Computational Civil Engineering Conference (CCE 2019)

IOP Conference Series: Materials Science and Engineering
Volume 586

Iasi, Romania
30 - 31 May 2019

Editors:

Gabriela Covatariu
Eugenio Oñate
Vasile-Mircea Venghiac
Petru Mihai

ISBN: 978-1-5108-9459-4
ISSN: 1757-8981

Printed from e-media with permission by:

Curran Associates, Inc.
57 Morehouse Lane
Red Hook, NY 12571

Some format issues inherent in the e-media version may also appear in this print version.

Institute of Physics
Dirac House, Temple Back
Bristol BS1 6BE UK

Phone: 44 1 17 929 7481
Fax: 44 1 17 920 0979

techtracking@iop.org

Additional copies of this publication are available from:

Curran Associates, Inc.
57 Morehouse Lane
Red Hook, NY 12571 USA
Phone: 845-758-0400
Fax: 845-758-2633
Email: curran@proceedings.com
Web: www.proceedings.com

Table of contents

Volume 586

Computational Civil Engineering Conference (CCE2019)

30–31 May 2019, Iasi, Romania

Accepted papers received: 4 July 2019
Published online: 10 September 2019

Preface

Papers

Computational Methods in Civil Engineering

Numerical Analyses of Civil Engineering Structures

Computational Fluid Dynamics

IOP Conf. Series: Materials Science and Engineering **586** (2019) 011001 doi:10.1088/1757-899X/586/1/011001

Preface

Over the last decades, we have witnessed an explosion in information generation related to all aspects of life, including all engineering disciplines. There has been an increase in active information collection to be used for solving critical engineering problems.

Computers are basic means for the efficient completion of construction projects from the design phase to the monitoring of constructions behaviour and facilities during exploitation.

The growing sophistication of computer programs, their availability, increased speed of computations and their ever-decreasing costs have already had a significant impact on Civil Engineering and Architecture and that can be considered a paradigm change.

The permanent development of the computing tools and the requirements for the delivery of complex and complete products / projects highlight the need to anticipate and, hence, the need to be at the current with the computational trends in constructions.

The Conference Computational Civil Engineering become traditional (since 2003).

The 15th edition CCE2019 was organized by Faculty of Civil Engineering and Building Services of Iasi in partnership with the International Centre for Numerical Methods in Engineering (CIMNE) Barcelona Spain, the “Matei Teiu-Botez” Academic Society Iasi and the "Anton Sesan" Civil Engineering Academic Society Iasi, Romania. The scientific event was held on May 30th-31st at Iasi, Romania.

CCE2019 aimed to bring together leading academic scientists, researchers, practitioners, numerical analysis programs developers to exchange / share their experiences to the Computational Civil Engineering domain to share their research findings to the theoretical and practical levels.

The proposed topics for CCE2019 edition of the conference was the following:

- *Computer Aided Design and Engineering*
- *Computational Methods in Civil Engineering*
- *Computational Fluid Dynamics*
- *Computer-Assisted Management in Civil Engineering*
- *Numerical Analyses of Civil Engineering Structures*
- *Computer Aided Training*
- *Civil Engineering Database*
- *Software for Civil Engineering*

In plenary sessions were presented very interesting and up-to-date exposures of the keynote speakers, well-known researchers or practicians in civil engineering domain:

- Jason Ingham (University of Auckland, New Zeeland) – *Multi- scale Multi-disciplinary Research to Mitigate the Seismic Hazard of URM Buildings*
- Neculai Tutos (Klaro Consulting, U.S.A.) - *Rethinking the Roadmap to Digital Age Performance in Construction*
- Konstantinos Tsavdaridis (University of Leeds, UK) - *Sustainable Seismic-Resistant Design of MRFS with the Use of Perforated Beams*
- Dan Palermo (York University, Canada) - *Shape Memory Alloy Reinforced Concrete Structures: From Experimental Testing to Advanced Nonlinear Modelling*

IOP Conf. Series: Materials Science and Engineering **586** (2019) 011001 doi:10.1088/1757-899X/586/1/011001

- Ioan Nistor (University of Ottawa, Canada) - *Extreme Hydrodynamic and Debris Loading on Structures - Forensic Engineering Field, Experimental and Numerical Studies*
- Alex H. Bărbat (CIMNE Spain) - *Evaluation of the failure pressure of the containment building of a nuclear power plant*

This year, at CCE2019 participate researchers and experts out of 7 countries distributed on the five topics of the conference.

The organizers want to express gratitude to members of Scientific Committee and welcome to the speakers, papers presenters and participants to this conference and to wish all of you a fruitful event

Finally, but not at the end, thank all sponsors and individuals who have contributed to the organization and success of this special event CCE2019.

More information about this scientific event can be found on the conference website www.cce.ci.tuiasi.ro.

Editors

IOP Conf. Series: Materials Science and Engineering 586 (2019) 011002 doi:10.1088/1757-899X/586/1/011002

The Committees of *CCE 2019 Conference*

Honorary Committee

Prof. Constantin Ionescu	"Gheorghe Asachi" Technical University of Iasi, Romania
Prof. Eugenio Oñate	International Centre for Numerical Methods in Engineering CIMNE Barcelona, Spain
Prof. Mihai Budescu	"Gheorghe Asachi" Technical University of Iasi, Romania
Prof. Nicolae Țăranu	"Gheorghe Asachi" Technical University of Iasi, Romania
Prof. Alex H. Bărbat	International Centre for Numerical Methods in Engineering CIMNE Barcelona, Spain
Prof. Dorina-Nicolina Isopescu	"Gheorghe Asachi" Technical University of Iasi, Romania
Prof. Daniela Manea	Technical University of Cluj-Napoca, Romania
Prof. Valeriu Stoian	"Politehnica" University of Timisoara, Romania
Prof. Radu Văcăreanu	Technical University of Civil Engineering Bucharest, Romania
Assoc.prof. Mihai Corneliu Drișcu	"Gheorghe Asachi" Technical University of Iasi, Romania

Scientific committee

Prof. Jason Maxwell Ingham	University of Auckland, New Zealand
Prof. Ioan Nistor	University of Ottawa, Canada
Prof. Dan Palermo	The York University, Toronto, Canada
Prof. Konstantinos Daniel Tsavdaridis	University of Leeds, UK
Prof. Thanasis C. Triantafillou	University of Patras, Greece
Prof. Hiroshi Yokota	Hokkaido University, Japan
Prof. João Carlos Gonçalves Lanzinha	University Beira Interior, Portugal
Prof. Daniel Aelenei	Universidade Nova de Lisboa, Portugal
Prof. Cristina Cosma	Wentworth Institute of Technology, USA
Prof. Cristina Mihaela Câmpian	Technical University Cuj Napoca, Romania
Prof. Sergio Oller	Natial University of Salta, Argentina
Prof. Loretta Batali	Technical University of Civil Engineering of Bucharest, Romania
Prof. Cedric D'Mello	City University of London, UK
Prof. Radu Pascu	Technical University of Civil Engineering of Bucharest, Romania
Prof. Irina Lungu	"Gheorghe Asachi" Technical University of Iasi, Romania
Assoc. prof. Filippo Giammaria Pratico	Universita Mediterranea di Reggio Calabria, Italy
Assoc. prof. Piero Colajanni	Università degli Studi di Palermo, Italy
Assoc. prof. Viorel Popa	Technical University of Civil Engineering of Bucharest, Romania
Assoc. prof. Mihail Iancovici	Technical University of Civil Engineering of Bucharest, Romania
Assoc. prof. Stefan Gutiu	Technical University of Cluj Napoca, Romania
Assoc. prof. Ferdinand-Zsongor Gobesz	Technical University of Cluj Napoca, Romania
Assoc. prof. Ligia Moga	Technical University of Cluj Napoca, Romania
Assoc. prof. Ioan Așchilean	Technical University of Cluj Napoca, Romania
Assoc. prof. Ioan Lucian Carstolovean	Transilvania University of Brasov, Romania
Assoc. prof. Carmen Ema Panaite	"Gheorghe Asachi" Technical University of Iasi, Romania
Assoc. prof. Carlos Chastre	Universidade NOVA de Lisboa, Portugal

IOP Conf. Series: Materials Science and Engineering **586** (2019) 011002 doi:10.1088/1757-899X/586/1/011002

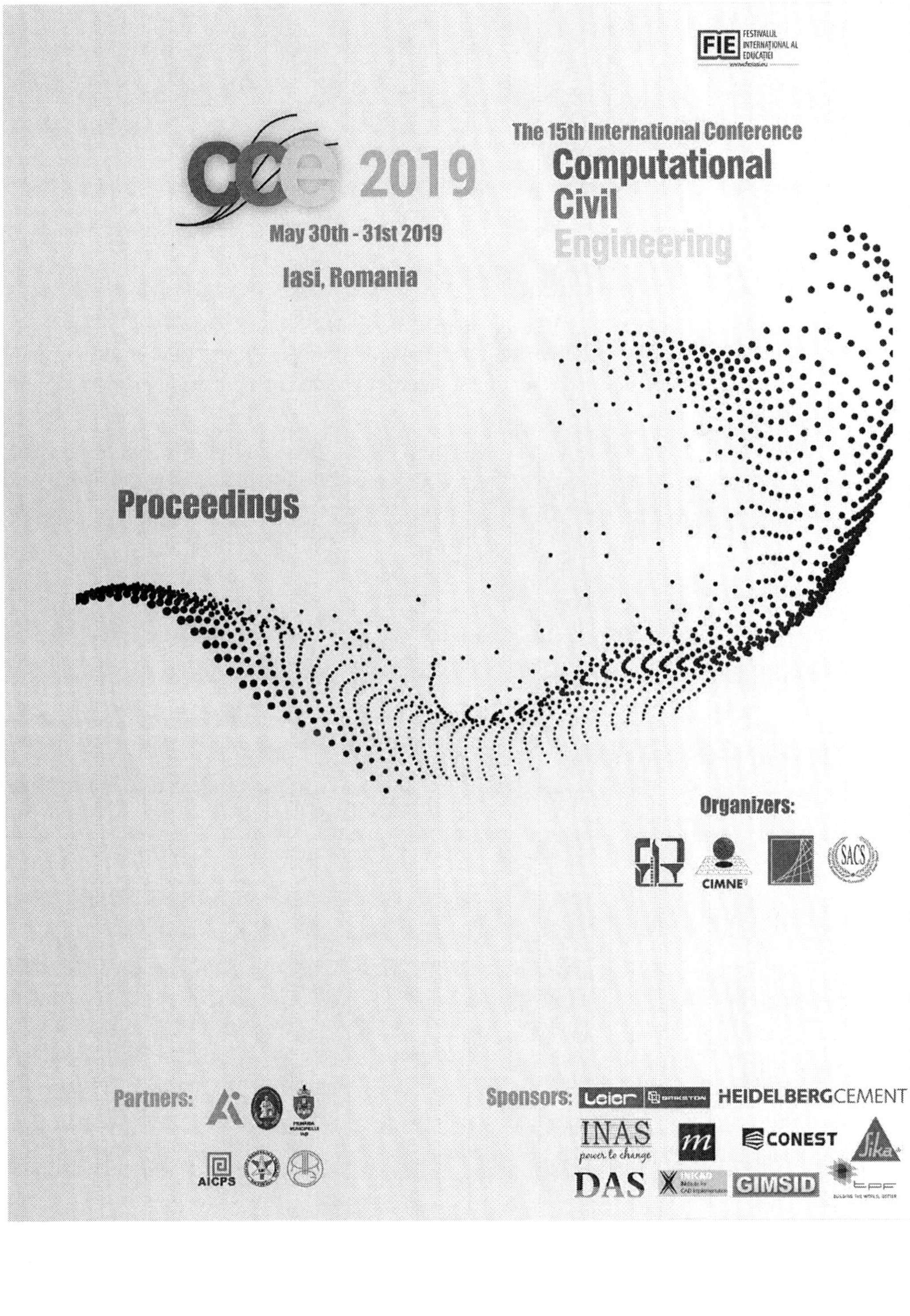
FIE FESTIVALUL INTERNAȚIONAL AL EDUCAȚIEI
The 15th International Conference
Computational
Civil
Engineering
CCE 2019
May 30th - 31st 2019
Iasi, Romania
Proceedings
Organizers:
CIMNE
SACS
Partners:
AICPS
Sponsors:
HEIDELBERGCEMENT
INAS
power to change
CONEST
Sika
DAS
GIMSID

Peer review statement

All papers published in this volume of *IOP Conference Series: Materials Science and Engineering* have been peer reviewed through processes administered by the proceedings Editors. Reviews were conducted by expert referees to the professional and scientific standards expected of a proceedings journal published by IOP Publishing.

IOP Conf. Series: Materials Science and Engineering **586** (2019) 012001 doi:10.1088/1757-899X/586/1/012001

Bearing capacity assessment for a timber girder when subject to a localised fire

K Kmiecik[1]
[1] Institute of Building Materials and Structures, Faculty of Civil Engineering, Cracow University of Technology, 31-155 Cracow, *Poland*,

E-mail: kamil.kmiecik@pk.edu.pl

Abstract. Timber girders are often used as roofs in large-area buildings. One of the most important issues concerning the design of timber structures are fire safety requirements. Safe design requires adequate knowledge and modelling of the physical and chemical reactions inside the timber members exposed to fire. Wood is a combustible material. At 300°C the pyrolysis takes place which cause loss in mass and decrease the strength and mechanical properties. European standards introduce simplified methods for assessing load-bearing capacity in fire conditions. The second approach is advanced numerical modelling. This paper presents results from numerical studies on fire resistance of timber girder when subject to a localised fire. The numerical simulations have been performed with application of the SAFIR computer program. The numerical model of a timber girder was made. European standards recommend estimating the fire resistance of a structure based on the ISO fire curve. The advantage of such an approach is simplicity. It can lead to overestimating load-bearing capacity. Therefore, the cost of construction and fire protection will be higher. Another approach is implementation of appropriate fire models. Then, determining the temperature-time relationship for the analysed structure members. In this paper localised and IS0 fire were assumed. The heat was applied for three sides of the beam. An uneven cross-section along the length was also assumed. The obtained results indicate that advanced numerical analysis allows for a more accurate estimation of the load-bearing capacity of timber elements exposed to fire.

1. Introduction

Glue laminated timber is widely used as a construction material for buildings and structures. It is used for public, residential, industrial, warehouse, and other buildings. The biggest disadvantage of using timber in constructions is combustibility. During the fire, the wood heats up and the moisture begins to evaporate. The pyrolysis takes place when the temperature of material is around 300°C. It causes loss in mass and reduces the strength and mechanical properties.

Timber elements have to fulfil fire resistance criteria when used in structures. The most essential in fire safety design is the estimation and selection of fire scenarios that may occur in buildings. Fire development depends on many factors, circumstances and events, which is why each fire scenario is unique. Fire scenarios determine the development of fire and the spread of combustion products throughout the building [1,2]. Therefore, the number of fire scenarios for a building can be very large. It is difficult to analyse every scenario of fire development. Therefore, the most probable cases that provide an acceptable level of safety should be considered.

To determine the load-bearing capacity of structures or members, European standards provide standard fire curve IS0 [3]. According to assumptions of the IS0 curve, the fire resistance of buildings and each element of the structure is estimated by a fully developed fire scenario [4]. Therefore, the load-bearing capacity of structure is relatively low. This approach is simple in calculations and provides a large safety reserve. The disadvantage is understatement of the load-bearing capacity, which leads to increased building costs. EN 1991-1-2 [5] also provides the possibility of using advanced calculation. Therefore, numerical models of fire development can be made for each building.

2. A description of the structure

2.1. Assumptions

This paper presents an investigation on bearing capacity assessment for a timber girder when subject to a localised fire. The analysis was subjected to wooden hall with the dimensions 36,00 x 90,00 meters and height to eave of 6,50 m. Wood properties at fire temperature were adopted in accordance with EN 1995-1-2 [5].

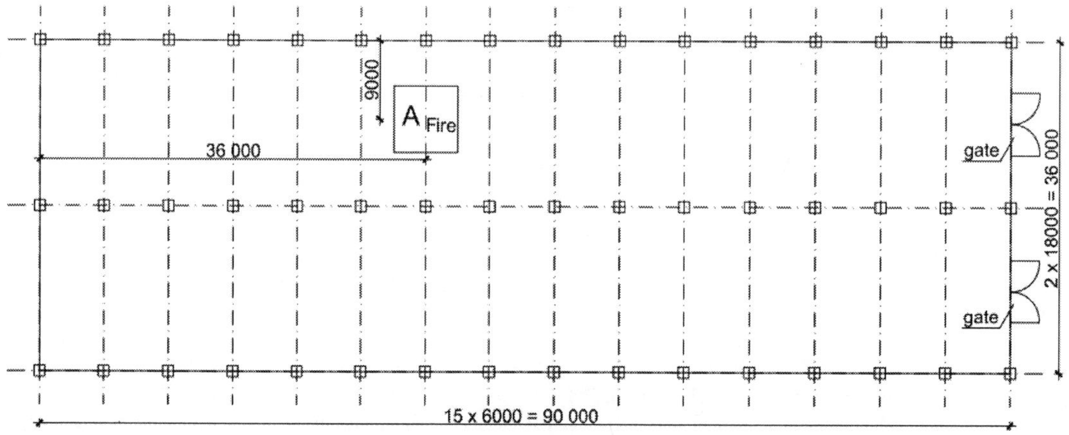

Figure 1. The floor plan of the building.

The location of the fire source is shown in Figure 2. A fire power Q_{max} = 25 MW was assumed. Heat release rate RHR_f was 500 kW/m^2. Fire area should be calculated using the relationship:

$$A_{fire} = \frac{Q_{\max}}{RHR_f} = \frac{25MW}{500kW/m^2}\ [\mathrm{m}^2] \tag{1}$$

The main structure of the roof are double tapered girders with a span of l = 18 meters. The height of the girder on the support h_l is 0.72 m, and in the ridge h_{ap} is 1.67 m. The beam width is fixed at b = 0.2 m.

The girders were made of glued laminated timber, class GL24h. Live and dead loads were assumed, which amounted 5 kN/m.

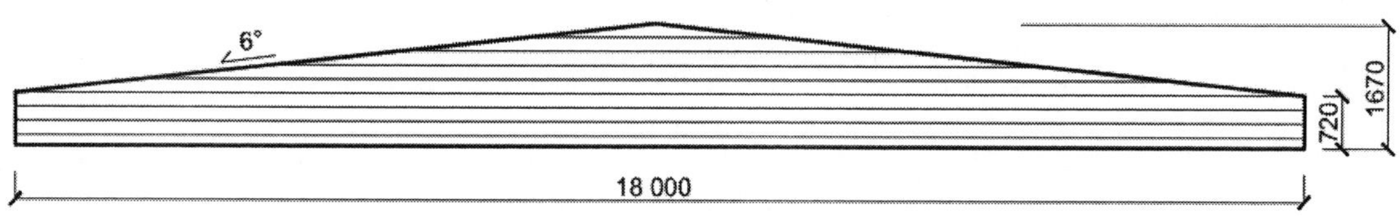

Figure 2. Double tapered girder.

2.2. Load-bearing capacity in normal conditions

To determine the load-bearing capacity of the girder in normal conditions, several conditions must be fulfilled. First, the shear capacity should be determined. Reduced shear force on the support is equal:

$$V_{d,red} = \frac{V_d}{l}\left(l - \left(\frac{l}{2} + h_1\right)\right) = 123\ [\mathrm{kN}] \tag{2}$$

where V_d is design shear force [kN]. Then a reduced cross-section area should be determined:

$$h_{red} = h_1 \cdot (h_{ap} - h_1) / \left(\frac{l}{2} - h_1 \right) + h_1 = 80,3 \text{ [cm]} \tag{3}$$

$$A_{red} = b \cdot h_{red} = 1610 \text{ [cm}^2\text{]} \tag{4}$$

The reduced shear stress is equal:

$$\tau_{d,red} = 1,5 \frac{V_{d,red}}{k_{cr} \cdot A_{red}} = 1,7 \text{ [MPa]} \tag{5}$$

Design shear strength is equal:

$$f_{v,d} = \frac{f_{v,k} \cdot k_{\text{mod}}}{\gamma_M} = 2,4 \text{ [MPa]} \tag{6}$$

where $f_{v,k}$ is characteristic shear strength [MPa], k_{mod} is modification factor, γ_M is partial factor for material properties. 0.2 m.
Other, necessary load-bearing capacity conditions were also checked. The results are shown in table 1.

Table 1. Ultimate Limit States for timber girder.

Ultimate Limit States	Design stress σ_i [MPa]	Design strength f_i [MPa]	$\frac{\sigma_i}{f_i}$
Shear	1,70	2,40	0,71
Bending	9,18	16,60	0,55
Bending at an angle α to the grain	9,18	14,90	0,62

3. Fire model

3.1. Heat transfer

To analyse the heat transfer within the timber girder in fire conditions, three modes of heat transfer should be considered: convective and radiative heat transfer from fire to element boundary and conduction in the element. The net heat flux in fire conditions, should consist of heat transfer by radiation and convection. The equation is given by:

$$q" = h_c(T_f - T_s) + \Phi \varepsilon_{eff} \sigma (T_f^4 - T_s^4) \tag{7}$$

where h_c (W/m^{2}K) is the convection coefficient, Φ is the configuration factor, σ is the Stefan-Boltzmann constant, ε_{eff} is the effective emissivity, T_s (K) is the surface of the member temperature and T_f (K) is the fire temperature. The convection coefficient h_c for the standard fire is 25 W/m^{2}K. The efficient emissivity ε_{eff} should be defined as:

$$\varepsilon_{eff} = \frac{\varepsilon_f \varepsilon_s}{\varepsilon_f + \varepsilon_s - \varepsilon_f \varepsilon_s} \tag{8}$$

where ε_f is emissivity of the fire source, ε_s is the surface emissivity.
In the literature are analytical models [6]. The governing partial differential equation should be defined:

$$\frac{\partial}{\partial x}\left(\lambda_x \frac{\partial T}{\partial x}\right)+\frac{\partial}{\partial y}\left(\lambda_y \frac{\partial T}{\partial y}\right)+\frac{\partial}{\partial z}\left(\lambda_z \frac{\partial T}{\partial z}\right)+\dot{Q}=\rho c\frac{\partial T}{\partial t} \tag{9}$$

where T(K) is temperature, $\lambda_{x,y,z}$ (W/mK) are thermal conductivities in x, y, z directions, Q(W/m^3) is internally generated heat, ρ (kg/m^3) is density, c (J/kgK) is specific heat, and t (s) is time.

3.2. Localised fire

In the estimation of fire resistance of structures, it is generally assumed that the scenario of the developed fire for which the flashover point has been achieved is the most suitable for the assessment. This means considering the even temperature distribution of exhaust gases in the entire volume of the load-carrying structures, and this leads to the conclusion that the basic structural members lose their load-bearing capacity relatively quickly. It also provides a safe estimation of the fire resistance of the structure. However, in many cases, this approach is too conservative. Flashover depends on many factors. First of all, on the type and distribution of the combustible materials accumulated in the building and the access of oxygen, which would sustain combustion. The risk of a flashover for a specific building can be estimated on the basis of survey of the building. In further analysis, characteristics describing the intensity of combustion and the mass of inventoried materials should be taken into account. The energy released during the entire duration of the fire will be sufficient only for a localised fire confined only to a part of the building's internal space. In many cases a localised fire can be considered as a reliable estimate of the fire resistance of the building. A localised fire has much less impact on the building structure than a fully developed fire.

The thermal action of localised fire can be assessed by using the expressions given in EN 1995-1-2 [5]. The flame lengths L_f of a localized fire is determined using the relationship:

$$L_f = -1.02D + 0.0148Q^{2/5} \text{ [m]} \tag{10}$$

where D is the diameter of the fire [m], Q is the rate of heat release [W] of the fire. When the flame is not impacting the ceiling ($L_f < H$), the temperature $T_m(z)$ in the plume along the symmetrical vertical flame axis is given by:

$$T_m(z) = 20 + 0.25Q_c^{2/3}(z - z_0)^{-5/3} \le 900 \text{ [°C]} \tag{11}$$

where Q_c is the convective part of the rate of heat release [W], z is the height [m] along the flame axis, H is the distance [m] between the fire source and the ceiling. The virtual origin z_o is given by:

$$z_0 = -1.02D + 0.00524Q^{2/5} \text{ [m]} \tag{12}$$

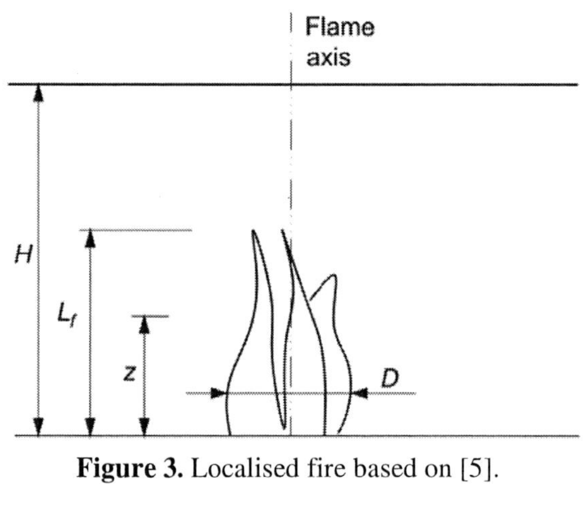

Figure 3. Localised fire based on [5].

4. Temperature distributions in a fire situation

4.1. Localised fire in Ozone V3 software

To estimate the temperature distribution, the Ozone V3 software was used. This program deals with two types of natural fire models: compartment fires and localised fires. In large-area buildings, where the flashover does not occur, localised fire should be considered. Localised fire in this software is based on the procedure where the thermal exchanges are drawn by convective fluxes are treated by application of the existing equations available in the EN 1991-1-2 [5].

4.2. Temperature distributions

The simulation results of the fire development are presented in the figure 4. The first curve shows the development of the standard fire. According to IS0-fire curve, temperature reaches about 700 °C after 10 minutes of fire. The temperature increases and after about one hour it is more than 900 °C. Considering the localised fire, the temperature increases to approximately 600 °C after 15 minutes of fire. Then the temperature decreases below 400 °C.

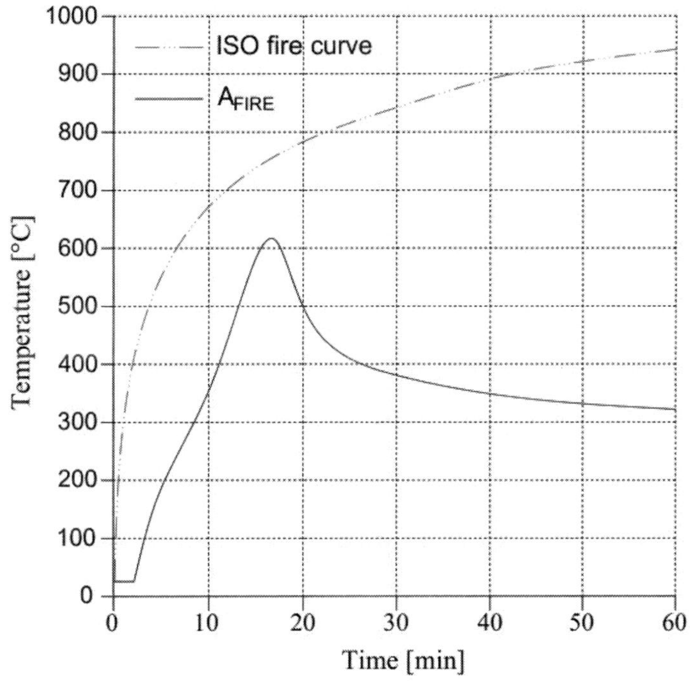

Figure 4. The standard ISO fire-curve and the temperature for A_{FIRE} under fire.

Figure 4 shows significant differences in temperature-time relationships depending on the fire scenario. Advanced numerical model that describes a local fire, determines much lower temperatures than those adopted according to the IS0 fire curve.

5. Heat transfer through timber girder

The numerical model of the timber double tapered girder was made in the SAFIR software. Material parameters were adopted in accordance with EN 1995-1-1 [7]. For numerical analysis of heat transfer through timber girder, the temperature-time relationship presented in figure 4 were used. The above curves have been implemented in the SAFIR software [8]. It was assumed that the fire acts on three sides of the beam. The results of the numerical simulations are presented in figure 5. Figure 5a shows the heat transfer through timber girder for the IS0 fire curve, while figure 5b for the localised fire at 15, 30, 45 and 60 min, respectively.

Many researchers have conducted studies on the properties of wood in a fire [9, 10]. When timber is subjected to elevated temperature, it undergoes chemical and physical changes. It was investigated that at a temperature of about 300 °C the pyrolysis takes place which cause loss in mass and reduce the strength and mechanical properties. A char layer is created that is not able to carry any loads. Comparing figure 5a and figure 5b it can be seen, that pyrolysis occurs faster and deeper in the case of the IS0-fire curve. In both cases the isothermal distributions turn into U-shape.

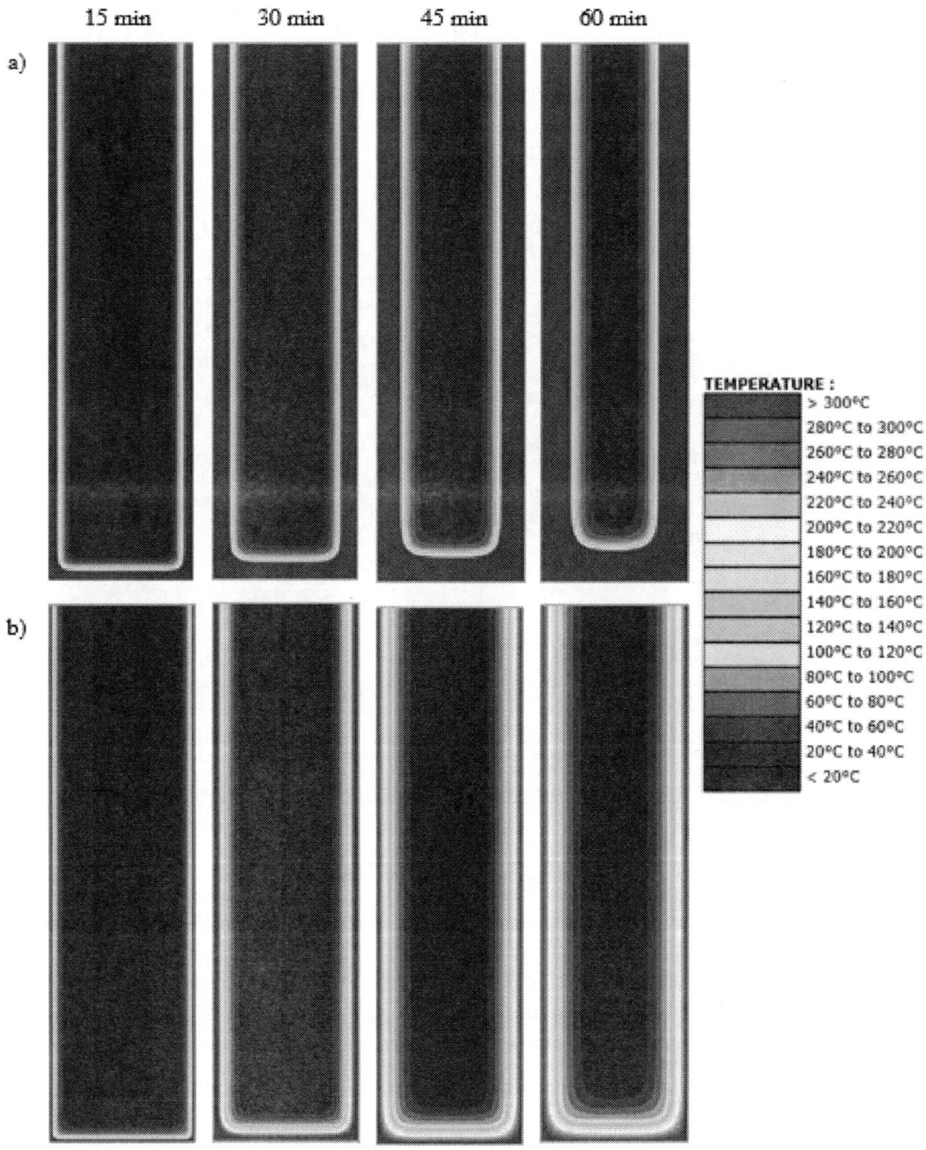

Figure 5. Temperature distribution of the timber girder exposed to: a) standard IS0-fire, b) localised fire A_{FIRE}.

Fire resistance of a girder depends on the residual cross-section, which is able to carry loads. Taking into consideration the standard IS0 fire, which acts on the girder, the cross-section is significantly reduced. After 30 minutes, the section was reduced by about 20% and after an hour only 50% of the initial cross-section was left. It is completely different in the case of a localized fire. After an hour of fire, the cross-section was reduced by approx. 13%. The residual cross-section was obtained from the SAFIR program based on the 300 °C isotherm. Figure 6 shows the residual section of the girder for two different fire scenarios, obtained as a result of numerical simulations.

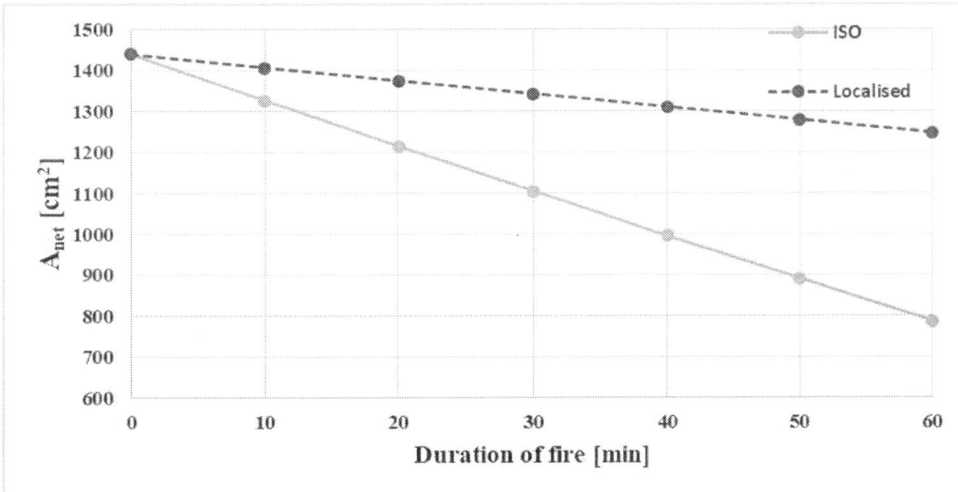

Figure 6. Residual cross-section of the girder in: a) IS0 fire, b) localised fire.

6. Load-bearing capacity in fire situation

According to EN 1995-1-2 [11], advanced calculation methods for determining mechanical strength should provide realistic analysis of structures exposed to fire. They should be based on basic physical behaviour of the element. Advanced calculation models should be applied for: the development and distribution of the temperature within structural members, the determination of the charring depth and the evaluation of structural behaviour of the structure. Calculations were made on the basis of decreasing cross-section in time. Therefore, at constant loads, the stresses in the residual cross-section were increased.

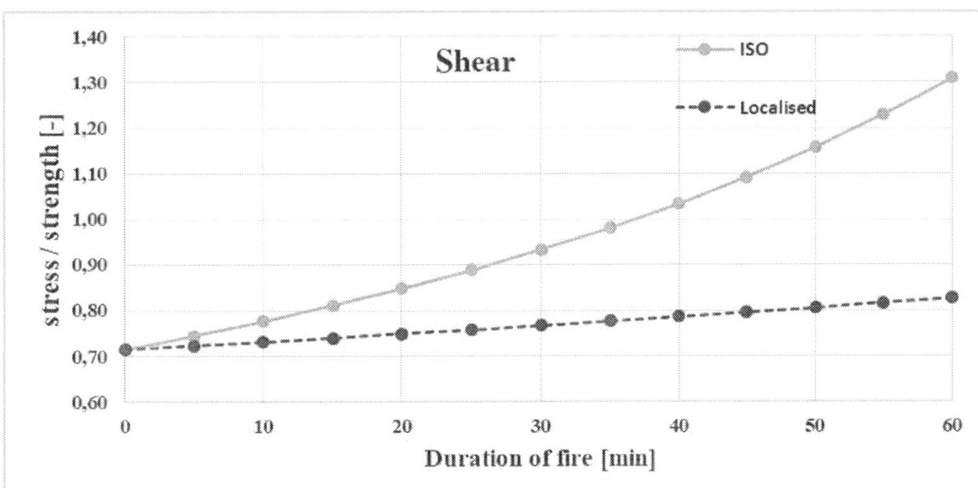

Figure 7. Shear strength of the girder in the IS0 and localised fire.

Figure 7 shows the decrease in shear capacity in two different fire scenarios: according to IS0 and localized. In the case of IS0 fire, after 35 min, the load capacity is completely lost. In the other case, fire resistance of the girder is ensured for 60 min of fire. Similar conclusions can be obtained by analyzing the bending (figure 8) and bending at an angle α to the grain (figure 9) capacity under fire. In these cases, the fire resistance of the beam is about 50 minutes. However, to consider the localised fire, after the hour of fire will still be 30% of the load capacity reserve.

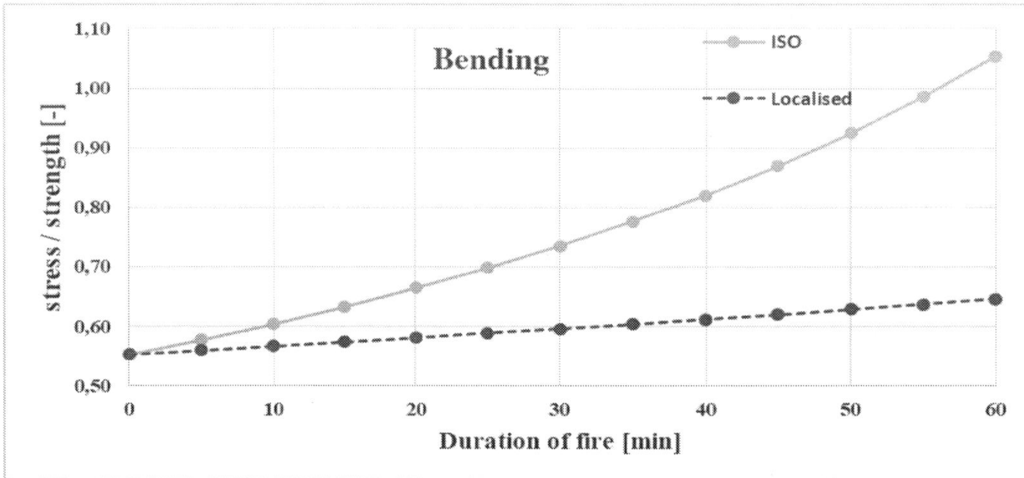

Figure 8. Bending strength of the girder in the IS0 and localised fire.

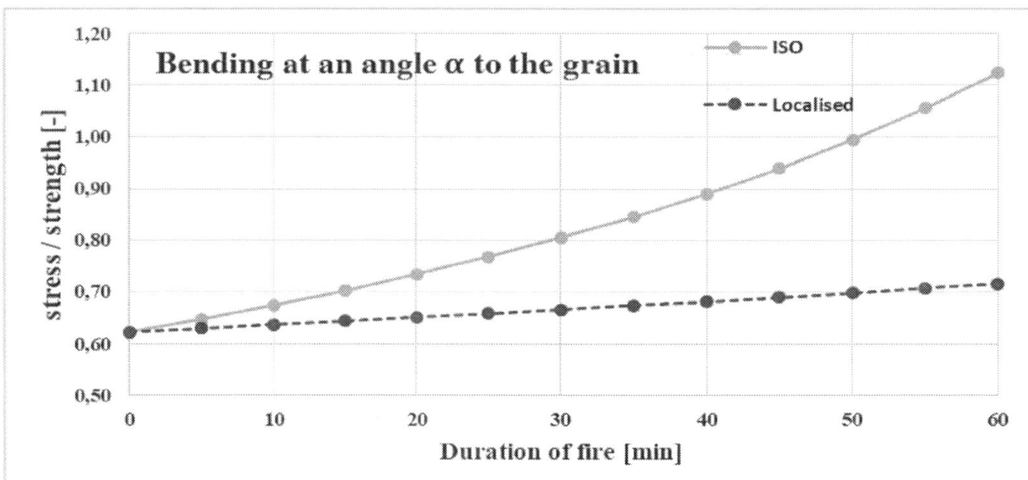

Figure 9. Bending strength at an angle α to the grain of the girder in the IS0 and localised fire.

7. Conclusions

Estimation of fire resistance of structural elements based on the IS0 standard curve is conservative. The disadvantage of this approach is understatement of the load-bearing capacity of the structure. This may lead to increased building costs. In this paper, numerical models were developed to analyse bearing capacity assessment for a timber girder when subject to a localised fire. Advanced computer numerical methods allow to estimate more precisely the development of the fire. The number of fire scenarios for the building is very large. It is important to choose the most probable fire scenarios, which provides acceptable levels of safety.

The temperature-time relationship is necessary to determine the heat flow through the element. Based on the temperature distribution in the cross-section, the speed and depth of char layer can be estimated in fire conditions. Then the fire resistance of the structural element can be determined. Numerical modelling of the fire, taking into consideration the fire source location is necessary for rational and economical design of timber structures. Correction coefficients should be introduced depending on the volume of the building as well as combustible materials that are in the building.

8. References

[1] Zhang J, Xu Q, Xu Y, Wang B and Shang J 2012 A numerical on fire endurance of wood beams exposed to three-side fire *Appl. Phys. & Eng.* **13** pp 491-505
[2] König J 2006 Effective thermal actions and thermal properties of timber members in natural fires *Fire Mater.* **30** pp 51-63
[3] ISO 834 1999 Fire-resistance tests - Elements of building construction – Part 1: General requirements
[4] Young S and Clancy P 2001 Compression mechanical properties of wood at temperatures simulating fire conditions *Fire Mater.* **25** pp 83-93
[5] EN 1991 1-2 2002 Eurocode 1: Actions on structures – Part 1-2: General actions – Actions on structures exposed to fire
[6] Fredlund B 1993 Modelling of heat and mass transfer in timber structures during fire *Fire Saf. J.* **20** pp 39-69
[7] EN 1995 1-1 2004 Eurocode 5: Design of timber structures – Part 1-1: General – Common rules and rules for buildings
[8] Franssen J and Gernay T 2017 Modeling structures in fire with SAFIR: theoretical background and capabilities *J. Struct. Fire Eng.* **8** pp 300-23
[9] Babrauskas V 2005 Charring rate of wood as a tool for fire investigations *Fire Saf. J.* **40** pp 528-54
[10] Friquin K 2011 Material properties and external factors influencing the charring rate of solid wood and glue-laminated timber *Fire Mat.* **35** pp 303-27
[11] EN 1995 1-2 2008 Eurocode 5: Design of timber structures – Part 1-2: General – Structural fire design

Acknowledgments
The authors acknowledge the support of the PROM programme no. PPI/PRO/2018/1/00013/U/001 which is co-financed by the European Social Fund under the Knowledge Education Development Operational Programme.

European Funds Knowledge Education Development
Republic of Poland
NAWA
European Union European Social Fund

Simulation of fire development in a large-area shopping hall

M Maslak[1], M Pazdanowski[1] , P Wozniczka[1]
[1] Cracow University of Technology, Warszawska 24, 31-155 Cracow, *Poland*

E-mail: mmaslak@pk.edu.pl

Abstract. The results obtained after numerical simulation of fire development in a large area shopping hall are presented and widely discussed. The analysis has been conducted for two types of halls differing in size and efficient ventilation capabilities. A fire development scenario has been forecast under the assumption that a sufficient number of smoke vents had been installed as required by the provisions of law. The obtained results have been related to the hypothetical case where there were no smoke vents whatsoever. The stacking height of the merchandise gathered in the hall has been differentiated during the analysis as well. The numerical simulations have been performed with application of the FDS computer program. It has been assumed, that the fire had been initiated by a single source uniquely defined and initially limited in spatial area of influence. The results obtained indicate that the spatial and temporal fire development is different for each of the considered scenarios. The key role of the possible limitations in the exchange of gases between the area affected by fire and the environment, influencing the way and speed with which the fire propagates, has been confirmed during analysis.

1. Introduction

In the general design practice, when an authoritative response of a complex bearing structure to the external actions including fire exposure is sought, commonly a fully developed fire scenario is considered, with fire plume gas temperature increasing in time but distributed evenly within the affected fire compartment. An assumption of this type, especially pertaining to the building structures of large area and large enclosed volume seems to constitute an unnecessary oversimplification with currently available computational models. In such cases fire load gathered in the fire zone, at the ventilation conditons existing in reality, only rarely is capable of generating a fire having sufficient combustion energy to, even if only locally, reach the stage of fully developed fire. Thus an assumption of a localized fire, characterized by the fire plume temperature varying in space and time, seems to be far more rational from both technical and economic point of view. Forecasting the potential fire development within the considered fire compartment, under the assumed initial conditions and unequivocally defined, in general known a'priori, boundary conditions thus becomes one of basic tasks facing the expert evaluating the safety level ensured to the user of the considered structure. Only following such, undoubtedly non trivial analysis, one may on the one hand verify the real level of risk, however refraining from overestimating it, as usually happened when the fully developed fire model has been applied, and subsequently, on the other hand determine the locally changing temperature field affecting, as the exceptional load accompanying other loads, the considered bearing structure.

2. Description of the halls and fire scenarios analyzed in detail in this paper

The considerations presented in this paper pertain to the fire development forecast in large area shopping halls. The specificity of fire compartments of this type lies in the fact, that at very large surface area they have relatively low height. In addition, due to the way of use, these halls are densely filled with storage racks, on which the merchandise is stockpiled. The merchandise, and in many cases the rack structure itself [1], constitute a potential fuel after the fire initiation, interpreted in the following analysis as an important component of the total fire load in the considered hall. The storage racks, filled with densely packed and often combustible merchandise, in themselves may constitute a potential fire source, bursting into flames in a sequence with fire plume temperature increasing in their direct vicinity. In addition, these racks obstruct the efficient ventilation of the hall. The ventilation, in the case of fire, is to assure the exchange of exhaust gasses with the surroundings. This on one side allows for more or less restricted delivery of combustion supporting oxygen [2], but on the other side is necessary to create at least temporarily the conditions for safe evacuation. In order to completely describe the restrictions affecting the ventilation of so characterized fire compartment one has to add that in most cases compartments of this type are completely devoid of windows and number and size of access gates is relatively small as well. Thus proper formation of such fire compartments, allowing for instance for gravitational evacuation of smoke gathering under the ceiling, is commonly required by the building codes. An application of smoke vents to evacuate the smoke to the outside may be required by the code as well.

Let us consider here two types of shopping halls, the first of which represents a hall of rather average usable area, while the second one repersents a typical large area shopping hall [3, 4]. The in plan dimensions of the smaller hall are 52.00m by 36.00m, with constant height equal 4.50m (Figures 1 and 2). The same dimensions of the large shopping hall are 135.00m by 60.00m with a height of 7.00m (Figure 3). As one can see, in both of these halls the roof was modelled as a flat one, because in a real object the slope angle was only about 5%. Furthermore, in both cases the numerical models assumed the existence of permanently open access gates. It was assumed, that in the smaller hall one large gate having the dimensions of 3.50m by 3.00m is located in the front, shorter wall, while two smaller gates having the dimensions of 1.00m by 2.50m are located in the side wall (Figures 1 and 2). Locations and dimensions of the access gates in the large hall are analogous to those assumed for the smaller hall. The only difference is in the fact, that instead of two, three small acces gates are present in the side wall (Figure 3). The sheathing is modeled as typical, made of sandwich panels with 15cm thick mineral wool core. The properties of the insulating material, dependent on the temperature affecting it, have been assumed according to [5]. It has been also assumed, that before the initiation of fire the internal temperature of the hall was constant and equal to 20°C. It has been assumed as well, that regularly spaced storage racks in the halls contain the stockpiled merchandise (a derivative of cellulose, i.e. mostly paper like materials). These represent the basic fire load in the fire compartment. In addition 3x5=15 evenly spaced smoke vents having the dimensions of 2.00m by 2.50m each (Figures 1-2) have been located in the roof over the smaller hall. These smoke vents are opened individually after the activation of the temperature sensor, with the activation temperature set to 74°C. Analogous vents, of the same type and size have been assumed to be installed in the roof over the large hall. Of course the number of the vents over the large hall has been much higher than before and was equal to 6x13=78 (Figure 3). For comparative reasons, the fire development scenario in both halls devoid of smoke vents has been considered as well.

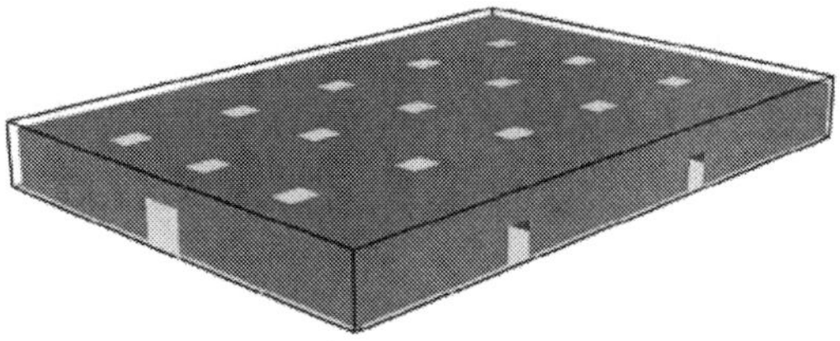

Figure 1. General view of the small shopping hall considered in the scenario A. The locations of the open gates and smoke vents are indicated.

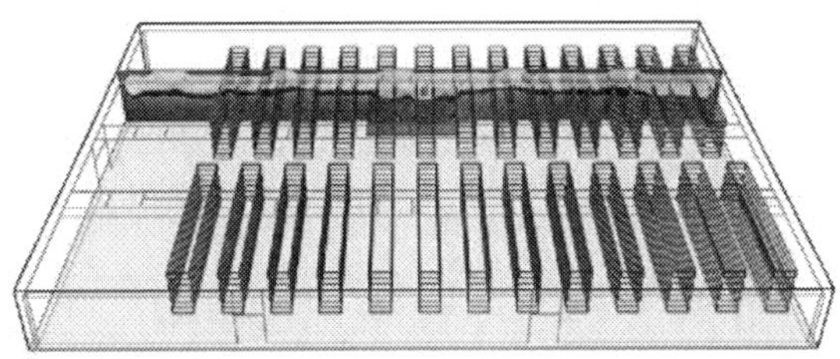

Figure 2. Distribution of the storage racks with paper like merchandise in the small hall considered in the scenario A. The location of fire source initiating the fire and the authoritative cross section depicting the distribution of fire plume temperature in the selected moment of fire exposure are indicated.

IOP Conf. Series: Materials Science and Engineering **586** (2019) 012002 doi:10.1088/1757-899X/586/1/012002

In the first two fire development scenarios simulated by the authors, of which the first one (scenario A) pertained to the smaller hall, while the second one (scenario B) pertained to the large one, the same stacking height of 3.50m has been assumed. This meant, that in the case of large hall and scenario B much more unobstructed space remained between the ceiling and the rack tops, than in the case of small hall and the scenario A (i.e. 3.50m and 1.00m respectively). In order to at least partially overcome the quantitative influence of this difference an additional scenario C has been considered (an alternative to the scenario B), where the stacking height has been increased from 3.50m to 5.50m while leaving other parameters of the scenario B unaffected. This means, that in the scenario C only 1.50m of the free space remained between the rack tops and the ceiling.

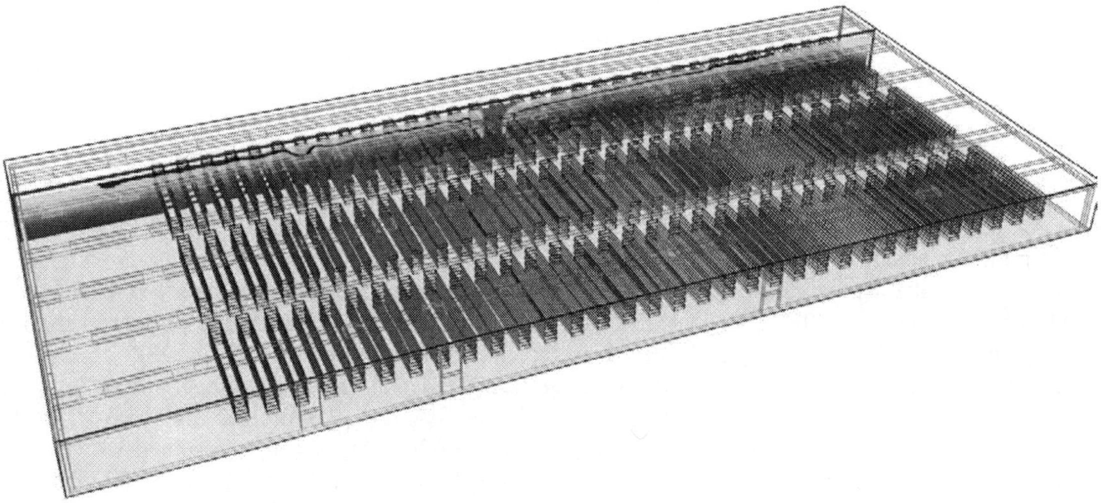

Figure 3. Distribution of the storage racks with paper like merchandise in the large hall considered in the scenario B, with low stacking height. The location of fire source initiating the fire and the authoritative cross section depicting the distribution of fire plume temperature in the selected moment of fire exposure are indicated.

Figure 4. Scenario A – Fire plume temperature map obtained in the authoritative cross section of the smaller hall after 10 minutes of fire exposure, under the assumption that the smoke vents have been activated sequentially when the temperature in their nearest vicinity reached 74°C. The isotherm depicted in black delimits the zone inside of which the temperature exceeded 100°C. Open smoke vents are visible.

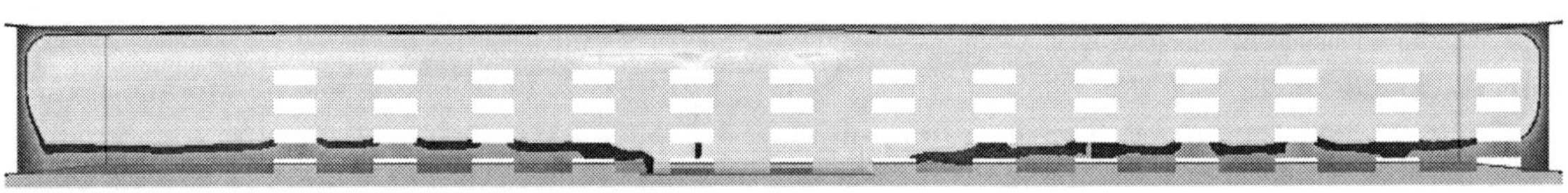

Figure 5. Scenario B – Fire plume temperature map analogous to the map depicted in figure 4, determined at the same cross section of the considered hall after 10 minutes of fire exposure, under the assumption that there are no smoke vents. The interpretation of the isotherm depicted in black is the same as in the previous figure.

The numerical simulation of fire development in the considered halls has been performed with application of Fire Dynamic Simulator (FDS) [6]. Convection, radiation and heat penetration through the partitions have been taken into account. A flame modeled as localized fire exhibiting the heat generation intensity of 500kW/m2 and total power of 25MW (recommended in [7] as the maximum for the steel structures and stockpiling of combustible materials) initiated the fire. The duration of this initial fire has been set to 120 seconds, and its location is indicated in Figures 2 and 3. The fire initiated in this manner developed further, in the scenario A engulfing, while in the scenario B not engulfing additional racks in the hall. The intensiy of fire development has been conditioned by the fire compartment geometry, the fire load gathered in the fire compartment and the available ventilation. The materials stockpiled on the racks have been modeled by the following physical properties: specific heat 1.36J/(kg·K), heat conductivity 0.25W/(m·K), volumetric weight 1100kg/m3. The flash point has been set to 250°C, and he chemical composition has been modeled in a simplified manner, as for cellulose, based on [5].

3. Development of the fire plume temperature during each of the considered fire scenarios

3.1. Scenario A

Detailed results of the fire plume temperature development, obtained for the fire forecast in the scenario A for the smaller of considered halls, are depicted in Figures 4, 5, 6 and 7. The location of source flame initiating the fire development is indicated in Figure 2.This location determines the location of the authoritative cross section, for which the fire plume temperature map is determined during the fire exposure. Let us consider the fire plume temperature map depicted in Figure 4 and obtained after 10 minutes of fire, in the hall equipped with operating smoke vents, as the reference one. The isotherm depicted in black in this picture refers to the temperature of 100°C. Thus the temperature in the zone surrounded by this isotherm is higher than 100°C, while the temperature outside of this zone is lower. Let us note, that this isotherm in this case is located relatively high above the floor level, and thus the zone containing relatively cool air after this long time of fire exposure is sufficiently thick to allow for safe evacuation. The limited thickness of fire plume under the roof is obviously the result of operating smoke vents evacuating hot combustion gases outside, as may be observed in Figure 4. Comparison of this picture with Figure 5, in which the analogous distribution of temperature determined in the same cross section of the hall and after the same fire exposure time is depicted, but this time for the hall in which the smoke vents did not activate, shows how much faster would the potential fire develop in a hall of the same fire compartment geometry and the same distribution of access gates ventilating the compartment, should one forego the additional obligatory equipment. As it is indicated in Figure 5, after 10 minutes of fire exposure the user of fire compartment stands no chance of evacuation. The temperature of fire plume in the whole height of the considered cross section, i.e. in its layer next to the floor as well, reaches now the level of 100°C.

A comparison of fire plume temperature maps obtained in the smaller hall in the horizontal cross section at the height of 4.00m above the floor level after 10, 15 and 20 minutes of fire exposure (Figure 6) may yield interesting conclusions. One may easily notice, that only after 20 minutes of the fire forecast for this hall the fire reaches the symptoms of fully developed one in large part of the considered fire compartment. In general all materials gathered in this area are on fire, and the temperature of the fire plume tends to equalize. This development is rather of local nature, as it does not affect the whole area of the considered hall. The graph depicting fire plume temperature in the authoritative cross section of the considered hall shown in Figure 2, at the height of 4.00m above floor level, seems to support this conclusion. The levels of temperature depicted in Figure 7, however, have been reached much later, after one hour of fire exposure. The results of simulations conducted for a hall equipped with operating smoke vents and the hypothetical one devoid of such vents are juxtaposed and compared here [8, 9].

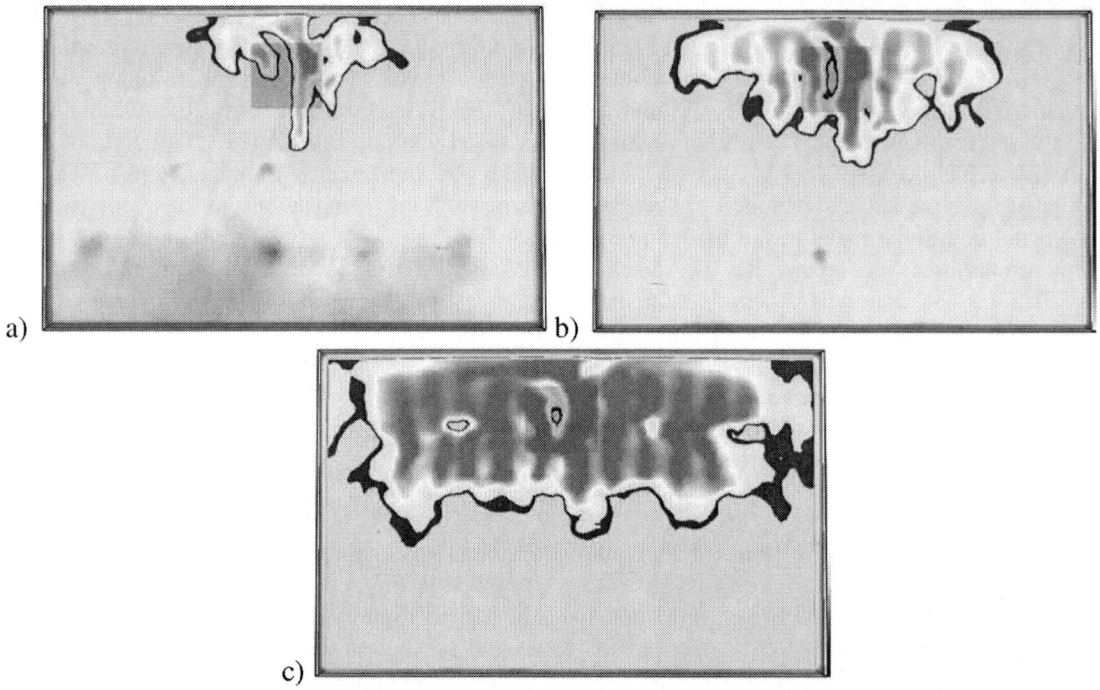

Figure 6. Scenario A – Fire plume temperature maps obtained in the horizontal cross section of the analyzed small hall at the height of 4.00m above the floor level, under the assumption, that the smoke vents operate as indicated above, after: a) 10 minutes, b) 15 minutes, c) 20 minutes of fire exposure, respectively. In this case the isotherm depicted in black delimits the zone inside of which the temperature exceeded 350°C. The location of the source of flame initiating the fire is indicated by the darkened square.

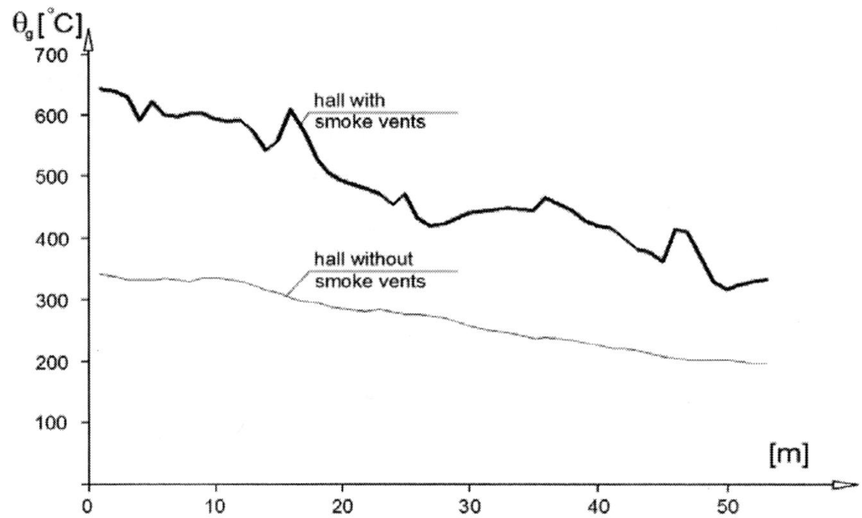

Figure 7. Scenario A – Graphs of fire plume temperatue obtained after one hour of fire exposure in the authoritative cross section of the smaller hall (depicted in figure 2), at the height of 4.00m above the floor level. The top graph refers to the hall equipped with operating smoke vents, while the bottom one to analogous hall devoid of vents.

3.2. Scenario B

The scenario A pertaining to the fire development simulation for the smaller of the considered halls is not typical for halls of the same type but having larger area. If the fire zone is sufficiently large (and this refers not only to the area, but also to the volume of the hall) the probability of fire progressing to the stage of fully developed one is negligibly small. For such a development to occur the availability of combustible materials, characterized by the sufficiently high burning efficiency, distributed evenly over the whole fire compartment and constituting the potential fire load would have to be accompanied by the sufficient availability of the combustion supporting oxygen, reaching the fire compartment via the numerous and permanently open gates. The analysis performed in this paper seems to indicate, that for large shopping halls a properly numerically modeled localized fire is more appropriate. This statement is in agreement with earlier recommendations of the authors, contained in [7]. A detailed analysis of fire plume temperature maps, similar to the maps presented in figures 4 and 5 above and obtained for the scenario B of fire development at low stacking height, after 60 minutes of fire exposure and related to the authoritative cross section of the larger of considered halls (figure 3), suffices to support this statement. The relevant maps are presented in Figures 8, 9 and 10, beginning with maps obtained for hall with operating smoke vents. The map depicted in Figure 8 shows the situation at the cross section located between the smoke vent axes (this location is shown as well in Figure 3), while the map shown in Figure 9 depicts the situation at the cross section located along the vent axes. Figure 10 presents the same maps for the analogous hall devoid of any smoke vents. The isotherm depicted in black denotes the temperature of 100°C.

Figure 8. Scenario B – map of fire plume temperature obtained after 60 minutes of fire exposure in the larger hall with operating smoke vents in the cross section located between the vent axes. The isotherm depicted in black denotes the temperature of 100°C.

Figure 9. Scenario B – map of fire plume temperature obtained after 60 minutes of fire exposure in the larger hall with operating smoke vents in the cross section located along the vent axes. The isotherm depicted in black denotes the temperature of 100°C.

Figure 10. Scenario B – map of fire plume temperature obtained after 60 minutes of fire exposure in the larger hall without smoke vents in the cross section located as in figure 8. The isotherm depicted in black denotes the temperature of 100°C.

Comparison of maps depicted in Figures 8, 9 and 10 with corresponding maps depicted above in Figures 4 and 5 should be made keeping in mind, that the larger hall considered in the scenario B is significantly higher (7.00m) than the smaller one considered in the scenario A (4.50m). With the same stacking height of combustible materials (derivatives of cellulose) assumed for both halls (3.50m) this results in a much larger open space, devoid of any obstacles such as shelves, located directly under the ceiling. Those 3.50m of open space under the ceiling in the scenario B kept the fire modeled in the larger hall during the whole 60 minutes of fire exposure at the stage of localized fire. The thickness of hot fire plume under the ceiling even after one hour of intense fire action proved to be insufficient to initiate ignition of combustible materials located on the racks below in the nearest vicinity of the fire source. Within such scenario of fire development the operation of smoke vents did not substantially

IOP Conf. Series: Materials Science and Engineering **586** (2019) 012002 doi:10.1088/1757-899X/586/1/012002

affect the temperature of air near the floor level, and thus it did not affect the forecast occupant evacuation time. The maps of fire plume temperature obtained after 60 minutes of fire exposure at the height of 6.00m above floor level seem to corroborate the conclusion that the fire remained at the stage of localized fire. These maps are analogous to the maps depicted in Figure 6 for the smaller hall and the scenario A of fire development. The first of the compared maps, depicted in Figure 11 pertains to the case with operating smoke vents, while the second one pertains to the analogous structure but devoid of smoke vents (Figure 12). Comparison of both maps seems to indicate a slightly larger fire affected zone in the case of hall devoid of smoke vents with respect to the case of a hall equipped with operating smoke vents, though the difference does not seem to be significant from the standpoint of the safety level ensured to the users of the considered hall.

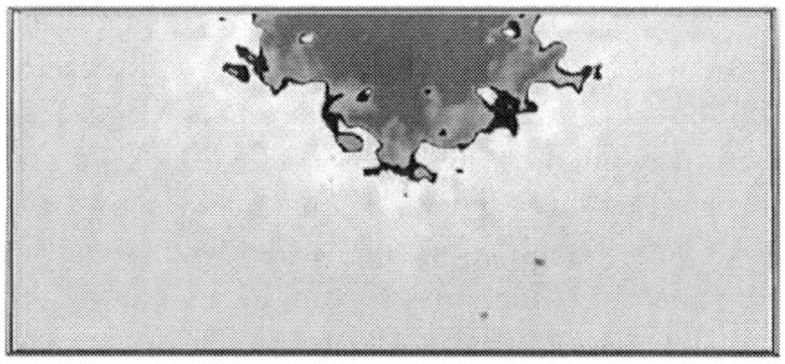

Figure 11. Scenario B – Fire plume temperature map obtained after 60 minutes of fire exposure at the height of 6.00m above the floor level in the larger of considered halls with operating smoke vents. The isotherm depicted in black denotes the temperature of 100°C.

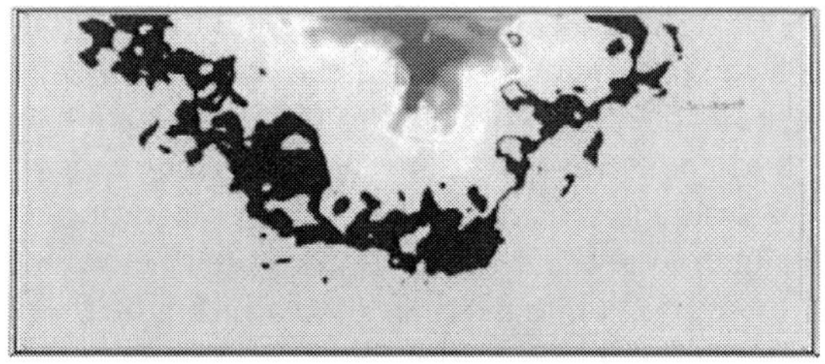

Figure 12. Scenario B – Fire plume temperature map obtained after 60 minutes of fire exposure at the height of 6.00m above the floor level in the larger of considered halls without smoke vents. The isotherm depicted in black denotes the temperature of 100°C.

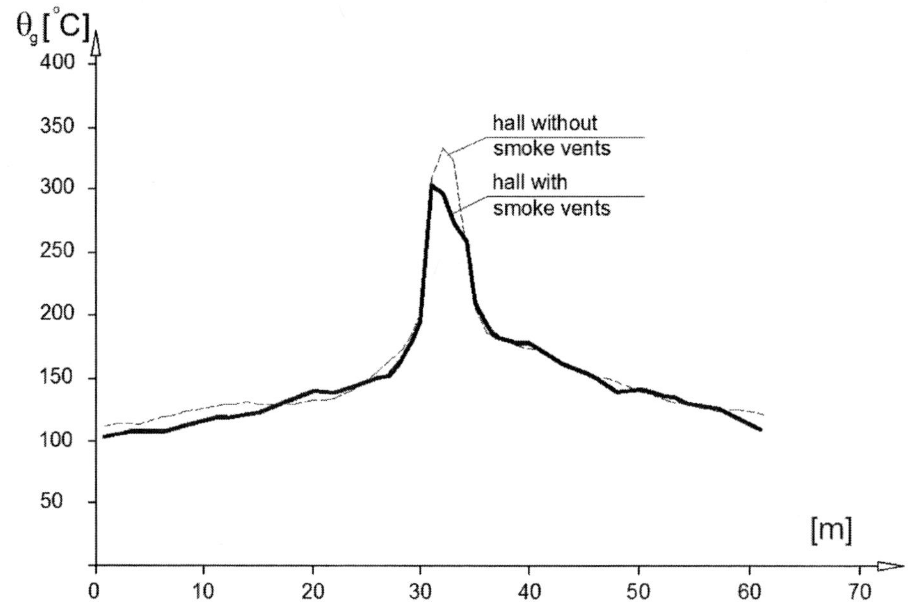

Figure 13. Scenario B – Distributions of fire plume temperature values obtained after 60 minutes of fire exposure in the cross section of the larger of considered halls at the height of 6.00m above floor level for the hall equipped with operating smoke vents and the hall devoid of such.

An additional confirmation of the locality of fire developing within the framework of scenario B after 60 minutes of fire exposure is offered by the analysis of fire plume temperature distribution obtained in the cross section of the large hall shown in Figure 3, at the height of 6.00m above the floor level. This distribution is depicted in Figure 13. The values of temperature forecast for the case of a hall equipped with operating smoke vents and for the same hall devoid of smoke vents seem to indicate, in both cases, a clearily defined maximum of the temperature in the nearest vicinity of the flame constituting the initial source of fire accompanied by rapid decrese of simulated fire plume temperature with increasing distance to the fire source. The shapes of both graphs are thus substantially different than the shapes of graphs depicted in Figure 7 and obtained for the scenario A and the smaller of the considered halls. Let us note, however, that the difference between the results obtained for the larger hall equipped with smoke vents, and the same hall devoid of such is now in general negligible [8, 9].

3.3. Scenario C
The scenario C differs from the scenario B analyzed for the large hall only in that the stacking height of the combustible materials has been changed from 3.50m to 5.50m. The spatial distribution of the racks has not been changed in any way. The parameters determining the ignition chances of goods placed on these shelves have not been changed in any way as well. This means that in this case the volume of empty space above the racks, where the fire plume may freely propagate is much more restricted. At the same time, due to the added amount of combusible materials, the fire load in the fire compartment has been substantially increased. Because of that, even at the same power of a fire source initiating the fire as in the scenario B, in the scenario C a local ignition of the combustible materials stockpiled in the hall is much more probable and in turn the cascading propagation of a fire to the neighbouring storage racks is more probable as well. In Figures 14 and 15 the fire plume temperature maps are depicted for the scenario C, after 60 minutes of fire exposure, obtained in the same cross sections of the considered hall as before. These maps correspond to the analogous maps depicted in Figures 8, 9 and 10. The 100°C isotherm is depicted in black in those pictures as well. The map from Figure 14 refers to the cross section located between the smoke vents, while the other one, from figure 15, refers to the cross section located along the axes of smoke vents. The scenario for a hall devoid of smoke vents has not been considered here. The maps of Figures 14 and 15 are accompanied by the maps of fire plume temperature distributions obtained for the scenario C and referring to the cross section at the height of 6.00m above the floor level. These maps are analogous to those depicted above in Figures 8, 9 and 10 referring to the scenario B. These maps are presented in Figures 16 and 17, to depict the simulated fire development in the scenario C after 30 minutes (Figure 16) and 60 minutes (Figure 17) of fire exposure. One may easily notice, that in this scenario, in spite of smoke vents acting to slow the progress of fire, only after 30 minutes of fire exposure the fire engulfed a substantial part of the affected hall. Thus the fire development happened to progress in a much faster pace than in the case identified for the scenario B.
Another important difference between the distribution of fire plume temperature obtained in the same cross section of the same large hall after the simulation of fire corresponding to the assumptions of the scenario C, with respect to the analogous distribution obtained previously for the scenario B is depicted in Figure 18. One may notice, after juxtaposition with the results depicted in Figure 13, that this time the maximum forecast fire plume temperature proved to be much higher than previously. Nevertheless, the simulated fire remained within the characteristics of a localized fire, with definite maximum at the axis of initial flame indicating the source of the fire. The higher forecast maximum temperature seems to be the result of increased stacking height with respect to the scenario B. In the scenario C the hot fire plume under the ceiling reaches the tops of storage racks much earlier than in the scenario B, and this in turn increases the possibility for cascading progress of fire to the new locations in the considered hall [10].

Figure 14. Scenario C - Exhaust plume gas temperature maps obtained after 60 minutes of a fire exposure in the case of a larger hall equipped with smoke vents, for the cross section of the hall located between the smoke vent axes. The isotherm depicted in black denotes the fire plume temperature of 100°C.

Figure 15. Scenario C - Exhaust plume gas temperature maps obtained after 60 minutes of a fire exposure in the case of a larger hall equipped with smoke vents, for the cross section of the hall located along the axes of smoke vents. The isotherm depicted in black denotes the fire plume temperature of 100°C.

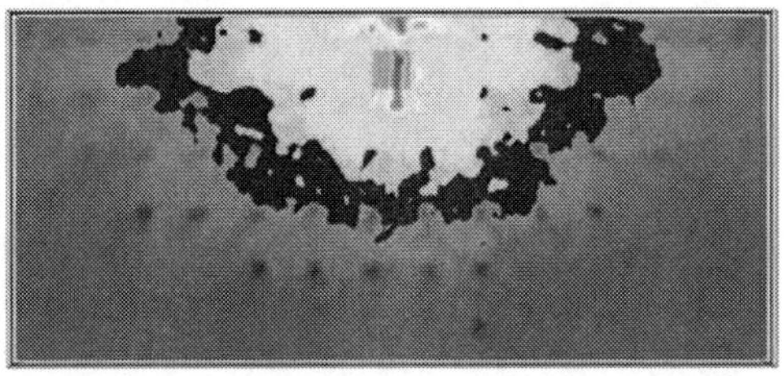

Figure 16. Scenario C – Map of fire plume temperature values obtained after 30 minutes of fire exposure in the horizontal cross section of the larger hall located at the height of 6.00m above floor level under the assumption, that the hall had been equipped with operating smoke vents. The isotherm depicted in black denotes the fire plume temperature of 100°C.

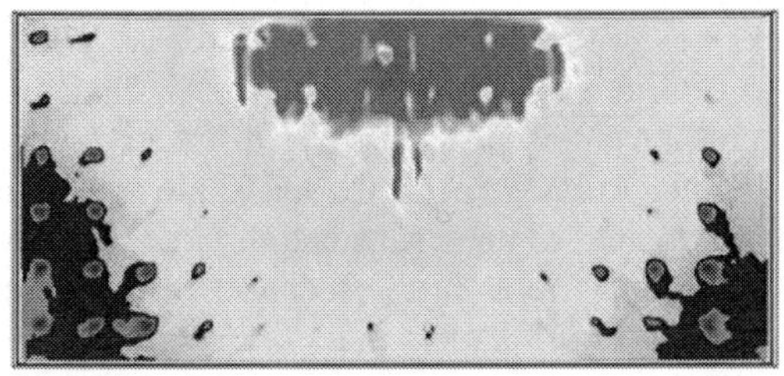

Figure 17. Scenario C – Map of fire plume temperature values obtained after 60 minutes of fire exposure in the horizontal cross section of the larger hall located at the height of 6.00m above floor level under the assumption, that the hall had been equipped with operating smoke vents. The isotherm depicted in black denotes the fire plume temperature of 100°C.

4. Infuence of the limited effective ventilation in the case of fire

The juxtaposition of graphs depicted together in Figure 7 unequivocally indicates that during simulation of fire in a small hall, corresponding to the scenario A, in the case of a hall equipped with operating smoke vents, the fire plume temperature at the considered level proved to be much higher than the analogous temperature forecast for the same hall devoid of any smoke vents. So strong differentiation may be attributed to the fact that in the second case the fire intensity was strongly affected by the limited availability of the combustion supporting oxygen [3]. The oxygen could, in this scenario, reach the fire zone only via the relatively distant constantly open access gates. Thus in this case the fire development was typical for the ventilation driven fires, as opposed to the case of the larger hall and the scenario B, where the fire was driven by the availability of the combustible material. This conclusion seems to be corroborated by the juxtaposition depicted in Figure 19. This figure presents the changes in the fire plume temperature measured at a point located at the distance of 3.00m from the initial source of fire and at the height of 4.00m above the floor level. This graph indicates, that in the case of a hall equipped with operating smoke vents the temperature averaged in the random process of burning progressed in general monotonically during the whole hour of analyzed fire exposure. The results obtained for the same hall, but devoid of vents indicate a completely different progress of fire. This time the fire plume temperature measured could not reach the sufficiently high level, and subsequently increase even further in analogy to the scenario considered previously, due to the limited availability of the oxygen extracted from the surroundings of the hall. Thus the temperature after initiation rapidly decreased, due to the momentary "suppression" of fire, and subsequently stabilized at the level of equilibrium conditioned by the available effective ventilation of the fire compartment [8, 9].

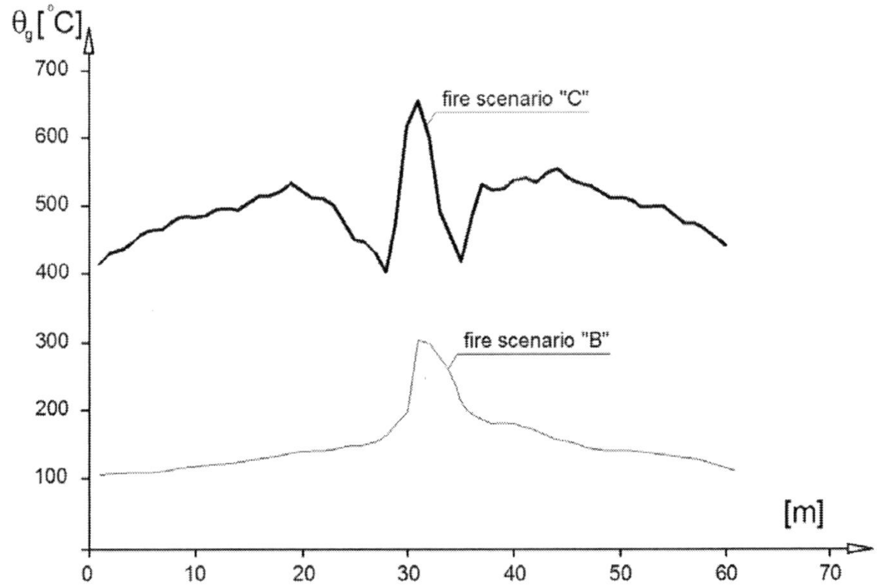

Figure 18. Distributions of fire plume temperature values obtained after 60 minutes of fire exposure in the cross section of the larger hall at the height of 6.00m above the floor level under the assumption, that the smoke vents operate efficiently. The lower graph pertains to the scenario B of the simulated fire, while the upper one pertains to the scenario C.

Analysis of analogous juxtaposition depicted in Figure 20, this time related to the scenario B of the fire and the larger of the considered halls, leads to a completely different conclusion. As one may observe, in the case of those simulations the forecast fire plume temperature increased monotonically both in the case when the hall was equipped with smoke vents, as well as when the hall was not so equipped. Thus here the suppression of fire did not occur. Anyway it would be difficult with so large volume of the fire compartment. The fire plume temperature increased rather slowly in this case. The faster increase in the temperature would be here a result of cascading progress of fire to adjacent storage racks containing combustible materials and located at the increasing distance to the initial source of flame. Therefore, if in the presented simulation there were no conditions for increasing fire development intensity, then it must mean that only the energy released in the localized fire, affecting the limited volume of the fire zone, contributed to the increase in the exhaust gas temperature [8, 9].
The juxtaposition depicted in Figure 21 illustrates another important difference in fire development modes in the larger of analyzed halls. This difference clearly distinguishes the development typical for the scenario C from the one observed when the scenario B is considered. The course of curves depicting the increase in the temperature of fire plume identified at the distance of 3.00m from the vertical axis indicating the location of the initial flame is different here. As far as the scenario B is considered opening of the smoke vents in general did not affect the progress of fire, as the generated heat proved to be insufficient to ignite the combustible materials gathered on the nearest storage racks adjacent to the initial flame source, while the fire corresponding to the scenario C, due to the increased stacking height, was capable of progressing from storage rack to the adjacent storage rack in a

cascading mode. In such case the fire in the considered location developed rapidly, in general in a monotonous manner, and the unlimited availability of the oxygen entering the fire affected area was assured via the opened smoke vents. Thus there was no question of possible fire suppression [10].

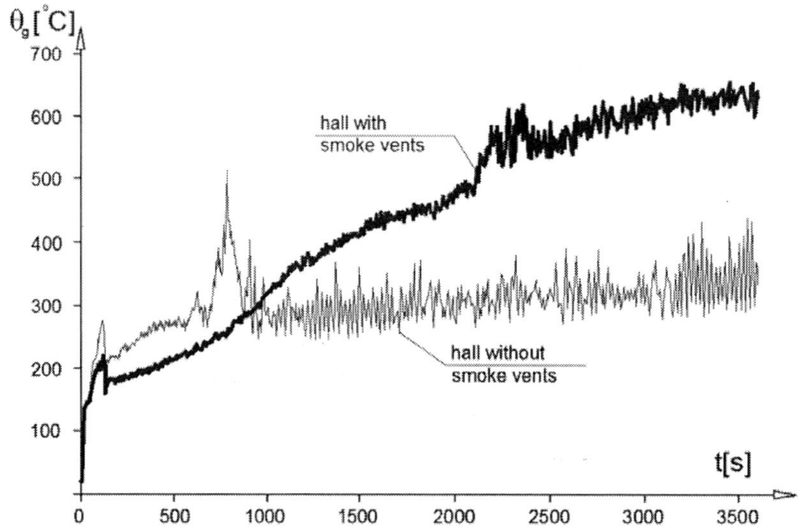

Figure 19. Scenario A – Changes in fire plume temperature determined for the fire in the smaller of considered halls at a point located 3.00m from the initial flame axis at the height of 4.00m above the floor level. In the case of a hall equipped with operating smoke vents the monotonous increase of averaged measured fire plume temperature is observed during the whole analyzed fire exposure time. For the hall devoid of vents the effects of suppressed combustion intensity typical for fires driven by ventillation conditions are visible.

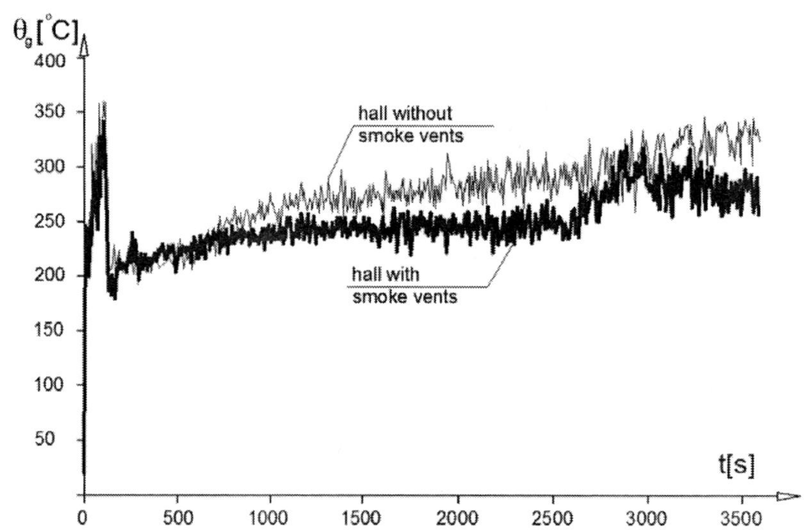

Figure 20. Scenario B – Changes in the exhaust plume gas temperature obtained inside the bigger of considered halls in a location at the distance of 3.00m from the initial fire axis and at the height of 6.00m above floor level.

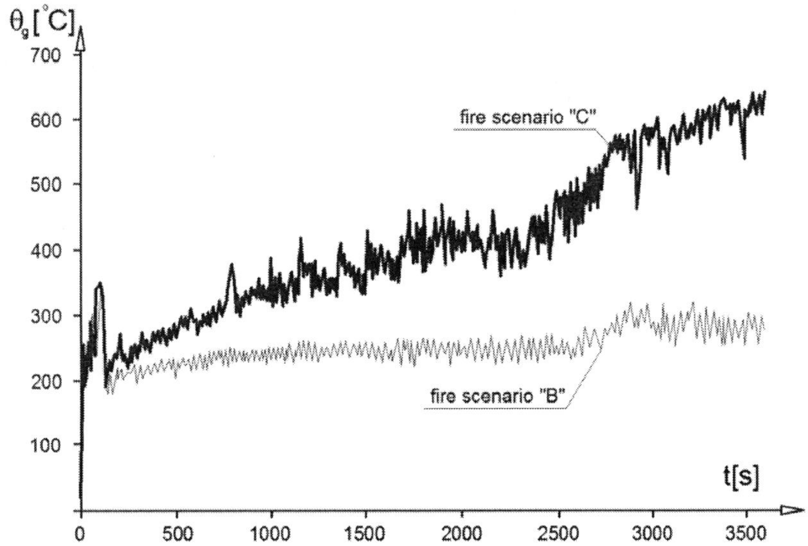

Figure 21. Changes in the exhaust plume gas temperature obtained inside the bigger of considered halls in a location at the distance of 3.00m from the initial fire axis and at the height of 6.00m above floor level – comparison of results obtained after simulation of fires developing under assumptions of scenarios B and C.

5. Concluding remarks

The results of simulations presented here seem to support the initial assumption, that for large area shopping halls, having a specific geometry, limited availability of efficient ventilation and more or less uniform distribution of combustible materials constituting the fire load, the localized fire of limited intensity may prove to be the authoritative one for identification of fire safety. Assumption of such reference fire instead of classical and commonly applied fully developed fire model, characterized by uniform distribution of temperature in the whole affected fire compartment, may allow for substantial savings gained in the process of rational selection of active and passive fire protection means. It has to be underlined here, that the more economical selection of the protection resources should be permitted only if the evaluator, despite recommended reductions, could justify the statement on guaranteeing the users of the hall the required safety level. Thus the Authors of this paper postulate, in the practically and economically justified cases, to approach in a less formal and more individualized manner the problem of specifying the required fire protection scheme. In our understanding this should be manifested by the selection of fire protection resources based on the qualitative and quantitative evaluation of risks supported by the assumption of a reliable and well justified authoritative fire development scenario followed by a detailed numerical analysis instead of schematically applying the conventional recommendations prepared for a wide spectrum of structures often substantially differing in the enclosed volume, shape and dimensions as well as mode of operation. Parameters of such a model are usually determined not only by the hall geometry, known a'priori, but also by the real hall ventilation capabilities, especially in the case of fire initiation, as well as by the quality and quantity of the accumulated merchandise representing the potential fuel and constituting the fire load. In the authors' opinion the numerical simulation of a fire development performed within the framework of FDS computer code presents an efficient computational tool allowing for reliable prediction of the most unfavourable, but at the same time probable fire scenario, which may be realized in the given design situation.

IOP Conf. Series: Materials Science and Engineering **586** (2019) 012002 doi:10.1088/1757-899X/586/1/012002

6. References

[1] Biczycki A., Maslak M., Kwasniewski L., Lukacz M., 2012, *Fire in a large-area shopping center*, in: Wald F. et al. (Eds.) – COST Action TU0904, Integrated Fire Engineering and Response – Case Studies, CTU Publishing - Production, Czech Technical University in Prague, March 2012, 131-139.

[2] Maslak M., Pazdanowski M., Wozniczka P., 2017, *Impact of the limited oxygen availability on the localized fire development in a large-area building compartment*, Proceedings of the International Scientific Conference „Fire Protection, Safety and Security 2017", Zvolen, Slovakia, May 3-5, 2017, 99-109.

[3] Maslak M., Wozniczka P., 2017, *Scenariusze rozwoju pożaru w wielkopowierzchniowej hali handlowej – część 1*, Nowoczesne Hale, 1/2017, 27-31.

[4] Maslak M., Wozniczka P., 2017, *Scenariusze rozwoju pożaru w wielkopowierzchniowej hali handlowej – część 2*, Nowoczesne Hale, 3/2017, 64-67.

[5] Wang Y., Burgess I., Wald F., Gillie M., 2014, *Performance-based fire engineering of structures*, CRC Press, London.

[6] McGrattan K., Hostikka S., McDermott R., Floyd J., Weinschenk C., Overholt K., 2013, *Fire Dynamics Simulator user's guide*, NIST, Gaithersburg, Maryland, USA.

[7] Fan Shen-Gang, Shu Gan-Ping, She Guang-Jun, Liew Richard J.Y., 2014, *Computational method and numerical simulation of temperature field for large-space steel structures in fire*, Advanced Steel Construction, 10(2), 2014, 151-178.

[8] Maslak M., Pazdanowski M., Wozniczka P., 2017, *Temperature distribution in a large-area shopping hall in the case of a localized fire*, in: Nigro E., Bilotta A. (Eds.) – Proceedings of the 2nd International Fire Safety Symposium (IFireSS 2017), Naples, Italy, June 7-9, 2017, 853-860.

[9] Maslak M., Wozniczka P., 2017, *The impact of the fire source location on fire development in a large-space steel commercial building*, Safety & Fire Techniques, 45(1), 2017, 154-169.

[10] Maslak M., Wozniczka P., Pazdanowski M., 2018, *Forecast of a fire spreading in a large-area shopping hall*, Technical Transactions, 12/2018, 61-76.

IOP Conf. Series: Materials Science and Engineering **586** (2019) 012003 doi:10.1088/1757-899X/586/1/012003

A posteriori modelling of a fire spreading in selected types of industrial halls

M Maslak[1], M Pazdanowski[1], M Suchodola[1], P Wozniczka[1]
[1] Cracow University of Technology, Warszawska 24, 31-155 Cracow, *Poland*

E-mail: mmaslak@pk.edu.pl

Abstract. Two different scenarios of fire development identified for two industrial halls are considered and discussed in detail. In both cases an attempt is made, using the appropriate numerical models, to describe a posteriori the spread of a fire that has already occurred in the past. The numerical model is implemented in a Fire Dynamic Simulator (i.e. FDS) computer program and used for simulation. The first considered hall is an one-storey two-aisled steel structure with an incomplete covering of its side walls. Recyclable materials were stored inside, mainly the plastic waste. The fire observed in this hall developed slowly, being repeatedly partly suppressed and then subjected to flashover again. The other building described in this contribution is an one-bay steel hall located in a transhipment terminal, in which the crude oil was stored in railroad tank cars. In this case, a fire initiated by the ignition of oil leaking from the damaged tank car and quickly intensified by the collapse of a part of the hall roof proceeded rapidly until the fuel supply ran out. The basic parameters of both models developed by the authors were calibrated on the basis of available material data as well as of the reports received from meteorological stations located in the vicinity of the destroyed halls.

1. Introduction

An inventory of the observed damage, i.e. deformations of structural components as well as their joints, plays an important role during the evaluation of the technical condition of steel bearing structure specified after fire. In general, simple visual observation combined with appropriately formulated material analysis program is insufficient to make an informed decision on possible extended service of selected structural components subjected to the episodes of at first monotonous heating due to the direct fire action of increasing intensity followed by intensive cooling during the firefighting period. In order to properly interpret and assess the state of damage after the fire it is advised to a posteriori replay the fire, which inflicted the observed damage [1]. It is not just about finding the values and distributions of internal forces actually generated in the structural members during the fire, which affected the considered structure. Thus, it is also important to verify, whether the a posteriori inventoried deformation state is formally in agreement with the a priori modelled fire plume temperature field in the nearest vicinity of structural members and the changes of basic parameters describing this field during the fire. In addition, a posteriori identification of possible local very intense heating or abrupt and nonuniform cooling episodes of the structural steel may give rise to suspicion, that adverse effects probably may have occurred in the microstructure of the steel, resulting in increased susceptibility to brittle failure during further service [2]. Changes of this type are extremely difficult to detect by direct approach, without anticipation due to the prior analysis. In this paper we intend to present two applications of the classical computer code – the *Fire Dynamic Simulator* (FDS) [3] to a posteriori model the development of two fires, which have occurred in real life and resulted in partial destruction of industrial steel halls [4]. Those two fires were completely qualitatively different, with different genesis, intensity and affected zone, though in both cases the obtained results proved to be useful not only in confirmation and justification of the conclusions drawn in appropriate expert appraisals but also in preparation of repair programs dedicated to both halls considered in this paper [5].

IOP Conf. Series: Materials Science and Engineering **586** (2019) 012003 doi:10.1088/1757-899X/586/1/012003

2. A fire in the steel hall of recycled materials warehouse

2.1. Description of the hall and the course of fire recreated based on eyewitness testimony

The considered hall was a single storey structure without basement, with incomplete sheathing of side walls (Figure 1). Its surface area was equal to 1314m^2 while the enclosed volume was 12337m^3. The basic dimensions of the hall within the system axes were 36.1m x 60.0m, the maximum height at the ridge 12.2m and the distance between the system frames 10.0m. The columns have been made of rolled steel, wall beams of rectangular hollow sections while purlins have been designed and made of multi span hot rolled I beams. The sheathing has been made of profiled trapezoidal sheet metal. The structure had been designed to store recyclable materials, mostly scraps of man-made materials (plastics) [4, 5].

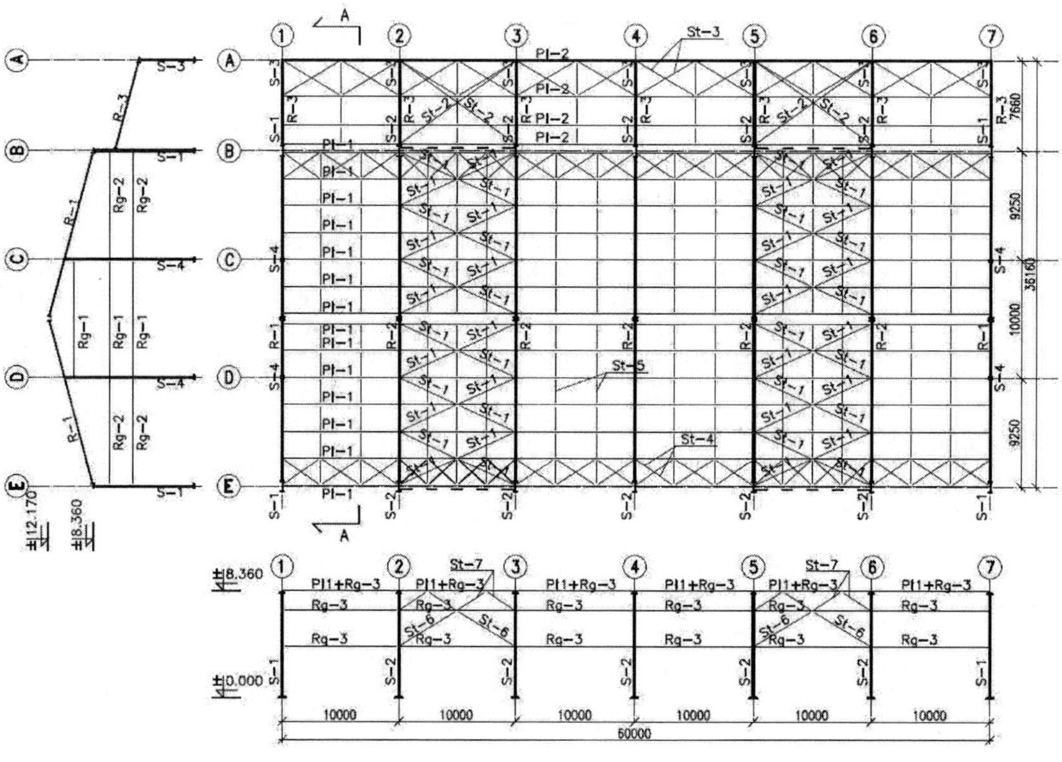

Figure 1. Statical scheme of the hall described in the text

The fire occurred during winter, while roof was free of snow cover, but the winds were strong. The eyewitness testimony indicated, that the fire lasted intermittently for approximately three days. The slowly burning recyclable materials faded for several times and after each fading flashed over again (Figure 2). The hall survived the fire, but the fire affected mostly the wall beams, braces and frames located at both ends of the hall. The columns of these frames sustained permanent damage in the form of bending (Figure 3). The remaining bearing elements, i.e. the roof purlins and the intermediate transverse frames have been deformed only locally or did not sustain any thermal deflections visible by naked eye. After the fire the hall has been reconstructed [4]. The a posteriori inventory indicated, that the average volumetric density of the materials stored in the hall at the time of fire was equal to 250kg/m^3, while the stacking height did not exceed 3.00m.

Figure 2. Described hall during the fire.

Figure 3. The state of the hall after the fire. Side view.

2.2. Parameters of the numerical model applied to a posteriori simulate the fire

Due to the nature of the stored material, the data corresponding to the properties of municipal waste have been used to simulate the fire load gathered in the hall, as it was recommended in [6]. It was a good approximation because these wastes consisted mainly of plastics intended for use as recyclable material. Based on the paper [6] the burning speed of such materials may be assumed in between 0.018–0.034 kg/(m^2·s). The heat of combustion of municipal waste usually is in between 16.7–29.0 MJ/kg. In the experimental analysis of shredded waste, composed of 62% man-made materials and 23% biomass (timber and other organic waste) of low humidity, the heat of combustion has been proved to be at the level of approximately 20 MJ/kg, as reported in the paper [6]. Considering the condition of the waste at the moment of fire initiation, and especially the lack of processing and partial exposure to precipitation the maximum rate of heat release has been estimated based on the formula:

$$RHR_f = Qv \tag{1}$$

where the heat of combustion Q has been assumed conservatively at the level of 16.7 MJ/kg, and the combustion speed v accordingly at the level of 0.018kg/(m^2·s). The value of RHR_f=300kW/m^3 estimated in this manner is close to the heat release rate assigned in the code [7] to offices and buildings of collective residence. The fire development scenario in this structure, according to the recommendations contained in [7], has been modelled as a t-square fire with subsequent phases described by the following equations (Figure 4):

$$Q(t) = \begin{cases} 10^3\, t/t_\alpha & \text{when} \quad 0\text{ s} \le t < 328\text{ s} \\ 300 & \text{when} \quad 328\text{ s} \le t < 29400\text{ s} \\ -0{,}012t + 652{,}8 & \text{when} \quad 29400\text{ s} \le t < 54400\text{ s} \end{cases} \tag{2}$$

In those equations the variable t represents the fire duration time measured in seconds, while the set value t_α stands for the time after which the heat release speed (in other words the fire intensity) reaches the threshold of 1MW.

The whole main bay area of the considered hall has been assumed as affected by the fire (burning fuel). As the information acquired from the eyewitnesses indicated that the fire burned for several days, the fire has been classified as a slow burning one in the sense of classification published in the code [7], this in turn means, that t_α=600s. Based on this it was estimated, that the growth phase took 328s, the fully developed fire phase took approximately 8 hours, while the fire extinguishing phase took additional 7 hours (Figure 4) [8]. According to the conventional t-square fire model it was assumed in addition, that the beginning of the fire extinguishing phase started after 70% of the combustible material burned out. The chemical composition of the burning material, needed to perform calculations, has been assumed in a simplified manner following the advice found in [3], with the values of functions responsible for generation of carbon monoxide and soot equal to: CO_YIELD=0.63 and SOOT_YIELD=0.163, respectively. Based on the archival meteorological data the weather conditions in the nearest vicinity of the fire have been recreated as well. A constant north-westerly wind has been modelled, with an average speed of 5.5m/s, and daily average air temperature at the beginning of numerical simulation has been assumed as equal to 0°C.

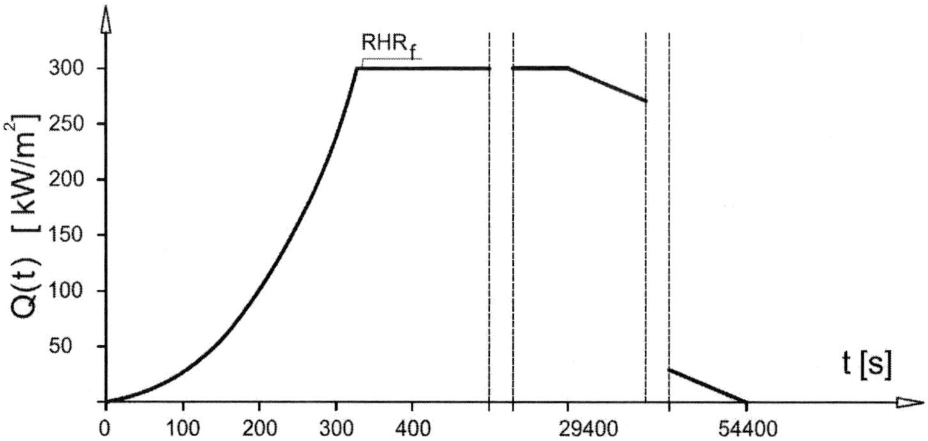

Figure 4. The course of a fire in the considered hall, simulated according to the *t*-square fire model in the coordinate system of: fire exposure time – released energy.

2.3. Discussion of the results obtained during numerical analysis
Results of the computer simulation performed by FDS [3] are shown in Figure 5. In this Figure the distribution of fire plume temperature recreated for the cross section located between the system axes 5 and 6 of the analyzed hall (Figure 1) is depicted after 30 minutes of the simulated fire. It is clearly visible, that the highest temperature has been reached along the longitudinal axis of the hall, directly under the ridge. This is, according to the Authors, a direct result of the hall shape, and precisely a ready and unlimited availability of combustion supporting oxygen from the surroundings due to the hall having incomplete sheathing of walls as well as stacking of the burning material into a more or less symmetrical prism with a ridge coinciding with the roof ridge axis. In Figure 6 the change of fire plume temperature during the simulated fire at the points located in the nearest vicinity of roof sheathing is depicted in detail. As one may easily observe, after reaching the fully developed fire stage, this temperature in both considered points climbed to reach the maximum value of approximately 400-450°C, which quickly stabilized and did not exhibit any significant growing trend in the following burning phase of the materials stockpiled in the hall. Similar value of the temperature observed at the same moments of fire at points located along the roof ridge and eaves indicates that the fire plume temperature evened out in the whole fire affected zone. However, in the simulation performed substantially different temperatures have been determined for the individual steel elements constituting components of the bearing structure. These differences are depicted in Figure 7, for the selected points on the main beam of the transverse frame and the adjacent wall beam, respectively. The differences in the heating speed of individual bearing components may be, therefore, attributed not only to the differences in the size of steel sections used to erect the hall but also to the differences in the location of measurement points, which directly affects the effective ventilation possible (the ridge well ventilated form the bottom versus badly ventilated eaves located in the corner between the wall and roof sheathing). The graphs juxtaposed in Figure 7 indicate, that in each of the compared points after 60 minutes of fire exposure the temperature did not exceed the level of 400°C. This in turn

means, that the yield limit of the steel of which the bearing structure has been made did not have to be reduced because of thermal action.

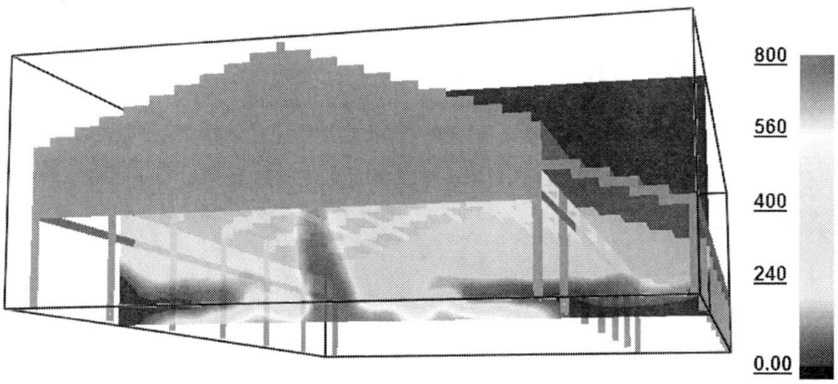

Figure 5. Distribution of fire plume temperature obtained in the considered hall after 30 minutes of the numerically simulated fire exposure.

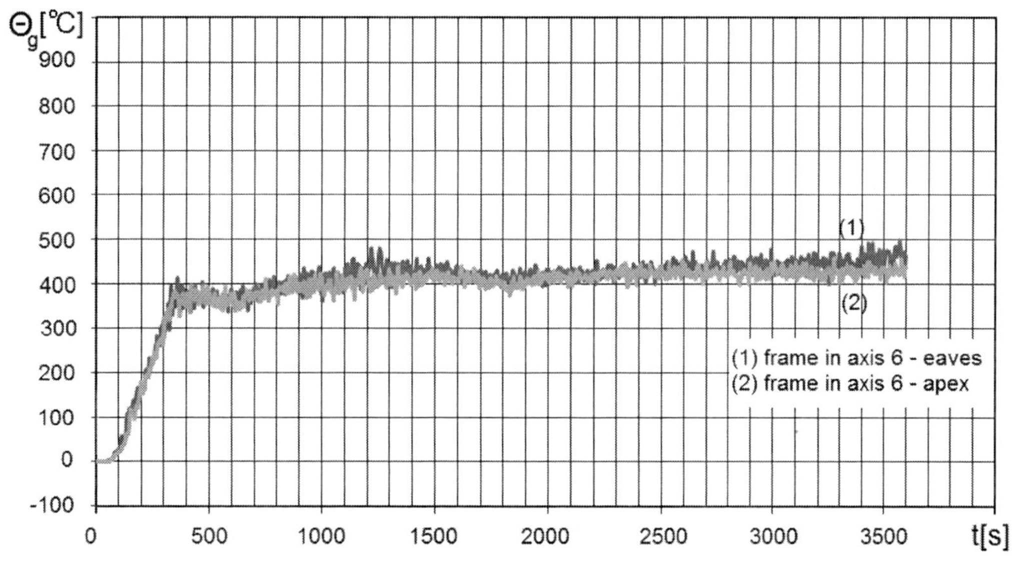

Figure 6. Development of the fire plume temperature in the numerically simulated fire exposure determined under the ceiling of the considered hall in the direct vicinity of the eaves and the ridge of the girder located at axis 6 (Figure1).

The above presented results of performed a posteriori numerical simulation of the fire seem to be formally in agreement with prior provisional conclusions, formulated ad hoc during taking the inventory of the deformation state of the structure after fire. These deformations, relatively limited with respect to both the number and the measured deviations from the initial shape (Figure 3), could be justified simply by the restrained thermal expansion capability of structural members during the fire, without the need to take into account the additional phenomenon of weakening the strength of material.

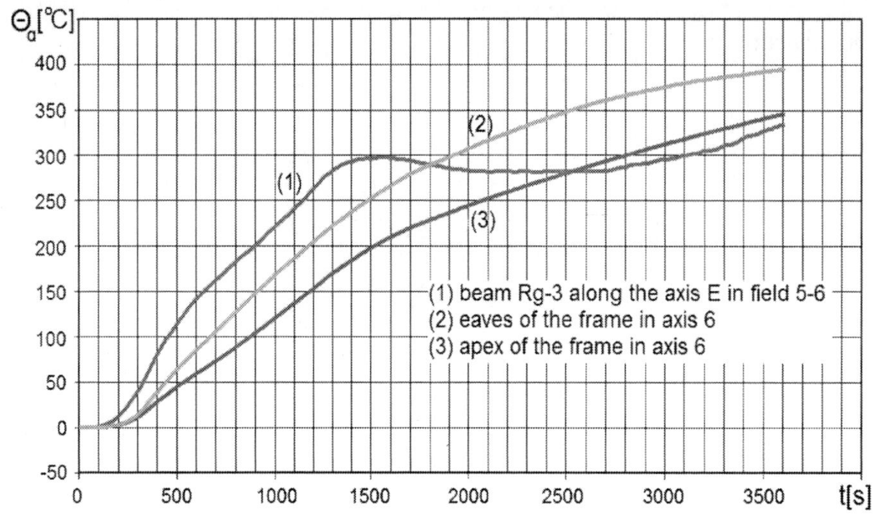

Figure 7. Changes in the steel temperature recorded during the numerically simulated fire exposure at the ridge and eaves of the main frame located at the axis 6 juxtaposed with corresponding changes in the temperature of the top steel beam located along the axis E in between the axes 5 and 6 (Figure 1).

3. Fire of the steel structure of a railway transhipment terminal

3.1. The cause and course of the fire

The fire described in this chapter occurred during summer in a hall of railway transhipment terminal, where the goods transported in railway cars conforming to gauge standard in European Union countries are transferred to the railway cars operating on wide gauge prevalent in the countries of Eastern Europe. This fire was initiated by the ignition of fuel spilled over a large area inside the terminal. This fuel leaked unnoticed for a long time from a depressurised railway tank car. The tank car contained raw oil, an easily inflammable material burning in a violent manner. The high intensity fire, according to eyewitness testimony, lasted for many hours, until the fuel burned completely. The bearing structure of the transhipment terminal consisted of a steel hall. The scheme of the hall is depicted in Figure 8. In the same picture the inventoried state of the deformed bearing members after the fire is presented in a simplified manner using exaggerated displacements [4]. The hall was one storey high, without basement, having the floor area of 770 m^2 and enclosed volume estimated at 5780m^3 (Figure 9). During the fire the roof of the hall was destroyed, falling in into the zone of intensive fire exposure while all the girders of the frames neighbouring with this zone sustained significant thermal deformations (Figure 10). Large part of the hall sheathing, including the supporting horizontal beams has been completely destroyed. Thus the sheathing ceased to function as a barrier between the inside of the hall and outside world and allowed for unlimited availability of combustion supporting oxygen. The main columns of transverse frames located outside the zone delimited by the destroyed roof survived the fire relatively intact, with only local deformations due to the thermal actions. As the fire occurred during summer, when the roof was loaded only by the covering material made of trapezoidal sheet metal panels, the 3D displacements depicted in Figure 8 could be attributed to the thermal loads only.

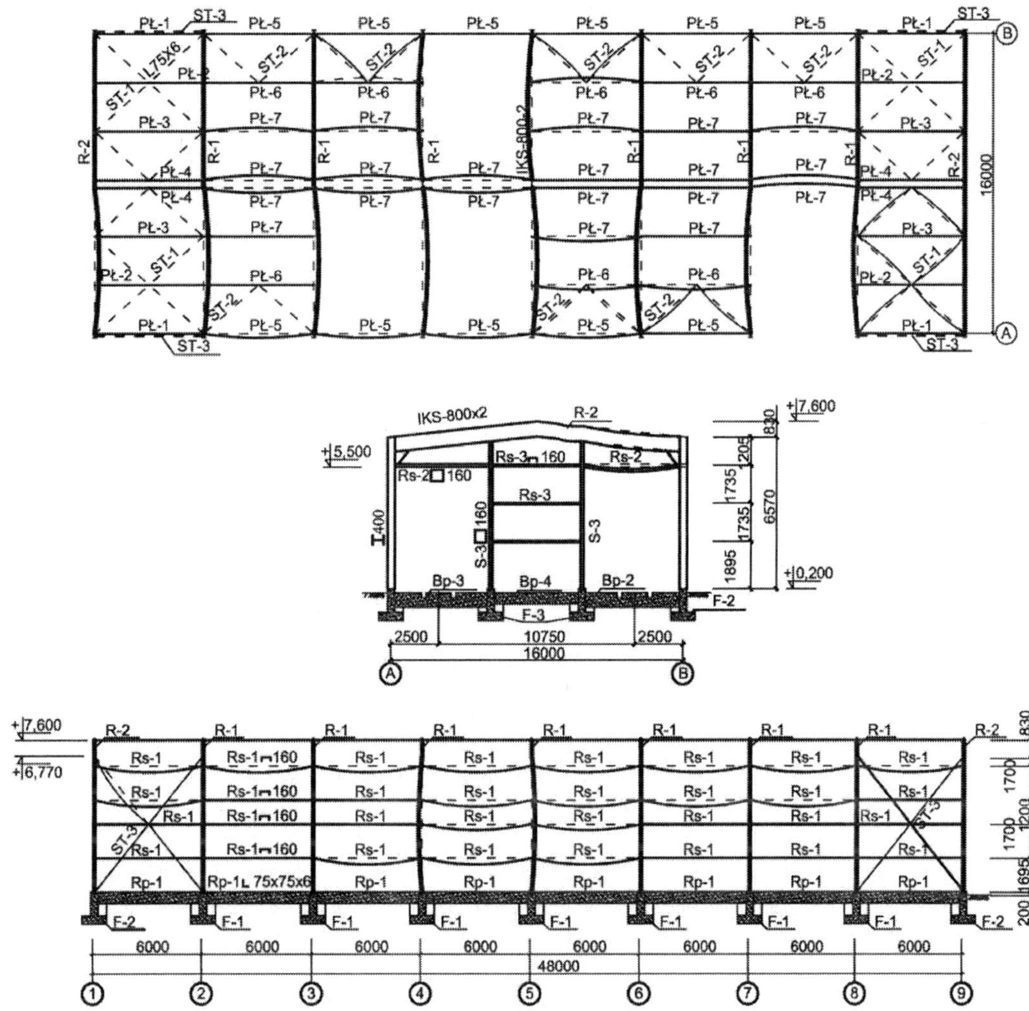

Figure 8. Scheme of the railway transhipment terminal hall described in this paper. The heat induced deformations of the bars inventoried after the fire are shown exaggerated.

3.2. The purpose and parameters of the a posteriori fire development simulation

During the post factum recreation of the considered fire it has been assumed, that because of the fuel initiating combustion, it followed the scenario of hydrocarbon fire [7]. The observation of this type leads to the conclusion, that if only the unlimited availability of oxygen from the outside to the fire area is ensured, the fire would almost immediately, in general during the first minute of fire exposure transform into the stage of fully developed fire, and the temperature of fire plume very quickly would reach the level exceeding 1000°C and remain more or less stable at this level until the fuel is completely exhausted. Such fire development scenario should in general result in complete destruction of the analysed structure, but this has not been observed on site after the inventory has been taken. The independently conducted a posteriori numerical fire development simulation precisely correlated with the real conditions determined for the terminal hall considered in this example should be able to dispel the doubts formulated in such a manner. This simulation has been performed using FDS code [3]. For the purpose of this analysis the heat release rate in the numerically simulated fire exposure has been assumed as equal to RHR_f=500kW/m^2, while the combustion speed ν=0.018kg/(m^2·s).

Figure 9. The terminal hall considered in this example after the fire has been extinguished.

Figure 10. Thermally generated deformations of beams in transverse frames observed a posteriori in the zone especially affected by the fire intensity.

3.3 Analysis of obtained results

In the numerical simulation performed a special emphasis has been laid upon a posteriori replication of the first 15 minutes of fire development, counted from the moment of initiation. According to the conventional hydrocarbon fire model after such fire exposure time the temperature of exhaust gases, uniform in the whole fire compartment, should have reached the level of:

$$\Theta_g = 1080\left(1 - 0,325e^{-0,167\cdot 15} - 0,675e^{-2,5\cdot 15}\right) + 20 = 1071\ ^{\circ}\mathrm{C} \tag{3}$$

Correlating this result with the results obtained via the numerically simulated for the considered hall fire plume temperature curve $\Theta_g(t)$ specified for the nearest vicinity of the eaves in the most intensely heated cross section of the bearing structure, depicted in Figure 11, one may confirm the thesis, that the so high level of the fire plume temperature has been actually reached in the structure. The significant difference with respect to the classical curve describing the hydrocarbon fire lies in the fact, that the level of the temperature Θ_g exceeding the value of 900°C has been obtained in the numerical model only after approximately 20 minutes of intensive burning of the spilled oil, instead of almost immediately as the standard model predicts. The temperature of the exhaust gases after 15 minutes (i.e. 900 seconds) of simulated fire exposure reached only about 500°C, i.e. much less than forecast by the formula (3). This in turn means, that in the case considered here the hydrocarbon fire model proved to be overly conservative. This conclusion seems to be corroborated by the curve depicting the evolution of the maximum steel temperature Θ_a determined at the eaves of the most intensely heated transverse frame in the considered hall. This curve, for the comparative purposes, has been shown in Figure 11. As one may see there, the temperature of the considered beam stabilizes at the level of approximately 900°C after more or less 22 minutes of interaction with hotter and hotter exhaust gases.

The temperature Θ_a distribution map shown in Figure 12, determined after 30 minutes of numerically simulated fire exposure for the sheathing panels on the roof, and the steel transverse bearing frames of the hall justifies the selection of the eaves cross section as authoritative for analysis. It is visible there, that the highest steel temperature values have been reached in the zones adjacent to the eaves. This may be attributable to the somewhat restricted ventilation in this zone, and thus limited cooling capability of the heated structural members. Air circulation in the nooks and crannies of this type is effectively restricted and its outflow limited on both sides by the sheathing of the hall. Of course in a real fire such temperature level will most probably never be recorded in steel sheathing panels, as these panels will sooner fall to the ground, as happened in reality during the fire analyzed here. Additionally, the map presented in Figure 12 confirms the limited reach of the intensive fire exposure zone. As it is visible there, this zone in general did not reach to the outermost transverse frames, and this is formally in agreement with the a posteriori inventoried structural member deformation state depicted in Figure 8.

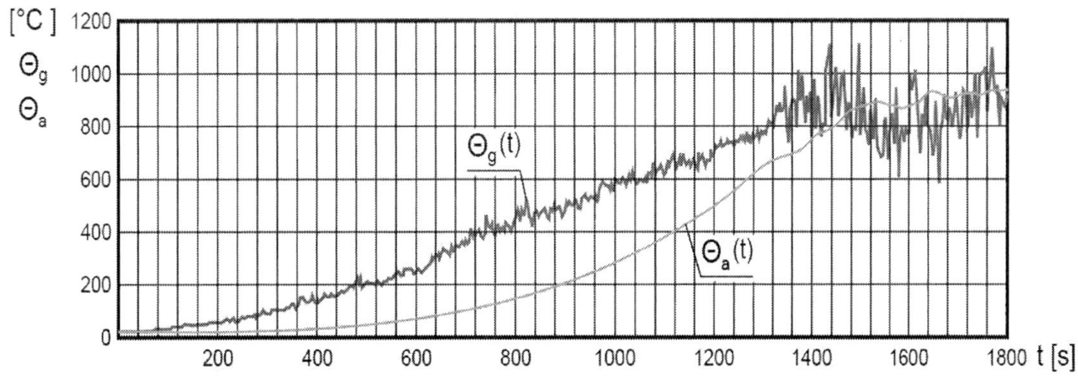

Figure 11. Development of the exhaust gas temperature Θ_g and maximum steel temperature Θ_a determined at the eaves of the most intensely heated beam of the transverse frame under the conditions of numerically simulated fire exposure for the transshipment terminal hall.

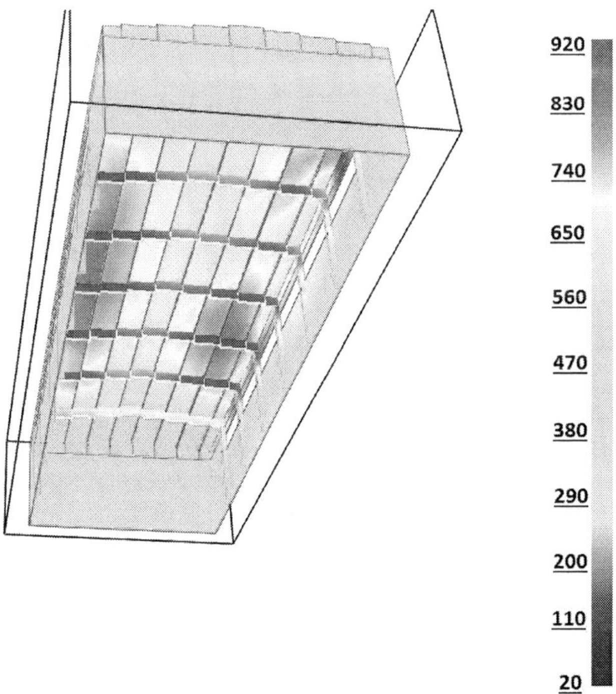

Figure 12. Temperature distribution map on the steel roof sheathing and steel bearing beams of the terminal hall obtained after 30 minutes of simulated fire exposure.

4. Concluding remarks

In this paper, based on the two selected examples related to a priori occurring fire development scenarios, we have tried to prove the usefulness of numerical simulation performed a posteriori and replicating the real conditions of fire exposure. According to our opinion the data obtained during such simulations, in both considered cases, significantly expanded the initial knowledge base of the evaluating expert, gathered in the conventional manner by visual identification and accompanied by additional supplementary material analyses. The results presented in this paper seem to prove, that the simple schemes and models of analysis applied initially have been often overly pessimistic with too wide margin of safety. The numerical analysis allowed for identification of weak spots in the structure, often difficult to discover when only the conventional methods of evaluation are applied, but never the less very important from the point of view of safety guaranteed to the user, should the decision be made to continue the use of the structure in its current state. The proposed approach, extended by numerical simulations yields the life of an expert making the decision on possible rebuilding works much easier, should those works be deemed necessary and possible. Of course, the reliability of the evidences obtained this way is limited by the diligence of the person conducting the numerical experiments and replicating the real conditions existing a priori within the considered structure at the moment of fire initiation. Of these conditions the parameters describing the potential combustible material, and especially its characteristics determining the burning speed and efficiency, as well as the conditions of ventilation prevalent in the fire affected area seem to play the most important role, as these conditions determine whether the fire will remain at the stage of localized fire with limited affected zone or will flare up and proceed into the stage of fully developed fire. The considerations presented in this paper are based on simulations performed using the computer code of the FDS type. Though commonly used in the engineering practice because of availability and ease of use this is only one of the programs, which may be applied for this purpose by the contemporary engineer.

5. References

[1] Maslak M., 2012, *Badania stali konstrukcyjnej po pożarze w kontekście oceny możliwości jej dalszego użytkowania w elementach nośnych ustrojów budowlanych*, Przegląd Budowlany, 6/2012, 48-51.

[2] Maslak M., Zwirski G., 2017, *Changes in structural steel microstructures following heating and cooling episodes in fires*, Safety & Fire Technique, 48(4), 2017, 34-52.

[3] McGrattan K., Hostikka S., McDermott R., Floyd J., Weinschenk C., Overholt K., 2013, *Fire Dynamics Simulator user's guide*, NIST, Gaithersburg, Maryland, USA.

[4] Gwozdz M., Suchodola M., 2016, *Bezpieczeństwo pożarowe budowlanych konstrukcji metalowych*, Cracow University of Technology Publishing House, Cracow, Poland.

[5] Gwozdz M., Tkaczyk A., Wozniczka P., 2016, *The reconstruction of fire-damaged industrial steel halls*, Safety & Fire Technique, 44(4), 2016, 51-66.

[6] Jodkowski W., Sitka A., Szumilo B, 2014, *Gasification of the oversize fraction of municipal waste with thermal power generation*, Archives of Waste Management and Environmental Protection, 3/2014, 45-52.

[7] EN 1991-1-2, Eurocode 1. Action on structures. Part 1-2. General actions – Actions on structures exposed to fire.

[8] Wozniczka P., 2017, *Modelowanie komputerowe pożaru stalowej hali magazynu surowców wtórnych*, Inżynieria i Budownictwo, 2/2017, 103-106.

IOP Conf. Series: Materials Science and Engineering **586** (2019) 012004 doi:10.1088/1757-899X/586/1/012004

Impact of wood species on the timber beam strength and stiffness under fire

K Kmiecik[1]
[1] Institute of Building Materials and Structures, Faculty of Civil Engineering, Cracow University of Technology, 31-155 Cracow, *Poland*

E-mail: kamil.kmiecik@pk.edu.pl

Abstract. A lot of different types of wood can be used to build the structure. Main constructions are usually built of solid wood or glued laminated timber. The division into design classes depends on the strength and mechanical properties of wood. One of the most important aspects to consider when designing timber structures is to provide fire resistance. In contrast to other construction materials such as concrete, steel, ceramics, wood is a combustible material. When timber member is subjected to fire, the surface of the wood initially ignites and burns rapidly. At 300°C the pyrolysis takes place which cause loss in mass and decrease the strength and mechanical properties. Charring rate and decreasing of properties depend on the wood species. As a consequence, it has a significant impact on the timber member strength in fire conditions. This article presents a summary from numerical simulations on impact of wood species on the timber beam strength and stiffness in fire conditions. Numerical models of timber beams were made in the SAFIR software. All modeled beams have the same dimensions. Each beam was loaded with a bending load. Also the fire load from four sides of the elements according to the ISO fire curve was given. Species and classes of wood were differentiated. Using the isotherm 300°C from the SAFIR program, time-dependent residual cross-sections were determined. As a result, the reduction of bending capacity and the increase in deflections of timber beams were obtained. It was noticed that higher class wood had a relatively lower load-bearing capacity reduction under fire.

1. Introduction

Wood is one of the oldest known materials used in construction and one of the most sustainable resources available. Timber has a very strength to weight ratio. It is used for variety of structural elements such as girders, columns, beams and trusses. Timber is a good construction material for many reasons: durability, high strength to weight ratio, good environmental influence, aesthetic values, high energy efficiency and quick erection time.

Timber for structural use is graded into strength classes. European strength class system is based on three grade determining properties: strength ($f_{m,k}$), modulus of elasticity parallel to the grain ($E_{0,mean}$) and density (ρ_k). Other strength and stiffness and density values are derived by empirical formulas from these primary properties [1]. Formulas for these derivative properties have been defined conservatively, for safety reasons. Provided in the current European standard EN 338 [2], the strength class system for softwood is based on edgewise bending properties.

Strength classes offer a number of advantages to both: the designer and the timber supplier. The designer is not obligated to check on the availability and price of a large number of species and grades that might be used. Suppliers can provide any grade that meet the strength class requirements specified in the documentation. EN 338 defines 18 strength classes: 12 for softwoods (C classes) and 6 for hardwoods (D classes). Table 1 presents strength and stiffness properties and density values for timber strength classes analysed in this paper.

As mentioned above, timber is a combustible material. Fire causes chemical and physical changes in timber structure [3, 4]. When the temperature of timber reaches 300°C the pyrolysis occurs. The result is mass loss and reduced strength. As a result of pyrolysis a char layer is formed. The charred layer

IOP Conf. Series: Materials Science and Engineering **586** (2019) 012004 doi:10.1088/1757-899X/586/1/012004

does not carry loads. Therefore, the cross-section of the member is reduced. In the residual cross-section the stress increases. EN 1995-1-2 [5] provides the constant charring rate.. In this paper, residual cross-sections were estimated using computer numerical methods.

Table 1. Strength and stiffness properties and density values for various timber strength classes.

Strength class [MPa]	Bending strength [MPa]	$E_{0,mean}$ [GPa]	Density ρ_k [kg/m^3]
C20	20	9,5	330
C30	30	12	380
C50	50	16	460
GL20h	20	9	340
GL30h	30	13	430
D30	30	10	530
D50	50	14	650

2. A description of the model

2.1. Assumptions for modeled beams

The main purpose of modelling the timber elements using computer numerical methods were to determine the impact of wood species on the timber beam strength and stiffness under fire. Figure 1 shows the geometry of the beams. A characteristic load was assumed, which was 5 kN/m. The beams were modeled using different wood species: softwood (C20, C30, C50), hardwood (D30, D50) and glued laminated timber (GL20h, GL30h). The samples were exposed to standard ISO fire [8] from four sides.

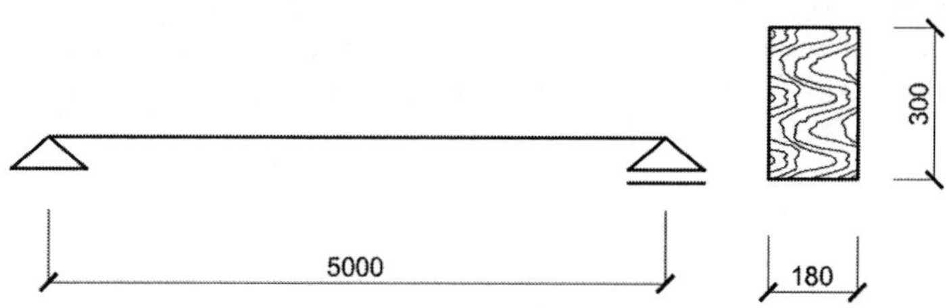

Figure 1. Geometry of the beams.

2.2. Heat transfer

Timber is a solid material, therefore the basic model is that in which heat is distributed in the structure essentially by conduction. To describe the geometry of the cross section, the fibre model is used. The cross section of the beam is subdivided into small fibres. The material behaviour of each fibre is calculated at the centre of the fibre and it is constant for the whole fibre. The general equation for two-dimensional heat conduction in wood is [6, 7]:

$$\left(k\frac{\partial^2 T}{\partial x^2}+k\frac{\partial^2 T}{\partial y^2}\right)+Q=\rho c\frac{\partial T}{\partial t} \quad (1)$$

where k is the thermal conductivity of the material, T is the temperature, Q is the amount heat generated in the material per unit volume, ρ is the density, c is the heat capacity, and t is time.
The heat flow should be defined as:

$$k\left(\frac{\partial T}{\partial x}n_x+\frac{\partial T}{\partial x}n_y\right)+q=0 \quad (2)$$

where q is the specified heat flow, and n_x and n_y are the components of the outward normal vector parallel to the x and y axes.

The net heat flux in fire conditions, should consist of heat transfer by radiation and convection. The equation is given by:

$$q" = h_c(T_f - T_s) + \Phi\varepsilon_{eff}\sigma(T_f^4 - T_s^4) \quad (3)$$

Where h_c (W/m^{2}K) is the convection coefficient, Φ is the configuration factor, σ is the Stefan-Boltzmann constant, ε_{eff} is the effective emissivity, T_s (K) is the surface of the member temperature and T_f (K) is the fire temperature. The convection coefficient h_c for the standard fire is 25 W/m2K. The efficient emissivity ε_{eff} should be defined as:

$$\varepsilon_{eff} = \frac{\varepsilon_f \varepsilon_s}{\varepsilon_f + \varepsilon_s - \varepsilon_f \varepsilon_s} \quad (4)$$

where ε_f is emissivity of the fire source, ε_s is the surface emissivity.
Other parameters necessary for the wood model are w – water content [kg/m^3], h_h - coefficient of convection on heated surfaces [W/m^{2}K], h_c - coefficient of convection on unheated surfaces [W/m^{2}K], and r – ratio of conductivity along the grain by conductivity perpendicular to the grain.

3. Residual cross-sections

Numerical models of heat flow through timber beams were made using the SAFIR software [9]. Material parameters were adopted in accordance with EN 338. Numerical simulations were conducted for seven different classes of wood. Results for two extreme cases are shown in figure 2. It was assumed that the cross-section that reached 300 °C is not able to carry loads.

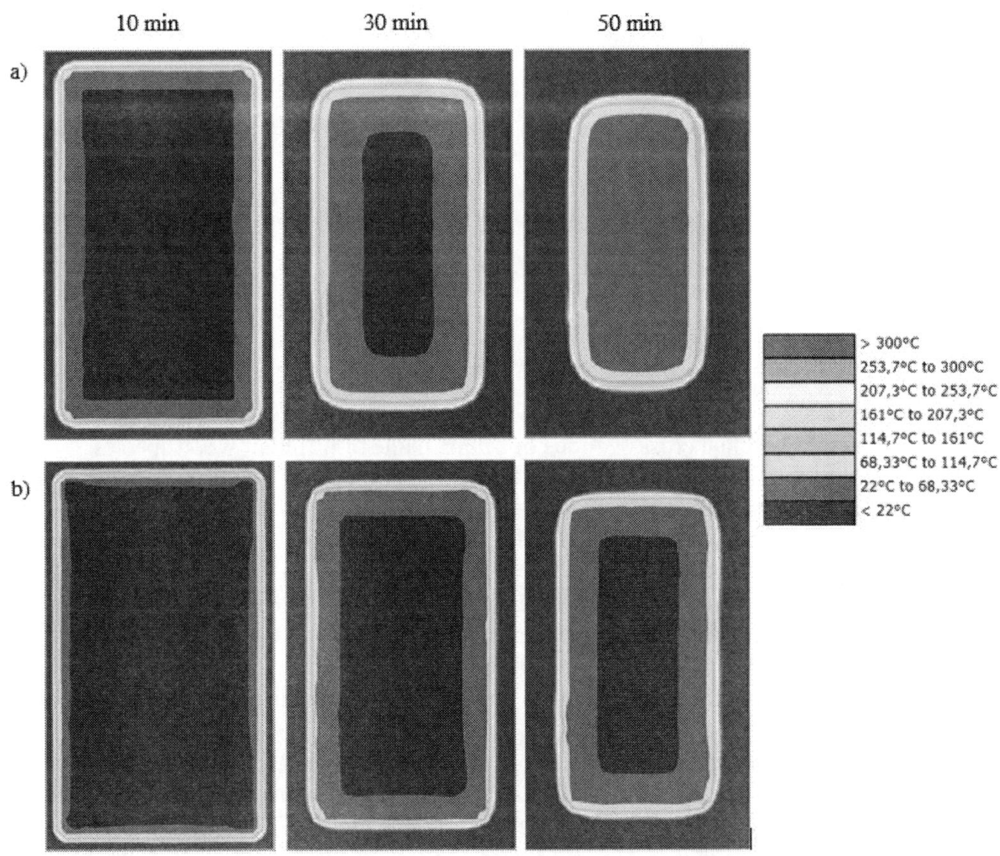

Figure 2. Temperature distribution of the timber beam under ISO fire a) C20 b) D50.

Residual cross-section was determined using the isotherm 300. Figure 2 shows temperature distribution through beams made of wood C20 (a) and D50 (b). It is easy to notice that the char layer is formed faster and deeper in cross-section made of C20. The moisture content of the wood has a huge impact on the charring. During fire, timber heats up and the moisture will begin to evaporate. The higher water content in the wood causes delay in pyrolysis. The models assume that all samples have 14% moisture.

Figure 3 shows the reduction of cross-sections during a fire for seven different classes of wood. The slowest formation of a char layer is in hardwood (D30, D50) while the fastest in the softwood (C20, C30, C50). There is a dependence that for a lower value of wood density, the cross-section reduces faster [10]. Comparing different wood species with a bending strength of 30 MPa, it can be noticed that after 60 minutes of fire duration, the cross-section of a beam made of wood class D30 decreased by about 50%. At the same time, for the glued timber beam GL30h, there was a reduction of about 60%, and for a beam made of softwood C30 by 66%.

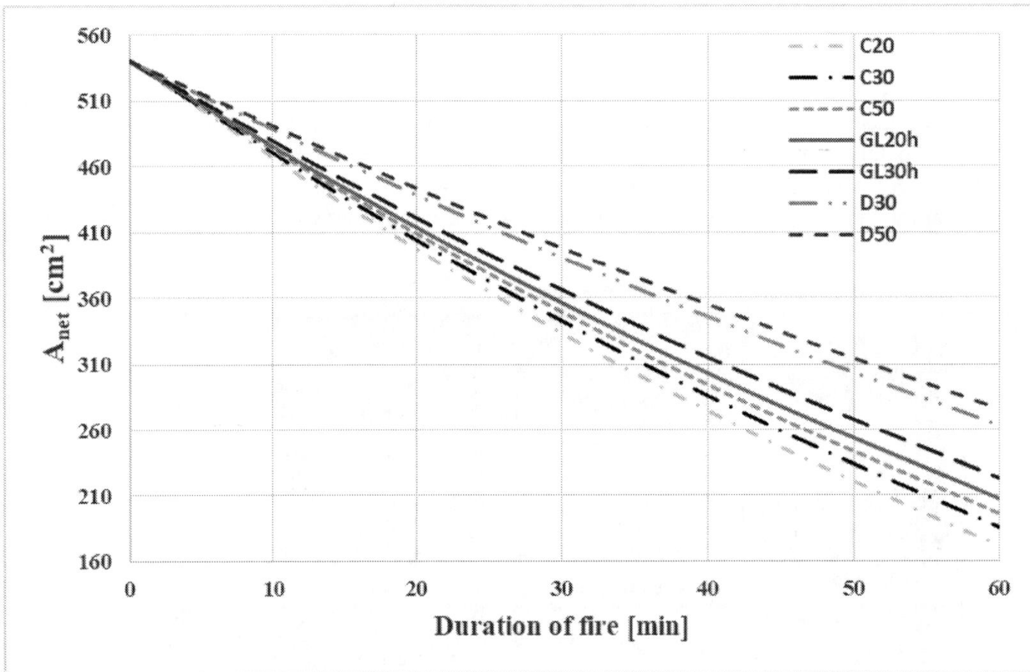

Figure 3. Residual cross-sections of beams made of different wood species.

4. Load-bearing capacity under fire

Load-bearing capacity of beams under fire depends, among other things, on stresses in residual cross-section. Due to the loss of cross-section, the stress in the element increases. Figure 4 shows the stress increment in beams during a fire. The largest increase in stress was noted in beams made of softwood. After 60 minutes of fire, stresses incremented about 5 times compared to the initial ones. On the other hand, the smallest increase in stresses is for hardwood beams. After the hour of fire, the stresses increased slightly more than 2 times compared to the initial ones. This is closely related to the charring rate.

Stresses in the cross-section are closely related to the load-bearing capacity of the beam. However, the greatest impact on the bearing capacity has bending strength. Beams made of wood of classes: D30, D50, C30 have fire resistance longer than 60 minutes. Beams made of C30 and GL30h provide load-bearing capacity for about 50 min of fire. Beams of the lowest classes (C20, GL20h) will lose their capacity after about 30 minutes of fire. A summary of the results is shown in figure 5.

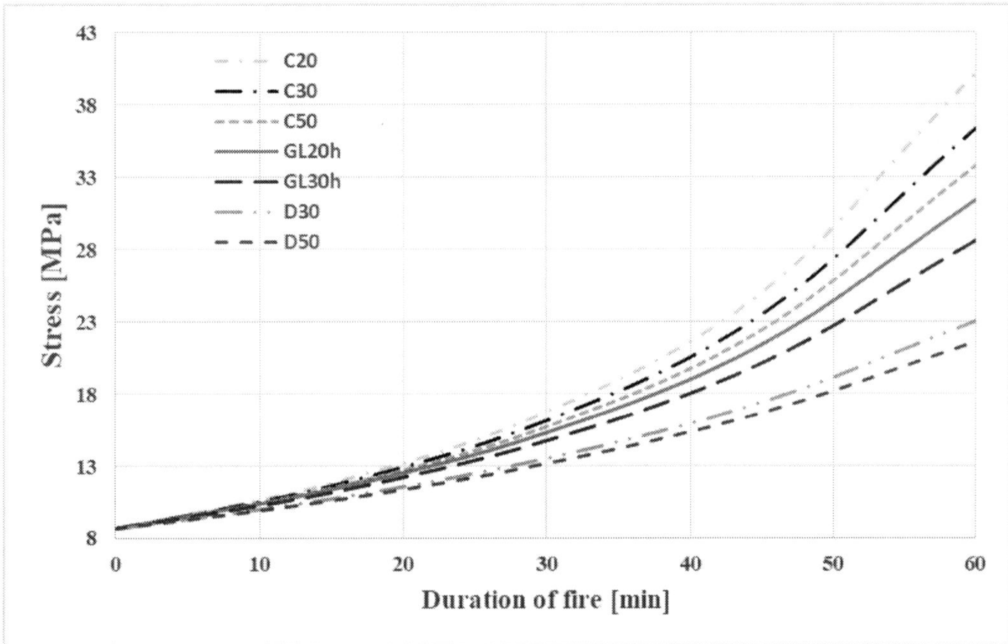

Figure 4. Stresses in residual cross-sections.

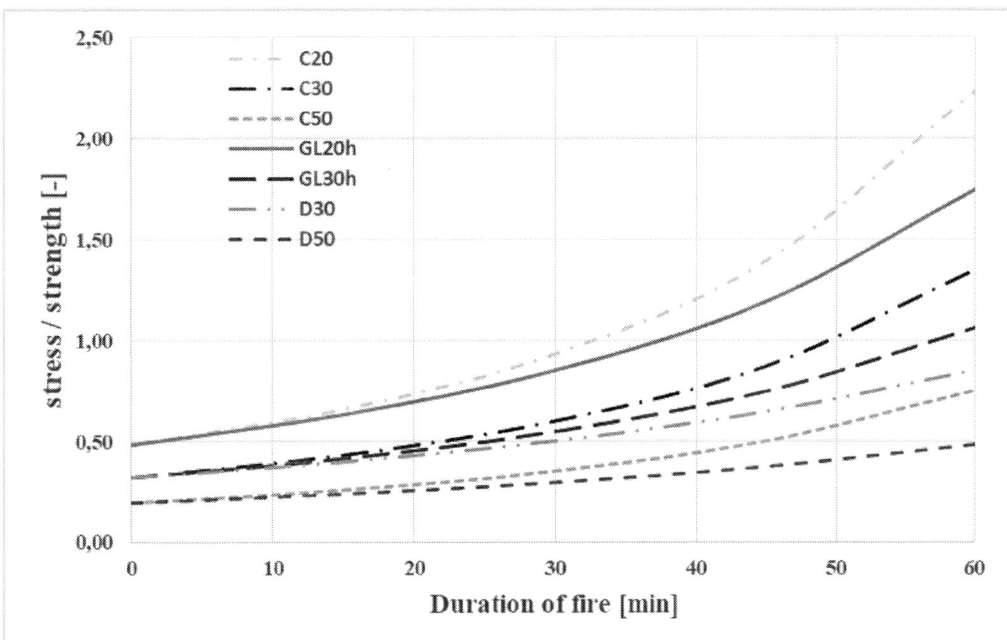

Figure 5. Load-bearing capacity (bending) for timber beams.

5. Deflections in fire conditions

Due to the reduction of the cross-section during a fire, the stiffness of the beam is decreased. The variability of the residual section during the fire was used to model the beam deflections. Figure 6 shows deflections of beams made of C20 (a) and D50 (b) after 30 minutes of fire. For each wood species analysed in this article, beam deflections were determined. Figure 7 shows the results of beam deflections during a fire, obtained from numerical modelling. Up to about 20 minutes of fire, the increase in deflections is relatively small for each case. For wood species with low density and low modulus of elasticity there are the greatest increases in deflection during a fire. For example, for C20 deflections have increased to about 700% after the hour of fire. On the other hand, for hardwood D50 at the same time, the deflections increased to 200% of the initial ones.

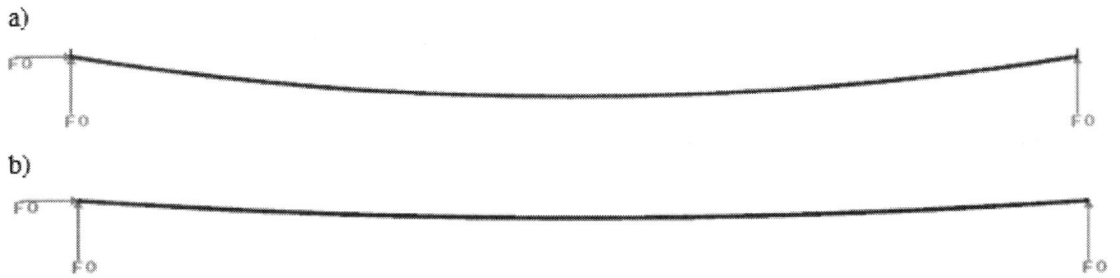

Figure 6. Deflection after 30 minutes of fire for timber beam: a) C20, b) D50.

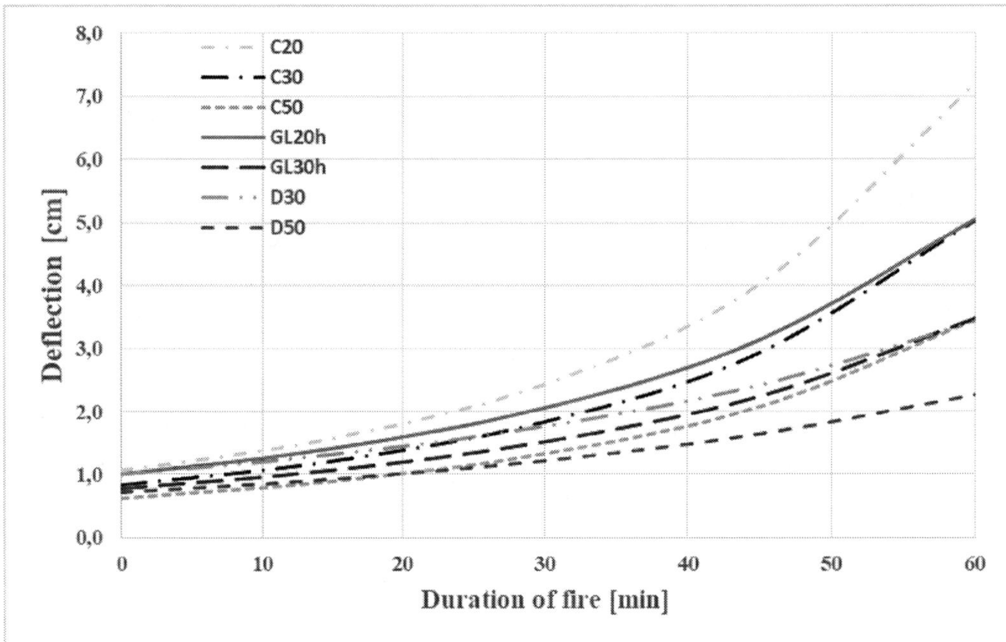

Figure 7. Deflections of beams under fire.

6. Conclusions

In this paper, numerical models were developed to analyse impact of wood species on the timber beam strength and stiffness under fire. In estimating fire resistance of beams, it is important to model the heat flow in the element. Then a char layer can be determined. Then stresses and stiffness can be obtained in the reduced cross-section.

As a result of numerical modelling, load-bearing capacities and deflections in fire conditions were obtained for beams made of various types of wood. It was noticed that hardwood had a relatively lower load-bearing capacity reduction under fire. The greatest loss of stiffness and strength was observed for beams made of low softwood classes. in numerical strength analyses, the reduction of glued laminated timber properties can be placed between hardwood and softwood.

7. References

[1] Spycher M, Schwarze F and Steiger R 2008 Assessment of resonance wood quality by comparing its physical and histological properties *Wood. Sci & Tech.* **42** pp 325-342
[2] EN 338 2011
[3] Babrauskas V 2005 Charring rate of wood as a tool for fire investigations *Fire Saf. J.* **40** pp 528-54
[4] König J 2006 Effective thermal actions and thermal properties of timber members in natural fires *Fire Mater.* **30** pp 51-63
[5] EN 1995 1-2 2008
[6] Zhang J, Xu Q, Xu Y, Wang B and Shang J 2012 A numerical on fire endurance of wood beams exposed to three-side fire *Appl. Phys. & Eng.* **13** pp 491-505
[7] Fredlund B 1993 Modelling of heat and mass transfer in timber structures during fire *Fire Saf. J.* **20** pp 39-69
[8] ISO 834 1999
[9] Franssen J and Gernay T 2017 Modeling structures in fire with SAFIR: theoretical background and capabilities *J. Struct. Fire Eng.* **8** pp 300-23
[10] Friquin K 2011 Material properties and external factors influencing the charring rate of solid wood and glue-laminated timber *Fire Mat.* **35** pp 303-27

Acknowledgments

The authors acknowledge the support of the PROM programme no. PPI/PRO/2018/1/00013/U/001 which is co-financed by the European Social Fund under the Knowledge Education Development Operational Programme.

European Funds Knowledge Education Development
Republic of Poland
NAWA
European Union European Social Fund

IOP Conf. Series: Materials Science and Engineering **586** (2019) 012005 doi:10.1088/1757-899X/586/1/012005

Experimental research of human evacuation in the case of a fire drill

Z C Grigoraș[1], M Ibănescu[1]

[1] "Gheorghe Asachi" Technical University, Faculty of Construction and Building services, Department of Structural Mechanics, Iași, 700050, *Romania*

E-mail: zeno.grigoras@gmail.com

Abstract. The main purpose of this study is to analyze the human evacuation in the case of a fire drill by two different means: experimental research - "real fire drill" and numerical simulation - "virtual fire drill".
The analyzed building can house a large number of people and its inner atrium can facilitate the spread of smoke and hot gases in the case of a real fire.
The evacuation drill was performed with about 50 students, located at the 2nd floor of the building, in two amphitheaters. They could evacuate themselves through two closed staircases to the ground floor of the building and then outside, through two exit doors.
The second purpose of this study is to establish, for the given case, if the used computer software can simulate the crowd movement and compute the travel times in accordance with the experimental research.
Another important subject is whether the computer software FDS+Evac - based on the CFD (Computational Fluid Dynamics) to simulate the crowd movement - can correctly identify the shortest escape path because one of the staircases, with its correspondent ground floor exit, is closer to the two analyzed amphitheaters.
Following the numerical analysis, it was concluded that for the considered scenario, the differences between the "real fire drill" and the "virtual fire drill" are acceptable from an engineering point of view.
FDS+Evac can simulate with high degree of credibility human movement in case of a fire drill.

KEYWORDS: fire drill, FDS+Evac, numerical simulation, CFD (Computational Fluid Dynamics), educational building, finite volume method.

1. Introduction

The purpose of this study is to analyze the human evacuation in case of a fire drill by two different means: experimental research and numerical simulation.
The experimental program presented in this paper (studied in detail by the author in his PhD Thesis [1]) addresses a direction of fire safety engineering less studied by researchers, considering the limited number of publications on human movement in case of real fires or in fire drill, as this process takes place in reality.
The numerical simulation – "virtual" fire drill – of the human movement in the analyzed building was performed and the results have been compared with those obtained from the "real" fire drill (evacuation time and escape route choosing).

2. Description of the experimental program

The considered building (Figure 1) is part of the Faculty of Construction and Building Services of the "Gh. Asachi" Technical University from Iasi, Romania.

Figure 1. Analyzed building – part of Faculty of Construction and Building Services from Iasi

The experimental program involves a fire evacuation drill from the building and monitoring the evacuation process with the help of the video surveillance system.

The occupants of the building are students and staff of the faculty (teaching staff and auxiliary staff). The users of the building are mainly young people (for students, the age range is between 20 and 25 years old), without psychic or motor disabilities and with a high-level of knowledge about the building.

The fire drill was planned, the users of the building being informed by posters of the date and time, but also, in particular, how to behave in case of such an event of evacuation.

The evacuation paths of the analyzed spaces are shown in Figure 2 for the first floor of the building and in Figure 3 for the ground floor of the building.

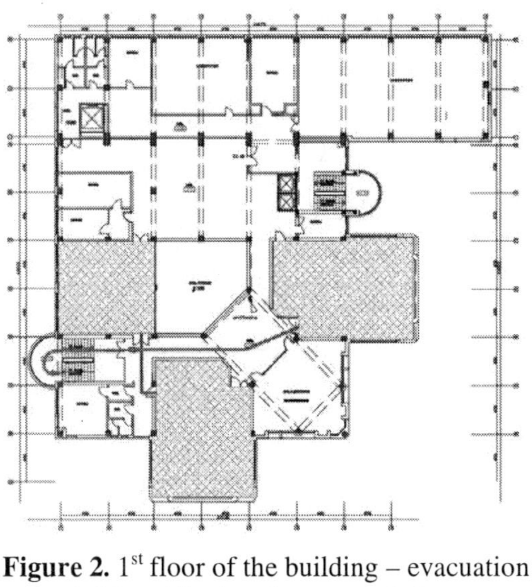

Figure 2. 1st floor of the building – evacuation paths (with red line) and evacuated classrooms (with green fill)

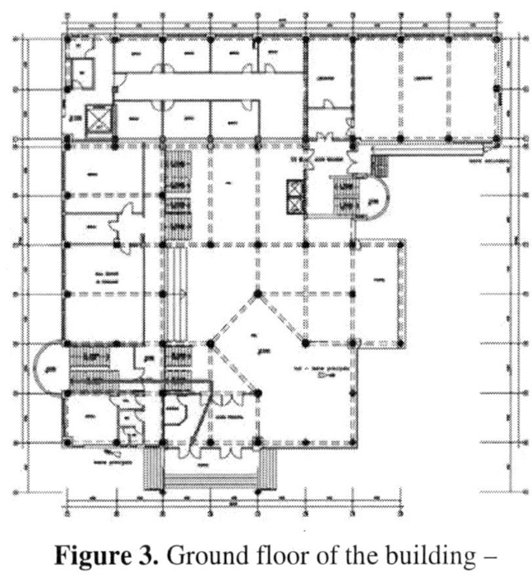

Figure 3. Ground floor of the building – evacuation paths (with red line)

3. The results of the experimental program

A total of 53 people participated in the evacuation drill, including 3 teachers and 50 students. The attending people were studying in two amphitheaters and a classroom, rooms located on the first floor of the building.

IOP Conf. Series: Materials Science and Engineering **586** (2019) 012005 doi:10.1088/1757-899X/586/1/012005

Figures 4÷9 show the movement of people, at different times, from the escape routes to the exterior of the building.

Figure 4. Amphitheaters exits

Figure 5. Main exit of the building

Figure 6. Main exit of the building

Figure 7. Secondary exit of the building

Figure 8. Main exit of the building

Figure 9. Main exit of the building

As a result of the evacuation drill, the following conclusions could be drawn:

- at the start of the alarm system, the participants responded firmly, gathering their personal belongings and leaving the rooms where they were studying;

- people travelled through the escape routes to the outside of the building in an orderly manner;
- only the main staircase and the main exit of the building have been used;
- the reaction time, measured from the moment when the alarm system was triggered till the first persons left the room, was 15 s;
- the movement time, measured from the moment when the first persons left the room till the last persons left the building, was 75 s;
- average movement speed, v=1.1÷1.7 m/s (uniform distribution);

FDS+Evac – short description

Fire Dynamics Simulation with Evacuation (FDS+Evac 2.5.0) is a human evacuation module implemented in the Fire Dynamics Simulator (FDS) software. The program is based on the analogy of crowd movement – fluid flow, considering the fire – people interaction.

FDS+Evac deals with each evacuee (called by the software "agent") as a single entity whose movement is controlled by a time-dependent equation. Each individual has his own properties and strategies for evacuation. The agents are subjected to moment and contact forces, deriving from psychological behaviour and displacement [2], [3].

People are modelled as agents moving in 2D geometries within the boundaries of the building floors (Figure 10 a). The plane size of each agent is estimated using 3 circles approximating the elliptical shape of the human body (Figure 10 b). FDS+Evac uses stochastic properties for agent characteristics. The motion equation, due to the granted degrees of freedom for translation and rotation, was solved by using specific methods for dynamics of dissipative particles. The model uses continuous time and space to trace agent trajectories (Figure 10 c).

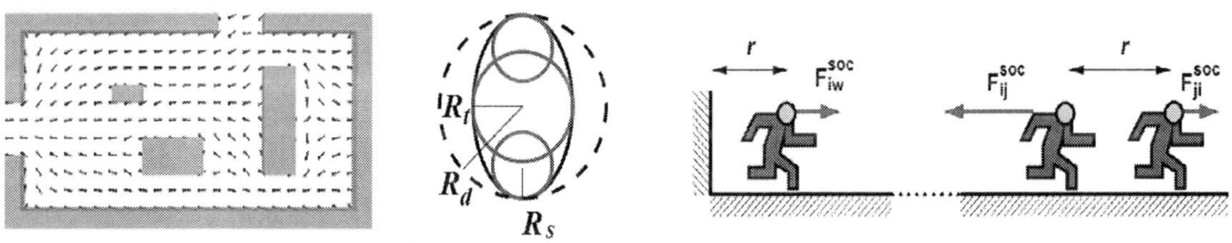

a) 2D flow field used to guide agents to the exit doors

b) human body shape

c) agents subject to forces

Figure 10. FDS+Evac agents movement modelling [3]

4. Numerical simulation

In the FDS+Evac the whole building was analyzed. The virtual model with finite volumes is presented in Figure 11. The location of the agents in the virtual computational domain is illustrated in Figure 12. Their location at time 0 (in the "virtual evacuation drill") corresponded to the moment when the "real evacuation drill" was started.

For generating the FDS and FDS+Evac input data, PyroSim (graphic user interface) was used.

In the numerical model, people were placed just like in the case of the "real" fire drill – 53 agents located in 3 classrooms on the first flood of the building. For the accuracy of the obtained results, the furniture in the classrooms has been modelled as obstacles for the moving agents.

The following characteristics, Figure 13, have been considered in the behaviour and agent modelling:

- detection time: 0 s (evacuation drill);
- reaction time: 15 s (taken from the "real" evacuation drill");
- displacement speed: 1.1÷1.7 m/s (taken from the "real" evacuation drill");
- software's default dimensions of the human body (correspond for "adult" type agents);

IOP Conf. Series: Materials Science and Engineering **586** (2019) 012005 doi:10.1088/1757-899X/586/1/012005

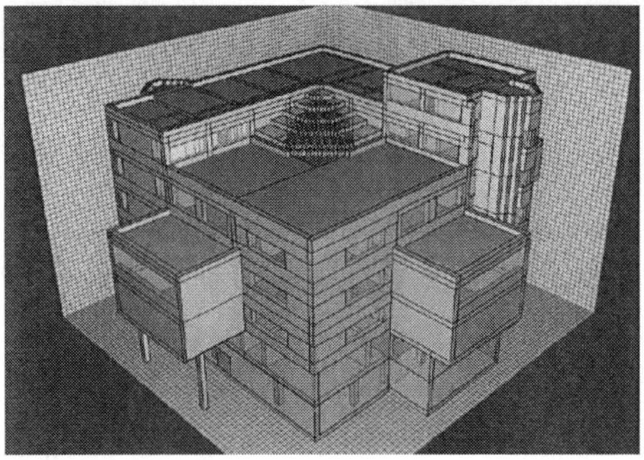

Figure 11. The virtual model of the analyzed building

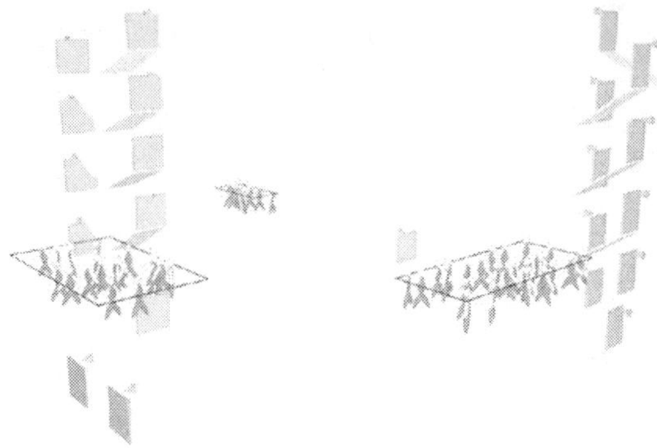

Figure 12. Location of agents within the computing field

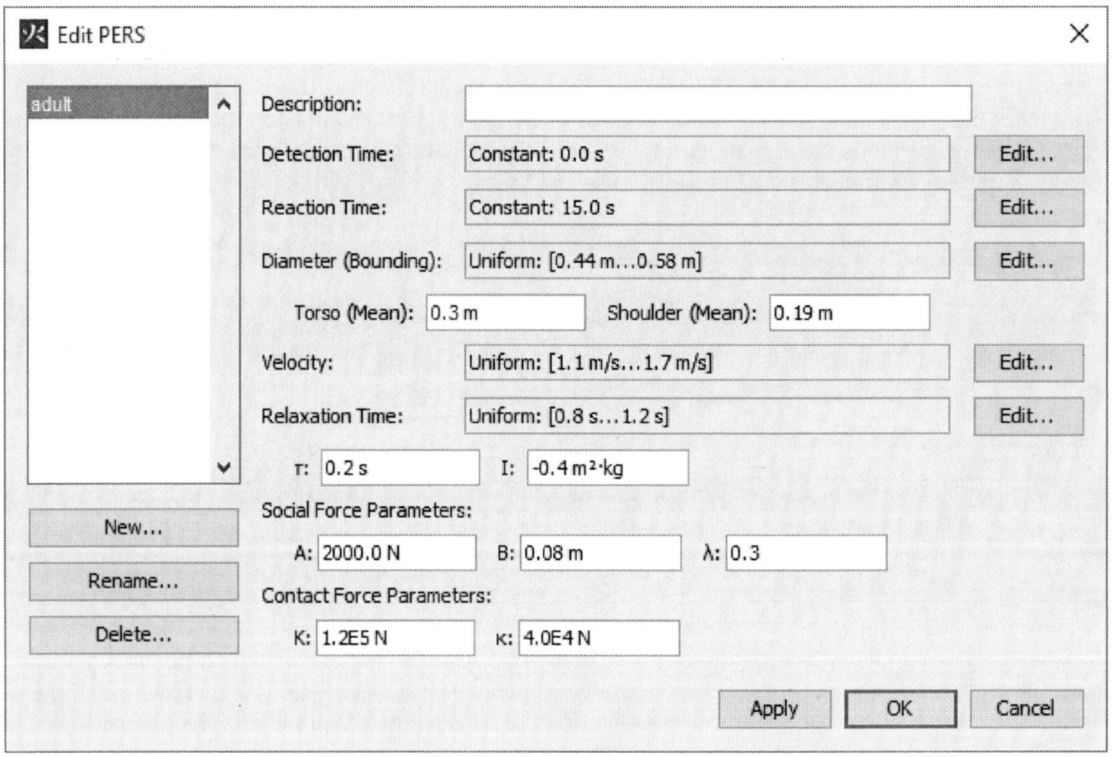

Figure 13. Input data for the agent movement model

The movement of agents inside the building is shown in Figure 14. The times when the first and the last person left the building are presented in Table 1:

Table 1. Characteristic times of the human evacuation process

	"real" evacuation drill	"virtual" evacuation drill
The time when the first person left the building [s]	40	45
The time when the last person left the building [s]	88	92

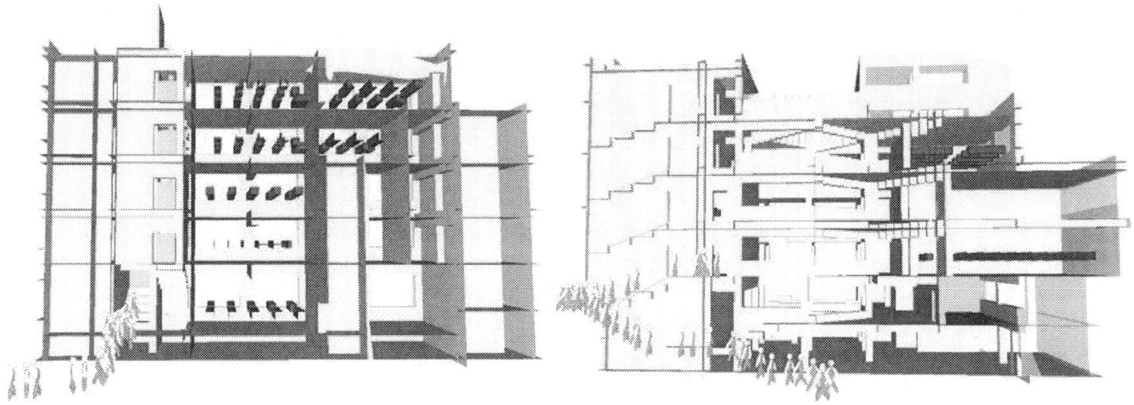

Figure 14. Agents movement inside the building

The graphs presented in Figure 15 show the variation in time of the number of people inside the building.

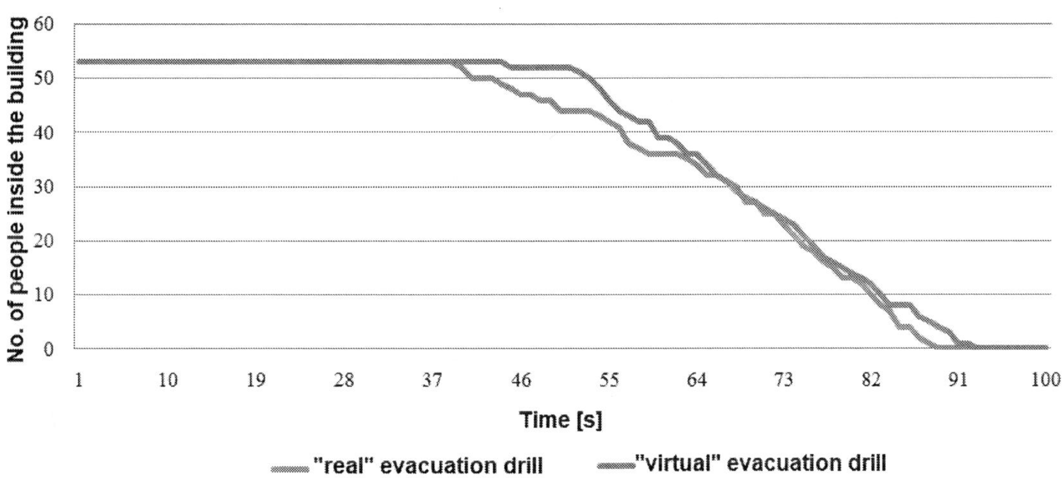

Figure 15. The variation in time of the number of people inside the building

After the comparative analysis of the results from the "real" and the "virtual" evacuation drill, it can be seen that the FDS+Evac software can approximate with a high degree of credibility the movement and behaviour of persons during an evacuation drill.

5. Comparative analysis of "real" vs. "virtual" evacuation drill

A conclusion of the comparison between the results obtained in the numerical analysis and in the in-situ experiment, is that there is a trend of results convergence: the time of evacuation in approximately the same (92 s for the numerical analysis and 88 s for the "real" evacuation drill), the movement of agents, from the software, is similar to the real movement of people in choosing the escape routes. It can be noticed that there is no turbulence in both situations: of numerical simulation and in-situ experiment.

The analysis of the differences between the evacuation times in the two situations, expressed in percents, shows that there is an increase of 12.5 % in the numerical simulation for the time when the first person has been evacuated and in increase of 4.5% in the numerical simulation, for the time when the last person has been evacuated.

FDS+Evac was tested by its developers only in small groups of people [3]: less the 100 individuals. It is possible that in case of a greater number of people the modelled situation does not coincide with reality. For this reason, we must avoid extrapolating the results from the situation where we operate with a small group of people (tens) to the situation where we operate with a large or very large number of people (hundreds or thousands).

The second situation, that of large groups of people, can be the subject of an in-depth, possibly parameterized study, in which the number of people is increased successively (in example: 50, 100, 150, 200, etc.), then the interpretation of the results will be performed by observing whether there is convergence or divergence. In this case the problem must be solved independently.

A detailed study, correlated with reality, is required when the FDS+Evac is significantly qualitatively influenced by the number of people with whom it operates. In this context, by qualitative it is understood the choice of other escape routes or the mode of completion the evacuation (as example: the number of deceased persons in the case of a fire). A quantitative difference is expected to increase the evacuation time.

6. References

[1] Grigoraş Z C 2016 *Posibilităţi de optimizare a evacuării umane în condiţii de siguranţă pentru cazul incendiilor din clădiri*, Facultatea de Construcţii şi Instalaţii, Universitatea Tehnică "Gheorghe Asachi" din Iaşi, România (in Romanian)

[2] Hostikka S, Korhonen T, Paloposki T, Rinne T, Matikainen K, Heliovaara S 2007 *Development of FDS+Evac for evacuation simulations, Project summary report, VTT Research Notes 2421*, VTT Technical Research Centre of Finland

[3] Korhonen T 2015 *Fire Dynamics Simulator with Evacuation: FDS+Evac, Technical Reference and User's Guide* (FDS 6.1.2, Evac 2.5.0, DRAFT), VTT Technical Research Centre of Finland

Acknowledgement

The authors are grateful to Thunderhead Engineering Consultants Inc. USA for providing the free education licence for PyroSim – graphic user interface for FDS and FDS+Evac.

THUNDERHEAD ENGINEERING

IOP Conf. Series: Materials Science and Engineering **586** (2019) 012006 doi:10.1088/1757-899X/586/1/012006

Thermal performance of insulated glazing units established by infrared thermography

L Moga[1]
[1] Technical University of Cluj-Napoca, 28 Memorandumului Str., *Romania*

E-mail: ligia.moga@ccm.utcluj.ro

Abstract. The targets for the building sector have been clearly defined starting with the Directive 2010/31/UE and ending with the long term strategy that implies the decarbonisation of the entire building sector. The aim consists in designing nearly Zero Energy Buildings (nZEB) starting from 2019 for public buildings and 2021 for all new buildings. For the energy retrofit of existing buildings, the last Directive 2018/844/UE on the energy performance of buildings sets a clear target for reaching nZEB also in the retrofit activity. Although that clear targets exists on a European level, some buildings are still poorly designed from the thermal point of view and are not able to meet these criteria. Thus, in many cases the thermal performance obtained in the design activity is not the one obtain in the operation phase of a building. That is why, not even buildings built after 2010, do not comply with the minimum recommended national thermal design criteria, and therefore need to be thermally retrofitted. Thus, the paper presents a thermography study done at a residential buildings, respectively the examination of the thermal performance of part of a façade. The focus is placed on a window-wall area that is examined by Infrared Thermography. The resulted thermograms are reconstructed by means of artificial intelligence in order to establish the thermal performance of the examined façade and building envelope components.

1. Introduction

In current practice, in the initial design stage of a building, the focus is placed more on structural aspects and less on the hygrothermal performance of a building, although that comfort is ensured based on an accurate hygrothermal design. A poor hygrothermal design, results in a poor energy performance of a building. As mentioned in literature, in some cases the final energy demand is in average about 20% higher compared to the designed value. [1] Thus, the Infrared Thermography (IRT) comes as a diagnosis solution for evaluating the thermal performance of the building envelope.

Infrared thermography is used as a non-invasive method at identifying thermal irregularities in the surface of the building envelope [2]. The irregularities are "shown" by the surface radiation temperature that varies on the thermographed surface. Thus, the heat flux directly associated with the infrared radiation emitted by the surface will offer an IR image called thermogram, which describes a thermal map of the investigated surface, indicating the weak parts of the building envelope, areas that need to be retrofitted. The thermal map can be expressed in vivid colours or using a black and white colour palette. In current practice IRT is mainly used in the thermos-energy retrofit activity with the aim of identifying: irregularities due to lack of thermal insulation, inhomogeneity and/or degradation of a material from the building envelope, weak thermal areas defined as thermal bridges, moisture issues in the mass of the envelope component, air leakages through the envelope and other issues regarding the envelope and the building systems. [3]

In the energy expertise and energy audit activity, specialist in the energy efficiency domain use IR cameras as an additional instrument in the auditing activity of a building. Nowadays the IR cameras available on the market are becoming popular considering that more producer exist on the market and thus a drop in prices took place. But the use of IR camera can also come in hand in other situations met in practice. In many cases of current practice, one of the next situations are found: the technical documentation of the building is not available, or if the technical documentation is available the information regarding the materials used for the building envelope are not accurate. The inaccuracy

comes from the fact that in the execution phase of the building, the constructive details (i.e. the materials) are not followed 100%, and sometimes the same happens with the constructive details of the building envelope described by the designer in the project. Thus, the real thermal performance of the building envelope is not the one obtained in the design phase and the impact is seen and felt in the maintenance costs of the building. Higher costs are showing the inaccuracy between the designed thermal performance of the building envelope, and the real thermal performance from the operation phase. In this case IRT can be a significant tool in diagnosing the real thermal performance of the building envelope.
The paper presents the development of a calculation algorithm and methodology and the computer program THERMOG that can be used to establish the thermal performance of an existing building by using IRT tools. The methodology and computer program can be used in both case, if technical documentation of the project exists or not. With the help of the program, the non-destructive measurement tool of IRT can offer a quick and reliable image on the thermal performance of a building component.

2. IR examination procedure

The IRT needs to follow several steps given in [2]. In order to prepare the surfaces that need to be themographed, one needs to heat the interior spaces of the examined building (i.e. space of the building). The boundary conditions of the measurements (i.e. wind speed, solar radiation, exterior and interior temperature, exterior and interior humidity, interior and exterior relative humidity, cloud cover) need to be registered to have an accurate interpretation of the acquired data. The data acquisition is recommended to be performed at sunrise or at sunset, on days with no solar radiation. [3, 4] The standard also recommends a temperature difference between the heated and the exterior environment of at least 10°C. That is why it is recommended that the measurement to be done during the cold period of the year, both in the exterior and in the interior environment of the building. Part of existing literature suggest measurement to be done in the early morning so that the effect of thermal inertia of a building can be avoided. [5]
In doing the thermography investigation, the emissivity of the materials defining the photographed surface (i.e. the finishing) must be defined or it can be taken as a standard value given by normed prescription or IR catalogues.
Although the thermographic non-invasive method is used lately as an adjacent resource for diagnose and testing of buildings [6], its accurate interpretation depends very much on the expertise of the specialist using it. Thus, by some practitioners the IR camera is presented as a tool that can calculate the thermo-energy performance of the building. Although that in some countries the IRT is an instrument in the energy auditing activity of buildings, the scientific literature mentions that errors in evaluating thermal performance by means of IRT are somewhere around 30%-100%. [5]

3. The calculation methodology and THERMOG program

In order to enable the use of the resulted thermograms as input data in establishing the thermo-energetic performance of buildings and also give an adjacent instrument that can be used with the IR camera for practitioners, a methodology and a computer program was developed in this purpose.
The developed algorithm is based on a mathematical model that reconstructs the thermographic image in order to identify the temperature value of each pixel (i.e. point) of the surface. The boundary conditions, the surface emissivity and the original IR thermogram, are used as inputs in the calculation algorithm. The thermogram reconstruction algorithm is based on an artificial intelligence (AI) approach. It is well known that AI implies different tools, starting from machine learning and data science, deep learning and other tools like unsupervised models, graphic models, reinforcement learning, and others. For developing the algorithm of reconstructing the thermogram, an AI feature is used that is trained for a process of retention and self-learning of the arrangements in the colour palette, so that the initial thermogram can be reconstructed in order to obtain all the temperature points identified on the surface of the examined building component.

The thermogram is a sequence of tens of thousands or hundreds of thousands of colour shades. To each colours shade corresponds a temperature contained between the minimum and maximum temperature values of the scale (i.e. legend). In the legend is usually presented an extract from the colour palette of the thermogram, somewhere between 100 and a maximum of 200 colours. The task of the algorithm, that includes the AI approach, is that to identify the correct order of the tens of thousands colour shades of the palette, having different nuances. Knowing the correct orderliness of the palette, through a ready at hand mathematical calculation, one can establish the temperature values belonging to the colour palette, values contained within the maximum and the minimum values from the colours legend. Therefore, the questions that are addressed are: how many different nuances exist and which is their correct order. The AI module is decomposing each colour in its RGB components allowing the AI algorithm to arrange the colour palette based on its increased or decreased intensity of various nuances, resulting in the end a colour palette as it can be seen in figure 3.a. For each domain of shades of colour, by nonlinear multi-criteria polynomial interpolation calculation formula are establish to determine the colour shades specific to each domain. A variant for the calculation formula can be as it follows:

$$CSh = a_o + a_1 \cdot R + a_2 \cdot G + a_3 \cdot B + a_4 \cdot R^2 + a_5 \cdot G^2 + a_6 \cdot B^2 + a_7 \cdot R \cdot G + a_8 \cdot R \cdot B + a_9 \cdot G \cdot B + a_{10} \cdot R \cdot G \cdot B + a_{11} \cdot \ln(R) + a_{12} \cdot \ln(G) + a_{13} \cdot \ln(B) \quad (1)$$

where CSh stands for color shade.

The validation of the algorithm is given by the new colour palette (i.e. colour legend/scale) described by the reconstructed themogram, colour palette that must be identical to the initial one. The reconstructed thermogram is than converted into heat fluxes values with the help of which the thermal performance of each building envelope component is established.

The problem of the mathematical formulation was also discussed by Vavilov [5] [7] who identify the necessity of reconstructing thermograms so that these can be used as a tool in expressing the thermal performance of the building envelope. The author offered a simple formula for estimating the heat losses based on the temperature differences between the exterior surface of a reference point of the building envelope component and the exterior air, the temperature difference between the average exterior temperature of the building component and the exterior air, the current heat flow density at a reference point from the interior surface of the building component, the duration of the heating season and the area for the obtained average results. Also the instrumental errors were discussed showing a possible error between 37%-110% in the measuring processes. [5]

Another approach in establishing the thermal performance of a building component was done by Danielski and Froling [8]. The study was divided in two parts, first to determine the heat transfer coefficient through convection by means of thermography and heat flux meters measurement. The second part includes the determination of the overall heat transfer coefficient of the building component by thermography. The results indicate that thermography is more efficient (i.e. about 11% higher) in giving accurate results compared to heat flux meters measurements.

In [9] the authors present an application of artificial neural networks analysis in the use themographic investigations. The developed program enables the analysis of a thermogram supplemented by a colour scheme. The analysis is done by a backward temperature field detection that is accompanied by several features, i.e. number of visual and statistic elements.

Thermography was also used in detecting thermal bridges by means of image processing approximation algorithms [10]. An enhancement of the thermographic images was done by optimizing the mathematical algorithm for digital image processing. The results obtained from the segmentation procedure give an improvement in the identification of the thermal bridges from the thermographic image and also provides the possibility of establishing heat losses using an incidence factor of thermal bridge denote by I_{tb}. The incidence factor is a parameter higher than one and its accuracy depends on the resolution of the thermal map.

IOP Conf. Series: Materials Science and Engineering **586** (2019) 012006 doi:10.1088/1757-899X/586/1/012006

4. The study case – application of the THERMOG program

In developing the assessment methodology and the calculation program for the thermal performance of a building envelope component, several cases from real practice were employed for its validation. The focus of this paper is on glazing units that are lately substituting the role of the opaque surface of the wall, due to the fact that beneficiaries want a greater connection with the exterior environment which is translated to larger glazing units at a building. It is well known that glazing units are weaker thermal areas compared to the opaque surface of the building envelope through which high heat losses occur [11]. Thus, a study case of a residential building located in Cluj-Napoca, Romania was chosen with the aim to demonstrate the usage of the developed methodology and computer program.

Figure 1. The thermographed area – west façade

The building is a block of flats with a height regime B+GF+2FL. Investigation were made at an apartment situated at the ground floor level of the building. The technical data provided by the project indicates that the building has a mixed structural system, reinforced concrete frame system. The exterior walls are made out of thermal efficient masonry of 30 cm on which is placed 15 cm of thermal insulation layers. For the glazing area, a double glazed PVC window was used. One of the thermographs taken for the building are evaluated and the thermal performance of the building envelope component is obtain.
The thermograph (i.e. see figure 4) chosen to illustrate the usage of the methodology and computer program has a dimension of 315x253 pixels, which gives a total number of 79.695 pixels. To each pixel corresponds a colour from the colour palette. From the total number of pixels, 38.300 have distinct colours, the rest being colours that appear at least two times on the surface of the thermograph. To each distinct colour corresponds a temperature value between the minimum and maximum value from the colour pallet (i.e. legend of colours), respectively the maximum 11.3°C and the minimum 4.4°C (see figure 2).
The colours from figure 2 shows all different shades of colours existent on the surface of the thermograph. In figure 3 one can see all distinct colours arranged coherently based on their shades. In order to obtain an accurate colour palette presented in figure 3.a, the calculation module was used so that through the process of retention and self-learning of the arrangements in the colour palette, the colour scale presented at 3.b was obtained.
To each colour shade having the coordinates (x,y) corresponds a surface temperature $\theta_{si,(x,y)}$. Starting from the temperature in each point from the surface, the thermograph was established with the colours generated by the colour palette. It can be seen that figure 4 and figure 5 are identical, therefor

demonstrating that the algorithm steps for the self-learning and restoration of all shades from the colour palette developed by the author, is validated.

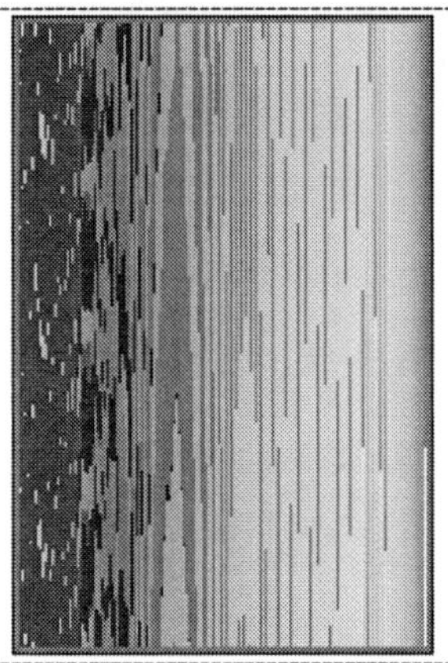

Figure 2. The resulted colour pallette based on the first processing on the thermography

Figure 3.a The final colour pallete of the thermography

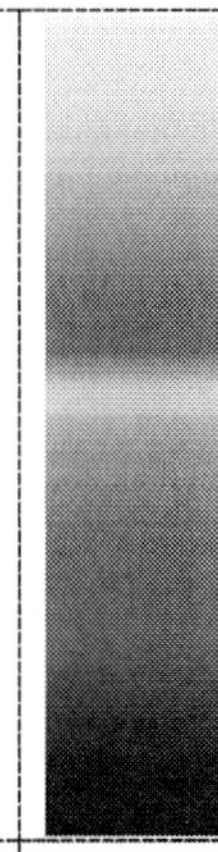

3.b

Based on the thermograph, the geometric characteristics for each type of surface were evaluated. The total surface, the opaque surface, the glazing surface divided by the frame are and the glass area. The next values were obtained: S_{total}=7.402 m^2, S_{opaque}=1.762 m^2, $S_{glazing}$=5.64 m^2 from which S_{frame}=1.462 m^2 and S_{glass}=4.178 m^2

Considering the formula given by nationals and internationals thermotechnical design norms [12] [13], the superficial thermal resistances in each point of the thermographed surface and the global average superficial thermal resistance on the surface of the thermograph was calculated $R_{se,\,av,\,T}$= 0.104 $m^2{\cdot}K/W$. Also, the values for each component of the surface were calculated: opaque surface $R_{se,av,opaque}$= 0.105 $m^2{\cdot}K/W$, $R_{se,\,av,\,glazing}$= 0.103 $m^2{\cdot}K/W$ global on the window surface, and on the frame $R_{se,\,av,\,frame}$= 0.104 $m^2{\cdot}K/W$ and on the surface of the glass $R_{se,\,av,\,glass}$= 0.103 $m^2{\cdot}K/W$.

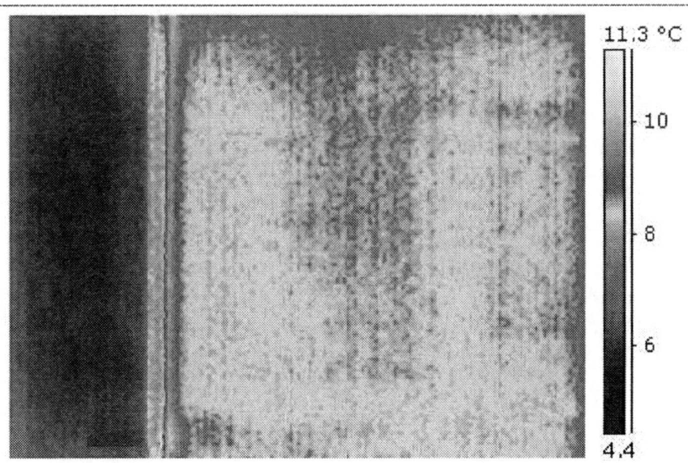

Figure 4. The initial thermogram corresponding to the window-wall area

Figure 5. The processed thermogram corresponding to the window-wall area

Knowing the exterior temperature and the temperature in each point of the surface, the heat flows that dissipate from each point of the building element to the exterior environment, were calculated. The sum of the heat flows on the entire surface of the element, and separately on each type of surface, was obtain as it follows:

- globally on the thermographed surface Φ_{global}=234.719 W
- on the opaque surface Φ_{opaque}=6.250 W
- on the glazing surface $\Phi_{glazing}$=228.469 W
- on the frame surface Φ_{frame}=52.168 W
- on the glass surface Φ_{glass}=176.301 W

With the help of the heat fluxes, the exterior superficial thermal resistances (i.e. average values), the surfaces for each distinct area, the specific thermal resistance for the analysed surface, the thermal resistance in the current field of each specific zone R_o [$m^2 \cdot K/W$] is calculated:

- globally on the thermpographed surface R_{oe}=0.468 [$m^2 \cdot K/W$]
- on the opaque surface - $R_{oe,opaque}$=4.185 [$m^2 \cdot K/W$]
- on the glazing surface - $R_{oe,glazing}$=0.366 [$m^2 \cdot K/W$]
- on the frame surface - $R_{oe,frame}$=0.416 [$m^2 \cdot K/W$]
- on the glass surface - $R_{oe,glass}$=0.352 [$m^2 \cdot K/W$]

After calculating these values, knowing the exterior air temperature, interior air temperature, the average superficial thermal resistances and the average unidirectional thermal resistances for each specific zone previously mentioned, by doing an elementary iterative calculation the average interior superficial thermal resistance for each distinct zone is calculated. Finding this value can give us the thermal behaviour of each analysed zone.

Other calculations can be done to identify if the thermotechanical properties of the used materials correspond to the ones mentioned in the technical data of the project. Thus, for the opaque zone of the thermographed area: we know that the opaque area is given by the interior rendering having a thickness of 2 cm with thermal conductivity λ=0.87 W/(m.K), aerated cellular concrete (i.e. ACC) of 30 cm and thermal insulation of expanded polystyrene with a thickness of 15 cm. Based on the ACC available at the beneficiary, the thermal conductivity was established, resulting a value λ_{ACC}=0.243 W/(m·K). Knowing the geometric characteristics of each layer and the thermotechnical characteristics of the materials, and knowing the unidirectional thermal resistance for the opaque area $R_{oe,opaque}$ the thermal resistance of the insulation layer is established being equal to $R_{therm.ins.}$=2.467 [$m^2 \cdot K/W$], corresponding to a thermal insulation having a value $\lambda_{therm.ins}$=0.041 W/(m·K), greater than the one mentioned in the technical documentation of the building, as being $\lambda_{therm.ins}$=0.036 W/(m·K).

IOP Conf. Series: Materials Science and Engineering **586** (2019) 012006 doi:10.1088/1757-899X/586/1/012006

Analysing the thermal performance of the glazing area for the frame, an average thermal resistance was obtained R_{frame} =0.416 [$m^{2}\cdot K/W$], which means a thermal transmittance U_{frame}=2.404 [$W/m^{2}\cdot K$]. Compared to U=1.2 [W/m^2K] that corresponds to a 6 chambers frame as mentioned in the technical documentation of the project, the value U_{frame}=2.404 [$W/m^{2}\cdot K$] corresponds to a PVC frame having three chambers [14], thus not meeting the data (i.e. the requirements) mentioned in the technical documentation of the project.

For the glass area of the window, an average value R_{glass}=0.352 [$m^{2}\cdot K/W$] was calculated, that gives a thermal resistance U_{glass}=2.841 [$W/m^{2}\cdot K$] which correspond to a glass area without thermoreflective surfaces, a different solution compared to the one mentioned in the technical documentation of the project which is a glass surface filled with Argon gas and thermoreflective surfaces. As a global value for the glazing surface, the overall values is $R_{glazing}$=0.366 [$m^{2}\cdot K/W$] which means half of the minimum recommended value $R_{min,glazing}$=0.77 [$m^{2}\cdot K/W$] as imposed by national thermotechnical design norms [15]

Similar calculations can be done for other thermograms, for both vertical and horizontal building envelope components. Thus, the non-destructive measurement tool of IRT together with the program THERMOG can offer a quick and reliable information regarding the thermal performance of a building component and of the entire building envelope.

5. Conclusions

It is well known that in design phase, decisions that are not properly chosen for the building envelope will have a direct negative impact on the lifespan of the building, which translate to high energy consumptions and large greenhouse gas emissions. The Infrared Termography is a tool mainly used for quality control of buildings envelope, although that its usage can also be for quantitative evaluation of the thermal performance of it. The calculation methodology and the computer program THERMOG elaborated by the author can be used both in the energy expertise activity of existing buildings and in the reception phase of new buildings. Considering the long path strategy of the European Union of decarbonizing all building sector [16], the energy retrofit of existing buildings and the fulfilment of the nZEB criteria for new and existing buildings is pushing the market on a path that implies accurate solutions in designing and retrofitting buildings. IRT together with the computer program THERMOG is a tool that can answer this call by helping practitioners in choosing solutions based on an accurate assessment of the real thermal behaviour of a building.

6. References

[1] Danielski I 2012 Large variations in specific final energy use in Swedish apartment buildings: Causes and solutions *Energy and Buildings* **49** 276-85

[2] ISO 6781:1983 Thermal insulation – Qualitative detection of thermal irregularities in building envelopes – Infrared method I. O. f. Standardization

[3] Pleşu R, Teodosiu G and Ţăranu G 2012 Infrared Thermography Application for Building Investigation *Buletinul Institutului Politehnic din Iasi* **LVIII** 157-68

[4] Jaworski J 2008 The infrared thermography of buildings proceeding its surrounding and their thermal performance in *9th International Conference on Quantitative InfraRed Thermography* Krakow

[5] Vavilov VP 2010 A Pessimistic View of the Energy Auditing of Building Structures with the Use of Infrared Thermography *Russian Journal of Nondestructive Testing* **46** 49-54

[6] Asdrubali F, Baldinelli G, Bianchi F, Costarelli D, Rotili A, Seracini M and Vinti G Detection of thermal bridges from thermographic images for the analysis of buildings energy performance

[7] Vavilova VP, Marinetti S and Nesteruk DA 2009 Evaluating the Thermal Resistance of Building Structures," *Russian Journal of Nondestructive Testing* **45** 50-61

[8] Danielski I and Frolling M 2015 Diagnosis of buildings' thermal performance - a quantitative method using thermography under non-steady state heat flow *Energy Procedia* **83** 320-29

[9] Koštialová Jančíková Z, Kostial P, Zimny O, Heger M and Spicka I Artificial Neural Networks Application in Thermography," in *13th International Conference on Heat Transfer, Fluid Mechanics and Thermodynamics* 408-11

[10] Asdrubali F, Baldinelli G, Bianchi F, Costarelli D, Rotili A, Seracini M and Vinti G 2018 Detection of thermal bridges from thermographic images by means of image processing approximation algorithms *Applied Mathematics and Computation* **317** 160-171

[11] Babotă F and Iernuţan R 2019 Calculation and analysis of corrected thermal resistance in a common dwelling building in Romania, in various thermal insulation options – The exterior wall *Procedia Manufacturing* **32** 442-49

[12] SR EN ISO 6946 2017 Componente si elemente de clădire. rezistenţa termică şi transmitanţă termică. metode de calcul.(Building components and building elements - thermal resistance and thermal transmittance - Calculation methods) Asociatia de Standardizare din Romania

[13] SR EN ISO 10211 2017 Punti termice in alcatuirea cladirilor. Fluxuri termice si temperaturi superficiale. Calcule detaliate. (Thermal bridges in building construction - heat flows and surface temperatures - Detailed calculations.) Asociatia de Standardizare din Romania

[14] SR EN ISO 10077-2 2018 Performanţa termică a ferestrelor, uşilor şi obloanelor. Calculul transmitanţei termice. Partea 2: Metoda numerică pentru profilurile de tâmplărie.(Calculation of the thermal performance of windows, doors and shutters- Part 2) Asociaţia de Standardizare din România

[15] Order no. 2641/2017 to amend and supplement the Technical Regulation "Calculation methodology of the energy performance of buildings", Romania: Ministry of Regional Development and Public Administration

[16] Directive (EU) 2018/844 amending Directive 2010/31/EU on the energy performance of buildings and Directive 2012/27/EU on energy efficiency Official Journal of the European Union

IOP Conf. Series: Materials Science and Engineering **586** (2019) 012007 doi:10.1088/1757-899X/586/1/012007

Green roof influence over the characteristics of the linear thermal bridges

I-R Baciu [1], D N Isopescu[1], N Taranu[1], S G Maxineasa[1]
[1] "Gheorghe Asachi" Technical University of Iasi, Faculty of Civil Engineering and Building Services, Department of Civil and Industrial Engineering, 1, Prof. Dr. docent Dimitrie Mangeron Blvd., Iasi, 700050, Romania

E-mail: ioanaroxana.baciu@yahoo.com

Abstract. Urban heat island (UHI) significantly affects the thermal-energy performance of the buildings. Moreover, urban materials absorb solar and infrared radiation, heat accumulation is dispersed in the atmosphere, a fact which increases air temperature. In this context, green roofs are the most suitable solution to resolve these vital issues. The expanding benefits of a green roof, such as energy saving, thermal insulation, and mitigating heat island effect emphasize the key role of this structure in overall thermal performance of buildings and microclimate conditions of indoor spaces. The main objective of the study is to analyse the influence of extensive green roof on the heat flow, in the thermal bridges developed structurally in buildings where they might replace the classical terrace roof. The influence is analysed through the comparison of the thermal impact of a classical terrace roof and that of an extensive green roof, through thermal characteristics of their components. The first part of the paper presents the structures of each type of roof and their thermic characteristics established and evaluated according to the normative regulation in force. In this context, the classical terrace roof and extensive green roof are examined from a thermic point of view, analysing the parameters of the types of thermal bridges usually met, in order to establish if the extensive green roof might have an influence on the overall heat flow. In conclusion, the unpredictable results obtained for the analysed thermal parameters let conclude that the extensive green roof solution presents a favourable environmental impact, promoting a "beautiful" added value to a building, in terms of both sustainability and aesthetics.

Keywords: thermal insulation; classical terrace roof; extensive green roof; heat flow; buildings.

1. Introduction

Nowadays, most of the world population lives in cities, and there is a rising tendency towards urban life year after year. A recent report of the United Nations shows that the population living in cities will have increased to 67% by 2050 [1]. This fact entails a significant growth of the environmental problems at a global scale and, also, affects the urbanization process. Issues such as urban heat island, greenhouse and a major reduction of energy sources can be assigned to urban agglomerations. In some researches of the urban heat island that analyse four areas in New York city, the reports showed that there is an average of 2^0C temperature difference between the intensive vegetated areas and less vegetated areas, that can be explained by the substitution of the greening areas with built ones. This is the reason why efficacious control of urbanization and urban heat island with energy-efficient and ecological buildings has vital importance. In some reports of the International Energy Agency, on a worldwide scale, the built sector is responsible for 30-40% of energy consumption. The rise of urban temperatures increases the overheating risk and indoor thermal discomfort, which is mostly influenced by the heat flux and energy storage of the materials used in constructions [2,3,4].
The solutions for creating more environmentally friendly construction and improving the energy efficient technology and designs are continually searched. A green or vegetated roof represents a structural design that joins nature and engineering for building a sustainable alternative to the classical

roof [5]. Green roof systems can be taken into account as efficient and esthetical tools for energy saving in buildings. The high characteristics which must be implemented on the building's envelope, regarding these systems are the effect of evapotranspiration and the large thermal inertia [6].
The biggest influence on these properties is given by external factors like weather conditions, greenery dynamics, substrate composition and also drainage layers. All layers of the structure are considered as thermal insulation layers regarding their physical characteristics and their capacity to properly retain water. Through the multifunctional benefits of the green roof identified until now, the enhanced building envelope thermodynamics has been one of the most important characteristics for decreasing energy consumption in the built sector [7,8].
A large number of research studies have been conducted to evaluate the performance and benefits of green roofs through experiments and modelling analysis. In this context, it is very important to compare the thermal performance of a classical terrace roof and an extensive green roof, in different climatic parameters, in order to establish the behaviour mode, the influence over the indoor temperature and the thermal resistance of each structure over the building. The classical solutions of the envelope for thermal protection are usually insufficient; this is the reason why there are necessary new solutions for insulation or for improving the thermal conditions in buildings. The climate conditions and the degree of thermal protection are the facts taken into account when establishing the thickness of the insulation layer.
The purpose of this paper is to observe and to compare the heat fluxes and the thermal effect of the classical terrace roof and the extensive green roof through computer simulations only. Therefore, the case studies have been conducted to assess the degree of thermal protection of the building in the case of the classical terrace roof and in the case of the extensive green roof. The analyses of the structural components and the materials properties in both cases are carried out separately for a better understanding and easier tracking of the results. The structure of each roof and the four thermal bridges taken into study (the connection between exterior wall and the terrace roof slab, the connection between interior wall and the terrace roof slab - without reinforced concrete beam, the connection between interior wall and the terrace roof slab - with reinforced concrete beam, the connection between reinforced concrete beam and the terrace roof slab) have been modelled in AutoCAD software [9] and the numerical modelling has been realized using RDM 6 software [10], to determine the effect of thermal bridges.
Finally, the set of calculation of unidirectional thermal resistance and linear thermal transfer coefficients for the classical terrace roof and also for the extensive green roof is made according to Romanian regulations. The results have been compared to assess the impact of each roof over the thermal performance. The boundary conditions for the location of the building is the 3rd climate zone, characterized by outdoor design temperature T_e = -18°C and indoor design temperature T_i = 20°C.

2. Case studies

2.1. Case study for thermal bridges in buildings with classical roof

In the simplest definition, the envelope of the building is the architectural element, acting as a physical separator between the conditioned and unconditioned environment of a building. The envelope of the building consists of the exterior walls and the terrace roof, (for the case studies presented in the paper), with all the components: ceiling, windows, doors, etc. The efficient use of the energy in a building is directly connected to the building envelope and can be presented as diminishing the energy demands for heating and cooling according to the structural characteristic of the envelope. The loss of energy is one of the important issues in energy performance and it is directly dependent on building type and age, climate and geographical conditions, the types of materials of the envelope and residents' behaviour [11]. The building envelope thermal efficiency can be defined by the achievement of a considerable reduction in urban heat island (UHI) effects. In this context, energy-efficient, low-cost and eco-friendly retrofits are necessary, to decrease these effects. For the analysis and calculation, different numbers have been considered for the thermal bridges in the performed assessment, which are presented in Table 1. At first, it is analysed the classical roof and its components. For a better understanding of the thermal performance of the classical roof, the structural components must be presented as they are indicated in Figure 1.

Table 1. The notations used for the thermal bridges in the performed assessment

Analysed thermal bridges	Notation of thermal bridge for the classical roof	Notation of thermal bridge for the extensive green roof
The connection between the exterior wall and the roof slab	1.1	2.1
The connection between the interior wall and the roof slab (without reinforced concrete beam)	1.2	2.2
The connection between the interior wall and the roof slab (with reinforced concrete beam)	1.3	2.3
The connection between the reinforced concrete beam and the roof slab	1.4	2.4

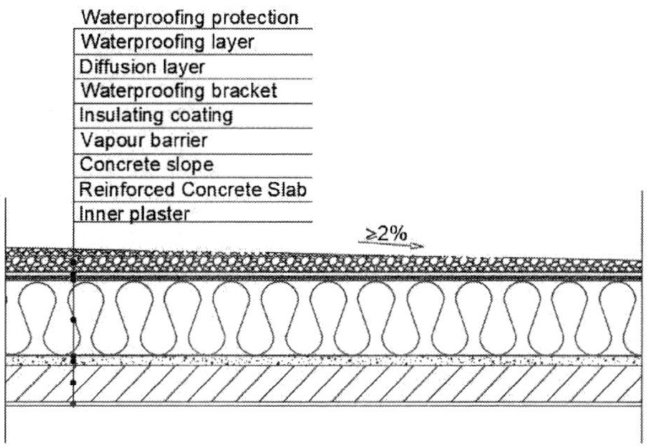

Figure 1. The structural components of the classical terrace roof.

In Table 2 the principal characteristics of the envelope elements are presented. These properties are the thickness of the materials, the bulk density, the coefficient of thermal conductibility and the heat mass capacity [12].

Table 2. Material properties (envelope with classical terrace roof)

	Envelope elements	Layer thickness [m]	Bulk density [kg/m^3]	The coefficient of thermal conductivity [W/m.K]	Heat mass capacity [J/kg.K]
1	Inner plaster	0.015	840	0.87	1700
2	Reinforced Concrete Slab	0.15	2500	1.74	840
3	Masonry (brick without gaps)	0.25	1800	0.80	870
4	Concrete slope	0.02	1800	0.93	840
5	Vapour barrier	0.01	1700	0.38	1460
6	Insulating coating	0.30	20	0.044	1460
7	Waterproofing bracket	0.01	1800	0.93	840
8	Diffusion layer	0.01	600	0.17	1460
9	Waterproofing layer	0.015	600	0.17	1460
10	Waterproofing protection	0.10	1800	0.7	840
11	Mortar	0.01	1800	0.93	840
12	Exterior plaster	0.025	840	0.87	1700

From the survey data, the most common types of thermal bridges found for the classical roof and taken for analysis, are those presented in Figure 2 ÷ 5.

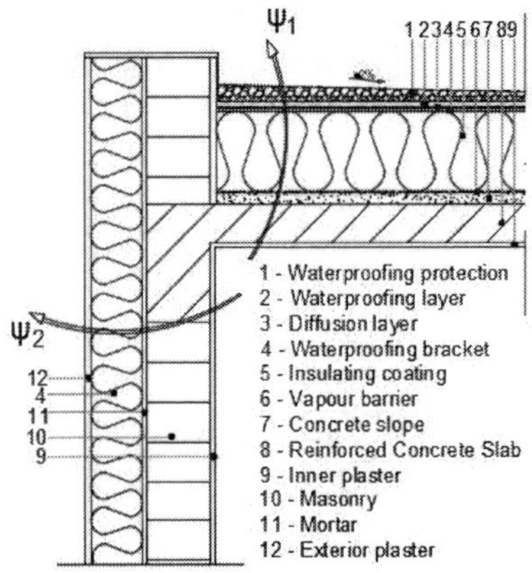

Figure 2. The connection between the exterior wall and the roof slab - Thermal bridge 1.1.

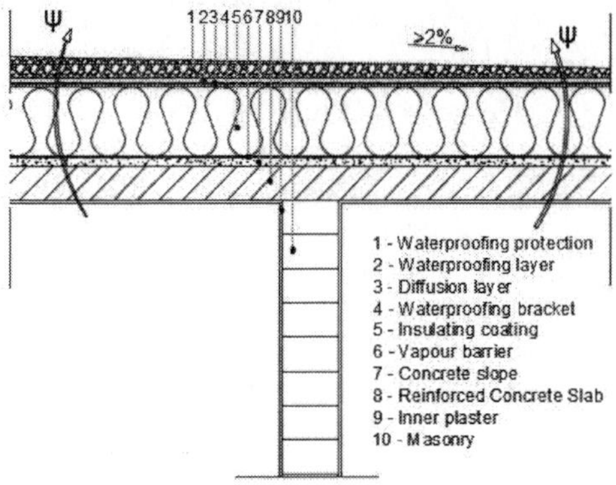

Figure 3. The connection between the interior wall and the roof slab (without reinforced concrete beam) - Thermal bridge 1.2.

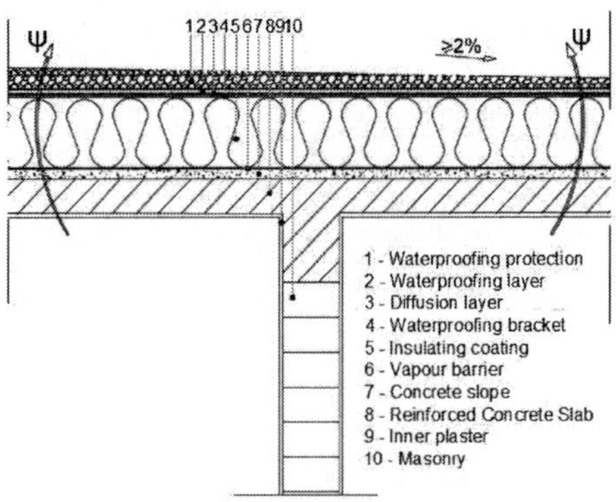

Figure 4. The connection between the interior wall and the roof slab (with reinforced concrete beam) - Thermal bridge 1.3.

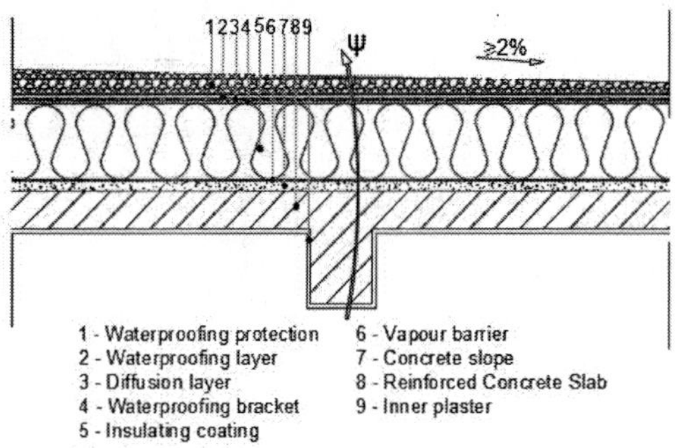

Figure 5. The connection between the reinforced concrete beam and the roof slab - Thermal bridge 1.4.

2.2. Case study for thermal bridges in buildings with an extensive green roof

This section presents the structure of the extensive green roof taken into consideration, the representative thermal bridges for this roof and the material components of the roof.

A green roof is composed of the vegetation, the soil layer, the gravel layer (drainage layer), the geotextile, the anti-root layer and profile membrane over the classical roof. The proposed green roof scenarios were analysed in terms of thermal resistance based on the selection of the components.

For a better understanding of the thermal bridges of the extensive green roof, the structural components must be presented as they are indicated in Figure 6.

Table 3 presents the principal characteristics of the extensive green roof elements. These properties are the thickness of the materials, the bulk density, the coefficient of thermal conductibility and their heat mass capacity [12].

From survey data, the most common types of thermal bridges for the extensive green roof are the same chosen for the classical terrace roof, which are presented in Figure 7÷10.

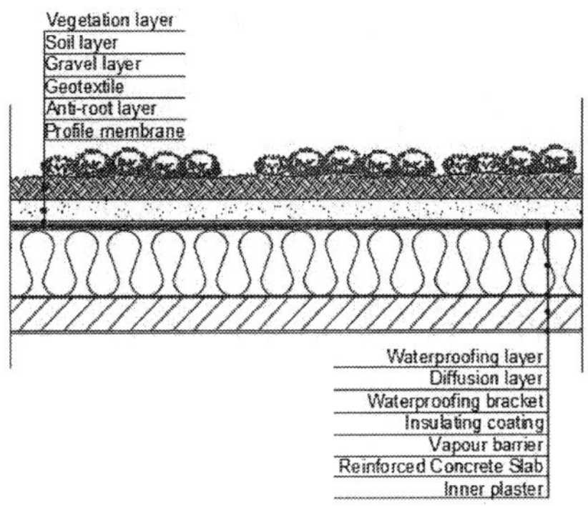

Figure 6. The structural components of the extensive green roof

Table 3. Material properties (envelope with extensive green roof)

	Envelope elements	Thickness of materials [m]	Bulk density [kg/m^3]	The coefficient of thermal conductivity [W/m.K]	Heat mass capacity [J/kg.K]
1	Inner plaster	0.015	840	0.87	1700
2	Reinforced Concrete Slab	0.15	2500	1.74	840
3	Vapour barrier	0.01	1700	0.38	1460
4	Insulating coating	0.30	20	0.044	1460
5	Waterproofing bracket	0.01	1800	0.93	840
6	Diffusion layer	0.01	600	0.17	1460
7	Waterproofing layer	0.015	600	0.17	1460
8	Profile membrane	0.01	150	0.7	1460
9	Anti-root layer	0.01	150	0.70	1700
10	Geotextile	0.2	150	0.10	1700
11	Gravel layer	0.15	1800	0.7	840
12	Soil layer	0.10	1800	1.16	840
13	Masonry (brick without gaps)	0.25	1800	0.80	870
14	Mortar	0.01	1800	0.93	840
15	Exterior plaster	0.025	840	0.87	1700
16	Cork layer	0.02	120	0.04	2500

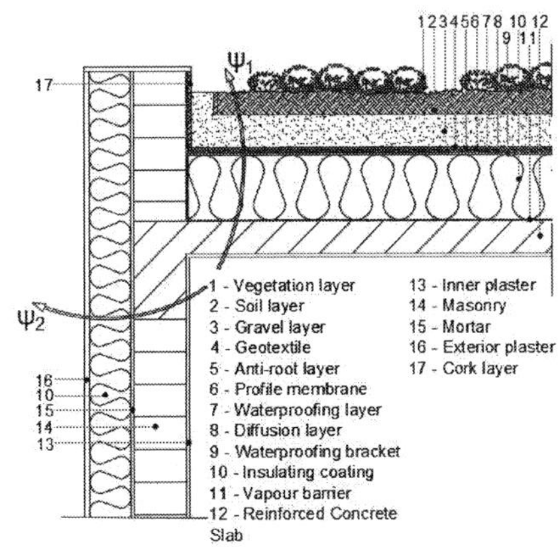

Figure 7. The connection between the exterior wall and the roof slab - Thermal bridge 2.1

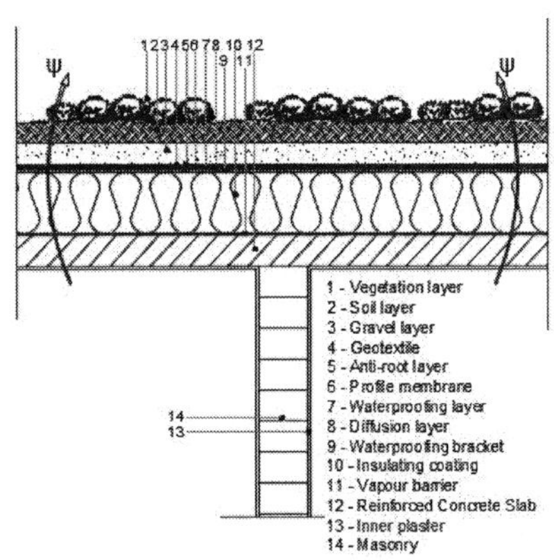

Figure 8. The connection between the interior wall and the roof slab (without reinforced concrete beam) - Thermal bridge 2.2

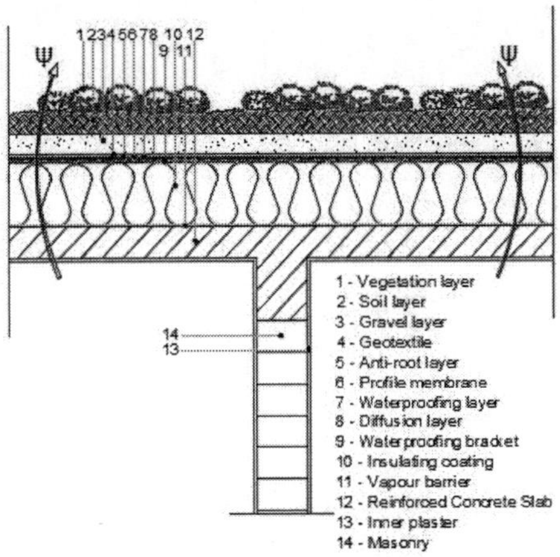

Figure 9. The connection between the interior wall and the roof slab (with reinforced concrete beam) - Thermal bridge 2.3

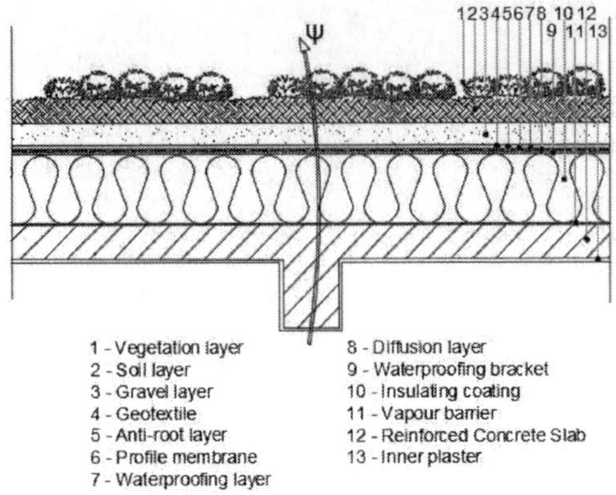

Figure 10. The connection between the reinforced concrete beam and the roof slab - Thermal bridge 2.4

3. Analyses and discussions

The linear thermal transfer coefficient, ψ, is the surplus flow $\Delta\phi$ transmitted by a linear bridge, relative to its length l and thermodynamic driving force for the flow of heat, ΔT (temperature difference).
This is the defining relation for the linear thermal transfer coefficient, ψ, but the practical calculation relationship is the mathematical formula given by Equation (1) [12]:

$$\psi = \frac{\phi}{\Delta T} - \frac{B}{R} \tag{1}$$

where ϕ is the thermal flow associated with a bridge with a width *B* and a length of 1 m (W/m); and *R* represents the unidirectional thermal resistance (m^2K/W).
The specific unidirectional thermal resistance is determined by the sum of the unidirectional resistances of the material layers, and the superficial resistance to the convective and radiative heat transfer of the interior and exterior surfaces, as shown in Equation (2) [12]

$$R = \frac{1}{\alpha_i} + \sum \frac{d}{\lambda} + \frac{1}{\alpha_e} \tag{2}$$

where α_i and α_e represent the convective and radiative heat transfer coefficients of the interior and exterior surfaces. In the calculus are used the following values for α_i, α_e and for corresponding thermal resistance:

α_i=8 R_i=0,125
α_e=24 Re=0,042

Table 4 presents the thermal resistance value for the classical terrace roof (R_c), determined based on the dimension of each layer (*d*) and the coefficient of thermal conductivity (λ) [10].

Table 4. Determination of Unidirectional Thermal Resistance for the structure of the classical terrace roof

Envelope elements	d [m]	The coefficient of thermal conductivity (λ) [W/m.K]	Thermal resistance (R_c) [m²K./W]
Inner plaster	0.015	0.87	7.455
Reinforced Concrete Slab	0.150	1.74	
Concrete slope	0.040	0.93	
Vapour barrier	0.010	0.38	
Insulating coating	0.300	0.044	
Waterproofing bracket	0.010	0.93	
Diffusion layer	0.010	0.17	
Waterproofing layer	0.015	0.17	
Waterproofing protection	0.100	0.70	

The thermal resistance value for the structure of the extensive green roof (R_g) determined based on the dimension of each layer (d) and the coefficient of thermal conductivity (λ), is presented in Tabel 5 [10].

Table 5. Determination of Unidirectional Thermal Resistance for the structure of the extensive green roof

Envelope elements	d [m]	The coefficient of thermal conductivity (λ) [W/m.K]	Thermal resistance (R_g) [m²K./W]
Inner plaster	0.015	0.87	7.789
Reinforced Concrete Slab	0.150	1.74	
Vapour barrier	0.010	0.38	
Insulating coating	0.300	0.044	
Waterproofing bracket	0.010	0.93	
Diffusion layer	0.010	0.17	
Waterproofing layer	0.015	0.17	
Profile membrane	0.002	0.38	
Anti-root layer	0.010	0.70	
Geotextile	0.020	0.10	
Gravel layer	0.150	0.70	
Soil layer	0.100	1.16	

Table 6 gives the thermal resistance value for the exterior wall (R_w), determined based on the dimension of each layer (d) and the coefficient of thermal conductivity for the material (λ) [10].

Table 6. Determination of Unidirectional Thermal Resistance for the exterior wall.

Envelope elements	d [m]	The coefficient of thermal conductivity (λ) [W/m.K]	Thermal resistance (R_w) [m²K./W]
Inner plaster	0.015	0.87	5.23
Masonry	0.250	0.80	
Mortar layer	0.015	0.93	
Insulating coating	0.200	0.044	
Exterior plaster	0.025	0.87	

In order to evaluate the thermodynamic performance for each type of roof taken into consideration in the study, the identification of the unidirectional thermal resistance and the linear thermal bridges coefficients have been established as parameters for comparison. All the thermal bridges have been analysed using the RDM 6 software as they are presented in Figure 11÷18. Table 7 presents the values

IOP Conf. Series: Materials Science and Engineering **586** (2019) 012007 doi:10.1088/1757-899X/586/1/012007

of linear thermal bridges coefficients, ψ, for both types of roof, and also, the thermal properties used in calculus, values being taken from the specialized literature and national design norms.

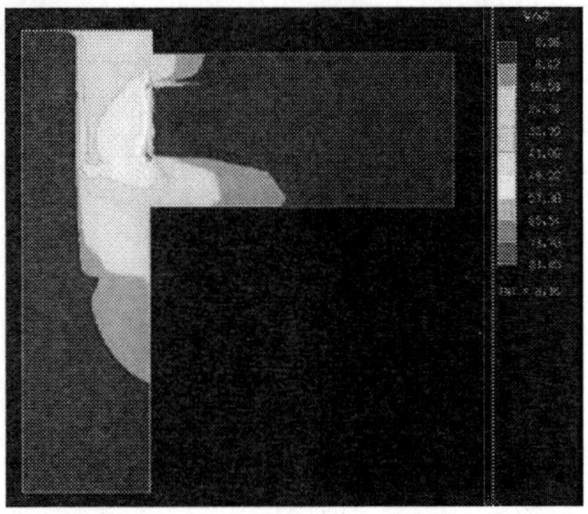

Figure 11. Isovalues of heat flow for the thermal bridge 1.1 of the classical terrace roof

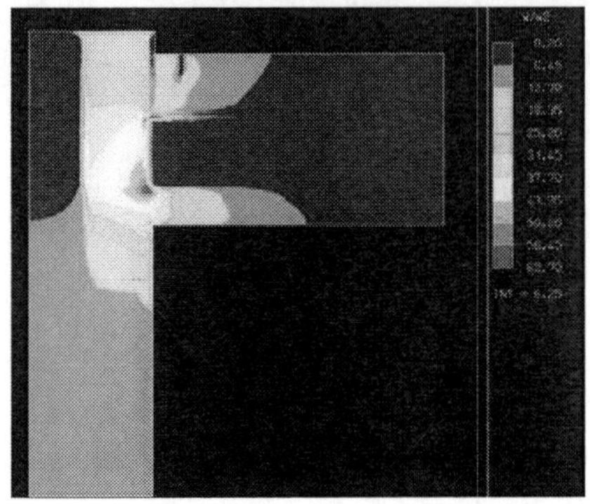

Figure 12. Isovalues of heat flow for the thermal bridge 2.1 of the extensive green roof

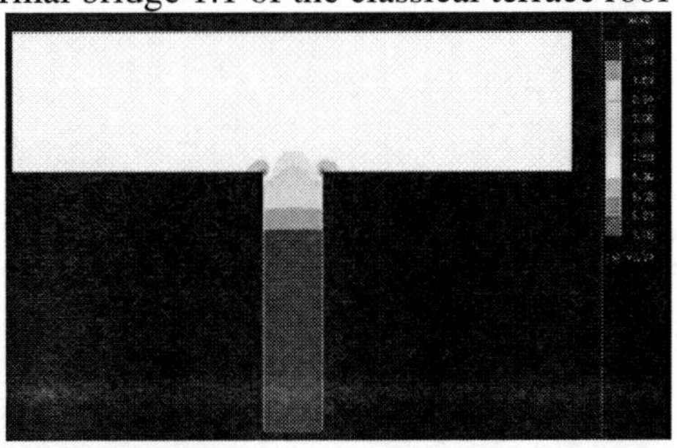

Figure 13. Isovalues of heat flow for the thermal bridge no 1.2 of the classical terrace roof

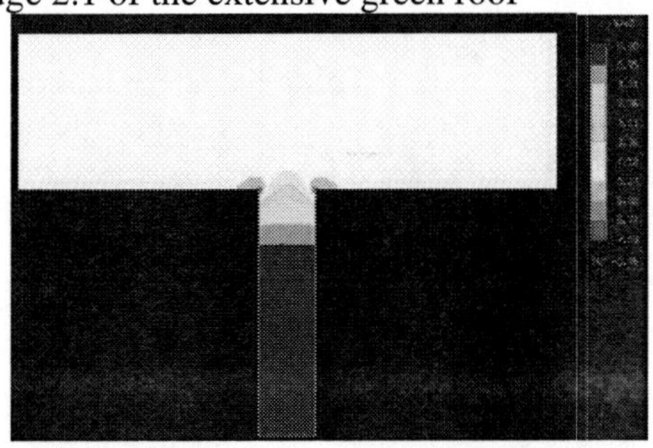

Figure 14. Isovalues of heat flow for the thermal bridge 2.2 of the extensive green roof

The results obtained for both roof cases analysed show that there is a positive influence in the case of the extensive green roof in comparison with terrace classical roof, Figures 11÷18 and Table 7. These positive influences are expressed as an increase of thermal resistance ($R_{element}$) exceeding 5%, for thermal bridge identified at the connection of roof with exterior wall, from 7.455 m^2K/W to 7.789 m^2K/W and a decreasing of linear thermal bridge coefficient (ψ) with almost 20%, from 0.1898 W/m^2C to 0.1549 W/m^2C, Figures 11÷12 and Table 7. In the cases of the other thermal bridges analysed, the variations of these parameters are in the same sense, but in the limits of 1-2%.

Figure 15. Isovalues of heat flow for the thermal bridge 1.3 of the classical terrace roof

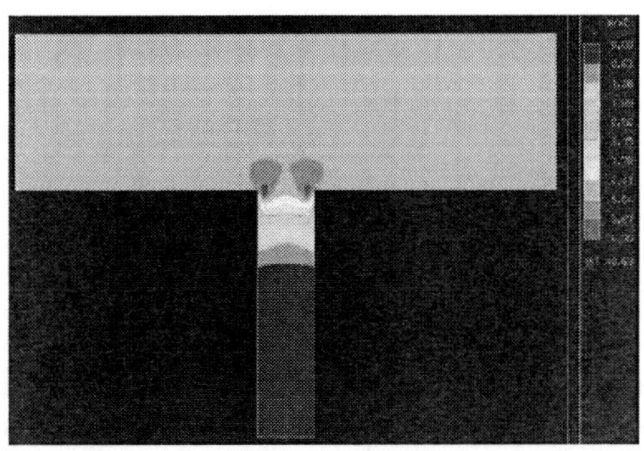

Figure 16. Isovalues of heat flow for the thermal bridge 2.3 of the extensive green roof

IOP Conf. Series: Materials Science and Engineering **586** (2019) 012007 doi:10.1088/1757-899X/586/1/012007

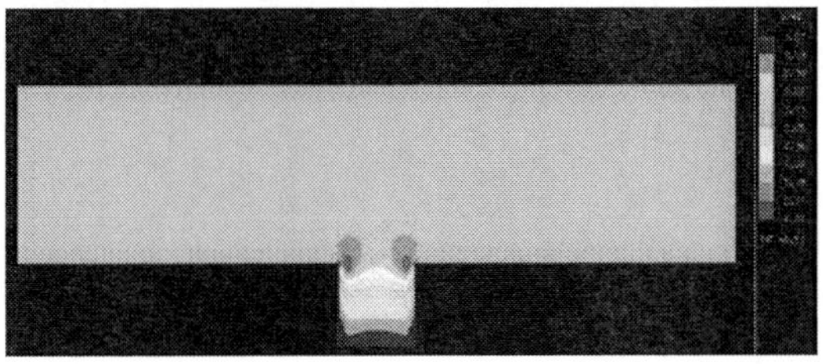

Figure 17. Isovalues of heat flow for the thermal bridge 1.4 of the classical terrace roof

Figure 18. Isovalues of heat flow for the thermal bridge 2.4 of the extensive green roof

Table 7. Linear thermal bridges coefficients, ψ

No.	Thermal bridge type	Thermal flow (ϕ) [W/m]	Temperature difference (ΔT) [°C]	The width of the thermal bridge (B) [m]	The unidirectional thermal resistance ($R_{element}$) [m²K/W]	Linear thermal bridge coefficient (ψ) [W/m²C]
1	1.1/figure 2 - classical terrace roof	13,33	38	1,20	7,455	0,1898
2	2.1/figure 5 – extensive green roof	11,74	38	1,20	7,789	0,1549
3	1.1/figure 2 - classical terrace roof (wall)	15,16	38	1,20	5,23	0,169
4	2.1/figure 5 – extensive green roof (wall)	15,24	38	1,20	5,23	0,172
5	1.2/figure 3 - classical terrace roof	6,78	38	1,20	7,455	-0,1435
6	2.2/figure 6 – extensive green roof	6,72	38	1,20	7,789	-0,1313
7	1.3/figure 4 - classical terrace roof	6,78	38	2,40	7,455	-0,1435
8	2.3/figure 7 – extensive green roof	6,70	38	2,40	7,789	-0,1318
9	1.4/figure 5 - classical terrace roof	13,57	38	3,75	7,455	-0,1459
10	2.4/figure 8 – extensive green roof	13,39	38	3,75	7,789	-0,1291

The results also emphasize the fact that the unidirectional thermal resistance in the case of the extensive green roof has an acceptable value, as it is presented in Table 5.

4. Conclusions

In the paper, the thermal resistance and the heat flow of the extensive green roof were analysed in order to establish the benefits of the green roof as a technique of insulating the buildings. Additionally, the components of these two roofs and the properties of the materials were selected and used to test their capacity to influence the heat flow intensity. Taking into account the complexity of this problem, the structural components and thermal analyses were conducted on the constructive details of thermal bridges by means of the RDM6 software.

Results from the thermal analyses show that there is an improvement for the extensive green roof in terms of thermal resistance and linear thermal bridge coefficient in the case of thermal bridge 1.1/2.1 and very low improvements for the other cases studied. This fact is confirmed by the numerical results and the modelling in the program.

In conclusion, the unpredictable results obtained for the analysed thermal parameters let conclude that, although there is no well determined thermo-technical advantage for the extensive green roof solution, anyhow, this solution presents a favourable environmental impact, promoting a "beautiful" added value to a building, in terms of both sustainability and aesthetics, as long as the building's structure has been built to accommodate it.

References

[1] Raji B, Tenpierik MJ, and van den Dobbelsteen A 2015 *J. Rene. Sustain. Energy Rev.* **45** 601
[2] Pyrgou A, Castaldo VL, Pisello AL, Cotana F, and Santamouris M 2017 *Sustain. Cities Soc.* **28** 187
[3] Salata F, Golasi I and Vollaro RDL 2015 *Energy Build.* **99** 32
[4] Santamouris M, Cartalis C, Synnefa A, and Kolokotsa D 2015 *J. Energy Build.* **98** 119-124
[5] Getter KL and Rowe DB 2006 *HortScience* **41** 1276
[6] Fabiani C, Coma J, Pisello AL, Perez G, and Cotana F 2018 *J. Energy Build.* **178** 140-153
[7] Zhao M and Srebric J 2012 *J Cent South Univ Technol* **19** 639
[8] Berardi U, Ghaffarianhoseini AH, and Ghaffarianhoseini A 2014 *J. Appl Energy* **115** 411
[9] Lockhart S 2016 *Tutorial Guide to AutoCad, 2D Drawing, 3D Modelling* SDC Publications
[10] Ştefănescu D 2012 *Hygrothermal Building Design Manual* (in Romanian) ed Soc. Acad. „Matei-Teiu Botez” (Iasi, Romania)
[11] Van der Hoeven M 2013 *Technology roadmap energy efficient building envelopes* International Energy Agency (Paris, France)
[12] *Normative Regarding the Calculation of Thermic Properties of Building Construction Elements 2011*(in Romanian) RC (C 107-2005) Revised Edition in 2011 Instal Engs Assoc of Romania (Bucharest Romania)

Comparison of current stiffness modulus values in national and international literature

V-I Iftode[1], I Olteanu-Dontov[1], P Mihai[2], M Budescu[1], I Sococol[2], I-D Dinu[1]
[1] "Gheorghe Asachi" Technical University of Jassy, Faculty of Civil Engineering and Building Services, Department of Structural Mechanics, Str. Prof. Dr. Doc.Dimitrie Mangeron, nr. 1, 700050, Romania,
[2] "Gheorghe Asachi" Technical University of Jassy, Faculty of Civil Engineering and Building Services, Department of Concrete Structures, Building Materials, Technology and Management, Str. Prof. Dr. Doc.Dimitrie Mangeron, nr. 1, 700050, Romania

E-mail: iftodevlad@yahoo.com

Abstract. The paper objective is to realize a comparative synthesis of the current coefficients used to reduce the stiffness modulus. For this, national and international codes were considered as well as reference papers on the subject. In the end, the paper presents a case study in which behaviour differences were observed for a reinforced concrete frame structure between national and international codes.

1. Introduction
Most of reinforced concrete structure are statically indeterminate structures, so the stiffness of the structural elements influences not only the displacements, but also the internal forces distribution in the elements.
The structural walls, columns and beams respond in working state II, cracked state. For this reason, in the structural computation is mandatory to consider the stiffness corresponding to working state II, the cracked one.
Admissible values are obtained for the displacement computation of a structure from the point of view of the real behaviour, if approximate values of the stiffness are used. These approximate values are computed by multiplying the stiffness modulus of the initial cross section with subunitary factors.
The subunitary factors recommended in national and international codes, reduce in fact the elasticity modulus or the geometric characteristics of the section.

2. Research methodology

2.1. National codes
In Romania, according to the design code for reinforced concrete wall structures, CR 2-1.1/2013, the stiffness values for reinforced concrete structural walls, coupled or not, are given by several computation hypothesis, presented below:

If
$$\upsilon_d = \frac{N_{Ed}}{A_c \cdot f_{cd}} = 0{,}4: \quad (1)$$

$$I_{eq} = 0{,}80 \cdot I_c \quad (2)$$

$$A_{eq} = 0{,}90 \cdot A_c \quad (3)$$

IOP Conf. Series: Materials Science and Engineering 586 (2019) 012008 doi:10.1088/1757-899X/586/1/012008

$$A_{eq,s} = 0{,}80 \cdot A_{c,s} \tag{4}$$

If

$$\upsilon_d = \frac{N_{Ed}}{A_c \cdot f_{cd}} = 0{,}0: \tag{5}$$

$$I_{eq} = 0{,}40 \cdot I_c \tag{6}$$

$$A_{eq} = 0{,}60 \cdot A_c \tag{7}$$

$$A_{eq,s} = 0{,}50 \cdot A_{c,s} \tag{8}$$

If

$$\upsilon_d = \frac{N_{Ed}}{A_c \cdot f_{cd}} = -0{,}2: \tag{9}$$

$$I_{eq} = 0{,}10 \cdot I_c \tag{10}$$

$$A_{eq} = 0{,}40 \cdot A_c \tag{11}$$

$$A_{eq,s} = 0{,}20 \cdot A_{c,s} \tag{12}$$

For intermediate values of υ_d ratio, the calculus values for the inertia moment I_{eq}, the area of the cross section A_{eq} and the area of the shearing section $A_{eq,s}$ are established through linear interpolation. I_c, A_c and $A_{c,s}$ values correspond to the section without cracks.

N_{Ed} and f_{cd} represents the design value for the axial force (positive for compression) in the section and the concrete compression strength value respectively.

The characteristics for the coupling beams are computed based on the reinforcement type:

- orthogonal bars reinforcing (longitudinal bars and stirrups):

$$I_{eq} = 0{,}20 \cdot I_c \tag{13}$$

$$A_{eq} = 0{,}20 \cdot A_c \tag{14}$$

- diagonally reinforced:

$$I_{eq} = 0{,}40 \cdot I_c \tag{15}$$

$$A_{eq} = 0{,}40 \cdot A_c \tag{16}$$

According to the Seismic design code, part I, P100-1/2013, the subunitary reduction factor of the stiffness modulus depends on the selected structural type (frame or structural walls). The recommended values for reinforced frame structures are given in Table 2.1. and take into consideration the connection types between non-structural components and the structural system.

- E_c - modulus of elasticity for the concrete
- I_g - inertia moment of the initial concrete section (without cracks).

IOP Conf. Series: Materials Science and Engineering **586** (2019) 012008 doi:10.1088/1757-899X/586/1/012008

Table 1. Design values for the stiffness modulus according P100-1/2013.

Structural type	Connection type between the non-structural component and the reinforced concrete structure	
	Non-structural elements influence the overall stiffness of the structure	Non-structural elements do not interact with the structure
Reinforced concrete structures		
Frames structures	$E_c I_g$	$0{,}50 \cdot E_c I_g$
Structural walls	$E_c I_g$	

2.2. International codes

The New Zealand standard, NZS 3101-1/2006, provides a computation equation (17), which takes into account the modulus of elasticity and the moment of inertia of the concrete section without cracks in relation to the normalized axial force. For the accuracy of the calculation, the characteristics of the reinforcement are also considered, equation 18.

$$EI = \frac{\left(E_c I_g / 2{,}5 \right)}{1 + \beta_d} \tag{17}$$

$$EI = \frac{\left(E_c I_g / 5 + E_e I_{se} \right)}{1 + \beta_d} \tag{18}$$

In which I_g is the inertia moment of the initial concrete section (without cracks), I_{se} - moment of inertia for the reinforcement, E_c - elasticity modulus for concrete, E_s - elasticity modulus for steel and β_d - ratio of design axial dead load to total axial design axial load of a column or pier:

$$\beta_d = \frac{N_{Ed}}{A_c \cdot f_{cd}} \tag{19}$$

- f_{cd} - design compression strength for concrete;
- A_c - initial concrete area section;
- N_{Ed} - design value for axial force.

Recommended values of effective section properties that may be used for the elastic analysis of structures subjected to seismic forces corresponding with the ultimate limit state are as shown in columns 2 & 3 of Table C6.6 from standard NZS 3101-2/2006.

Notes from table 2:

- (§) With these values the E value should be the elastic modulus for concrete with strength of 40 MPa regardless of the actual concrete strength.
- (‡) The values in brackets apply to columns which have a high level of protection against plastic hinge formation in the ultimate limit state.
- (¶) For additional flexibility, within joint zones and for conventionally reinforced coupling beams refer to the text.

Table 2. Effective section properties, I_e

Type of member	Ultimate limit state		Serviceability limit state		
	$f_y = 300MPa$	$f_y = 500MPa$	$\mu = 1,25$	$\mu = 3,00$	$\mu = 6,00$
Beams					
Rectangular¶	$0,40I_g$ (use with E_{40})§	$0,32I_g$ (use with E_{40})§	I_g	$0,70I_g$	$0,40I_g$ (use with E_{40})§
T or L Beams¶	$0,40I_g$ (use with E_{40})§	$0,32I_g$ (use with E_{40})§	I_g	$0,70I_g$	$0,40I_g$ (use with E_{40})§
Columns					
$N^*/A_g f_c' > 0,50$	$0,80I_g\left(1,00I_g\right)^‡$	$0,80I_g\left(1,00I_g\right)^‡$	I_g	$1,00I_g$	As for the ultimate limit state value in brackets
$N^*/A_g f_c' = 0,20$	$0,55I_g\left(0,66I_g\right)^‡$	$0,50I_g\left(0,66I_g\right)^‡$	I_g	$0,80I_g$	
$N^*/A_g f_c' = 0,00$	$0,40I_g\left(0,45I_g\right)^‡$	$0,30I_g\left(0,35I_g\right)^‡$	I_g	$0,70I_g$	
Walls¶					
$N^*/A_g f_c' > 0,50$	$0,48I_g$	$0,40I_g$	I_g	$0,70I_g$	As for the ultimate limit state value
$N^*/A_g f_c' = 0,20$	$0,40I_g$	$0,33I_g$	I_g	$0,60I_g$	
$N^*/A_g f_c' = 0,00$	$0,32I_g$	$0,25I_g$	I_g	$0,70I_g$	
Diagonally reinforced coupling beams	$0,60I_g$ – for flexure Shear area, A_{shear}		I_g $1,50A_{shear}$ for ULS	$0,75I_g$ $1,25A_{shear}$ for ULS	As for ultimate limit

In the Japanese code, JSCE-No. 15-2010, a detailed procedure for the reduction of the stiffness modulus assessment is presented.

If the effective bending stiffness is considered as a function of the bending moment, the evaluation relation is:

$$E_e I_e = \left(\frac{M_{crd}}{M_d}\right)^4 \cdot \frac{E_e I_g}{1-\dfrac{\Delta M_{csg}}{M_d - P\left(d_p - c_g\right)}} + \left[1-\left(\frac{M_{crd}}{M_d}\right)^4\right] \cdot \frac{E_e I_{cr}}{1-\dfrac{\Delta M_{csg,r}}{M_d - P\left(d_p - c_g\right)}} \tag{20}$$

If the effective bending stiffness is considered constant on longitudinal direction, the evaluation relation becomes:

$$E_e I_e = \left(\frac{M_{crd}}{M_{d,\max}}\right)^3 \cdot \frac{E_e I_g}{1-\dfrac{\Delta M_{csg}}{M_{d,\max} - P\left(d_p - c_g\right)}} + \left[1-\left(\frac{M_{crd}}{M_{d,\max}}\right)^3\right] \cdot \frac{E_e I_{cr}}{1-\dfrac{\Delta M_{csg,r}}{M_{d,\max} - P\left(d_p - c_g\right)}} \tag{21}$$

where:
E_e - effective elasticity modulus for concrete and is computed by 22 relation.

$$E_e = \frac{E_{ct}}{1+\varphi} = \frac{E_{ct}}{1+\left(\frac{E_{ct}}{E_c}\right)\cdot \varphi_{28}} \quad (22)$$

The following notations were made: E_{ct} - elasticity modulus for the concrete from permanent loading moment, φ - creep coefficient, φ_{28} - creep coefficient at 28 days, I_g - moment of inertia of gross section, I_{cr} - moment of inertia for the section with cracks, I_e - effective moment of inertia, M_{crd} - critical moment for bending when a crack appears in the transversal section, M_d - design bending moment, P - axial force or the pretension force and ΔM_{csg} - apparent bending moment produces by the contraction in the section without cracks taking into consideration the reinforcement, computed by 23 relation:

$$\Delta M_{csg} = \left(\frac{I'_{sg}}{c_g - d'} - \frac{I_{sg}}{d - c_g} - \frac{I_{pg}}{d_p - c_g}\right)\cdot \varepsilon'_{cs} \quad (23)$$

$\Delta M_{csg,r}$ - apparent bending moment produces by the contraction in the section with cracks taking into consideration the reinforcement, computed by 24 relation:

$$\Delta M_{csg,r} = \left(\frac{I'_{scr}}{c_{sr} - d'} - \frac{I_{scr}}{d - c_{cr}} - \frac{I_{pcr}}{d_p - c_{cr}}\right)\cdot \varepsilon'_{cs} \quad (24)$$

The creed coefficient depends on the concrete grade and the reinforcement percentage from the elements. Its values are given in table 3, when $E_{ct} = E_c$.

Table 3. Values for the creed coefficient

Creed coefficient for concrete					
Exposure conditions	Concrete age for the pretension or loading applications				
	4-7 days	14 days	28 days	3 months	1 year
Exterior	2,7	1,7	1,5	1,3	1,1
Interior	2,4	1,7	1,5	1,3	1,1
Creed coefficient for light concrete					
	4-7 days	14 days	28 days	3 months	1 year
Exterior	2,0	1,3	1,1	1,0	0,8
Interior	1,8	1,3	1,1	1,0	0,8
Creed coefficient for concrete with 1% reinforcement					
	4-7 days	14 days	28 days	3 months	1 year
Exterior	2,1	1,4	1,2	1,1	0,9
Interior	1,9	1,4	1,2	1,1	0,9

Where: I'_{sg} - moment of inertia of the compressed reinforcement of the section without cracks, I_{sg} - moment of inertia of the tensioned reinforcement of the section without cracks, I_{pg} - moment of inertia of the prestressing reinforcement of the section without cracks, I'_{scr} - moment of inertia of the

IOP Conf. Series: Materials Science and Engineering **586** (2019) 012008 doi:10.1088/1757-899X/586/1/012008

compressed reinforcement of the section with cracks, I_{scr} - moment of inertia of the tensioned reinforcement of the section with cracks, I_{pcr} - moment of inertia of the prestressing reinforcement of the section with cracks, c_g - distance between the compressed face to the centroid of the section without cracks, c_{cr} - distance between the compressed face to the centroid of the section with cracks, d' - distance from the compressed fibre to the compressed reinforcement, d - distance from the compressed fibre to the tensioned reinforcement, d_p - distance from the compressed fibre to the prestressing reinforcement and ε'_{cs} - contraction strain.

2.3. In reference works

T. Paulay and M. Priestley, in their book "Seismic design of reinforced concrete structures and masonry buildings" from 1992, give an interval of values which depends on the constituent elements of the structure, and on the recommended values, table 4.

Table 4. Design values for the stiffness modulus according to Paulay and Priestley/1992

Element	Interval	Recommended value
Beams	$0{,}30 \div 0{,}50 \cdot I_g$	$0{,}40 \cdot I_g$
Beams T, L	$0{,}25 \div 0{,}45 \cdot I_g$	$0{,}35 \cdot I_g$
Columns, $P \geq 0{,}50 \cdot A_g f_c$	$0{,}70 \div 0{,}90 \cdot I_g$	$0{,}80 \cdot I_g$
Columns, $P = 0{,}20 \cdot A_g f_c$	$0{,}50 \div 0{,}70 \cdot I_g$	$0{,}60 \cdot I_g$
Columns, $P = -0{,}05 \cdot A_g f_c$	$0{,}30 \div 0{,}50 \cdot I_g$	$0{,}40 \cdot I_g$

I_e depends on the structural element type, so

- for structural walls:

$$I_e = \left(\frac{100}{f_y} + \frac{P_u}{A_g f} \right) \cdot I_g \tag{25}$$

- for coupling beams with diagonal reinforcements:

$$I_e = \frac{0{,}40 \cdot I_g}{\left[1 + 3 \cdot \left(\frac{h}{l_n} \right)^2 \right]} \tag{26}$$

- for coupling beams with orthogonal bars reinforcements:

$$I_e = \frac{0{,}20 \cdot I_g}{\left[1 + 3 \cdot \left(\frac{h}{l_n} \right)^2 \right]} \tag{27}$$

Where f_c is the compressions strength for the concrete, f_y - elasticity limit of the steel in case of reinforced concrete, P_u - axial force obtained from the seismic combination, h - section height and l_n - coupling beam length.

3. Quantitative results, qualitative result and discussions

For the case study a reinforced concrete frame structure was considered with one opening of 6m and height of 3m. Columns of 50x50 cm^2 were considered, reinforced with 12 longitudinal bars of Ø16 diameter and stirrups of Ø8 at a distance of 15cm. 30x60 cm^2 cross section beams were considered reinforced top and bottom with 2 bars of Ø16 and 1 bar of Ø10, and stirrups of Ø8 at a distance of 15cm. A uniformly distributed load of 50kN/m is applied on the beams as live load and 50kN/m for permanent load.

A modal and a static nonlinear analysis were performed in SAP2000. The software is based on the finite element method. The stiffness modulus of the frame elements was reduced according to the synthesis from chapter 2.

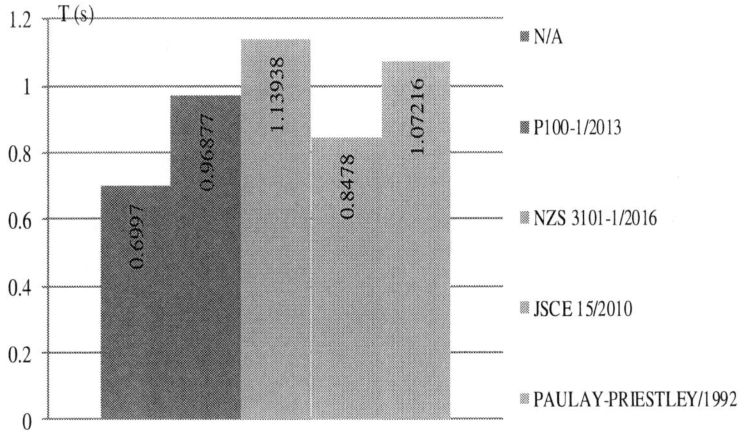

Figure 1. Results synthesis – comparison of the fundamental period of vibration

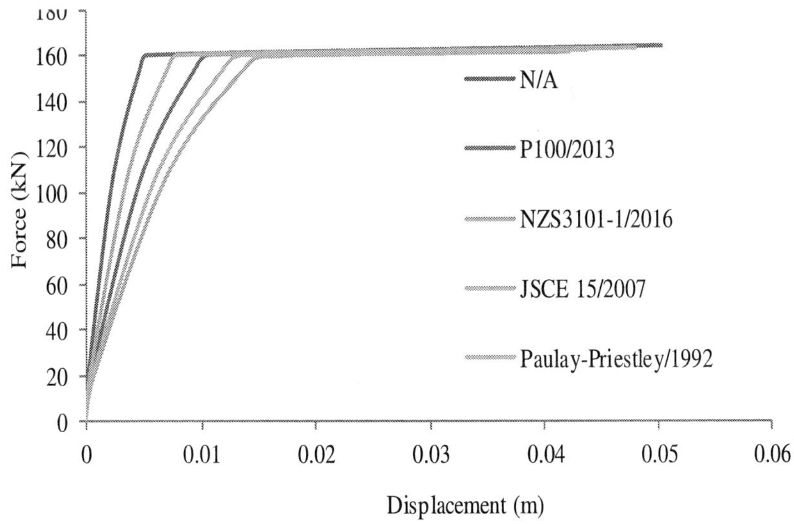

Figure 2. Results synthesis – comparison of the capacity (pushover) curve

4. Conclusion

The reinforced concrete frame structures are flexible one for which as long as their period of vibration is greater, they experience the maximum dynamic displacement. According to the Romanian code, the period of vibration is greater with 27.8% with respect to the initial structure. The highest increase of the period of vibration is observed according to the New Zeeland code, of 38.6% with respect to the neutral state. Based on the results obtained from the nonlinear static analysis, for the maximum recorded lateral force a significant difference is observed by applying the New Zeeland code. The lateral force is smaller with 1.5% and the displacement with 19.2% smaller with respect to the neutral state.

Further studies should be carried out to establish the optimum reduction factor for the stiffness modulus.

5. References

[1] Seismic design code indicative P100-1 2013 *Design provisions for buildings*

[2] Seismic design code indicative P100-3 2018 *Provisions for seismic assessment of existing buildings*

[3] Paulay T and Priestley M J N 1992 *Seismic design of reinforced concrete and masonry buildings* ISBN 0-471-54915-0

[4] New Zeeland Standard NZS 3101-1 English version 2006 *Concrete structures standard - The design of concrete structures* ISBN 1-86975-043-8

[5] New Zeeland Standard NZS 3101-2 English version 2006 *Concrete structures standard – Commentary on the Design of Concrete Structures* ISBN 1-86975-043-8

[6] National code indicative CR 2–1–1.1 2013 *Design code for structures with reinforced structural walls*

[7] Japan Society of Civil Engineers (JSCE) 2010 Guidelines for concrete No. 15 *Standard specifications for concrete structures – Design* ISBN 978-4-8106-0752-9

IOP Conf. Series: Materials Science and Engineering **586** (2019) 012009 doi:10.1088/1757-899X/586/1/012009

Improving elastic properties of bricks by polyurethane matrices mean

V-I Iftode[1], I Olteanu-Dontov[1], P Mihai[2], M Budescu[1], M Harja[3]
[1] "Gheorghe Asachi" Technical University of Jassy, Faculty of Civil Engineering and Building Services, Department of Structural Mechanics, Str. Prof. Dr. Doc.Dimitrie Mangeron, nr. 1, 700050, Romania,
[2] "Gheorghe Asachi" Technical University of Jassy, Faculty of Civil Engineering and Building Services, Department of Concrete Structures, Building Materials, Technology and Management, Str. Prof. Dr. Doc.Dimitrie Mangeron, nr. 1, 700050, Romania
[3] "Gheorghe Asachi" Technical University of Jassy, Faculty of Chemical Engineering and Environmental Protection, Department of Chemical Engineering, Str. Prof. Dr. Doc.Dimitrie Mangeron, nr. 73, 700050, Romania

E-mail: iftodevlad@yahoo.com

Abstract. The usage of polyurethane was considered in order to obtain an innovative brick (polyurethane brick) for the envelope of reinforced concrete frame structures and also for partition walls. The following properties were studied: self-weight, compression strength and bending strength. In order to increase its mechanical properties, polyurethane materials can be reinforced with recycled or organic materials, reducing costs in this manner (crashed glass, straw, chopped rubber). Using these innovative infill materials, the heating costs are reduced and the execution time is decrease. The polyurethane bricks are easily adapted to any climatic condition and allow water vapour transfer from inside to the exterior side of the masonry, preventing condensation. Compressive strength, flexural strength and split tensile strength were experimentally determined. From the laboratory results it was observed that an increase of 32.6% occurred for the compressive strength for the sample with glass pieces. Overall improvements of the mechanical properties were noticed for the samples with smeared reinforcement.

1. Introduction

Buildings are intended to provide people the indispensable conditions for their activities. The society's degree of development is reflected by the technical level in construction. Each construction or construction component must meet a set of technical requirements and conditions, concerning durability, fire resistance, toughness and stability of the construction, architectural and economically-organizational ones. Every building has a well-established destination and satisfies a range of complex and varied functions.

Polyurethane is a resilient, flexible and durable material that can replace rubber, metal and wood in many applications. It can be produced in different colour, take any shape or size. The polyurethane properties depend on the type and percentage of the necessary components: isocyanate and polyols. A piece of polyurethane can be considered a giant molecule, as a consequence, typical polyurethanes do not soften or melt when heated. Isocyanates and polyols available options, compared to other additives and processing conditions, allow the polyurethane to have a wide range of properties that make it such a widely used polymer [4].

Polyurethane is used in construction since 1950 and its main destination is building insulation, in the shape of insulating roof panels, walls, ceilings and floors. One of the desirable attributes of polyurethanes is their ability to be converted to foam. Polyurethane foam sandwich panel is fit for projects which have requirements regarding constant high temperature or strict hygiene maintenance,

because polyurethane core material is considered to be the best material in keeping warm and behaving as thermal insulation.
Metal-faced polyurethane sandwich panels are the system of choice today for large industrial buildings, office blocks, exhibition halls, schools etc. Prefabricated sandwich walls and lightweight roofing systems which consist of metal facings bonded tightly together by a core of rigid polyurethane foam. Polyurethane sandwich panels come complete with specially formed tongue-and-groove joints ensuring a perfect fit and maximum integrity [12].
Polyurethane bricks (PUB) are made using a mould and a special machine with a dosing device, in order to obtain the required mechanical characteristics. PUB can be produced in a variety of sizes, depending on the construction site, the destination of the building and the size of the frames structure [6].
Ventilated facades are currently made of: HPL tiles (High Pressure Laminate), cement, ceramics, clinker, terracotta, artificial stones, recomposed one (marble, granite, travertine), composite wood, aluminium, decorative glass, steel, natural stone.
By its insulating qualities, for both low and height temperature, a ventilated facade made of polyurethane, has the economic advantage of reducing the costs for air conditioner or heating. Using a ventilated facade, the building structure may represent the most modern and attractive line, its maintenance being simple, without major problems.

2. Material and method

2.1. Overview

The polyurethane brick is a mixture of chemical substances. The percentage of each component, the environment conditions and the technological process significantly influences the final product. Isocyanate is the functional group having the formula $R - N = C = O$.
Di-isocyanates are made from reactions with polyols in the production of polyurethanes. Isocyanates are electrophiles, so they are reactive among a variety of nucleophiles, including alcohols, amines or even water. The treatment with alcohols leads to a urethane bond. If a di-isocyanate is treated with a compound that contains two or more hydroxyl groups, such as a diol or polyol, a polymer chain called polyurethane is formed.
Monomeric polyols, made from the polymerization of ethylene oxide and propylene oxide, such as glycerol, pentaerythritol ethylene glycol, are often used as point of departure for polymer polyols and the initiation of the polymerization reaction. Polymer polyols are typically polyether or polyester. Hydroxyl terminated polybutadiene polyol is used to produce the polyurethane.
Characteristics of the mixture are:

- mixture ratio: 100:120
- free density: 100 ÷ 350 Kg/m3
- working temperature: 25 ° C
- time to expand: 16-18 sec
- moulding time: max. 15 minutes.

Polyurethane bricks are made using a mould filled in by a special machine with a dosing device, in order to obtain the essential density. During the manufacturing process it is desired to use the best quality of the material components, compliance with technological parameters and quality of finished products, in order to ensure proper operating behaviour [11].
Various sizes of polyurethane bricks can be produced, depending on the construction site, the destination of the building and the size of the reinforced concrete frame structures. In figure 1 it is shown a model for a polyurethane brick that has fibre cement boards on the lateral sides, in order to increase the mechanical strength [6].

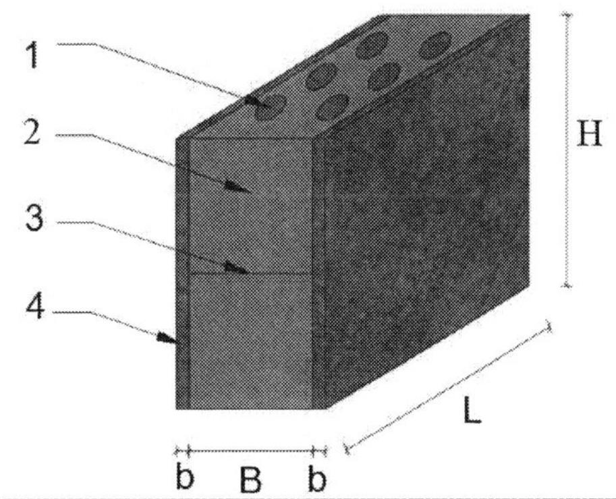

Figure 1. PUB: 1. Gaps (ceramic, extruded polystyrene, wood); 2. Rigid polyurethane; 3. Reinforcement mesh; 4. Cement slabs

The gaps presented in this prototype can be left empty or filled with various materials – ceramics, polystyrene, and wood, in order to reduce the costs and increase the mechanical strength.
Considering the available size of a classic bricks that are already on the market – ceramic, lightweight concrete, Table 1 presents some suggestions for PUB possible dimensions. For each case the polyurethane and the cement dimensions are presented. Models 11 and 12 are proposed for implementing new dimensions.
Large size products are possible due to the low density of the polyurethane, meaning as well as low weight.

Table 1. Possible dimensions for polyurethane bricks

Crt. No.	Brick	L [mm]	H [mm]	B [mm]	b [mm]
1	Module 1*	240	115	43	10
2	Module 2*	240	115	68	10
3	Module 3*	240	115	118	10
4	Module 4*	240	115	168	10
5	Module 5*	240	115	158	15
6	Module 6*	290	140	68	10
7	Module 7*	290	140	118	10
8	Module 8*	290	140	168	10
9	Module 9*	380	250	168	10
10	Module 10*	380	250	158	15
11	Module 11 (Concept)	500	250	150	10
12	Module 12 (Concept)	500	250	150	20

As said previously, if creating appropriate moulds, polyurethane can even take the shape of a round section as half log, figure 2. Polyurethane concrete block can represent a proper replacement for logs used for the construction of country houses, the environmentally friendly properties being a major advantage for this brick. Table 2 presents three different possible dimensions that could be used for moulds, but these are not necessarily the only ones. Their dimensions can be adapted according the owner request or the construction requirements.

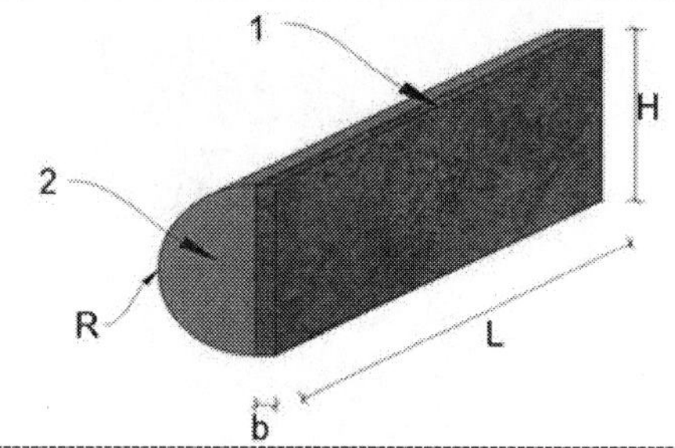

Figure 2. PUB: 1. Cement slabs; 2. Rigid polyurethane

Table 2. Possible dimensions for polyurethane bricks

Crt. No.	Brick	L [mm]	H [mm]	R [mm]	b [mm]
1	Module 1	2000	200	150	20
2	Module 2	2000	150	100	20
3	Module 3	1000	100	50	20

2.2. Reinforcement

Fibre reinforcement of construction materials has been used for a long time in order to increase the tensile capacity. Unburned bricks (adobe) were reinforced with chopped straw or animal hairs are considered, in order to prevent cracking and to provide a high strength to tearing and moisture represent a novelty. The extrapolation was made from clay to cement and from straw and animal hair to fibres. Due to the progressive increase of the steel – concrete price on the world market, and as a result to the technical and economic studies that were made, polymer fibres were chosen for dispersed reinforcement [3].

Within the past few years, there has been a significant increase in the use of natural fibres, such as leaves from flax, jute, hemp, pineapple, and sisal, for making a new type of environmentally-friendly composite. In general, two types of natural fibres are identified for making fibre-reinforcement polymer: plant-based and animal-based fibres. For the former, due to their abundant supply in the natural environment, the raw material cost is relatively low and can be completed with synthetic fibres [1].

The comparison between the physical and mechanical properties of this fibres and the steel ones have led to the exponential growth in the use and demand of this type of material in the construction world market.

To improve the quality of the polyurethane bricks and to increase compressive and flexure strength, reinforcement can be used for PUB, also. In addition to this, the amount of polyurethane will be reduced, and thus, the costs [2].

In order to find the ideal reinforcement material, bricks with different reinforcement type have been tested in three points to bending and compression tests. The materials used for this reinforcement are: fibreglass mesh, fibre rubber, Geogrid mesh, metal mesh, glass shards shape. Section 3 presents a synthesis of the laboratory tests and the conclusions.

2.3. PUB masonry

In order to do the infill of reinforced concrete structures, only bricks of good quality will be used, with no cracks. A polyurethane adhesive is recommended in order to obtain a homogenous infill wall. Wall thickness will correspond to the architectural drawings.

For execution, labour skilled workers will be used, along with masons and related masonry workers. Both, flatness and verticality will be watched; according to actual code. A deviation of ± 0.5 cm is allowed both vertically and horizontally, measured toward a straight edge of 3 m length.

The following operations must be strictly controlled:

- good adhesion between polyurethane bricks and adhesive;
- horizontal and vertical joints shall be well filled with adhesive all over;
- in the vertical joints reinforcing mesh fibreglass can be placed by geogrid etc.;
- vertical joints will be weave so that the height superposition of two successive rows, both in the field and at the corner's intersection, to make the minimum of ¼ from the brick length, and ½ of its thickness. Weaving is required in each row;
- the horizontality of the brick's rows will be watched;
- the interruption of the masonry work will be done in steps;
- the connections between walls, corners, junctions and branches will be made alternatively;
- the anchoring of the infill masonry is made with steel whiskers of 8 mm diameter and 50 cm long or creating strep to achieve weaving ancient masonry;
- corrosion protection of the anchors will be provided;
- partition walls (bricks edge) are stiffen by weaving and anchoring steel bars of 6 mm diameter every 3-4 rows in horizontal joints, according to the P2-85 Standard.

3. Laboratory test and results

3.1. 3 point bending test

To sustain the efficiency of polyurethane masonry, laboratory tests were performed in order to obtain and compare the physical and mechanical characteristics of four types of materials: autoclaved aerated concrete (AAC), brick, wood, and respectively polyurethane brick.
For the bending tests, a simple scheme of loading is considered. The test tube, having the shape of a circular or rectangular section bars, will be simply supported, and the load will be applied in the midway of the supports, until the time of rupture, as shown in figure 3. Tension in the test tube it is not even, due to the variation of the bending moment in the longitudinal axis, and because of a given cross-section in elastic areas, tension varies linearly.

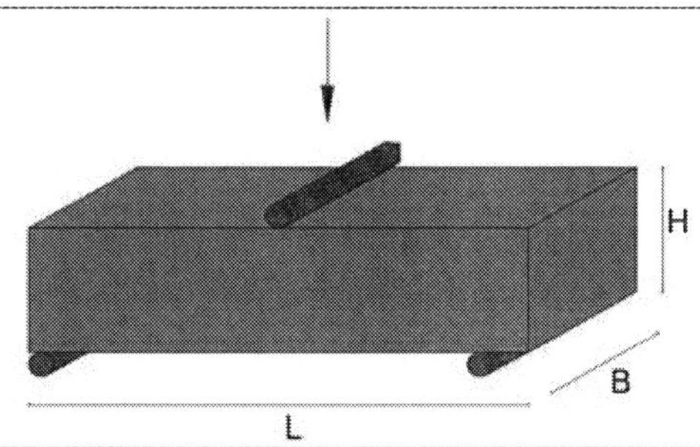

Figure 3. Testing specimen for 3 point bending test

Six testing tubes were considered – 2 made of autoclaved aerated concrete (A.A.C.), 2 made of clay (usual brick) and 2 of PUB. The samples had the following dimensions 40x40x160 all in mm. Traditional bricks and A.A.C. had behaved similarly, while polyurethane brick's behaviour was significantly different, see figure 4.
The displacements obtained for polyurethane brick were significantly higher than the other ones. Also, the maximum force was 1.15 kN for the most resistant traditional brick versus 1.35 kN for polyurethane brick. One of the PUB sample has been reinforced using fibreglass mesh. An increase of about 60% for the reinforced brick was reached, compared to the simple brick, concerning the maximum force, and an increase of 40% for the maximum displacement.

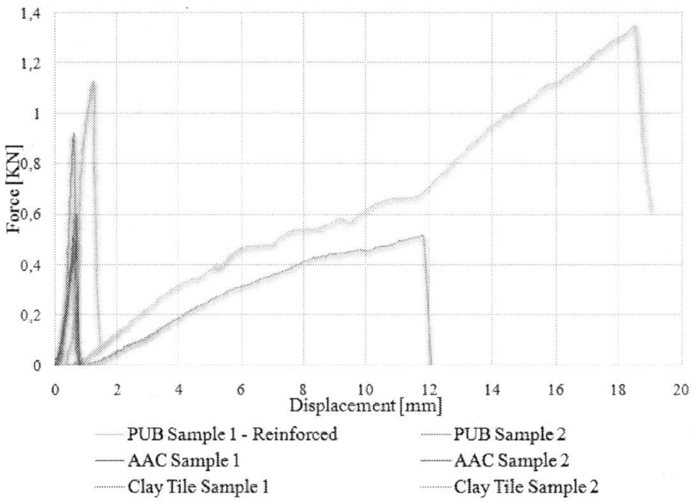

Figure 4. Results for 3 point bending test

From the 3-point bending test carried out in the faculty laboratory, it can be concluded that polyurethane brick has an adequate behaviour in comparison with current used materials. It was noticed that its capacity is significantly enhanced when reinforcement solutions are considered. figure 5 presents the sample taken during the testing.

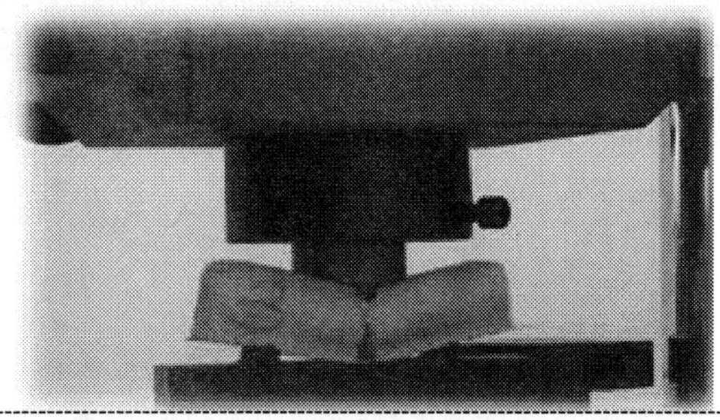

Figure 5. Photo from the 3 point bending test

3.2. Compressive test

The compression test is usually performed for determining the mechanical properties of brittle materials (when it is cold or warm), which present lower tensile strength (cast iron, metals or non-ferrous alloys etc.). This test allows to establish the compressive stress, according to STAS 1552-78 (STAS 1552-78).

The main issue that this laboratory test is facing is the homogeneous tension inside the test tube. Another problem is that tenacious materials cannot be damaged by this experiment, furthermore a plastic deformation can appear along with the increasing of the load. Rigid metals present an elastic deformation, followed by a plastic one that leads to damage. After deformation, the samples look like a barrel.

For this test, six polyurethane bricks samples were considered, which differed by the reinforcement type. Two types of reinforcement were considered, horizontal, perpendicular to the loading action and smeared reinforcement. Samples 1 and 5 were longitudinally reinforced with fibreglass mesh, sample 2 was using geogrid mesh of 40 mm and dispersed rubber particles, sample 3 was reinforced with

broken glass, sample 4 was longitudinally reinforced with metal mesh of 5 mm, and sample 6 was simple, made at a specialized factory, with a special dosing machine [9].
Figure 6 shows photo take during the compression test on some of the samples.

Figure 6. Photo from the compressive test

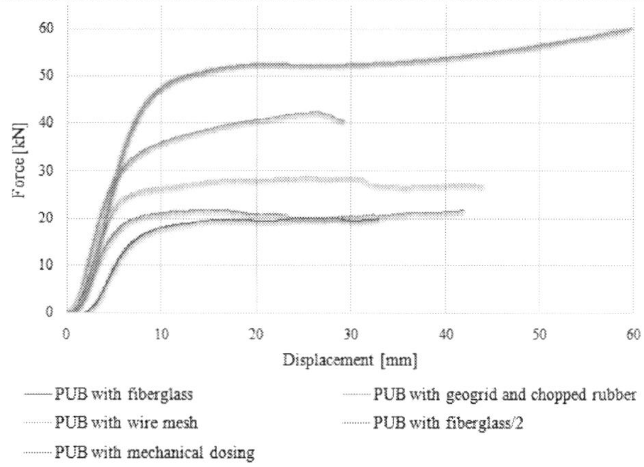

Figure 7. Compressive test results

Figure 7 is a synthesis of the results obtained in the compression test. The main conclusion of this test is that the use of mixed reinforcement – longitudinal and smeared one, lead to significantly better results. Although it was initially assumed that broken glass reinforcement would have a good behaviour, laboratory tests had shown otherwise.
It can also be concluded that the execution technology of the components is essential. Relatively large differences were observed between the samples hand made in our laboratory and those obtained by automatic dosing and mixing the components at constant pressure. This sustains the idea that the process is essential in reaching the top quality of polyurethane brick elements.

4. Conclusions

Polyurethane is not a new material, on the contrary is being widely utilized in different industries. In the construction field is generally used for sandwich panels, as building insulator, as adhesive, etc. The new use, that of masonry block, has numerous advantages. Firstly, good results obtained in the laboratory for the mechanical characteristics. Among the analysed properties were studied: self-weight, compression strength, bending strength and thermal strength. Also, it was established that in order to increase its mechanical properties, PUB can be reinforced with recycled or organic materials, reducing the costs in this way (crashed glass, straw, chopped rubber). The PUB masonry used as infill wall for reinforced concrete frame structures overcome the failure mechanism due to stiffness

differences between the components. Secondly, due to its good thermal characteristics, the masonry made of PUB elements lead to a superior thermal strength of the element in comparison with those frequently used (ACC, clay). A consequence of this property is represented by the cost reduction for the energy required to heat the interior spaces. The polyurethane bricks are easily adapted to any climatic condition and allow the water vapour transfer from inside to the exterior side of the masonry, preventing condensation. Thirdly, the technological process needed to put in place PUB elements is fast and requires minimum resources, to be exact only electric energy. According to the U.E. Directives on the classification, packaging and labelling of dangerous substances, there are no special requirements for this material. In addition to this, polyurethane corresponds to protocols relating to ozone layer in Montreal and Kyoto Summit. From the point of view of health, raw materials are treated in such a way that, under normal conditions, does not emit noxious fumes and are not toxic by touching. Environmentally speaking, the products do not emit harmful gases over their lifetime, nor degrade the environment. As one of the most effective insulators in construction, it actively contributes to energy saving, greatly reducing carbon dioxide emissions.
Last aspects which need to be considered are that this material can be recycled and used again in the construction field. Such an example is to mix small polyurethane pieces in the concrete in order to obtain thermal insulated slabs.
Further studies need to be performed in order to have a complete understanding on how the product behaves in particular situations, for example extreme weather conditions, fire, etc.

5. References

[1] Alan Kin-tak Lau, Ada Pui-Yan Hung 2017 Natural Fiber-Reinforced Biodegradable and Bioresorbable Polymer Composites, Woodhead Publishing
[2] Barbuta M, Harja M, Ciobanu G 2014 Mechanical properties of polymer concrete containing tire waste powder, *Journal of Food, Agriculture & Environment*, vol. 12, nr 2, 1186-1190
[3] Diaconescu R M, Barbuta M, Harja M 2013 Prediction of mechanical Properties of Polymer Concrete with tyre rubber Using Neural Networks, *Materials Science Engineering* B, 178, 1259-1267
[4] Hepburn C 1991 Polyurethane Elastomers, *2nd Ed. Elsevier Appl Sci*, London.15
[5] Olteanu I, Budescu M, Iftode V I 2014 Case Study on Failure Mechanism for Reinforced Concrete Frame Structure with Different Infill Material, *Proceedings of the 5th International Conference Advanced Materials and Systems*, ICAMS2014, Bucharest, Romania
[6] Olteanu I, Budescu M, Iftode V I 2014 Flexybrick – Revolutionary Solution for Polyurethane Application, *Proceedings of the 5th International Conference Advanced Materials and Systems, ICAMS2014*, Bucharest, Romania
[7] Olteanu I, Toma A M 2012 Applicability of vulnerability index of reinforced concrete frame structure in seimic, *4th International Conference on Advanced Materials and Systems*, Bucharest, Romania
[8] Ngapeya Gelen Gael Chewe, Waldmann D, Scholzen F 2018 Impact of the height imperfections of masonry blocks on the load bearing capacity of dry-stack masonry walls, Construction and Building Materials, Elsevier, 165, 898-913
[9] Pope K 1973 Method of fabricating a building panel containing cellular polyurethane, *U.S. Patent No 3,715,417*
[10] STAS 1552-78, Tests of metals. Compression test
[11] Septevani A A, Evans D A, Annamalai P K, Martin D J 2017 The use of cellulose nanocrystals to enhance the thermal insulation properties and sustainability of rigid polyurethane foam. *Industrial Crops and Products*, 107, 114-121
[12] Woods G 1982 Flexible Polyurethane Foams, *Chemistry and Technology*, *Appl Sci Pub Ltd*; Essex, England

IOP Conf. Series: Materials Science and Engineering **586** (2019) 012010 doi:10.1088/1757-899X/586/1/012010

Analysis of contemporary solar house in terms of sustainability criteria

D N Isopescu[1], A I Dumitraşcu[2], C G Corduban[2] and C Ţâşnei[3]
[1] Civil and Industrial Constructions Department, Faculty of Civil Engineering and Building Services, "Gh. Asachi" Technical University, Iaşi, 700050, Romania
[2] Urbanism Department, Faculty of Architecture, "Gh. Asachi" Technical University, Iaşi, 700050, Romania
[3] SC Energy Logistic Park, Romania

E-mail: alin_corduban@yahoo.com

Abstract. When discussing the importance of sustainability principles for both architecture and construction fields, the selection and analysis of design directions (from concept to implementation phase) represent a major topic of the whole integrated process. The constructions sector has already established itself worldwide as an important consumer of both energy and resources. Therefore, the long life of the edifices implies that initial design decisions (conceptual phase) have a significant, amplified effect over time. This paper presents a proposal for a contemporary solar house, with a design in accordance to sustainability criteria. The research highlights the methods applied in the conceptual phase, as well as a comparative study in terms of embodied energy, with two scenarios for the structure: wooden frame versus masonry. Life cycle assessment methods (LCA) have played an important role in optimizing the proposed model in order to deliver a product that is both environmentally and economically efficient.

1. Introduction

When discussing the future of sustainable development, the overall consensus is that finding new solutions to energy production and consumption issues should be of fundamental importance. Currently, the maintenance and further progress of living standards, in the economic, technical and social sectors depend mainly on improving the energy efficiency of buildings, the technological systems and especially ensuring the use of sustainable energy sources and materials with minimal impact on the environment [1]. The future of housing design will definitely take into account environmental issues, which are already at the core of the policy for governments, industry and non-profit organisations worldwide. The most important aspects to be considered are: life-cycles assessments, production technologies analysis, energy efficiency and regional impact studies [2].

1.1. Sustainability principles incorporated in the design stage

Integrated design is essential in respect to sustainability criteria and involves several aspects that need to be taken into account from the conception phase (house orientation, materials, energy efficiency strategies, etc.) and the collaboration of different fields (architecture, construction, sociology, etc.). An important component is represented by passive solar design, a cumulative method of organizing the premises of a building that exploits and controls radiation input. The goal is to maximize positive inputs in the cold season while reducing the effects of overheating during the warm season [3].

The house presented in this paper incorporates integrated passive design strategies, analyzed in the conceptual phase. Life cycle assessment (LCA) and energy efficiency evaluations have been performed in order to support the premises and to optimize the model.

The semi-detached house (cumulating a total living area of 160m^2), figure 1, has been designed as part of a research study focused on developing ecological and affordable dwellings for a residential

IOP Conf. Series: Materials Science and Engineering **586** (2019) 012010 doi:10.1088/1757-899X/586/1/012010

development comprising approximately 100 housing units, on the outskirts of Iasi, Romania. The task was to establish a balance between ecological design principles and economical solutions, specific to regional technical building practice. In this respect, the following strategies have been implemented:

- *optimal orientation.* The houses overall geometry coupled with the opportunity of double access/entrance (from the front and the back) determined an unconventional approach in order to maximise solar exposure for both houses that compose the semi-detached unit. Instead of the traditional linear distribution that would have provided southern exposure only to one house, thus leaving the other exposed mainly to the North, a different juxtaposition solution was adopted (back-wall adjacency). Thus, both houses can benefit from southern exposure, even if not in equal terms, exploiting south-east and south-west directions.
- *passive solar design strategies.* The functional distribution of rooms and the subsequent glazing ratio of the facades follow the passive design strategies, with southern orientation of living area and north orientation for technical spaces. The generous south-facing windows are shaded during summer by either volumes or roof eaves, while allowing for passive solar gains in winter, when the sun's trajectory is lower. The indoor open space permits north to south cross-ventilation.
- *materials.* The environmental benefits of wood are universally acknowledged, as listed forward: renewability, good strength-to-weight ratio, low- embedded energy, low carbon impact. Also, it is an aesthetically pleasing material, thus being a desirable choice for the housing program [4]. The wooden platform framing system has been analysed in a LCA comparison with a reinforced concrete and brick structure, the latter representing a commonly used solution in Romanian building practice.
- *energy efficiency of envelope elements.* The optimal solution selection strategy for energy efficiency in residential buildings synthetically consists of: listing of above mentioned technical solutions, designed to increase energy efficiency; optimal solutions selection in accordance with restrictions arising from specific analysis and impact assessment.
- *active solar design strategies.* The implementation of solar systems for housing units technologies remains a priority, Iasi being located in an area with annual solar energy fluxes ranging from 1120-1490 kWh/m2 per year, applying the Photovoltaic Geographical Informational system (PVGIS) [5].

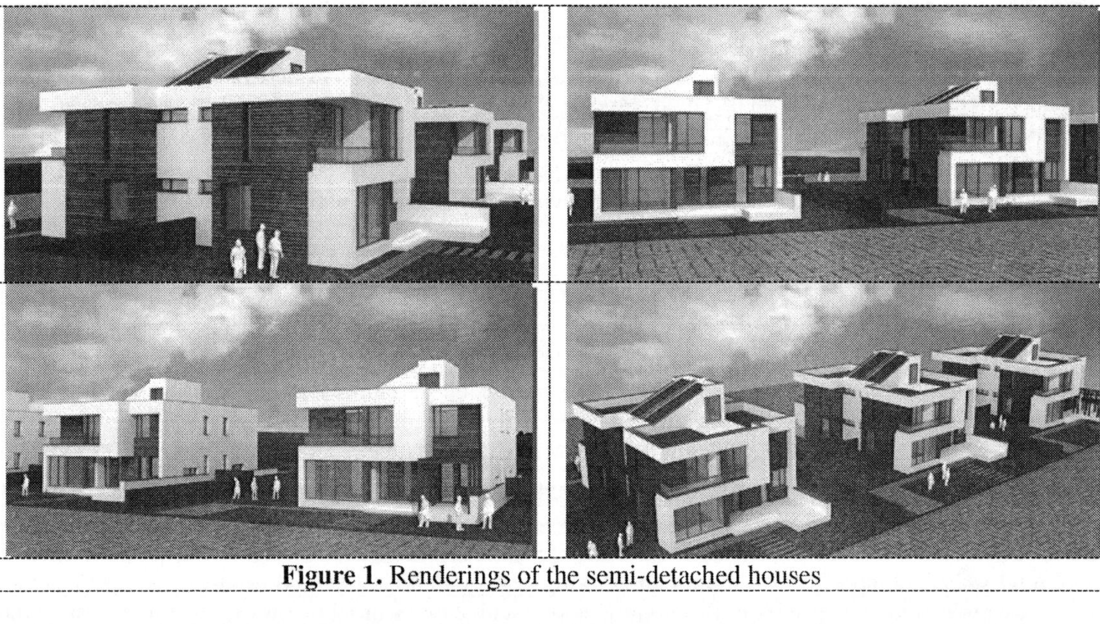

Figure 1. Renderings of the semi-detached houses

1.2. Investigation of structural and envelope scenarios

For the same architectural solution, two scenarios for the structure and enclosures have been considered, both well suited in terms of spans and loads: *wood platform framing* and *reinforced concrete post and beam* with vertical perforation bricks masonry framed walls, figure 1. The wooden structure consists of 200 mm wide studs, spaced at 400 mm, stiffened at the exterior with OSB boards and protected by an outer layer, 100 mm thick, of rock wool insulation. In between studs insulation is placed, with vapour barrier and gypsum boarding to the interior. The slabs consist of 240 mm thick beams, strengthened with OSB boards and the same median layout and to the interior as the walls. The concrete system consists of 300x300 mm columns and 300x450 mm beams, with 290 mm thick masonry, plaster to the interior and 100 mm thick rock wool insulation to the exterior, figure 2. The reinforced concrete slabs are 150 mm thick, with a 100 mm thick ground slab that was implemented in both solutions. The pre-dimensioning of elements had to take into account the extra loads given by the extensive green roof proposed in both cases.

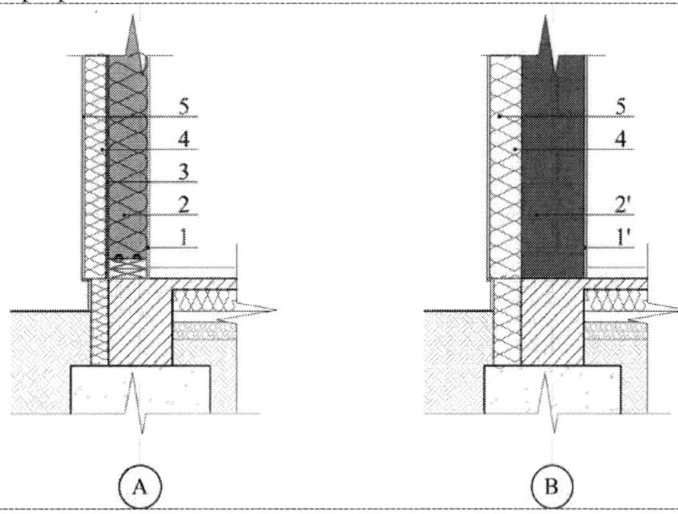

Figure 2. Wall detail, wood platform framing (A) and reinforced concrete post and beam (B). 1- gypsum board, 2- vertical studs/180 mm rock wool insulation, 3- OSB boards, 4- 100mm rock wool insulation, 5- exterior plaster; 1'-interior plaster, 2'- 290mm vertical perforation bricks

2. Life Cycle Assessment comparison in the design stage

Life Cycle Assessment (LCA) methodology evaluates inputs and outputs of a system, in particular a construction, over a life cycle, in terms of system interaction with the natural environment. Systemic modelling takes into account all inputs, including raw material extraction, energy and water consumption, among the most important and all outputs, including mainly air emissions, water and soil pollution values [6].

2.1. Methodology of the environmental assessment

Based on analysis of available software and study of scientific literature, ATHENA Impact Estimator was chosen, a program that allows complex modelling and is also suitable for comparative evaluations in early design stages. The software can be used for LEED global assessments and is based on the life cycle standard: *ISO 14040, Environmental Management, Life Cycle Assessment, Principles and Framework.* ATHENA Impact Estimator [7] is a useful tool in making decisions about the impact of materials on the environment, in cradle to grave assessments, incorporating input data related to production, energy and transport processes [8]. Jrade and Jalaei, 2013, have developed an integrated building evaluation model using ATHENA Impact Estimator, which they consider to be an international tool for LCA evaluation, which allows users to implement various project modifications, replace materials, and make comparisons to optimize any model [9]. For the model evaluation, the information related to the proposed dwelling and a detailed extract of materials, in both scenarios, were used. The estimated life expectancy is of 60 years, from material extraction phase up to demolition and recycling. Since the software uses the energy balance specific only to cities and states in North America (not from Europe), a comparison was made between Romania's energy balance and all states in the program database, opting for Portland, Oregon, with the closest values to those of our country. The data for each building element was entered separately, based on predefined program elements, modelled to match the proposed stratifications. Since the program has certain limitations, which do not allow for the introduction of more complex elements, each building component (floor, walls, roof) was added, using the *Extra Materials* option, with additional materials not included in the predefined configurations, resulting in a model matching the proposed design. For the windows, project-specific data was introduced, modelling the opaque – glazed ratio for each wall, opting for triple glazed aluminium frame. Two simulations have been performed for each of the scenarios for the structure and enclosures, wood platform framing (WPF) and reinforced concrete post and beam (CPM).

2.2. Results and discussions

After entering all the necessary information, ATHENA Impact Estimator provides a series of reports on:

- Primary Energy Consumption;
- Acidification Potential;
- Global Warming Potential;
- Human Health (HH) Respiratory Effects Potential;
- Ozone Depletion Potential;
- Photo- chemical Smog Potential;
- Eutrophication Potential, table1.

LCA Measures	Unit	Product		Process		End Of Life		**Total**	
		WPF	CPM	**WPF**	CPM	**WPF**	CPM	**WPF**	CPM
Global Warming Potential	kg CO_2 eq	**3.24E+04**	5.47E+04	**5.67E+03**	6.40E+03	**1.80E+03**	2.95E+03	**5.44E+04**	7.86E+04
Acidification Potential	kg SO_2 eq	**2.23E+02**	2.97E+02	**6.01E+01**	6.44E+01	**1.64E+01**	3.02E+01	**4.01E+02**	4.95E+02
HH Particulate	kg PM2.5 eq	**5.83E+01**	8.23E+01	**4.83E+00**	4.87E+00	**8.06E-01**	1.62E+00	**1.06E+02**	1.30E+02
Eutrophication Potential	kg N eq	**1.71E+01**	3.74E+01	**3.93E+00**	4.69E+00	**8.97E-01**	1.75E+00	**2.45E+01**	4.65E+01
Ozone Depletion Potential	kg CFC-11 eq	**7.94E-04**	1.03E-03	**4.85E-05**	5.05E-05	**1.10E-07**	1.55E-07	**9.29E-04**	1.17E-03
Smog Potential	kg O_3 eq	**2.57E+03**	3.57E+03	**1.68E+03**	1.87E+03	**4.61E+02**	9.05E+02	**5.48E+03**	7.21E+03
Total Primary Energy	MJ	**5.43E+05**	7.21E+05	**8.35E+04**	8.95E+04	**2.52E+04**	4.25E+04	**9.65E+05**	1.17E+06
Non-Renewable Energy	MJ	**4.54E+05**	6.58E+05	**7.76E+04**	8.69E+04	**2.51E+04**	4.20E+04	**7.91E+05**	1.02E+06
Fossil Fuel Consumption	MJ	**4.25E+05**	5.84E+05	**7.68E+04**	8.55E+04	**2.46E+04**	4.15E+04	**7.46E+05**	9.33E+05

IOP Conf. Series: Materials Science and Engineering **586** (2019) 012010 doi:10.1088/1757-899X/586/1/012010

From the comparative results presented in the table above, we can conclude that the post and beam solution has a greater impact on the environment, aspect attributed in part to the higher embodied energy in the reinforced concrete elements and the difficulties related to the recycling of these materials. The overall difference is about 45% higher in the case of post and beam structure in terms of global warming potential, 21% for the total primary energy. These differences in the calculated values for the global warming potential can be largely attributed to the use of wood, which stores CO_2 through the photosynthesis process, during the tree's life, which afterwards becomes "captive" in the products, elements or structures that use this material. As shown in figure 3, the main differences occur within the elements employing the most reinforced steel and concrete (post and beam, floors, where the total energy difference is almost 2.0E+5MJ) and less in respect to elements utilising masonry (for walls the difference is negligible). The estimated design life of only 60 years is a simplified approach that does not give enough credit to the main sustainability quality of concrete and masonry, and that is their durability.

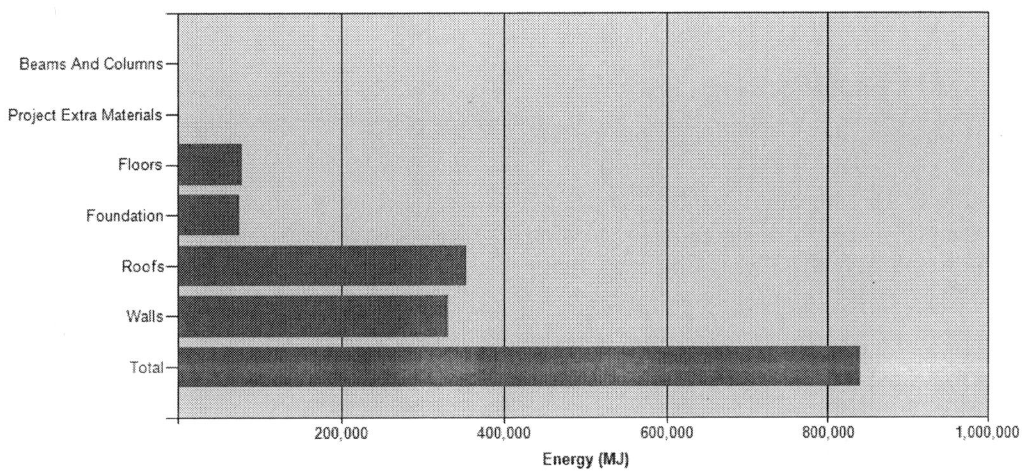

wood platform framing (WPF)

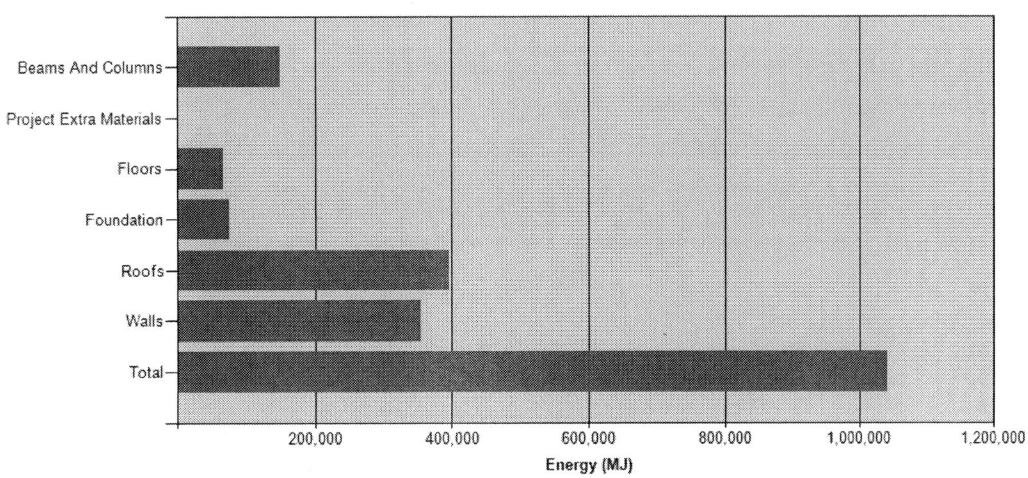

reinforced concrete post and beam (CPM)

Figure 3. Total primary energy LCA measure

IOP Conf. Series: Materials Science and Engineering **586** (2019) 012010 doi:10.1088/1757-899X/586/1/012010

3. Energy efficiency assessment

In order to determine the best scenario in terms of thermal efficiency, the house has been evaluated through numerical simulations using specialized programs, in particular CASAnova software [10]. The inputs are based on the computed values of the adjusted thermal resistances for each specific element for the wooden platform framing (WPF) and concrete post and beam (CPM) solutions, table 2. Adjusted thermal resistances (*R'*), expressed in m^2 *K/W*, represents the corrected value of thermal resistance by the influence of the thermal bridges and tends to equal the value of the real characteristic of the constructive element [11].

Table 2. Adjusted thermal resistances (R′) for constructive elements specific to the wooden platform framing (WPF) and concrete post and beam (CPM) scenarios.

Element	R′ value WPF [m^2K/W]	R′ value CPM [m^2K/W]
Ground slab	6,083	6,083
Exterior wall	7,416	3,910
Roof	7,495	6,188

The results of the comparative study indicate a high level of energy efficiency, characterised by low specific annual energy consumption for heating (q_i) in both scenarios, aspect largely due to solar gains, as presented in the Sankey diagrams, figure 4.

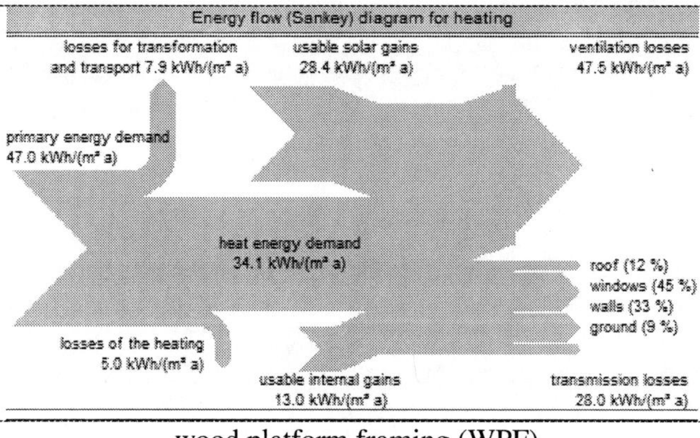

wood platform framing (WPF)

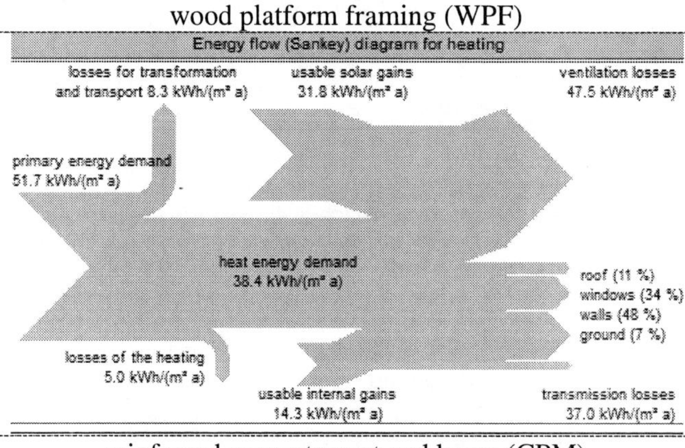

reinforced concrete post and beam (CPM)

Figure 4. Sankey diagrams for heating energy demand

4. Conclusions

Implementation of passive design strategies in the conceptual phase of a project is as important as design optimization through numerical simulations, in terms of embedded energy and thermal efficiency. With respect to sustainability criteria, the comparative study presented in this paper underlines that wooden structure is a better solution than reinforced concrete and masonry. Both life cycle assessment and energy efficiency simulations prove that wooden platform framing can be a more appropriate sustainable option. Thus, in terms of life cycle assessment, the comparative study all sustainable parameters indicate a less impact on the environment in the case of the wood structure. The most relevant is the Total Global Warming Potential that is 7.86 E+04 Kg CO^2eq in the case of reinforced concrete, compared to 5.44 E+04 Kg CO^2eq in the case of wood platform framing (a 70% difference). In terms of heating and cooling energy demand, primary energy for heating is smaller in the case of wood platform framing (47.0 kWh/m^2a compared to 51.7 kWh/m^2a for concrete post and beam).

5. References

[1] Ochinciuc V 2002 *Conceptul dezvoltării durabile în arhitectură. Proiectarea integrată* (Editura Universitară "Ion Mincu", Bucureşti) p 145-150;

[2] Steele K 2010 *Life Cycle Assessment and environmental profiling of building materials, Materials for energy efficiency and thermal comfort in buildings* (Woodhead Publishing Limited, Cambridge, UK) p 245-278;

[3] Chiras D 2006 *The Homeowner's Guide to Renewable Energy* (New Society Publishers, BC) p 102-103;

[4] Khatib J M 2016 *Sustainability of Construction Materials, Second Edition* (Woodhead Publishing, Elsevier, US) 129-130;

[5] PVGIS © European Communities, 2001-2017 data download, available free at http://re.jrc.ec.europa.eu/pvgis/

[6] Zeumer M, Hegger M, Fuchs M and Stark T 2007 *Atlante della Sostenabilita e della efficienza energetica degli edifici* (UTET Scienze Tecniche, Milano) p 36-42, 164-178, 191;

[7] Athena Impact Estimator for Buildings, available free at http://www.athenasmi.org;

[8] Srinivasan R, Ingwersen W, Trucco C, Ries R and Campbell D M S 2014 Building and Environment **79** Elsevier, 138-151;

[9] Jrade A and Jalaei F 2013 Build Simul **6** Tsinghua University Press and Springer-Verlag Berlin Heidelberg 429–444;

[10] CASAnova software, available for free at: http://nesa1.uni-siegen.de/index.htm?/softlab/casanova_e.htm;

[11] Secu A and Romila C 2014 The Bulletin of the Polytechnic Institute of Jassy. Constructions. Architecture Section **t. LX (LXIV), f. 3** Universitatea Tehnică „Gheorghe Asachi" din Iaşi 67-77.

Acknowledgments

This paper was elaborated with the support of the "Ecoinnovative Products and Technologies for Energy Efficiency in Constructions – EFECON" re-search grant, project ID P_40_295/105524, Program co-financed by the European Regional Development Fund through Operational Program Competitiveness 2014-2020.

IOP Conf. Series: Materials Science and Engineering **586** (2019) 012011 doi:10.1088/1757-899X/586/1/012011

Comparative Analysis of Hygrothermal Behaviour of the Exterior Walls in Transient Regime

R A Pescaru[1], L Dumitrescu[1], I Baran[1] and C M Grădinaru[1]
[1]"Gheorghe Asachi" Technical University from Iaşi, Bulevardul Profesor Dimitrie Mangeron 65, Iași 700259, *România*

E-mail: irina.baran9@gmail.com

Abstract. Energy conservation and an intelligent use of resources are among primary objectives in the contemporary society. The decrease of energy consumption in buildings has many benefits both on operating costs and environment preservation too. During the buildings life-cycle, the energy consumption for operation, in winter through heating and in summer through ventilation, is more important than the energy used in the construction phase, requiring passive measures from the design stage. The paper presents the results of a study of the behaviour, in a variable regime, of the external wall with different structures, to the mass transfer and temperature variations. The influence of the placement of interior or exterior thermal insulation layers on the hygrothermal behaviour of the envelope element was determined by means of WUFI® Pro software, which takes into account the built-in moisture, driving rain, solar radiation, long-wave radiation, capillary transport, and summer condensation.

1. Introduction

Massive walls exposed to the natural climate, without special rain protection, show a dynamic moisture balance influenced by the alternating events of rain and solar radiation. The moisture reduces the thermal insulation capacity of the wall. In this case, thermal rehabilitation solutions are necessary to improve the building performances. But, if existing walls are usually breathable and they can withstand rain and snow as they will dry out during the year, when they are additional insulated, the whole dynamic of the wall is changed, because the vapours cannot pass through the wall and the heat cannot dry out, having negative consequences on the wall structure.

When a building envelope wall has to be thermally insulated, it is advisable to apply the thermal insulation layer to the external face of the façade (besides avoiding the most of the thermal bridges, it increases internal thermal inertia, maintains building structure at constant temperature and diminishes the interstitial condensation risk in building elements). However, sometimes may be too costly or undesirable because the original façade should remain unaltered for aesthetic reasons. In such cases, the only possible alternative is the insulation of the indoor face of the wall.

In this situation are the patrimony buildings which are often made of load bearing masonry walls. In this situation, besides increasing the thermal bridge effect, the decrease in temperature of the masonry element, caused by this type of thermal insulation will reduce the drying process of precipitation moisture. Elevated moisture contents, due to the impact of driving rain on exposed walls, can also lead to frost damage or accelerated degradations.

In addition to accounting for the thermal response of buildings and building components, it is also important to know the level of the moisture and the effects of humidity on building elements. Long-term exposure to high moisture level can cause degradation in the building's closing elements, and the development of mould on surfaces where condensation occurs may lead to significant health problems. Heat and mass transport in buildings elements are coupled. It is well known that high moisture levels result in higher heat losses, and the temperature conditions in building elements influence the mass

IOP Conf. Series: Materials Science and Engineering **586** (2019) 012011 doi:10.1088/1757-899X/586/1/012011

transfer phenomenon. The analysis of the coupling of heat and mass transfer is known as "hygrothermics".
During time, different models for the coupled heat, air, moisture and salt transport have been developed, beginning with old manual analysis tools as "dew point method" and "Glaser diagram" to more accurate modern HAM models (heat, air and moisture) which combine the flow equations with the mass and energy balances [1-3].

1.1. The simplified graphical and numerical method: Glaser

The traditional method for assessing the moisture balance of a building component has been the Glaser method (described in the DIN 4108 standard) which analyses the vapour diffusion transport in the building component, and is based on the following diffusion equation:

$$w=-(\delta'/\mu')\Delta p/d \tag{1}$$

where:
w = vapour flow per unit of area, [kg/m^2·s];
δ' = diffusion coefficient of water vapour in air, [s];
μ' = diffusion resistance factor of the material, [-];
p = vapour pressure [Pa];
d = flow path or thickness of the material, [m].
However, this method does not take into account the mass transfer through capillarity and the absorption capacity of the material, both of them reducing the risk of degradation due to condensation.
Furthermore, because the Glaser method takes into account only the steady-state mass transfer phenomenon under simplified boundary conditions, it cannot have in view short-term events or to consider the rain and solar radiation. It was intended to offer a general assessment of the hygrothermal behaviour of a building element, not to give a simulation of the real heat and moisture transfer into a building element exposed to location specific climatic conditions at a certain location.

1.2. Simultaneous heat and moisture transfer in building elements

Advanced hygrothermal computer models available today are based on the latest knowledge and deeper understanding of the combined heat, air and moisture transfer in building materials and components. The main advantage of modelling is that it can predict the long-term hygrothermal behaviour of the building envelope elements under different climatic and interior conditions and the effect of various energy retrofit methods on the building lastingness [2].
The equations proposed by Künzel [3] for the simultaneous heat and moisture transfer are:

$$\frac{dH}{dT}\frac{dT}{dt} = \nabla(\lambda \nabla T) + h_v \nabla\left(\delta_p \nabla(\varphi p_{sat})\right) \tag{2}$$

$$\frac{dw}{d\varphi}\frac{d\varphi}{dt} = \nabla\left(D_\varphi \nabla \varphi + \delta_p \nabla(\varphi p_{sat})\right) \tag{3}$$

where:
H = enthalpy, [J/m^3];
T = temperature, [K];
h_v = latent heat of phase change, [J/kg];
δ_p = water vapour permeability of the building material, [kg/m·s·Pa];
p_{sat} = water vapour saturation pressure, [Pa];
λ = thermal conductivity of the moist building material, [W/m·K];
φ = relative humidity, [-];

IOP Conf. Series: Materials Science and Engineering **586** (2019) 012011 doi:10.1088/1757-899X/586/1/012011

D_φ = liquid conduction coefficient of the building material, [kg/m·s];
w = water content [kg/m^3];
t = time, [s].

During the last decades, many hygrothermal simulation methods have been developed [4, 5] to analyse the physical phenomena that influence heat and moisture transfer within building materials. There are also many experimental studies signalled in literature which aim to compare the results from numerical models with experimental measurements and the general trend shows that discrepancies are observed most particularly for highly hygroscopic or bio-based materials than for traditional construction materials such as concrete or brick [6].

WUFI is a computer program for the calculation of the simultaneous heat and moisture transfer in multi-layer building components, developed by the Department of Hygrothermics at Fraunhofer IBP. WUFI® is an acronym for Wärme Und Feuchte Instationär - which means heat and moisture transiency. It has repeatedly been validated by comparison with experimental results and provide realistic simulation of hygrothermal conditions in building elements and buildings under actual climate conditions.

2. Case study

One of the constructive solutions for the thermal insulation of the external wall of the buildings, which is preferred due to its simplicity, is the ETICS (External Thermal Insulation Composite System), and consists in the attachment of a thermal insulating material layer directly on the outer face of the structural wall. Often the structural wall is made of brickwork masonry, and sometimes due to particular restrictive conditions (as heritage building façade preservation), the thermal insulating layer is fixed on inside part of the envelope wall.

By means of experimentally-validated heat and moisture transfer calculation software, the effect of different thermal insulation constructive solutions on the hygrothermal behaviour of brick wall will be studied in this research. The study approach is the behaviour, in time and in a variable hygrothermal regime, of the external wall to the mass transfer and temperature variations.

The 25 cm thick solid brick masonry external wall is insulated in three constructive versions, which have been modelled in the experiment. In the first case (a.) it is considered a 10 cm thick EPS insulation layer fixed on the outer face of the wall, in the second case (b.) it is considered a 10 cm thick EPS insulation layer fixed on the inside part of the wall, and in the third case (c.) it is considered a 10 cm thick mineral wool insulation layer fixed on the outer face of the wall.

The main physical and geometrical characteristics (ρ - material density, c - specific heat capacity, λ - thermal conductivity, thickness) of the constitutive materials used in each wall element are given in the tabular and graphical manner presented in Figure 1. The thermal resistance of all layers for each of the three studied cases is comprised between 2.87 m^{2}K/W – see Figure 1.b - and 3.05 m^{2}K/W – see Figure 1.a - values that recommend them as effective solutions for energy conservation.

As outdoor climate data, the hourly values of air temperature, humidity and solar radiation on a south-east oriented façade and the driving rain recorded during an average year in Graz (Austria) (the nearest location to Romania included in database of the software) were used.

Indoor air temperature for dwelling buildings, according to EN15026 standard is T_i=+20°C during winter (heating season) and T_i=+25 °C (cooling season). The indoor relative humidity varies between 30% (winter) and 60% (summer).

For new reinforced concrete frame building constructions, the constructive materials were exposed to the open air during storage and construction time, thus having attained equilibrium moisture. In this case, the initial moisture (water content) can be set to the equilibrium water content of the material at 80% RH (relative humidity). In Romania, this is the annual mean outdoor RH, if sheltering of the material from the rain and direct sunshine action is assumed.

Homogenous layers

Thermal resistance: 3,05 m²K/W

Heat transfer coefficient(U-value): 0,31 W/m²K

Thickness: 0,389 m

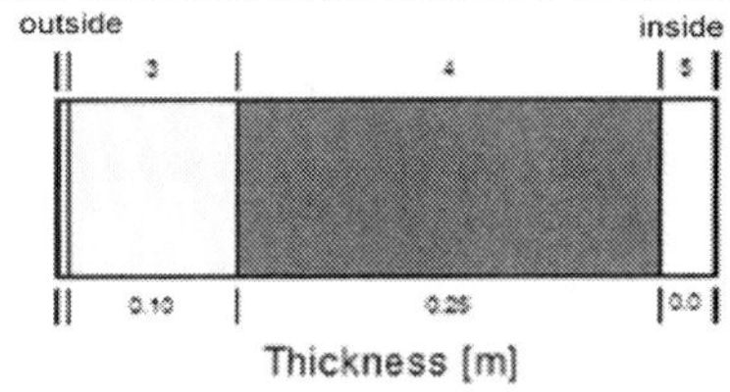

Nr.	Material/Layer (from outside to inside)	ρ [kg/m³]	c [J/kgK]	λ [W/mK]	Thickness [m]	Color
1	Mineral Finishing Coat	1482	850	0,954	0,002	
2	Mineral Plaster (stucco, A-value: 0.1 kg/m2h0.5)	1900	850	0,8	0,005	
3	EPS (heat cond.: 0.04 W/mK - density: 15 kg/m³)	15	1500	0,04	0,1	
4	Solid Brick Wienerberger	1786	889	0,519	0,25	
5	Lime Cement Mortar, fine	1880	850	0,6	0,03	
6	Interior Plaster (Gypsum Plaster)	850	850	0,2	0,002	

a. external wall with polystyrene insulation on outer face

Homogenous layers

Thermal resistance: 2,876 m²K/W

Heat transfer coefficient(U-value): 0,33 W/m²K

Thickness: 0,389 m

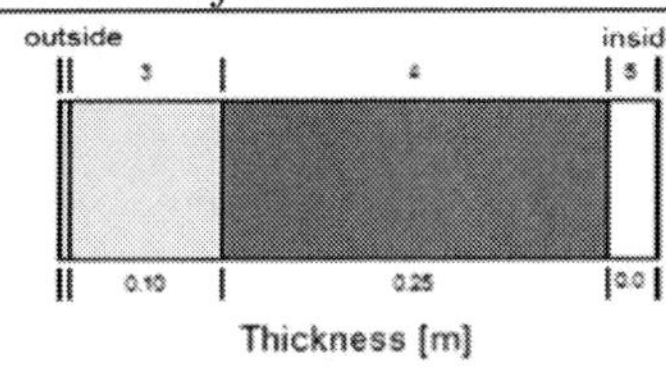

Nr.	Material/Layer (from outside to inside)	ρ [kg/m³]	c [J/kgK]	λ [W/mK]	Thickness [m]	Color
1	Mineral Finishing Coat	1482	850	0,954	0,002	
2	Mineral Plaster (stucco, A-value: 0.1 kg/m2h0.5)	1900	850	0,8	0,005	
3	Mineral Insulation Board	115	850	0,043	0,1	
4	Solid Brick Wienerberger	1786	889	0,519	0,25	
5	Lime Cement Mortar, fine	1880	850	0,6	0,03	
6	Interior Plaster (Gypsum Plaster)	850	850	0,2	0,002	

b. external wall with mineral wool insulation on outer face

Homogenous layers

Thermal resistance: 3,022 m²K/W

Heat transfer coefficient(U-value): 0,31 W/m²K

Thickness: 0,394 m

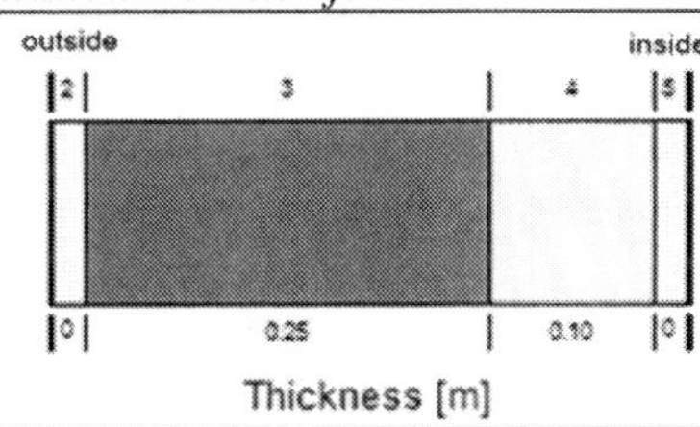

Nr.	Material/Layer (from outside to inside)	ρ [kg/m³]	c [J/kgK]	λ [W/mK]	Thickness [m]	Color
1	Mineral Finishing Coat	1482	850	0,954	0,002	
2	Cement Lime Plaster (stucco, A-value: 2.0 kg/m2h0.5)	1900	850	0,8	0,02	
3	Wienerberger Solid Brick	1744	889	0,5438	0,25	
4	EPS (heat cond.: 0.04 W/mK - density: 15 kg/m³)	15	1500	0,04	0,1	
5	Cement Lime Plaster (stucco, A-value: 2.0 kg/m2h0.5)	1900	850	0,8	0,02	
6	Interior Plaster (Gypsum Plaster)	850	850	0,2	0,002	

c. external wall with polystyrene insulation on inner face

Figure 1. The structure and main hygrothermal characteristics of the studied walls

IOP Conf. Series: Materials Science and Engineering **586** (2019) 012011 doi:10.1088/1757-899X/586/1/012011

3. Results and discussion

Some or the result of the computer simulation are described in this section. Form the entire period of simulation of three years, the results from the middle of last summer and winter (in the third year of analyse) are presented in Figures 2 and 3.

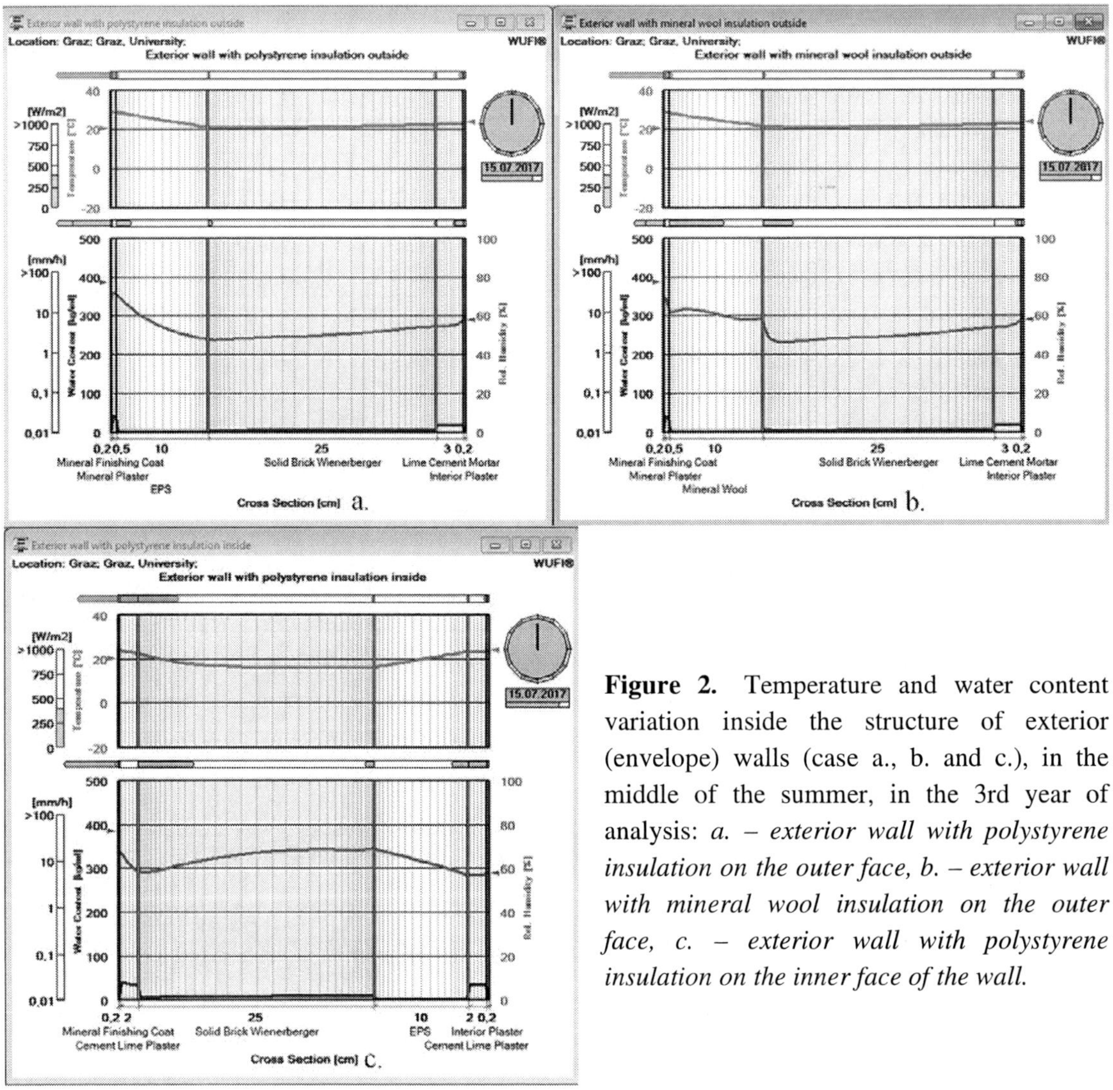

Figure 2. Temperature and water content variation inside the structure of exterior (envelope) walls (case a., b. and c.), in the middle of the summer, in the 3rd year of analysis: *a. – exterior wall with polystyrene insulation on the outer face, b. – exterior wall with mineral wool insulation on the outer face, c. – exterior wall with polystyrene insulation on the inner face of the wall.*

A time history simulation performed by the model is presented as a film file (Figure 3), which allows the display, as an animation, the thermal and hygric processes in the walls, hour by hour, during the middle of last winter season. The temperature profile from outdoors (left) to indoors (right) is presented in the upper graph. The red line shows the level of the temperature at twelve o'clock on 1 January in the third year of simulation. The pink area shows all the temperatures values reached in the various layers. The graph below with the green and blue lines and areas represents the moisture analysis. Relative humidity is in green colour and water content is represented in blue (the lines refer also for twelve o'clock on 1 January in the third year of simulation).

It can be observed that, in case c. with thermal insulation on the inner face, the masonry wall is submitted to a greater range of negative temperatures, with an undesirable impact on the mechanical characteristics and durability of this structural element.

IOP Conf. Series: Materials Science and Engineering **586** (2019) 012011 doi:10.1088/1757-899X/586/1/012011

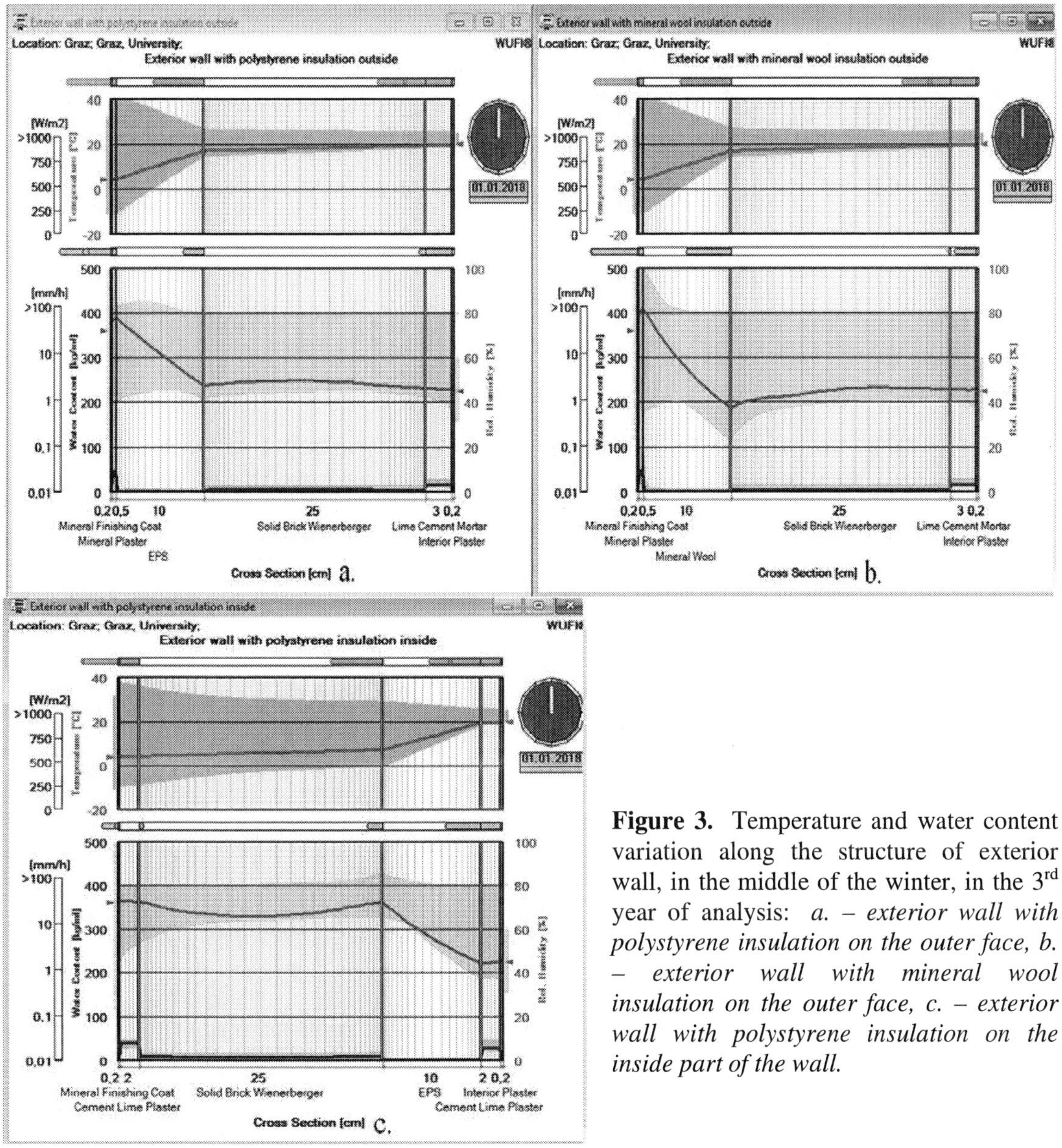

Figure 3. Temperature and water content variation along the structure of exterior wall, in the middle of the winter, in the 3rd year of analysis: *a. – exterior wall with polystyrene insulation on the outer face, b. – exterior wall with mineral wool insulation on the outer face, c. – exterior wall with polystyrene insulation on the inside part of the wall.*

The fluxes of heat and moisture across the layer interfaces and the component surfaces are presented by the red and blue arrows above each graph. On the left side of the screen are shown the current amounts of solar radiation and rain.

The film is useful for gaining insights into the hygrothermal processes in the component. In this way, the reaction of different building elements to the changing of climatic conditions could be seen. The moisture migration in different seasons and the construction parts which might be critical with respect to relative humidity are also shown.

The placement of thermal insulation materials with low vapour permeability on the external face of the wall can significantly reduce the drying capacity of a wall system (especially during winter season). This constructive solution could be appropriate if it is sufficient and available the drying of the wall to the interior space of the building without affecting the indoor climate / comfort. Exterior position of the thermal insulating materials with high vapour permeability facilitates drying to the outdoor space as well as wall systems with no exterior insulation.

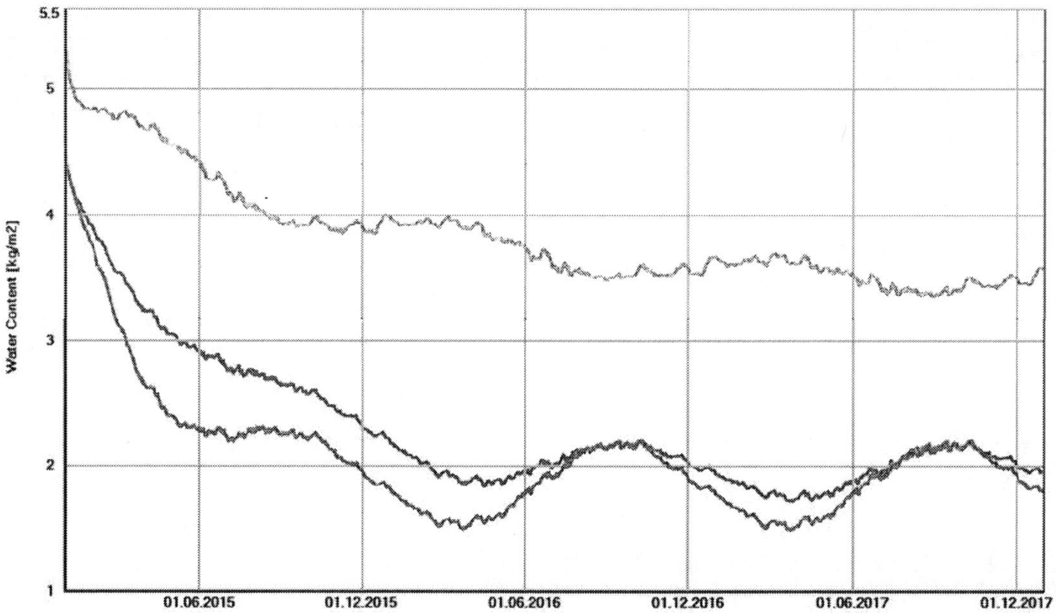

Figure 4. Variation of water content: red line - external wall with polystyrene insulation on the outside face, green line - external wall with polystyrene insulation on the inside face, blue line - external wall with mineral wool insulation on the outside face.

The interior insulation leads to higher average water content in the masonry (see Figure 4). In order to improve the moisture situation induced by the interior insulation, a water-repellent impregnation may be applied to the façade.
Due to higher vapour diffusion resistance, the EPS insulated wall dries somewhat more slowly than the wall insulated with mineral wool, but the effect is not significant because the main moisture flux is directed to the outside of the wall.
A mineral wool insulation system leads to a faster drying of the masonry underneath than do insulation systems with polystyrene. There may be, however, one danger: a frost damage risk if a thermal insulation system with mineral wool is applied to a rather wet wall during a period when frost is imminent. Due to the almost unhindered vapour diffusion through the mineral wool, moisture can accumulate behind the surface plastering, especially if this rendering layer material has an elevated vapour diffusion resistance. This risk can be avoided by using polystyrene or by installing a cladding instead of a rendering.
In cases where an exterior insulation cannot be installed, a water-repellent impregnation or paint coat should be applied to the façade prior to the interior insulation; otherwise, there will be an increase in the risk of frost damage due to influence of the insulation on the drying capacity of the masonry, along with a slight decrease in thermal resistance of the masonry. It is recommended to improve the rain protection of the façade some time before installing the interior insulation. If the vapour diffusion resistance of the exterior wall surface is negligible, any insulation material can be used inside, as long as convection of the room's air beneath the insulation layer can be safely avoided.

4. Conclusion

Modern calculation methods are a valuable tool to assess the hygrothermal behaviour of wall constructions. As a general rule, it is recommended the exterior insulation method of the external

walls, mainly because is the most effective method to avoid the thermal bridges from the building structure, which can result in low surfaces temperatures values, increasing the risk of mould.
Simulation tools to estimate moisture movement in building elements (new or retrofitted by adding internal or external thermal insulation) are very useful for estimating moisture accumulation and interstitial condensation risk, contributing to the choice of the best design option. Together with experimental testing results, simulation studies conclusions should be included in norms and codes of practice in construction, for the guidance of building practitioners.

5. References

[1] Glaser H 1959 "Graphisches Verfahren zur Untersuchung von Diffusionsvorgängen"(Graphical Procedure for Investigating Diffusion) . Kältetechn. Vol.11, pp.345-349.
[2] Karagiozis A N, Künzel H M, Holm A 2001 "WUFI-ORNL/IBP - A North American Hygrothermal Model." Proc. Performance of Exterior Envelopes of Whole Buildings VIII, Dec. 2-7 2001, Clearwater Beach, Florida
[3] Künzel H M 1995 Simultaneous Heat and Moisture Transport in Building Components. One- and two-dimensional calculation using simple parameters. Fraunhofer IRB Verlag, pp.39-40
[4] Woloszyn M, Rode C 2008 Tools for performance simulation of heat, air and moisture conditions of whole buildings, Building Simulation, 5–24
[5] Delgado J, Ramos N M, Barreira E, De Freitas V 2010 A critical review of hygrothermal models used in porous building materials, Journal of Porous Media 13 (3).
[6] Busser T, Berger J, Piot A, Pailha M, Woloszyn M 2018 Experimental validation of hygrothermal models for building materials and walls: an analysis of recent trends. <hal-01678857>

IOP Conf. Series: Materials Science and Engineering **586** (2019) 012012 doi:10.1088/1757-899X/586/1/012012

Evaluation of the swelling pressure for expansive soils

M Aniculaesi[1] , I Lungu[2]
[1] "Gheorghe Asachi" Technical University of Iaşi, Bd. Dimitrie Mangeron nr. 67, 700050 Iaşi, România,
[2] "Gheorghe Asachi" Technical University of Iaşi, Bd. Dimitrie Mangeron nr. 67, 700050 Iaşi, România

E-mail: aniculaesi.mircea@gmail.com

Abstract. The interaction between light structures and expansive soils as mainly active clays was studied since the 1950's. The first stage in foundation design on this problematic soil is the evaluation of the swelling potential, based on which one identifies the clay's activity and its correlation with the value of the swelling pressure (p_s) for the decisional support of the geotechnical design ($p_{effective} > p_s$). In the literature there are many formulas that can be used to predict the swelling pressure based on the correlation between various geotechnical parameters. This paper provides the swelling pressure as correlation with the consistency index (CI) and the liquid limit (LL), by linear regression of the experimental results for the 56 active clays from seven countries. Using the analytical methods, one can predict the total heave of a floor slab founded on expansive clay. The validity of the proposed correlation has been tested in the prediction of the total heave in the centre of a floor slab for a light industrial building in North-Central Regina, Saskatchewan, Canada.

1. Introduction

It is a well-known phenomenon that unsaturated expansive clays change their volume with the change of their water content. The water content variation at shallow depth is mainly due the seasonal climate change. As it triggers activity of expansive soils, the depth up to which the water content variation is influenced by the seasonal climate change is called the active zone. This depth can be approximated following the guidelines given in the Engineer Manual EM 1110-1-1904 [1].

When the soil heave is impeded by an engineering structure, the soil develops a swelling pressure against the contact structural element, foundation or slab. Many light weight constructions founded on these soils disregarding the swelling pressure occurrence can suffer structural degradations with limitations on the functionality and additional costs for their rehabilitation. Consequently, it is very important to accurately estimate the swelling potential and the correlated swelling pressure in order to select and implement the optimum measures to minimise the soil heave.

The evaluation methods of the swelling pressure can be divided in two categories: direct or indirect ones [2]. The direct methods are based on the oedometer test, using the readings from the vertical swell of a sample after saturation [3][4]. The indirect methods relate the swelling pressure and swelling potential to one or several physical soil parameters such as liquid limit, clay content, plasticity index, natural water content. In table 1 the typical empirical correlations between the physical properties and the swelling pressure are presented.

When focusing on the light weight constructions interacting with expansive soils one can separate them in three categories such as: (a) road infrastructures; (b) slabs for outdoor parking spaces; (c) houses. Although all of them are founded within the active zone, the first two categories benefit from large budgets for soil improvement to reduce its heave, the third category owners seldom accept such costs. For houses, the interest stays with counteracting the swelling pressure of the natural soil with net pressure from the house itself hence an efficient foundation design represents the main aim. The task becomes the estimation of the swelling pressure at reasonable geotechnical investigation costs. Following that task, the prediction of the total heave against measurements from documented case studies may reinforce the confidence of practitioners in controlling the heave effect.

This paper proposes a new correlation between the swelling pressure and two geotechnical indices: liquid limit and consistency index. The values obtained with the new empirical correlation were compared with the measured values from direct tests for 56 samples from seven countries (table 2). The values obtained using the proposed empirical equation have been used to predict the total heave, by replacing the swelling pressure values derived from the basic formula for the constant volume oedometer test based method [1] [5]. The results were verified using the data collected by the Prairie Regional Station of the Division of Building Research (DBR), in Western Canada, for a light industrial building in North-Central Regina, Saskatchewan [6].

Table 1. Prediction equations for the swelling pressure value based on empirical correlations.

Equation	*Author*
$logp_s = -2.132 + 0.0208 \cdot LL + 0.000665 \cdot \gamma_d - 0.0269 \cdot w_0 \; [daN/cm^2]$	Komornik and David [5]
$p_s = 2.5 \cdot 10^{-1} \cdot PI^{1.12} \cdot C^2/w_0^2 + 25 \; [kN/m^2]$	Nayak and Christensen [5] [7]
$p_s = -227.27 + 2.14 \cdot w_o + 1.54 \cdot LL + 72.49 \cdot \gamma_d \; [N/cm^2]$	Erguler and Ulusay [5]
$logp_s = -4.812 + 0.01405 \cdot PI + 2.394 \cdot \gamma_d - 0.0163 \cdot w_0 \; [daN/cm^2]$	Erzin and Erol [5]
$logp_s = -5.02 + 0.01383 \cdot PI + 2.356 \cdot \gamma_d \; [daN/cm^2]$	Erzin and Erol [5]
$p_s = 135 + 2 \cdot (C + PI - w_O) \; [kN/m^2]$	Sabtan [5]
$log_{10}p_s = (0.4 \cdot LL - w_O - 0.4)/12$	Vijayvergiya and Ghazzaly [7] [19]
$log_{10}p_s = (\gamma_d + 0.65 \cdot LL - 139.5)/19.5$	Vijayvergiya and Ghazzaly [7]

Notes: p_s – swelling pressure; LL – liquid limit; PI – plasticity index; w_0 – water content; γ_d – dry unit weight; C – clay content

2. Indirect correlation of the swelling pressure with index properties

Using the physical properties of soil samples collected from the literature (table 2) an indirect equation for the swelling pressure has been proposed. This equation correlates the liquid limit (LL) and the consistency index (CI) with the swelling pressure, indirectly but clearly correlated with the initial water content:

$$p_s = (3.71 \cdot \mathrm{LL} - 125)/CI \quad (1)$$

where: p_s - swelling pressure [kPa]; LL – liquid limit (%); CI – consistency index

Table 2. Physical properties and swelling pressures of the samples used in correlation.

	No. of samples	Liquid limit (%)	Consistency index (%)	Natural moisture content (%)	Swelling pressure (kPa)	References
Ethiopia	12	96 ÷ 110	0.87 ÷ 1.15	32 ÷ 42.1	199 ÷ 420	[8]
Turkey	7	50 ÷ 67	0.86 ÷ 1.42	14.4 ÷ 32.7	18.8 ÷ 72.7	[9]
Iraq	6	35 ÷ 51	1.3 ÷ 3.7	16 ÷ 19.7	5.5 ÷ 19.5	[10]
Brazil	5	51.5 ÷ 55	1.38 ÷ 2.1	12.2 ÷ 15.8	24.4 ÷ 45	[11]
USA	4	54 ÷ 92	0.93 ÷ 1.21	20.6 ÷ 27.1	75 ÷ 263	[12]
India	1	64	1.24	18	135	[13]
Jordan	1	83	1.38	25	245	[14]
Other soils	20	41.4÷ 129.2	0.99 ÷ 1.34	14 ÷ 23.3	46.54 ÷ 223.4	[15] [16] [17]

The comparison between the experimental data of the swelling pressure and the predicted values based on equation 1 is shown in figure 1. The equation capacity in offering the results closed to the ones obtained by laboratory testing can be made using the performance index like R^2 (determination coefficient). The value of R^2 using the linear regression between the predicted and measured data indicated a strong coefficient of correlation (0.70). If R^2 is 1 the prediction model is accepted as excellent [18]. The proposed equation for the swelling pressure evaluation exhibited a high performance for predicting the swelling pressure of the soils according to the R^2 value.

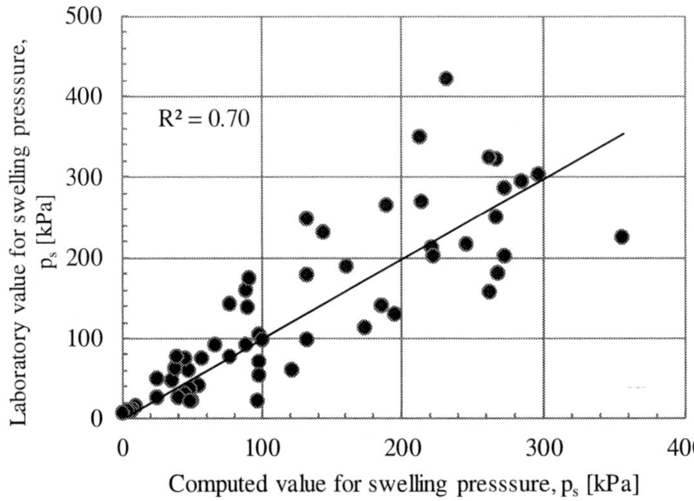

Figure 1. The linear regression as correlation between computed values of the swelling pressure and laboratory test values.

3. Soil heave analysis

The anticipated heave can be calculated using the equation presented in [1], [5]:

$$S_{max} = \sum_{j=1}^{n} \frac{c_{sj}}{1+e_{0j}} \cdot log_{10} \frac{p_{sj}}{\sigma'_{fj}} \cdot H_j \tag{2}$$

where: S_{max} - maximum potential vertical heave [m]; n - number of strata within the depth of heaving soil; c_{sj} – swelling index of stratum j, as the slope of the *e-log(σ-u)* loading line; p_{sj} – swelling pressure of stratum j [kPa]; σ'_{fj} – final or equilibrium average effective vertical pressure of stratum j [kPa]; e_{0j} – initial void ratio of stratum j.

The total vertical heave is calculated using the summation of heave on each sublayer (n) up to the active zone limit (or critical depth). The swelling pressure p_{sj} in each stratum can be obtained using the direct methods (oedometer tests) or by introducing the swelling pressure obtained from the indirect equation. By replacing the swelling pressure as from equation 1 in equation 2 the maximum potential heave can be evaluated by:

$$S_{max} = \sum_{j=1}^{n} \frac{c_{sj}}{1+e_{0j}} \cdot log_{10} \frac{(3.71 \cdot \mathrm{LL} - 125)/CI}{\sigma'_{fj}} \cdot H_j \tag{3}$$

3.1. Estimating of total heave for a slab on grade floor on Regina clay, Canada

In 1961 a light industrial building was constructed in the North of the city of Regina, Saskatcewan, Canada, and in 1962 the building owner noticed heave and cracks in the floor slab. The average values of geotechnical parameters for this soil are [5] [6]: clay content = 50%; liquid limit = 77%; plastic limit = 33%; average water content = 29%; unit weight = $18.88 kN/m^3$; initial void ratio = 0.962; swelling index = 0.09; swelling pressure = 325kPa. In analyses, a surcharge of 5.75kPa is applied, which represents the weight of the concrete slab and the weight of the fill under the slab. The measured heave under the center was 105mm. Yoshida et al. [6], estimates the soil heave of Regina clay considering different situation for the final pore water pressure: case 1 – final pore water pressure equal to zero; case 2 - final pore water pressure is negative, and case 3 - final pore water pressure is hydrostatic. For the benefit of testing the validity of the simple estimation of the swelling pressure with the new proposed correlation, the predicted values of heave are given in table 3, only for the assumption of final pore-water pressure as zero and thus considering that initially negative pore-water pressure increase to zero as water content increases during the wet season.

Table 3. Prediction of heave assuming final pore-water pressure is zero (case 1) [6].

Depth (m)	0.15	0.45	0.75	1.05	1.35	1.65	1.95	2.25
w_f (%)	40.1	39.2	38.5	37.9	37.3	36.7	35.9	34.6
S_{max} (mm)	118	92	70	51	35	21	10	2

Notes: $e_0 = 0.962$ – initial void ratio; $C_s = 0.090$ – swelling index; $S = 100\%$ – degree of saturation

The estimated values obtained with the equation 3 were compared with laboratory tests results for the expansive clay from the city of Regina. Assuming that the final pore water pressure will be equal with zero, as the case 1 from Yoshida et al [6], the final equilibrium average effective vertical pressure will be $\sigma'_{fj} = \sigma_{\gamma zj} + \Delta\sigma_j$. The soil heave under the slab was calculated using equation 3, assuming different values for the swelling pressure with the depth due to the water content variation (table 4).

Table 4. Prediction of heave using equation 3 assuming final pore-water pressure is zero.

Depth (m)	0.15	0.45	0.75	1.05	1.35	1.65	1.95	2.25
w_f (%)	40.1	39.2	38.5	37.9	37.3	36.7	35.9	34.6
S_{max} (mm)	109.5	99.7	82.9	68.7	56	44.8	34.5	25.4

Notes: $e_0 = 0.962$ – initial void ratio; $C_s = 0.090$ – swelling index; $S = 100\%$ – degree of saturation

Figure 2 present the variation with depth of the measured swell, calculated by Yoshida et al. [6], and by the proposed equation (equation 3).

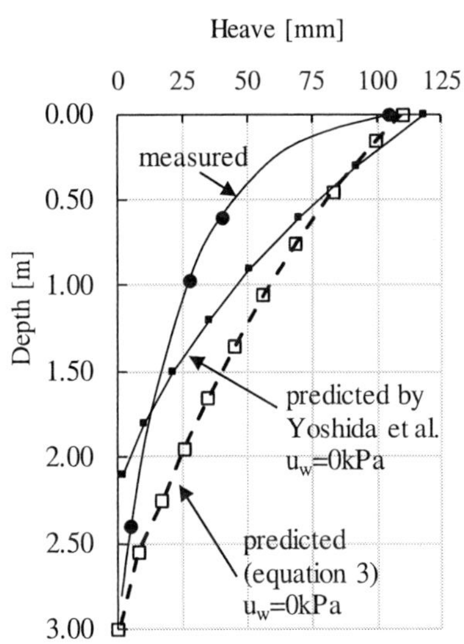

Figure 2. Predicted and measured ground heave under the centre of the slab

IOP Conf. Series: Materials Science and Engineering **586** (2019) 012012 doi:10.1088/1757-899X/586/1/012012

The maximum value of heave computed by the proposed equation is 109.5mm, which is very close to the measured one 105mm. As one compares the values of the heave at different depths up to the active zone (z_a = 3m), the calculated values differ from the measured ones (figure 2). The same trend can be noticed also in the paper published by Yoshida et al. [6] for the case 1 (final pore-water pressure equals zero) as in figure 2.

4. Conclusions

Soil improvement techniques for expansive soils to reduce the swelling potential and consequently the swelling pressure for the benefit of undamaged light constructions during their service time have been proved efficient based on extensive research for many years. As every construction together with its site conditions represent a unique situation, generalizations of simpler approaches for the estimation of the swelling pressure should be applied with caution. However, the budget allocated to proper geotechnical investigations in such cases plays a very important role. In case of houses, especially ground-floor only structures are light weight and their foundation should be geotechnically designed to balance the swelling pressure. Its estimation is essential since the final pore-pressure can be safely assumed to increase from a negative value up to zero. This paper proposes a new equation which correlates the swelling pressure with the liquid limit and consistency index. By comparing the computed results with the measured ones, the R^2 performance indices was found to be equal with 0.7, which mean that there is a relatively good correlation between the measured and predicted values. The validation has been made by predicting a total heave of the floor slab from a well-documented case for a light industrial building in North-Central Regina, Saskatchewan, Canada. The proposed correlation for the swelling pressure resulted in a total maximum heave of 109.5mm, which is with only 4% higher than the measured one (105mm) and thus in good accordance with the practical evidence.

5. References

[1] U.S. Army Corps of Engineers 1990 *Engineering and Design. Settlement analysis* Engineer Manual No. 1110-1-1904

[2] Al-Shamrani M A and Dhowian A W 2003 Experimental study of lateral restraint effects on the potential heave of expansive soils *Engineering Geology 69 (2003) 63–81*

[3] ASTM D 4546 – 03 *Standard Test Methods for One-Dimensional Swell or Settlement Potential of Cohesive Soils*

[4] STAS 1913/12-88 *Soil foundation. The evaluation of physical and mechanical properties of expansive soils (in Romanian)*

[5] Arafat H and Ebid A M 2015 Optimum Replacement Depth to Control Heave of Swelling Clays *International Journal of Engineering and Innovative Technology (IJEIT) Volume 4, Issue 9*

[6] Yoshida R T, Fredlun D G and Hamilton J J 1983 The prediction of total heave of a slab-on-grade floor on Regina clay *Canadian Geotechnical Journal 20: 69-81, NRC Research Press*

[7] Al-Rawas A and Goosen M F A 2006 *Expansive Soils: Recent Advances in Characterization and Treatment* CRC Press ISBN 9780415396813 - CAT# SW6816

[8] Teklu D 2003 Examining the Swelling Pressure of Addis Ababa Expansive Soil *A Thesis Submitted to The School of Graduate studies of Addis Ababa University*

[9] Ozer R, Ulusay R and Isik S N 2012 Evaluation of damage to light structures erected on a fill material rich in expansive soil *Bulletin of Engineering Geology and the Environment* Volume 71, Issue 1, pp 21–36

[10] Al-Khashab M N and Abdulnafaa M T 2008 Treatment of Expansive Clayey Soil with Crushed Limeston *Engineering & Technology Journal.* 26. 376-386

[11] Lew B 2010 Structure Damage Due to Expansive Soils: a Case Study *Electronic Journal of Geotechnical Engineering.* 15. 1317-1324

[12] Lin B and Cerato A B 2012 Prediction of expansive soil swelling based on four micro-scale properties *Eng Geol Environ* 71:71–78

[13] Ramesh P, Rao N A V and Murthy K N 2012 Efficacy of sodium carbonate and calcium carbonate in stabilizing a black cotton soil *International Journal of Emerging Technology and Advanced Engineering* ISSN 2250-2459, Volume 2, Issue 10, pp 197-201

[14] Nusier O Kh and Alawneh A 2007 Micropiles Reinforcement for Expansive Soils: Large-Scale Experimental Investigation *Ground Improvement* 11(2):55-60

[15] Nayak N V and Christensen R W 1971 Swelling characteristics of compacted, expansive soils *Clays and Clay Minerals,* Vol. 19,pp. 251-261. Pergamon Press

[16] Abdullah W N and Alsharqi A S 2011 Rehabilitation of Medium Expansive Soil Using Cement Treatment *Jordan Journal of Civil Engineering,* Volume 5, No. 3 pp 343-356

[17] Tripathy S and Rao K S S 2009 Cyclic Swell–Shrink Behaviour of a Compacted Expansive Soil *Geotechnical and Geological Engineering 27(1):89-103*

[18] Yilmaz I 2006 Indirect estimation of the swelling percent and a new classification of soils depending on liquid limit and cation exchange capacity *Engineering Geology, 85, 295-301*

[19] Elbadry H 2017 Simplified reliable prediction method for determining the volume change of expansive soils based on simply physical tests *Housing and Building National Research Center Journal 13, pp 353-360*

IOP Conf. Series: Materials Science and Engineering **586** (2019) 012013 doi:10.1088/1757-899X/586/1/012013

Study regarding the stiffness influence of slab to beams for a plan structural reinforced concrete frame system in seismic zones

I Sococol[1], P Mihai[1], V I Iftode[1], I Olteanu-Dontov[1]

[1]"Gheorghe Asachi" Technical University of Iași, Faculty of Civil Engineering and Building Services, 700050 Iași, Romania

E-mail: ion.sococol@tuiasi.ro

Abstract. For monolithic reinforced concrete structures, it is known that beams and slabs form a common body, so that the stiffness of the dissipative elements (beams) increases significantly. Also, it is known the fact that the beams are the principal structural dissipative elements. In these circumstances, it will try through numerical simulations (nonlinear calculation) a theoretical reproduction of a recently executed structure, so as to take into consideration the effect of excess rigidity brought to the horizontal dissipative structural elements (beams). It will be pursued the dissipation mode of seismic energy through plastic deformations (formation the punctual plastic hinges at the end zones of the beams and especially at the end regions of the columns).

1. Introduction
The design of reinforced concrete frame systems concerns to use the ductile design concept, for which it is necessary to activate the plastic energy dissipation zones by deformations in the nonlinear domain. These areas are considered to be the end regions of the beams and the end zones of the columns at the base of the structures. As beams become through this design concept the essential dissipative elements, the interest of their seismic response is primordial.
Because of the fact that in the last decade have occurred a significant number of severe earthquakes, it can be observed the real response of the designed and executed reinforced concrete frame structures in the corresponding seismic zones. In the considerable majority of cases it was observed the difference between the real structural seismic response and the critical areas of the structural elements considered by the designers. In these conditions, the beams are not seismic energy dissipators and the columns work intensively in the nonlinear domain with important deformations in the critical zones not only at the base of the structure.
Thus, through by this study it is desired a more realistic knowledge of the seismic response for the monolithic reinforced concrete frame systems, following the stiffness influence of slab to beams and structure. It also follows the dissipation mode of seismic energy through plastic deformations.

2. General data
In this case study is presented biographical calculation (nonlinear static) and time-history method for a planar reinforced concrete structure: Axis 2, figure 1(a), figure 1(b) from a hotel with the P+3E height regime located in Bucharest, performed with SAP2000 program [14].
Description of component parts from structural system:
- Reinforced concrete frame structure;
- Frame opening from structural system: 5.1 (m);
- The height of the floors: 3 (m);
- Marginal columns reinforcement (C_{01}): figure 2(a);
- Central columns reinforcement (C_{02}): figure 2(b);
- Beams reinforcement (B_{01}): figure 2(c);

IOP Conf. Series: Materials Science and Engineering **586** (2019) 012013 doi:10.1088/1757-899X/586/1/012013

Materials used: Concrete C20/25, Steel S500 (trade name: B500C) [12].

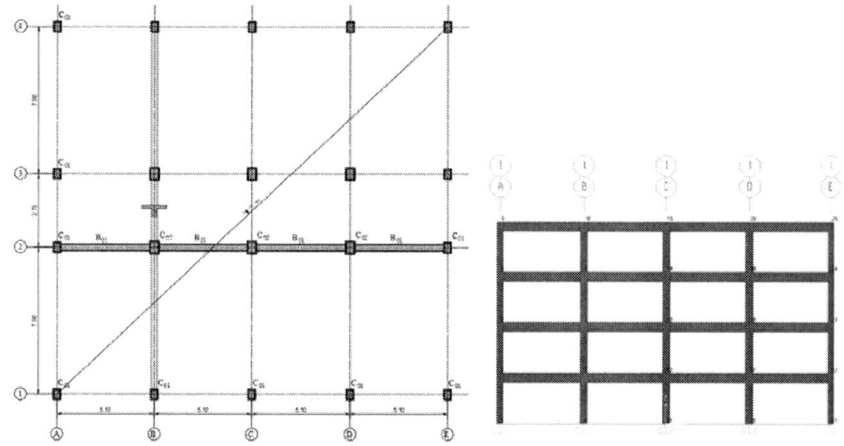

(a) (b)

Figure 1. (a) Current level plan with representation of axis 2; (b) Structural representation of the frame system in X-Z plan [14].

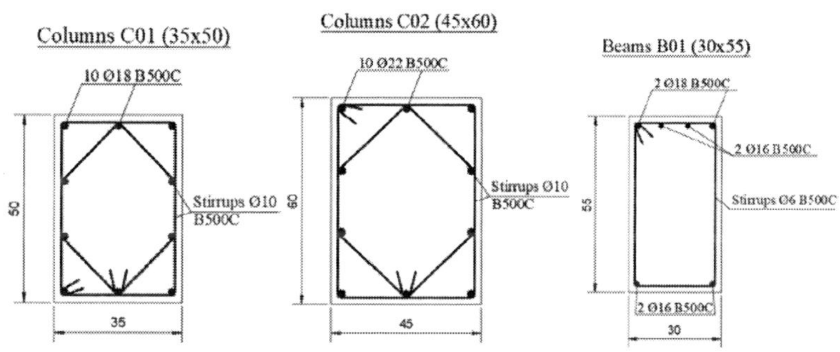

(a) (b) (c)

Figure 2. Columns reinforcement: (a) marginal, (b) central; (c) beams reinforcement.

Seismic loads:

- Used loads: table 1;
- Class of importance for structure: III;
- The structure was designed for high ductility class: DCH;
- Seismic parameters for Bucharest city: according to P100-1;
- Hysteretic degradation model: Takeda bilinear.

Accelerograms and elastic acceleration spectra corresponding to them:
Six accelerograms were used for non-linear Time-History analysis:
-two synthetic accelerograms generated by a specialized software: QKE_1 figure 3(a), QKE_3 figure 3(b) [2];
-two natural accelerograms (recorded): INCERC Bucharest 1977 N-S figure 3(c) and INCERC Bucharest 1986 E-W figure 3(d), scaled to maximum acceleration (PGA=0.30g) [13];
-the same natural accelerograms but compatibilized figure 3(e), figure 3(f) (scaled such that the elastic acceleration spectrum of the code corresponds to the elastic acceleration spectrum for the fundamental vibration period) [2].

In this order, lower is presented 6 accelerograms and 6 elastic acceleration spectra figure 4(a)-(f) corresponding to them [2],[3],[10],[13].

Table 1. Used loads.

Load Pattern	Story	Values	Units
Dead	GF; S1; S2	1.2	kN/m^2
	S3	2.5	kN/m^2
Live	GF; S1; S2; S3	1.5	kN/m^2
Snow	S3	1.6	kN/m^2
Seismic	-	-	-

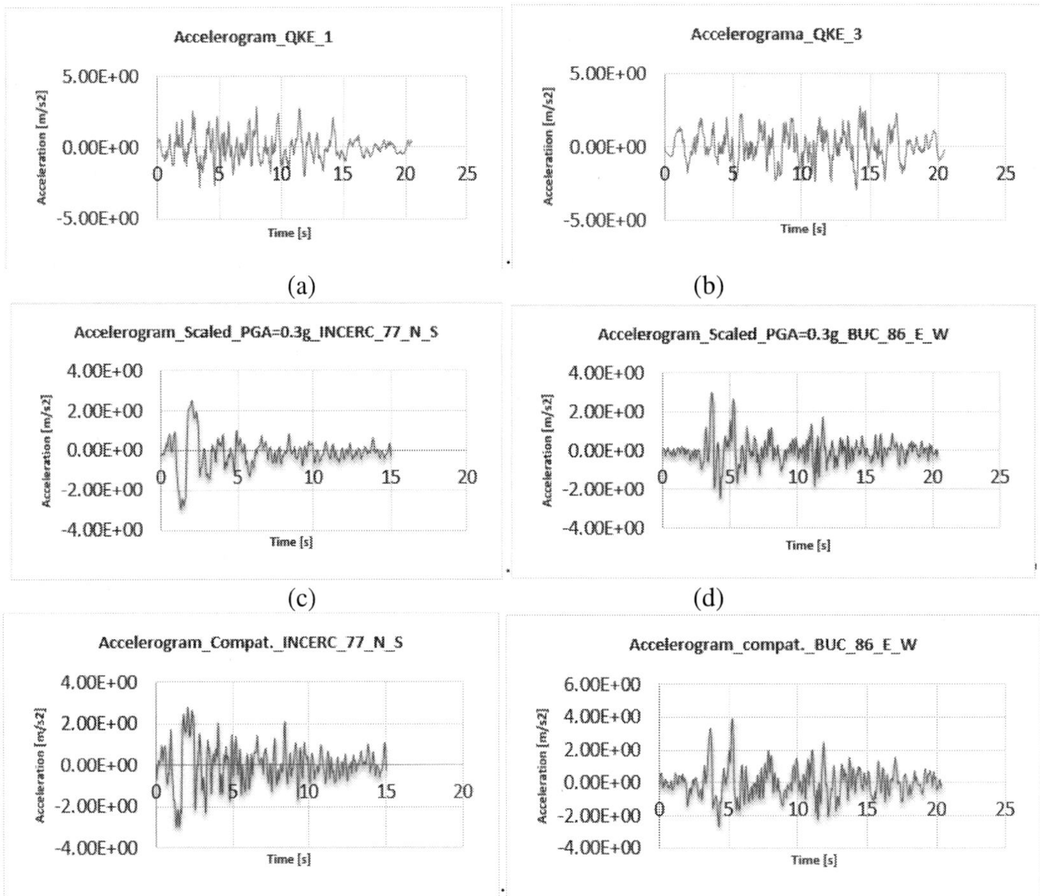

(a) (b)
(c) (d)
(e) (f)

Figure 3. Accelerograms:(a) QKE_1; (b) QKE_3; (c) INCERC_77_N_S_Scaled_PGA=0.30g; (d) BUC_86_E_W_Scaled_PGA=0.30g; (e) INCERC_77_N_S_Compatibilized by spectral design value criteria (PGA=0,304g); (f) BUC_86_E_W_Compatibilized by spectral design value criteria (PGA=0,397g).

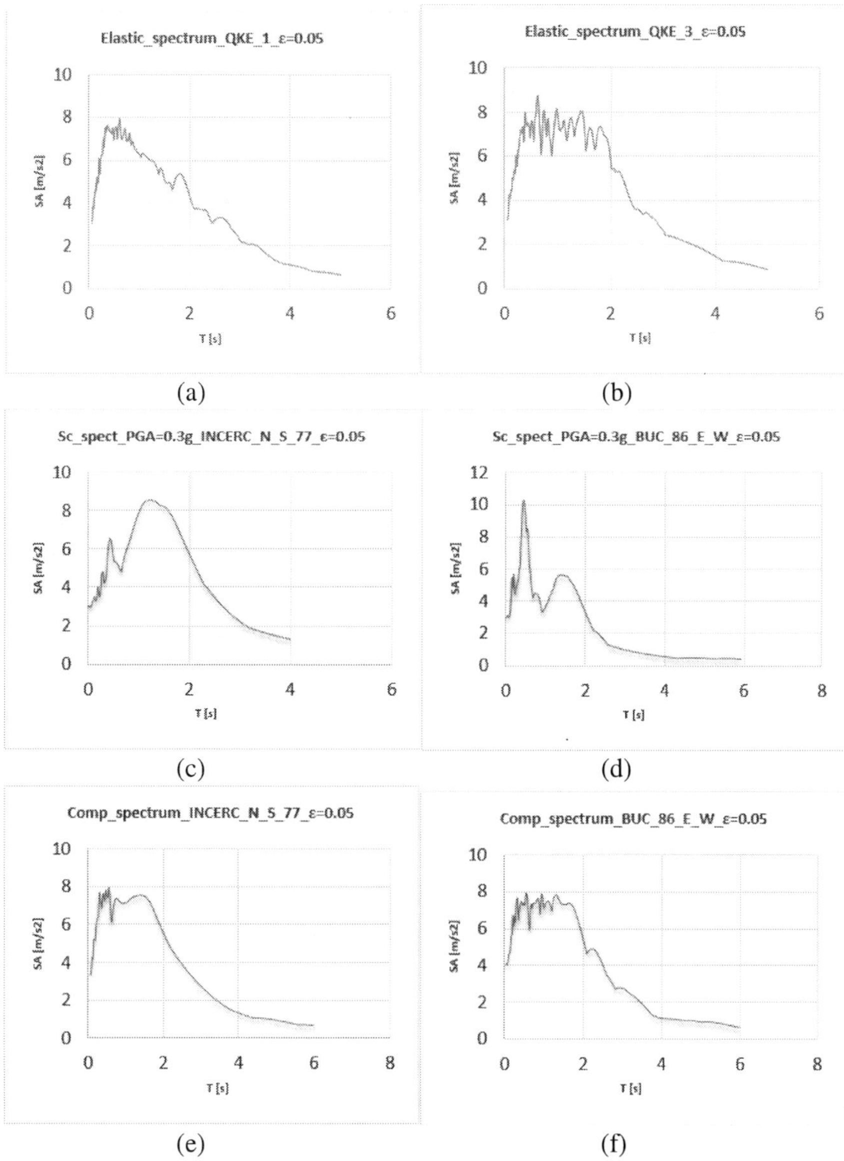

Figure 4. The elastic acceleration spectrum corresponding to the accelerogram: (a) QKE_1; (b) QKE_3; (c) INCERC_77_N_S_Scaled; (d) BUC_86_E_W_Scaled; (e) INCERC_77_N_S_Compatibilized; (f) BUC_86_E_W_Compatibilized.

The study mode and the influence parameter of the seismic response for structural system
Four study cases were considered with the flexural stiffness of the beams as variable:1EI; 1.5EI; 1.7EI; 2EI. The mode of considering the beams stiffness is idealized, reducing the stiffness of the beams-slab assembly to the stiffness of the "pure" section of the beams.

Hypotheses regarding the seismic response of the structural system:

- The rigidity of the beams-slab assembly influences the seismic response of the structural system with incursions in the inelastic domain through degradation and formation of plastic hinges in columns. This phenomenon can produce a floor (storey) mechanism [8],[9];

- The rigidity of the beams-slab assembly influences the seismic response of the structural system with incursions in the inelastic domain through degradation and formation of plastic hinges in beams [4],[5]. This structural response in the nonlinear domain corresponds to the design concept of "weak beams-strong columns": according to the current norms [10],[11] etc.

The veracity of these hypotheses can be demonstrated (interpreted) by the results obtained from nonlinear analysis (Push-Over and Time-History).

3. Push-Over analysis

For the analyzed case, two load distributions were used: (1) uniform distribution of lateral forces; (2) distribution corresponding to the fundamental vibration mode, [3],[10]. Following the bilinearization of the lateral force-displacement curve, have resulted the variation of the displacement requirement represented in figure 5(a), the variation of the lateral force corresponding to the structural system yielding figure 5(b), the variation of the fundamental period figure 6, having as a parameter the bending stiffness of the beams.

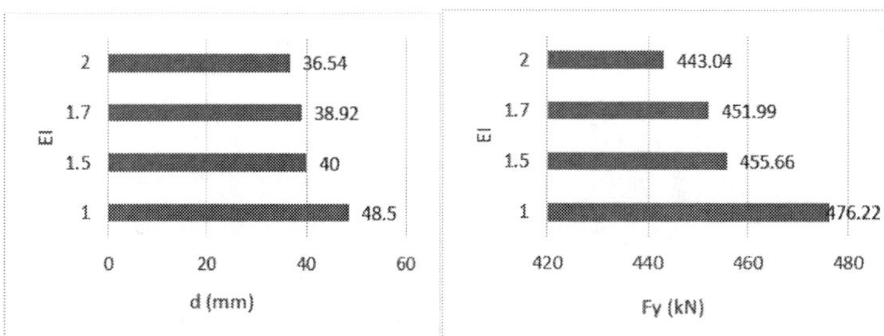

(a) (b)

Figure 5. (a) Displacement requirement variation for structural systems according to the stiffness variation of the beams; (b) The lateral force variation corresponding to the yield of the structural systems, depending on the bending stiffness of the beams.

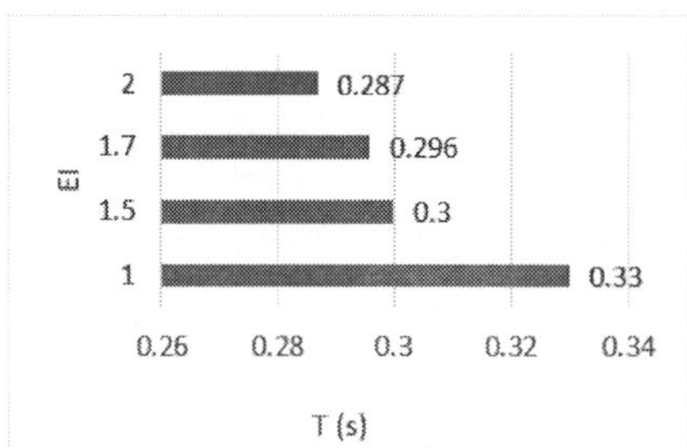

Figure 6. Variation of the fundamental period for structural systems according to the variation of bending stiffness of the beams.

The lateral force-displacement curves bilinearizated according with calculation based on corrected elastic spectra from P100-1 [10], function of bending stiffness *EI* for beams are shown in figure 7(a) and figure 7(b).

With all this data, it is possible to easily determine the stiffness influence of the beams to rigidity of the structure (figure 8), the total seismic load reduction factor (q), the seismic force reduction factor appropriate to structure ductility (q_{μ}), structure over-strength (q_S), design over-strength (q_{Sd}), redundancy or capacity for plastic redistribution of efforts (q_R) in figure 9, variation of seismic design force (F_d), elastic force (F_e), force corresponding to formation of the first plastic hinge (F_1), yielding force (F_y), depending on the bending stiffness *EI* of the beams (figure 10).

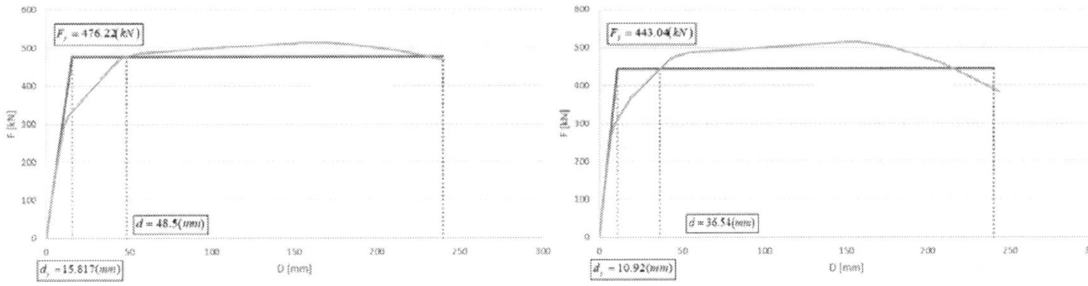

(a) (b)

Figure 7. Lateral force-peak displacement curves: (a) EI; (b) 2EI.

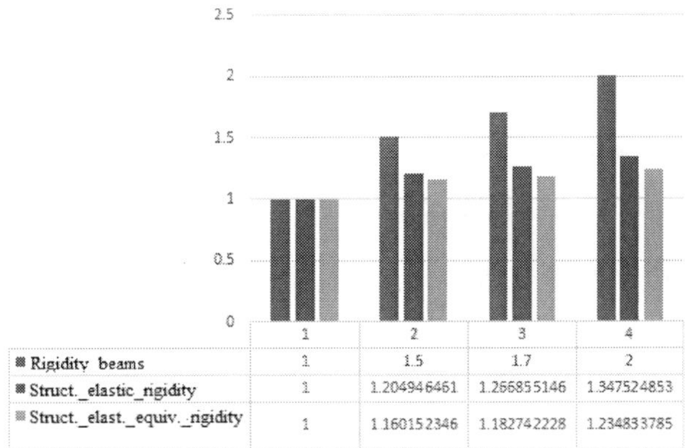

Figure 8. The bending stiffness influence to global rigidity of the structure.

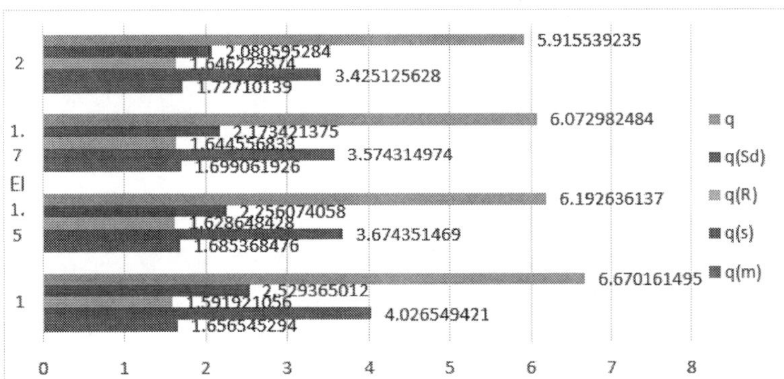

Figure 9. Variation of seismic load reduction factors.

Structure response to static actions according to the Push-Over mode of action can also be represented graphically through color mode of plastic hinge formation for each loading step in all four cases of bending stiffness. In this way, it is presented lower figure 11(a),(b) structural response for triangular loading, step 5 for *EI* and *2EI* rigidity. It was chosen step 5 of the load because in this step is touched the yield force for structural system and the displacement requirement. Thus, it can be seen what happens with the structure for maximum horizontal load.

IOP Conf. Series: Materials Science and Engineering **586** (2019) 012013 doi:10.1088/1757-899X/586/1/012013

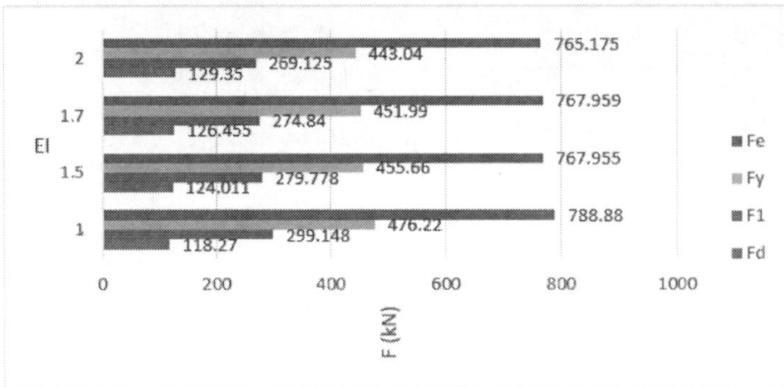

Figure 10. Variation of elastic forces, yielding forces, forces corresponding to formation of the first plastic hinge, designing forces.

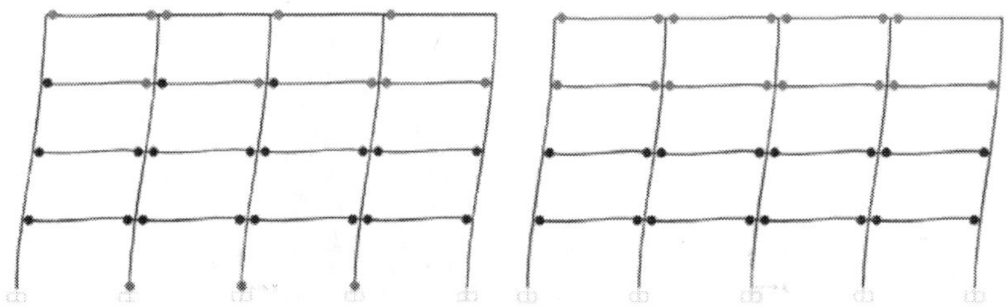

Figure 11. The seismic response of the structural system through formation the plastic hinges (Step 5, Push_Triang.): (a) EI; (b) 2EI, [14].

The colors of the plastic hinges correspond to the American codes [6],[7] who take in consideration four levels of performance in seismic design figure12, figure 13.

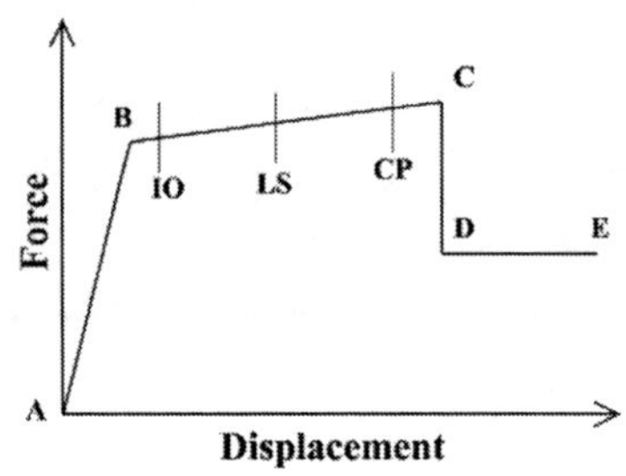

Figure 12. Lateral force-displacement curve according to FEMA 356 [6].

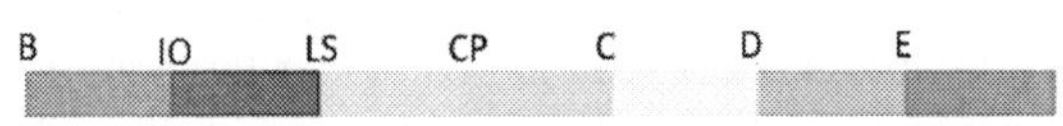

Figure 13. The colors representing each level of performance [14].

Thus, according to FEMA 356 [6] the translation of the performance levels in figure 12 and figure 13 are:

O - Operational Level;
IO - Immediate Occupancy Structural Performance;
LS - Life Safety Structural Performance Level;
CP - Collapse Prevention Structural Performance Level.

IOP Conf. Series: Materials Science and Engineering **586** (2019) 012013 doi:10.1088/1757-899X/586/1/012013

4. Non-linear Time-History analysis

To perform the most efficient evaluation methods for seismic response of structures, it is necessary to establish the accelerogram, hysteretic degradation laws of the material, the damping properties of the structure etc. [3]. Compared with linear time-history analysis, the nonlinear analysis takes into account rigidity parameters and degradation effects. Thus, the real-time response of the structure can be monitored and can be known a multitude of requirements (displacements, rotations, accelerations etc.) on a structural assembly and element [1],[11].

Further, will be presented the hysteretic acceleration-displacement curves for last level (node 5) only for beams with *EI* (figure 14) and *2EI* stiffness (figure 15). Thus, it is possible to accentuate the structural response in displacements:

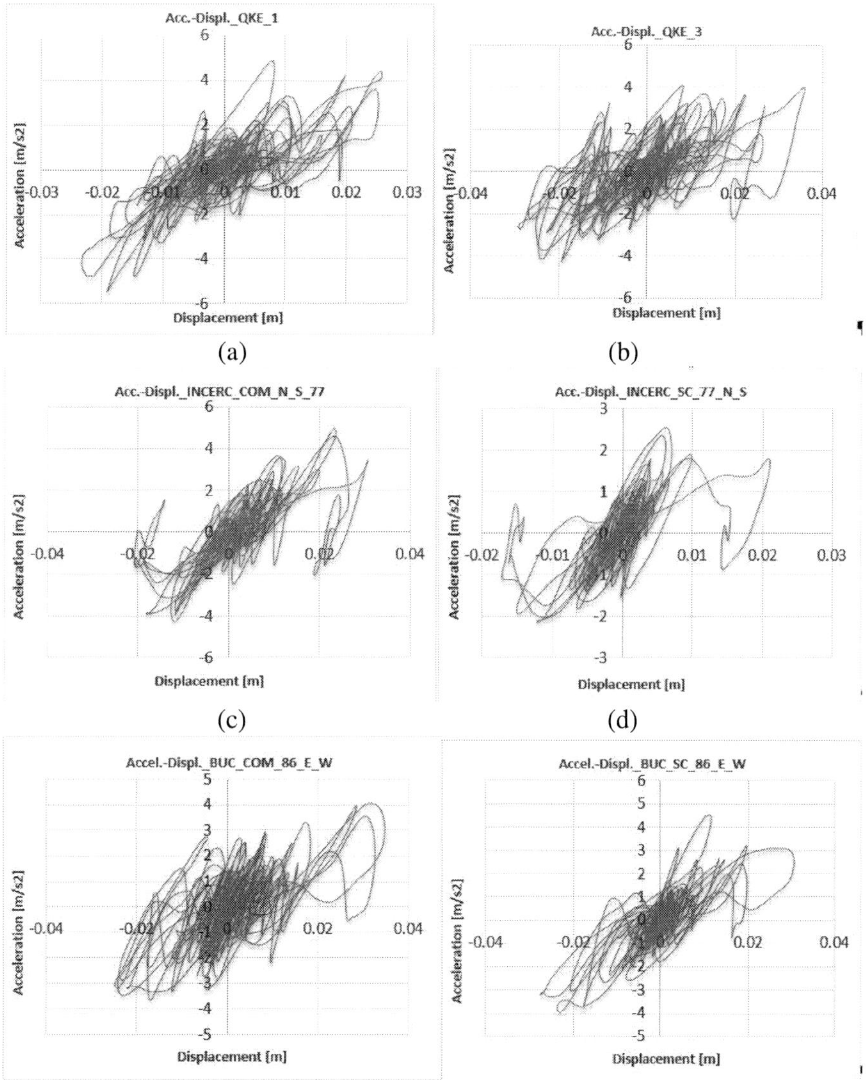

Figure 14. Hysteresis curves corresponding to Non-linear Time-History analysis for beams with EI stiffness: (a) QKE_1; (b) QKE_3; (c) INCERC_77_N_S_Compatib.; (d) INCERC_77_N_S_Scaled; (e) BUC_86_E_W_Compatib.; (f) BUC_86_E_W_Scaled.

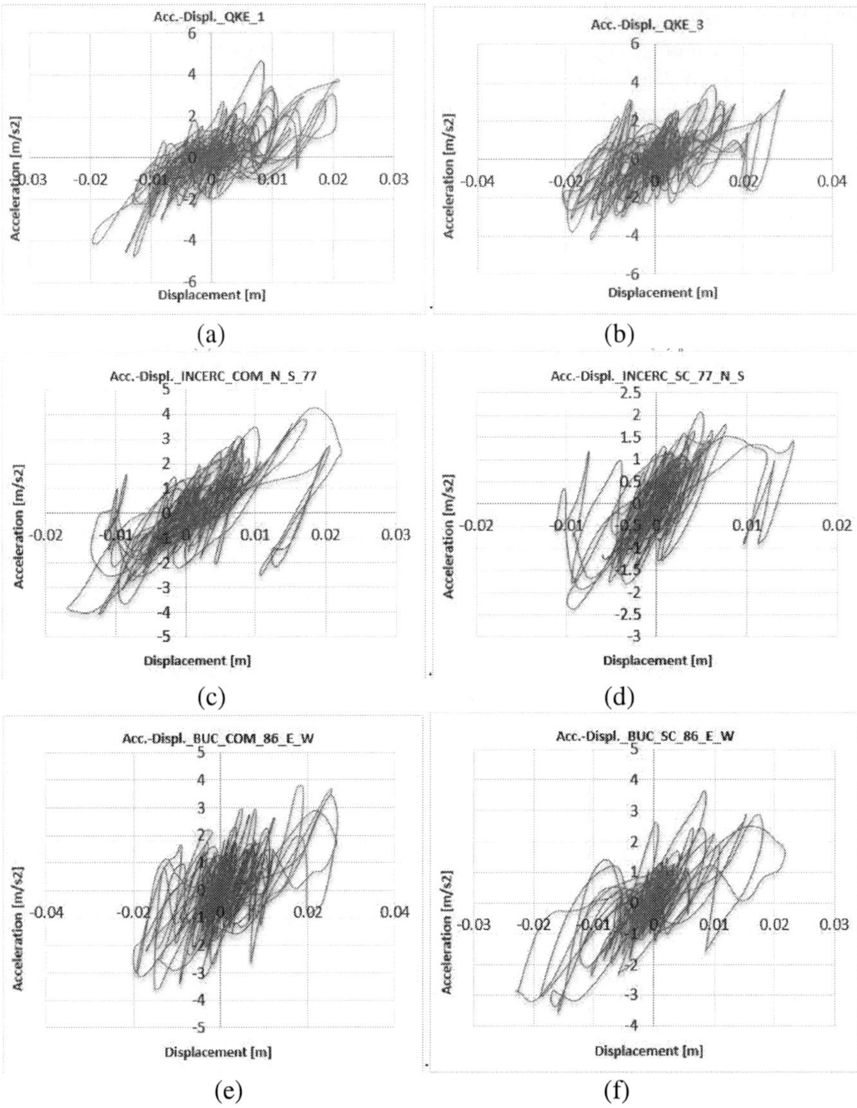

Figure 15. Hysteresis curves corresponding to Non-linear Time-History analysis for beams with 2EI stiffness: (a) QKE_1; (b) QKE_3; (c) INCERC_77_N_S_Compatib.; (d) INCERC_77_N_S_Sc.; (e) BUC_86_E_W_Compatib.; (f) BUC_86_E_W_Sc.

5. Conclusions

Push-Over analysis

(1) Increasing the bending stiffness of the beams, results a decrease with 25% of the displacement requirements for structural system (in the case if the maximum rigidity is considered to be 2EI) and a small decrease of the lateral force corresponding to the yield of the system. These phenomena are explained by a development of the structural rigidity imposed by the stiffness of the beams. Thus, for 2EI bending stiffness of the beams, the elastic rigidity of the structure increase with 35%. This effect can be verified by gradually decreasing of the fundamental vibration period for structural system.

(2) With increasing stiffness of the beams (structural dissipative elements) the global seismic load reduction factor (the behavior factor q) decreases (from 6.67 for EI to 5.9 for 2EI). In addition to this process, can be noted the decrease of structural over-strength (from 4.02 for EI to 3.42

for 2EI). The same phenomenon is also produced for elastic seismic forces, yielding forces, forces corresponding to formation of the first plastic hinge or designing seismic forces, but the differences are very small (which implies not considering them).

(3) This information helps us to understand the significant contribution of slab rigidity to structural stiffness. This can be better understood through graphical representation of the structure's response by forming plastic hinges in the considered dissipative areas (end zones for beams and columns). Thus, it is visible how the number of beams who touch the LS (Life Safety) performance decreases at the same time with developing bending stiffness. Of course, in this case the number of beams affected by plastic deformations decreases, but unfortunately the problem must be seen differently, because the seismic designing forces, yielding forces or elastic forces do not vary much, their values are close. In these conditions, it is evident how the structure is loaded approximately with the same level of horizontal excitation (grace to its maximum capacity to load with seismic forces). Thus, if plastic deformations do not occur in the dissipative elements (beams), then these deformations occur in vertical elements (columns) who works theoretically in elastic domain. In these circumstances, some design principles need to be revised.

Non-linear Time-History analysis

(1) For non-linear time-history analysis, it can be observed through hysteretic curves a close response of the structure following synthetic and compatibilized accelerograms. Thus, the compatibility method based on spectral design values is more efficient than the accelerogram scaling method for maximum acceleration (our PGA case =0.30g).

(2) Also, according to the acceleration-top displacement curves (response lines) of the structure, it is visible an extensive range of their activity, due to deformations in the considered dissipative elements. In addition to this, displacement requirements decrease in the same time with developing bending stiffness.

(3) Through non-linear dynamic analysis it was possible to confirm the final conclusion from the Push-Over analysis with regard to the influence of the plastic deformation capacity of the beams.

(4) Non-linear time-history analysis demonstrate how the structure possesses an important redistribution capacity and more over-strength than was considered to be. Proof is the values of the base shear forces, which are much higher compared to the design seismic forces. In this way, it was possible to appreciate the structure's capacity to deform in potentially plastic areas.

6. References

[1] Budescu M and Ciongradi I 2014 *Inginerie Seismică*, Ed. Politehnium, Iasi

[2] Damian I 2012 *Electiva I, Accelerograme artificiale compatibile cu un spectru de accelerații predefinit. Metode de generare și agresivitatea acestora,* București

[3] Postelnicu T 2012 *Proiectarea structurilor de beton armat în zone seismice,* București MarLink

[4] Paulay T and Priestley MJN 1992 *Seismic design of reinforced concrete and masonry buildings*, John Wiley & Sons, USA

[5] Park R and Paulay T 1975 *Reinforced concrete structures,* John Wiley & Sons USA

[6] FEMA 356 2000 *Prestandard and Commentary for the Seismic Rehabilitation of Buildings,* Washington, USA

[7] FEMA 273 1997 *NEHRP Guidelines for the seismic rehabilitation of buildings* Washington, USA

[8] FIB 24 2003 *Seismic assessment and retrofit of reinforced concrete buildings,* Sprint-Digital-Druck, Stuttgart:

IOP Conf. Series: Materials Science and Engineering **586** (2019) 012013 doi:10.1088/1757-899X/586/1/012013

[9] FIB 25 2003 *Displacement-based seismic design of reinforced concrete buildings,* Stuttgart: Sprint-Digital-Druck, Stuttgart
[10] P100-1 2013 *Cod de proiectare seismică, Partea-I, Prevederi de proiectare pentru clădiri,* Bucuresti
[11] SR EN 1998-1 2004 *Proiectarea structurilor pentru rezistența la cutremur. Partea 1: Reguli generale, acțiuni seismice și reguli pentru clădiri,* Bucuresti
[12] SR EN 1992-1-1 2004 *Proiectarea structurilor de beton. Partea 1-1: Reguli generale și reguli pentru clădiri,* Bucuresti
[13] PRISM for Earthquake Engineering (South Korea: INHA University), 2011
[14] SAP2000 *Computers and Structures* (www.csiamerica.com)

IOP Conf. Series: Materials Science and Engineering **586** (2019) 012014 doi:10.1088/1757-899X/586/1/012014

Parametric study in composite joints configuration

M Pop[1], C Campian[1], N Chira[1]
[1] Technical University of Cluj Napoca, Baritiu Street no 25, ZIP 400027, Romania

E-mail: maria.pop@dst.utcluj.ro

Abstract. Composite structures showed higher stiffness than steel structures with similar ductility and also better performance to fire action. The global behaviour of the composite steel-concrete structure depends mostly on the behaviour of the joints. Many studies have demonstrated that composite joints exhibit adequate mechanical properties, such as strength, ductility and energy dissipation. The present paper is based on the study of the behaviour of the joints and their influence on the structure. The study involves a sensitivity analysis in the finite element program Abaqus 6.13. The analysis focuses on determining the parameters that significantly contributed both on the strength and stiffness of the joints and quantifying their weight in the final results. The analyses also provide information on the ductility of the connections under monotonic loads.

1. Introduction

Multi-storey buildings have been a favourite investor option for centuries and a challenge for engineers involved in materializing these projects. Due to the growth of the service sector, tourism and other sectors specific to the epoch, on the one hand, and due to acute managerial influences, such as the cost-benefit ratio, on the other hand, the demand for this type of structures is growing. Analysing the types of efficient, technical and economic structures for high-rise multi-storey buildings is therefore a major, permanent engineering concern.

Following the earthquakes in recent years, numerous studies have been carried out on various steel-concrete composite structures. Over time, joints have shown an important role in seismic applications. The lack of data for seismic design of composite joints leads to the situation where the designer has to choose whether or not to take into account the steel-concrete composite action.

Compared to pure steelwork, composite joints provide greater moment resistance in a braced frame, through the tensile action of the slab reinforcement. However, they usually remain partial strength relative to the composite beam, and tests show that the rotation capacity at the enhanced moment is reduced. The rotation capacity is curtailed by several possible modes of failure, some are peculiar to composite joints, others are more likely because the tensile action of the reinforcement increases the balancing compression around the bottom flange of the steel section, thereby encouraging buckling [1].

The effect of behaviour of the joints on the distribution of internal forces and moments within a structure, and on the overall deformations of the structure, may generally be neglected, but where such effects are significant, they should be taken into account [2].

The joint connection inside the composite joint is numerical simulated in the finite element modelling program [3]. The complex method approach the design of steel and composite joints to the proven efficiency in terms of resistance and rotation capacity.

This parametric study is based into a FE model, calibrated on experimental program on these type of joints, made at the Civil Engineering Faculty of TUCN [4].

1.1. Steel-concrete composite joints

The main objective of this work is to obtain a functional model by using the numerical modelling analysis, which reflects as accurate as possible the behaviour of the joint, by using the physic-mechanical characteristics of the materials.

As part of the component modelling in the FEM program, a fine mesh refinement was used, especially in the interconnection area, in order to better obtain the stress concentration and deformations of the assembly.
The joints were modelled by using the finite element program Abaqus 6.13 [3] in which the physical-mechanical characteristics of the materials and the node geometry were taken into account. We highlighted the joints characteristics, meshing modes, contour conditions, and the chosen analysis method.

Figure 1. The joints geometry

In the finite element analysis of the joints the purpose is to determine the resistance characteristics, the ductility, as well as the local effects occurring in joints subjected to hogging bending (Figure 2).
The study was carried out to determine the bearing capacity of the joint, which is given by the strength of the weakest component. Considering the correct way in which the numeric model captures the joint behaviour, a parametric analysis was performed. In order to determine their inflection to the bearing capacity, rigidity and breaking ability of the composite steel-concrete joint configuration, we have to take into consideration the behaviour of each component, in particular.
The parameters considered were: the diameter of the joint, the thickness of the anchor plate, the steel profile, the quality of the steel and the number of connectors. The influence of their variation on the resistivity, stiffness and rotation ability characteristics was determined.

Figure 2. Test setup of hybrid joints

IOP Conf. Series: Materials Science and Engineering **586** (2019) 012014 doi:10.1088/1757-899X/586/1/012014

2. Numerical modelling

The geometrical nonlinearity of the materials of all components making up the joints was taken into account as well as the effect of imperfections. In the complex model the modelling of the end plate, the steel profile, the connectors, and the concrete slab with three-dimensional finite elements were taken into consideration. The analysed joints were loaded increasing monotonous, the force being applied from the free zone of the composite beam (base model Figure 3).
The interaction between the parts was of the type General Explicit by having the following property: "hard contact", on the surface and in the tangential direction (in the surface plane) and also a tangential behaviour with friction coefficient (μ =0, 45).
The Abaqus 6.13 [3] finite element program uses numerical integration methods using the Gaussian quadrature for most elements in all of the element integration nodes. Elements can have normal or reduced integration, options that can influence the accuracy of the solution. The integration modes type C3D8 (8 nodes) are used for low integration to obtain stress and deformation, finite elements of order I or II. with partial or total interaction.
The type of analysis chosen for the numerical study is the one-step dynamic load analysis that simulates a quasi-static analysis. To diminish the dynamic effects of the model, we controlled the test by the duration and the mode of application of the imposed displacement, using the "Smooth step" amplitude curve
An optimization process was made for each element in part, that makes up the joint, to determine a proper mesh discretization. In the analysis, the concept of Fracture energy was taken into account [3].
Element type S4R was used for modelling the behaviour of the reinforced concrete slab. This is a general-purpose conventional shell element which allows for transverse shear deformation to be taken into account. It also accounts for finite strains and arbitrarily large rotations [5].
The quasi-static analysis used in this paper is a particular case of dynamic analysis where the inertial effect is greatly diminished by applying the modelling rules. The most important parameters in this direction are:

- Application time (force, displacement),
- The amplitude of this request over the set time.

The application time must be long enough to avoid the shock effect. In order to monitor this, it is sufficient to check the total system energy and internal energy by percentage comparison. If the internal energy percentage is less than 3-5% of the total energy then the dynamic effect in the system can be considered as negligible.
The "concrete damage plasticity" (CDP) model of Abaqus [3] is used to describe the nonlinear material behaviour of concrete. The suitable materials and parameters were determined experimental. The purpose of this model is to present the effects of degradation associated with the failure mechanisms that are caused by cracking and crushing of the concrete.

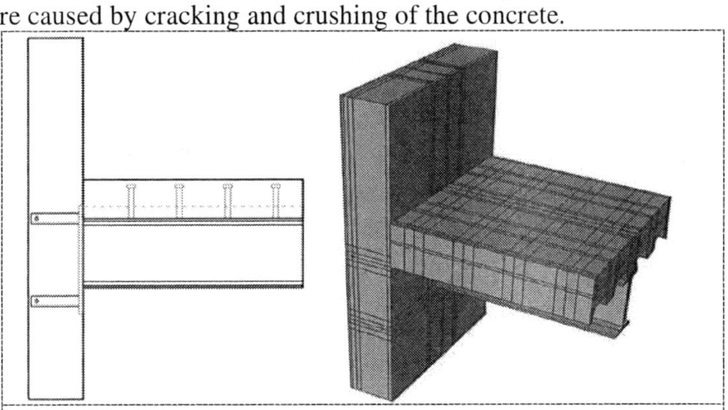

Figure 3. Numerical modelling of the joint.

Table 1. CDP (Concrete Damage Plasticity) parameters

Sinification	ψ	ε	f_{t0}/f_{c0}	k	μ
Parameter	Dilatation angle				viscosity
Predefined value	-	0,1	1,16	0.667	0
Value considered	38	0	1,6	0,667	0

The steel characteristics are used from the different types of elements that makes up the joint obtained from the experimental tests of the specimens.
The modulus of elasticity (E) has a value of 210000N/mm^2, Poisson's coefficient (0,3) elastic level. A design response that calculates the von Misses stress was use at the maximum value of the stress within a region of the model [3].

Table 2. Steel properties

Density	Modulus of elasticity E(N/mm^2)	Poisson coefficient	f_u/f_y
7800	20000	0.3	1.2
7850	21000	0.3	0.80

- *The results obtained by the FEM analysis are showed in Figure 4.* Stress concentration in the joint configuration

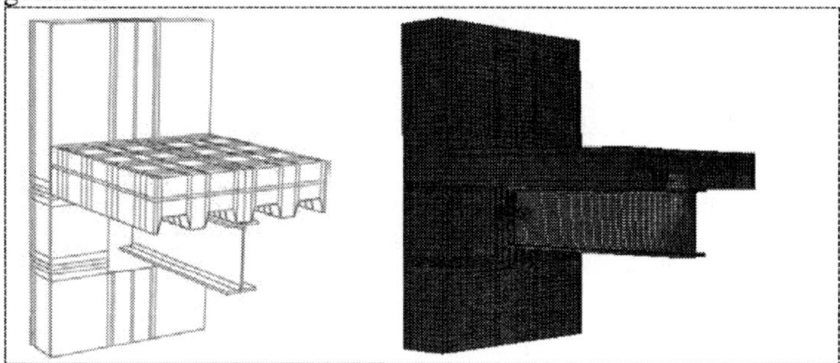

Figure 4. Joints stress concentration.

- *The results obtained by the FEM analysis are showed in Figure 5.* The ultimate bearing capacity and also the state of stress concentration in each element.

The joints deformations and stress concentrations, observed on the numerical analysis (Figure 5), allow a better perception of the behavior of these joints. Due to the obtained results the stress concentration is highlighted by transmitting the effort between the connected parts.

IOP Conf. Series: Materials Science and Engineering **586** (2019) 012014 doi:10.1088/1757-899X/586/1/012014

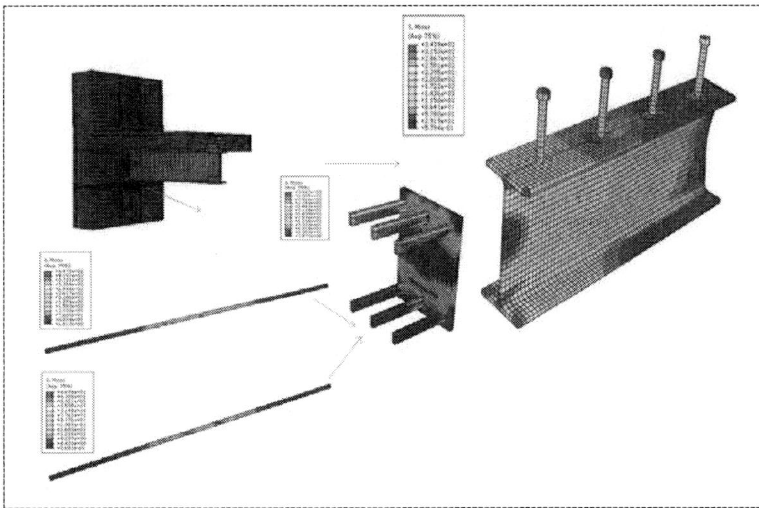

Figure 5. Stress concentration in elements

3. Parametric study

A parametric analysis in the finite element program Abaqus 6.13 [3] was conducted for this study. The analysis focuses on determining the parameters that significantly contributed both to the strength and stiffness of the joints. The main purpose is to minimize the weight of an element of the joint staying lower than allowable stress and displacement criteria.
The parameters that make up the moment-relative rotation curve define its behaviour (Figure 6).

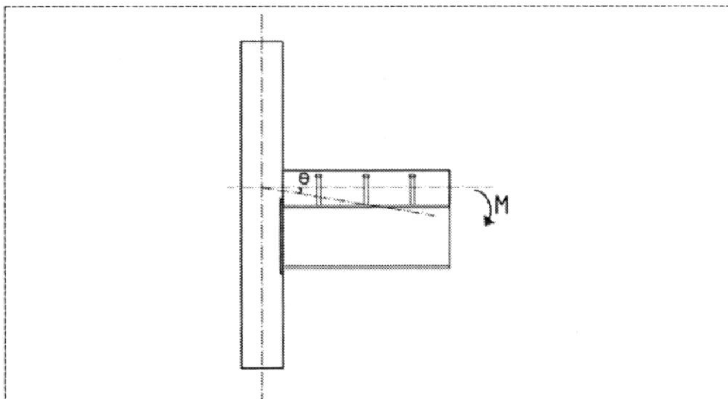

Figure 6. t Moment-relative rotation curve

The first component that started the parametric study is the longitudinal reinforcement of the composite steel-concrete slab.
The ability of the rotation of the joint is obviously limited by the elastic and mechanical properties that make up the connection. These properties define the ultimate strength of the joint that controls the imposed rotation to which it can grow.
The results obtained after the numerical analysis is presented in Figure 7, in terms of Moment-rotation curves.
The parameters considered were:

- Reinforcement bar from the steel-concrete composite beam
- Steel profile

- Number of connectors

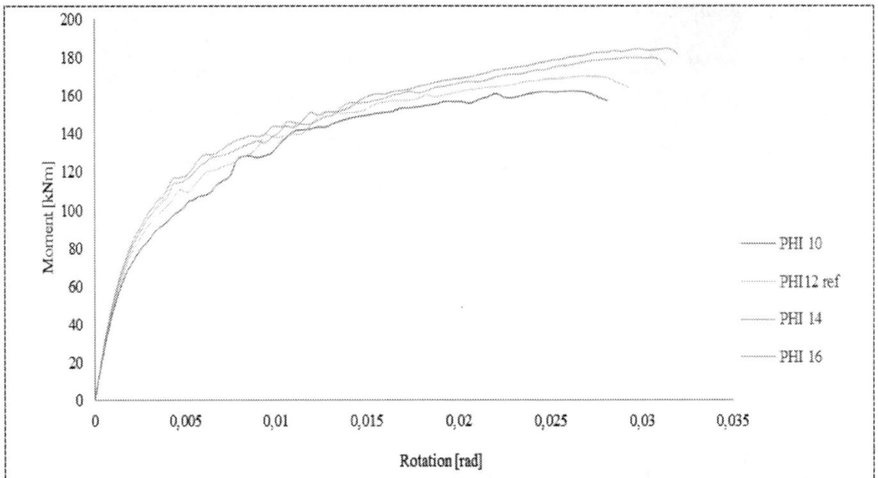

Figure 7. Moment –rotation curve obtained by parameterizing the Reinforcement bar.

An increase in maximum rotation , as observed in Figure 8, compared to the joints made with the diameter of the Ø12 mm, in percentage is :

- With 4.89% using the diameter of the Ø10 mm reinforcement bar
- With 10.85% using the diameter of the bar Ø14 mm
- With 13.65% using the diameter of the bar Ø16 mm

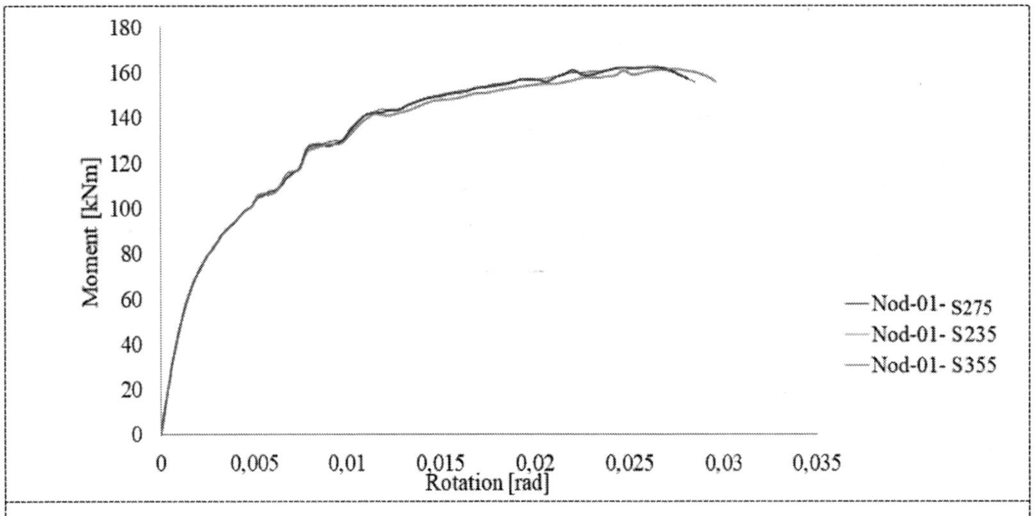

Figure 8. Moment –rotation curve obtained by parameterizing the steel profile..

In Figure 9 is showed the increase in terms moment-rotation with:

- 3.02% using S235
- 3.62% using S355

compared to the S275

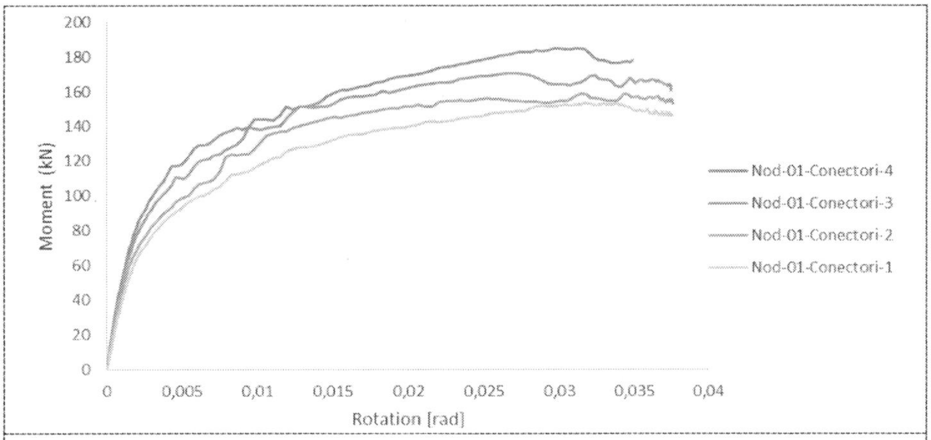

Figure 9. Moment –rotation curve obtained by parameterizing the number of connectors.

4. Conclusions

The presented paper is based on the study of the behaviour of the joints and their influence on the structure. The study involves a sensitivity analysis in the finite element program Abaqus 6.13. The analysis focuses on determining the parameters that significantly contributed both on the strength and stiffness of the joints and quantifying their weight in the final results. The analyses also provide information on the ductility of the connections under monotonic loads.

The failure mode of the joints appears by the fracture of the reinforcement, loss of anchorage (stress concentration in the end plates) failure of the shear connection in the superior part (concrete part) of the composite slab by bearing, and local buckling of the steel profile.

This type of joints is acting strongly related on the one hand, to the geometric configuration and on the other, to the particular behaviour of the materials.

The most pronounced influences on the joint response curves are given by the diameter of the reinforcement and the size of the steel profile. Also, large variations were obtained by parameterizing the number of connectors, the reference model being one with a stiffer reinforcement.

References

[1] Najafi Ali. A and Anderson D 1997 *Ductile steel-concrete composite joins, IABSE, Zurich*

[2] Eurocode 2004 Design of composite steel and concrete structures Part 1-1: General rules and rules for buildings

[3] ABAQUS - Explicit user's manual, version 6.13. Pawtucket: Hibbit, Karlsson & Sorensen, Inc., 2013

[4] Pop M 2017 Contributions regarding the use of steel-concrete elements in multi-storey buildings, Phd Thesis, Cluj Napoca, Romania

[5] Vasdravellis G, Valente C A Castiglioni 2009 Dynamic response of composite frames with different shear connection degree, Journal of Constructional Steel Research

[6] Eurocode 3 2005 Design of steel strucutes- Part 1-1: General rules and rules for buildings

[7] Henriques J, da Silva JS, Valente I 2013 Design model for composite beam-to-reinforced concrete wall joints. Steel Construction 6, Berlin

[8] EN 10002-1 2001 European Standard Metallic materials - Tensile testing - Part 1: Method of test at ambient temperature

[9] Henriques J, da Silva LS, Valente IB 2013 Numerical modeling of composite beam to reinforced concrete wall joints, Part I: Calibration of joint components. ELSEVIER

IOP Conf. Series: Materials Science and Engineering **586** (2019) 012014 doi:10.1088/1757-899X/586/1/012014

[10] Gracia J, Bayoa E, Ferrario F, Bursi O, Braconi A, Salvatore W 2009 The seismic performance of a semi-rigid composite joint with a double-sided extended end-plate. Part I: Experimental research, Elsevier
[11] ECCS 1986 Recommended testing procedure for assessing the behaviour of structural steel elements under cyclic loads. Doc. no 45 of European Convention for Constructional Steelwork, TWG 1.3
[12] Puskas A, Moga L M 2015 Sustenability of reinforced concrete frame structure – a case study, International of Sustainable Development and Planning, Volume 2

IOP Conf. Series: Materials Science and Engineering **586** (2019) 012015 doi:10.1088/1757-899X/586/1/012015

Simulation on implementing noise mitigation measures in a typical urban area

I Oancea[1] and G Covatariu[2]
[1] Department of Architecture, "G.M. Cantacuzino" Faculty of Architecture, "Gheorghe Asachi" Technical University of Iasi, Iasi, Romania
[2] Department of Structural Mechanics, Faculty of Civil Engineering and Building Services, "Gheorghe Asachi" Technical University of Iasi, Romania

E-mail: irina.oancea@tuiasi.ro

Abstract. Starting from the analysis of a real situation in which the equivalent noise level was measured in a residential area, an urban area with the same characteristics was simulated. The purpose of these tests was not to determine exactly the noise level in the area, but to simulate the introduction of noise mitigation measures, and to highlight the difference between the noise level at the edge of the sidewalk and the level next to the building, at the pedestrian height. Four series of simulations were made. The first series starts from the simple situation (street with 4 traffic bands and tram line), without measures of noise reduction, for which a green area with a height of 1.50 m was implemented, a noise barrier and then both measures. The second series of simulations tests the impact that the reduced capacity of the traffic in the area may have, transforming the street into one with only two traffic bands, the remaining space being given to the sidewalk. To optimize noise mitigation, the same situations were simulated, but with the implementation of a green area of 3.00m height. Thus, it is found that the noise level decreases from 74-76 dB (A) to 58-60 dB (A), meaning a decrease of 16 dB (A). From these simulations it can be concluded that in order to reach a comfortable noise level that respects the maximum admitted level, it is recommended to decrease traffic intensity (through different interventions on road surface capacity, promoting of public transport, etc.), to implement anti-noise barriers of materials with a high sound absorption capacity and a green space with a height that exceeds the height of the anti-noise barrier and has a high density of leaves.

1. Introduction

Ambiental acoustics or environmental acoustics study the management and protection against noise produced by fixed sources (industry, education facilities, entertainment, parking, etc.) and / or mobile (underground, land and air traffic) and the possibility of creating urban areas with a strong ecological character, based on the sensations felt by people in these spaces.

Even though the vast majority of the population do not concern about the influence of noise on their health, this being acknowledged by specialists in the field, noise has negative implications on quality of life [1,2] and on the "health budget".

To reduce the level of noise experienced by humans, a series of measures can be taken, regarding:

- *reducing the noise emission* - decrease of the level of noise generated by sources, by stimulating the evolution of the transport vehicles technology; the use of electric vehicles, silent tires, etc.;
- *increasing sound absorption* on the distance between source and receiver by using sound absorbing surfaces with high noise reduction potential;
- *protecting the receiver* by urban planning measures such as:
  - placing artificial or natural screens between noise sources and pedestrians;
  - restricting and rethinking traffic on urban traffic roads;
  - creating an acoustic protection zone for the pedestrians by rehabilitating the cross-section of the streets, this protection being made of vegetal elements, where the situation allows it, or artificial ones [3].

IOP Conf. Series: Materials Science and Engineering **586** (2019) 012015 doi:10.1088/1757-899X/586/1/012015

In terms of *urban road traffic*, a number of measures have been taken in recent years in various European cities, including:

- *reducing heavy freight traffic* within the city by developing logistics centres outside the cities and using low noise emission vehicles for the distribution of goods in dense urban areas;
- *directing the traffic to less sensitive areas*: limiting streets that allow agglomerations of cars to cross residential areas, those being directed to mainly industrial areas;
- *improvement and homogenization of traffic by*: designing of appropriately dimensioned roundabouts, imposing speed restrictions, implementation of a traffic light management system, synchronized with the volume of traffic, so that stops and starts are limited, traffic becoming fluidized;
- *promoting public transportation*, by creating special bus lanes, implementing a traffic light system with priority for public transport at intersections, setting up optimized routes so that public transport would become much easier than the use of one's own vehicle, which will lead to a decrease in the number of cars in the urban environment;
- *implementation of a "park & ride" system* - parking spaces located on the main entry roads into the cities, directly linked to the main points of interests by means of public transportation;
- *encouraging the use of bicycles* - by designing suitable bicycle paths, separated from roads and pedestrian traffic, for reasons of safety and fluidity of transport;
- *redesigning the cross-section of streets* - the Copenhagen model - limiting the number of lanes for cars, which are transformed into public transportation areas, bicycle lanes or green spaces, thus increasing the distance between the noise sources and the pedestrians. This measure can lead to the decrease of traffic volume, the transformation of the street from a traffic-only street into a space dedicated to people with a high degree of comfort [4].

This study aims to analyse the impact of some noise mitigation measures on its level as it is perceived by the pedestrians at the street level.

2. Description of the study

The study presented in this paper is based on the analysis of a real situation, presented in the paper [4]. The studied area is located between the intersection of Nicolina Street and Aleea Rozelor (figure 1), in the city of Iasi, Romania. Nicolina Street is a 2nd category street, according to STAS 10144 / 1-90 [5], with four lanes, and it is bordered by collective houses on both sides, of five storeys high. Road traffic is continuously pulsating, the number of vehicles registered over one day being over 11 000. According to the noise map of Iasi, the noise level is very high near the buildings on this street (L_{zsn}> 75dB) [6]. The measurement for the noise map study were made at a time when there was no noise barrier in the area. In 2015, along with the street rehabilitation, a protective barrier was mounted on the sidewalk, with the purpose of limiting the propagation of traffic noise.

Figure 1. Studied area in Iasi, Romania [4].

According to the measurements made in this area, the following values of the equivalent noise level were determined:

- at the road edge – 74.17 dB(A)
- behind the acoustic panel – 65.62 dB(A)
- next to the building (at a distance of 2.00m) – 56.75dB(A)

It can be noted that the noise levels exceed the legal limits, both at the road edge (legal limit of 70 dB(A) according to STAS 10009-88 [7]), and also next to the facade of the residential building (legal limit of 50 dB(A) according to C 125-133 [8]).

Starting from the analysis of the real situation presented above, an urban area with the same characteristics was simulated:

- buildings have a height of 16 m, the same as a 5-storey block of flats;
- the distance between the two opposite buildings is 40 m;
- the street was considered to be of 2nd category, connecting road, with four lanes and tram line;
- the distance between the street and the buildings is 12.00 m.

According to the Technical Norms of 27.01.1998, regarding the design and construction of streets in urban areas, a traffic intensity of 360-600 automobiles / hour / band [9] is considered for a 2nd category street. The simulations were made using SoundPlan. To calculate the emission level, the software needs some input factors: the number of lanes and their width, the type of road surface, the number of light and heavy vehicles that pass throughout the street, the type of trams that are used (witch you can chose from a library) and the maximum legal speed [10]. The simulations took into account the minimum number of cars scheduled (360). The number of trams passing through the area was considered to be 1 tram / 30 minutes. The simulations only took into account the day period (07:00 – 19:00).

The purpose of these tests was not to determine exactly the noise level in the area but to simulate the introduction of noise mitigation measures and to highlight the difference between the noise level at the pedestrian height of 1.70 m, near the building, for each of the tests. The pedestrian was considered at a distance of 2.00 m from the building.

3. Tests results

Figure 2 shows the simple street situation with 4 lanes and a tram line without any noise mitigation measures. It is noted that the noise level recorded near buildings is between 74 and 76 dB(A). The following test implies the placement of a green space with a width of 5 m and a height of 1.50 m. A decrease of the noise level is observed, having values between 70 and 72 dB(A) next to the building (figure 3).

The third test takes into consideration the setup of a noise barrier at the edge of the pavement, with a height of 2.00m. The program has the option of choosing a slightly absorbent barrier with a noise reduction level of 4 dB(A), which corresponds to an absorption coefficient $\alpha = 0.6$. In this test, there is a decrease in the noise level recorded near the building, which is between 68 and 70 dB(A) (Figure 4).

The fourth test takes into account the implementation of both noise mitigation measures (1.50m green space and 2.00m noise barrier) resulting in a noise level between 66 and 68 dB(A). It can be noticed that the implementation of both noise protection measures has the greatest impact on it, the decrease of the noise level being of approx. 8 dB(A) (Figure 5).

The second series of simulations tests the impact that the reduced traffic capacity in the area might have, by transforming the street into one with only two lanes, the freed space being given to the sidewalk (figure 6). This change results in a noise level near the building of 70-72 dB(A), with 4 dB (A) less than in the 4-lanes situation. As in the first series of tests, the following noise mitigation options were implemented:

- a green space with a width of 5 m and a height of 1.50m (figure 7);
- noise barrier with the height of 2.00m (figure 8);
- both measures of noise mitigation (figure 9).

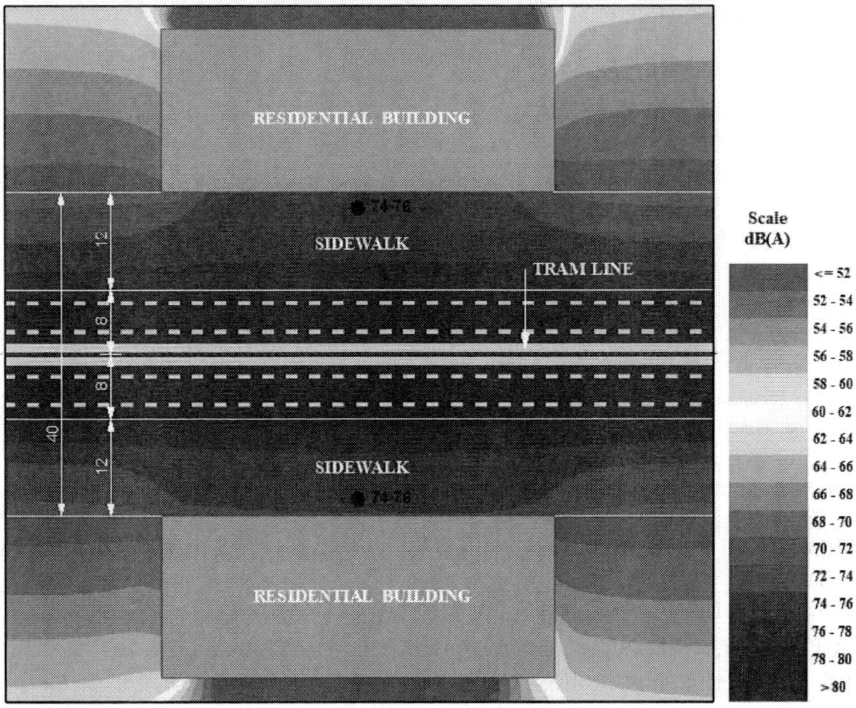

Figure 2. Noise level simulation for a street with 4 lanes and tramway, without noise mitigation measures.

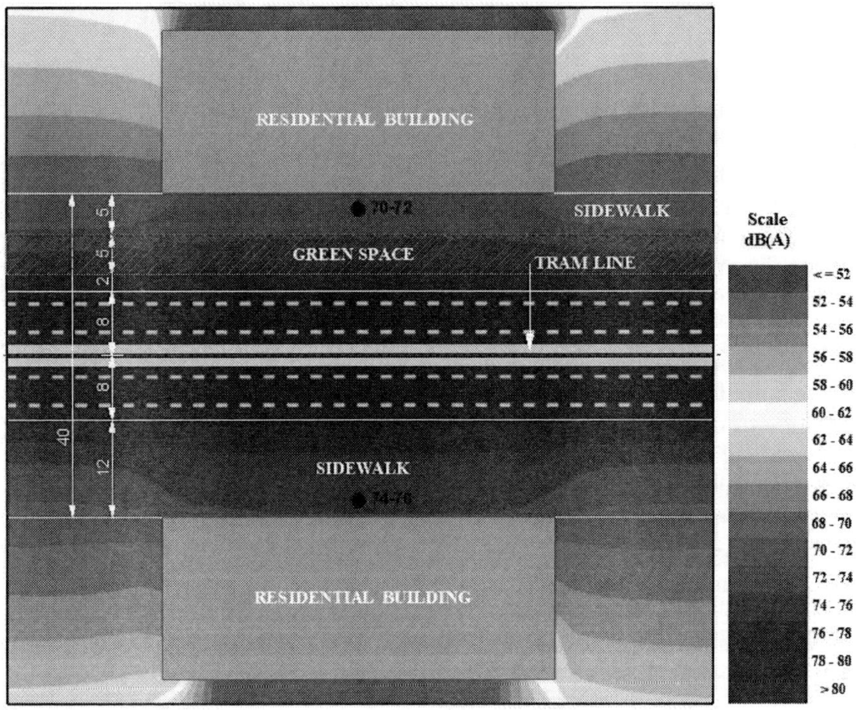

Figure 3. Noise level simulation for a street with 4 lanes, tramway and green space (h = 1.50m).

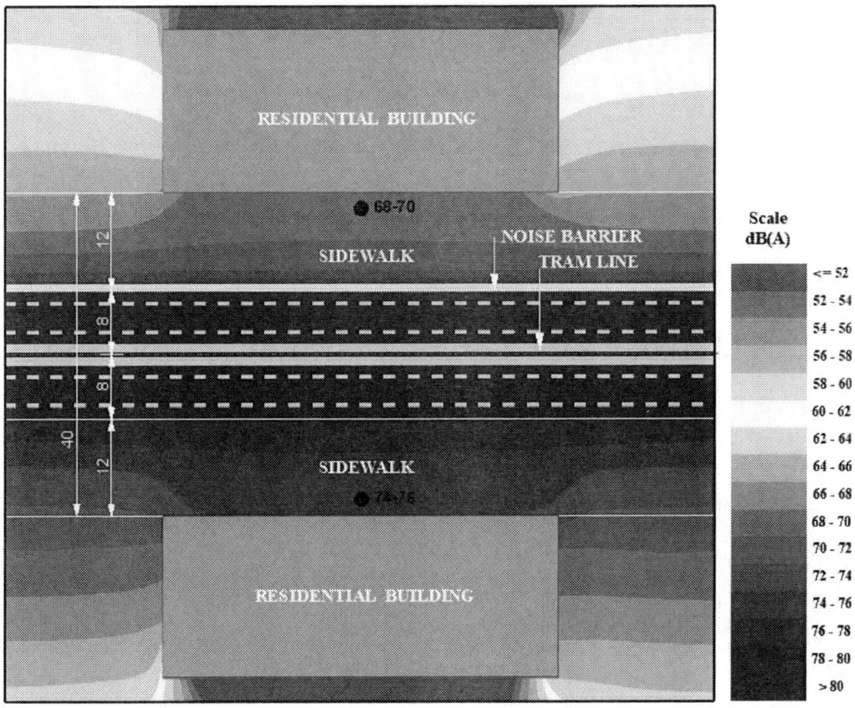

Figure 4. Noise level simulation for a street with 4 lanes, tramway and noise barrier (h = 2m).

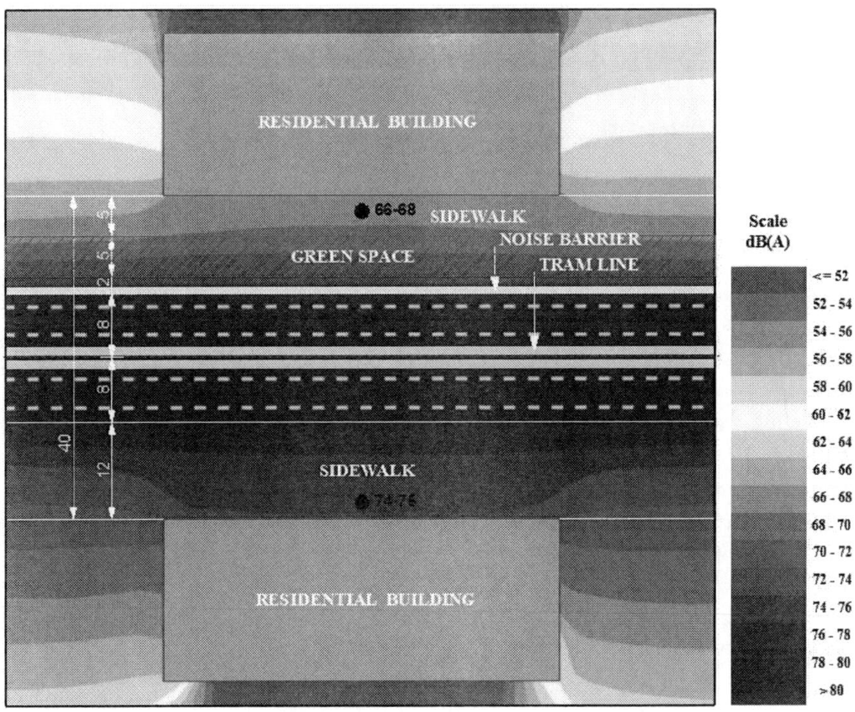

Figure 5. Noise level simulation for a street with 4 lanes, tramway, green space (h = 1.50m) and noise barrier (h = 2m).

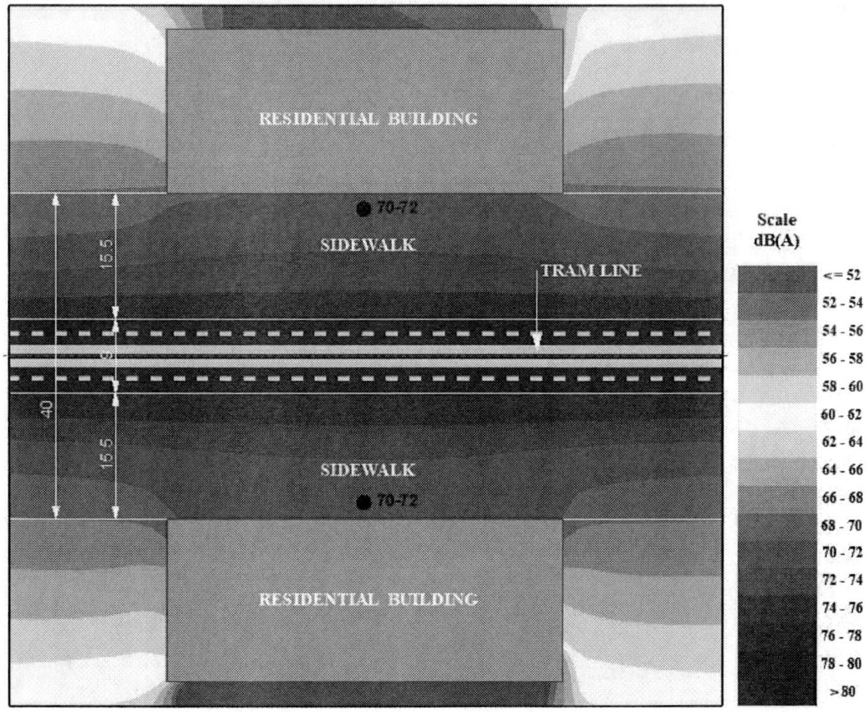

Figure 6. Noise level simulation for a street with 2 lanes and tramway, without noise mitigation measures.

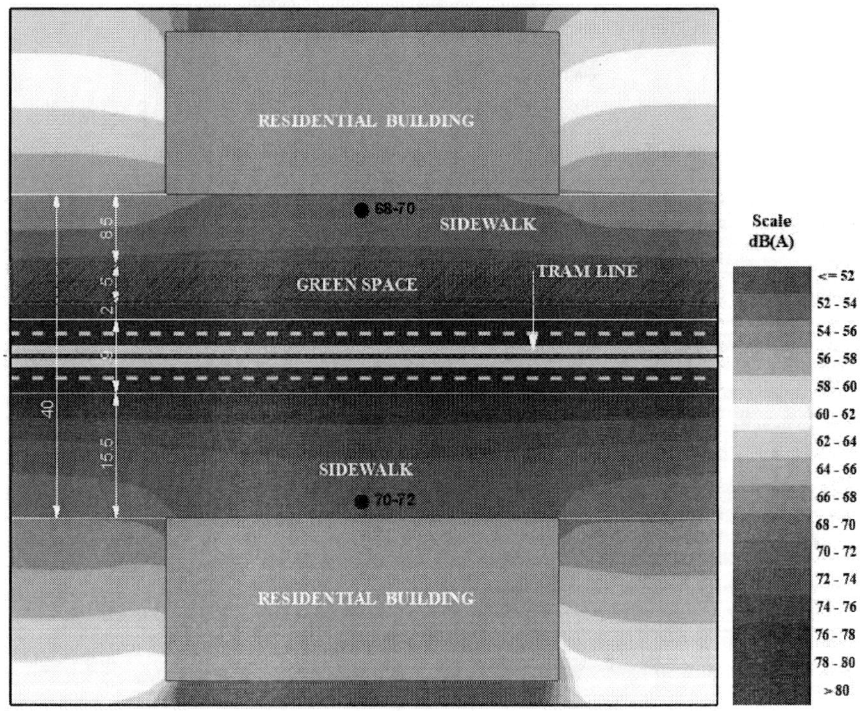

Figure 7. Noise level simulation for a street with 2 lanes, tramway and green space (h = 1.50m).

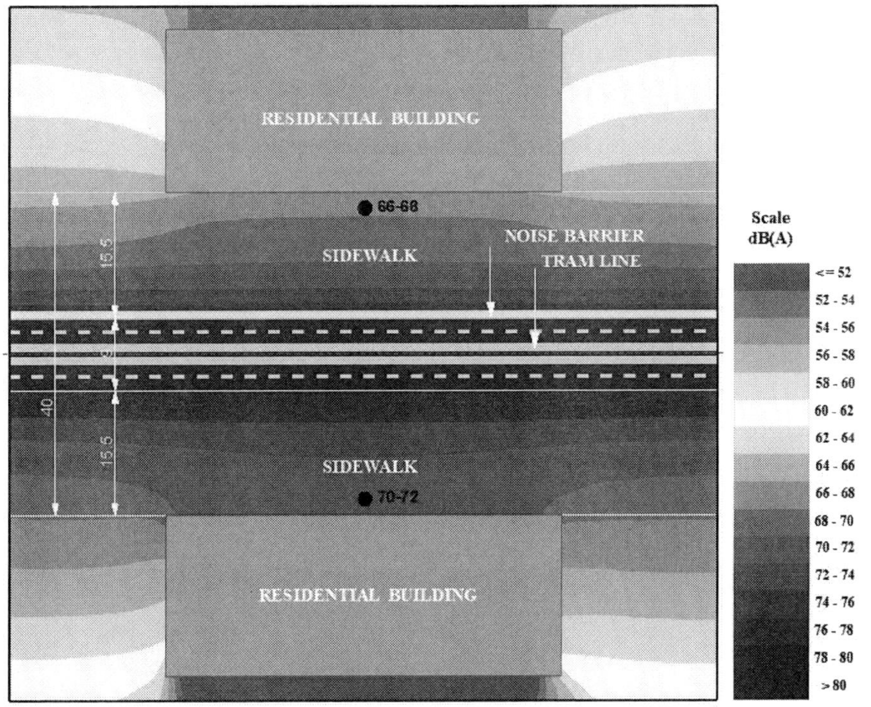

Figure 8. Noise level simulation for a street with 2 lanes, tramway and noise barrier (h = 2m).

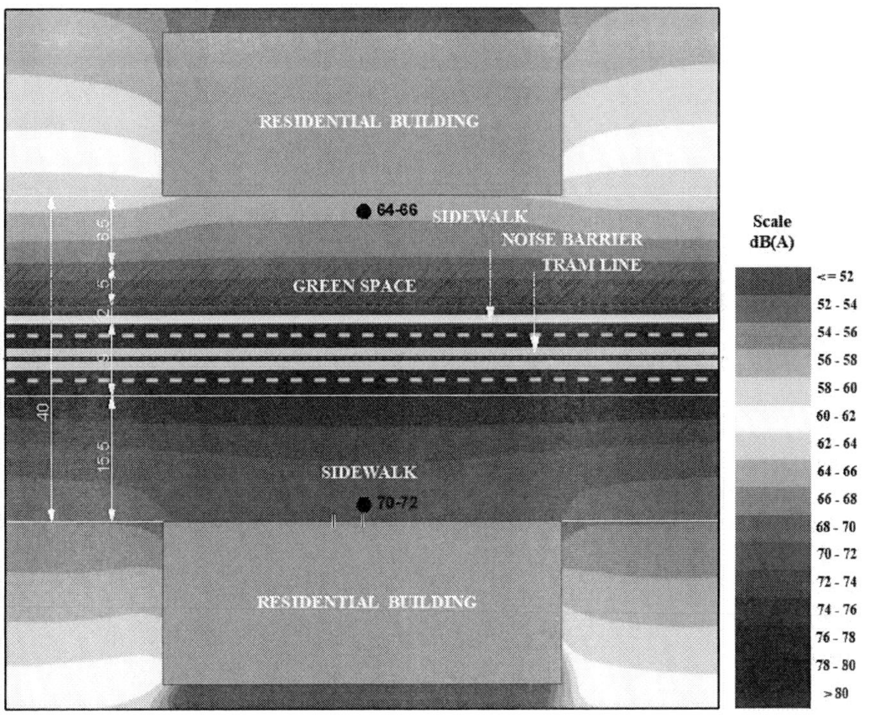

Figure 9. Noise level simulation for a street with 2 lanes, tramway, green space (h = 1.50m) and noise barrier (h = 2m).

By drawing a cross-sectional profile of the street, it can be noted the decrease of the noise level due to the implementation of the noise barrier and the green space. Figure 10 shows the street situation with 4 lanes, and Figure 11 shows the street with 2 lanes. The smallest noise level is 64 dB(A), obtained in the situation of the street with only two lanes, anti-noise barrier and green space with a height of 1.50m. However, the values obtained are above the limit allowed by Norm C 125 - 2013 of 50 dB(A), measured at 2.00 m distance from the residential building.
From these tests it can be concluded that if we would consider the reduction of the road surface to only two lanes, than the recorded level (without protection measures) would be 4 dB(A) lower than in the case of the street with 4 lanes. In case of implementing the green space and the noise barrier, the recorded pedestrian level would be 64-66 dB(A).

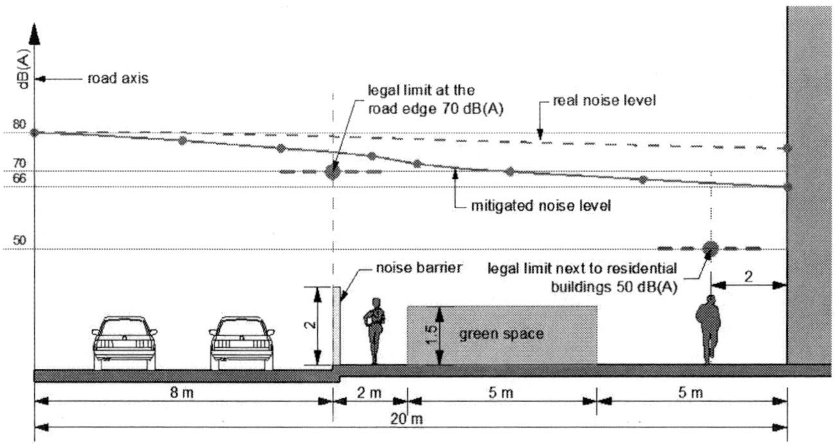

Figure 10. Highlighting the noise level - cross-sectional profile of 4 lanes street with barrier and green space (h = 1.50m).

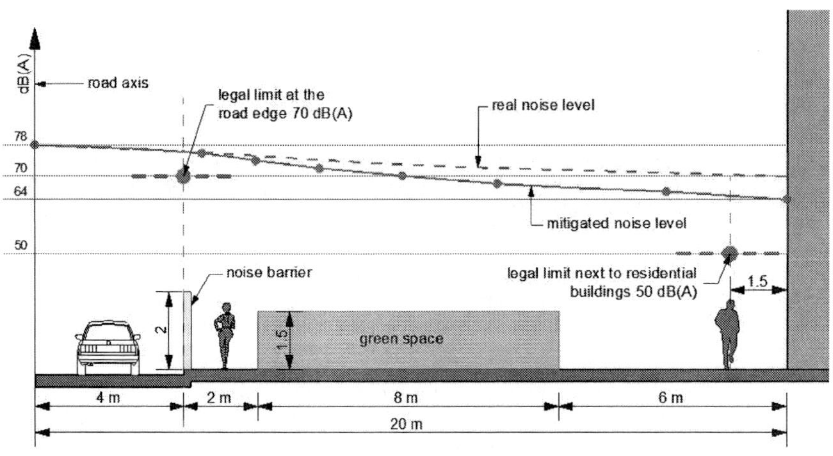

Figure 11. Highlighting the noise level - cross-sectional profile of 2 lanes street with barrier and green space (h = 1.50m).

To optimize noise attenuation, it was decided to simulate the same situations, but with the implementation of a green space of 3.00m height. This change led to the next distribution of the noise level presented separately for the 4-lanes street (Figure 12) and the 2-lanes street (Figure 13).
By drawing the cross-sections of these situations, a minimum noise level of 62 dB(A), for the 4-lane street and 58 dB(A), for the 2-lane street (Figures 14 and 15), is observed. Figure 16 presents

comparatively all the cross-sectional profiles of this road, with the highlighting of the noise level obtained in the simulations.

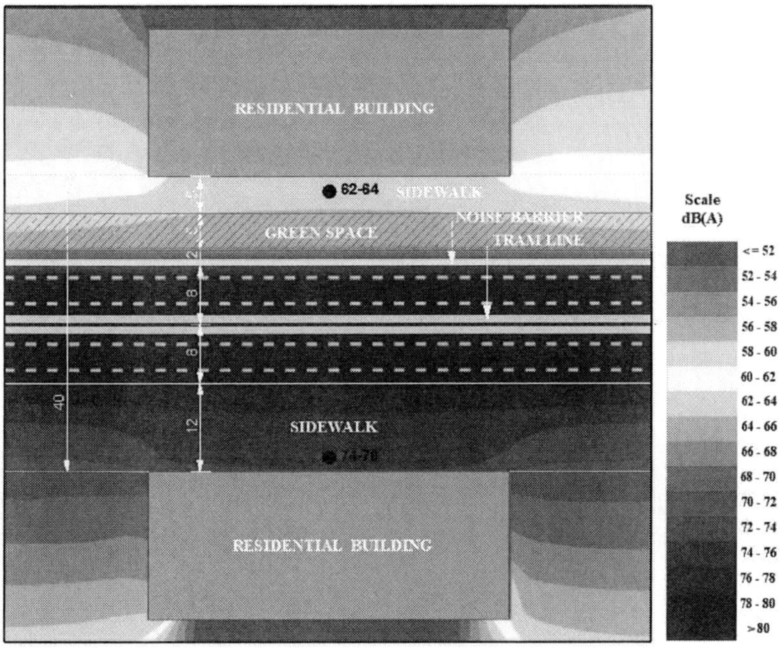

Figure 12. Noise level simulation for a street with 4 lanes, tramway, green space (h = 3.00m) and noise barrier (h = 2m).

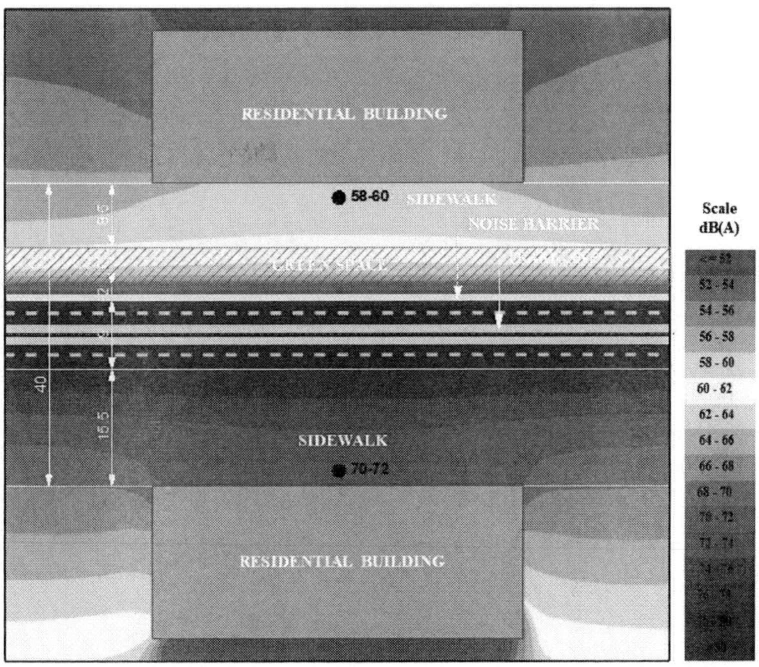

Figure 13. Noise level simulation for a street with 2 lanes, tramway, green space (h = 3.00m) and noise barrier (h = 2m).

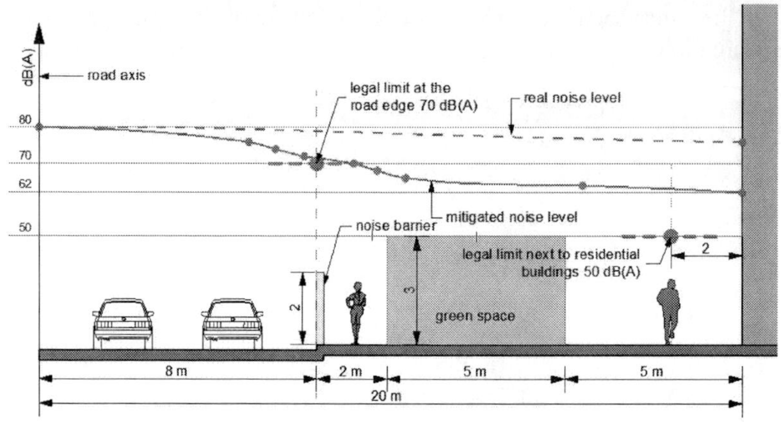

Figure 14. Highlighting the noise level - cross-sectional profile of 4 lanes street with barrier and green space (h = 3.00m).

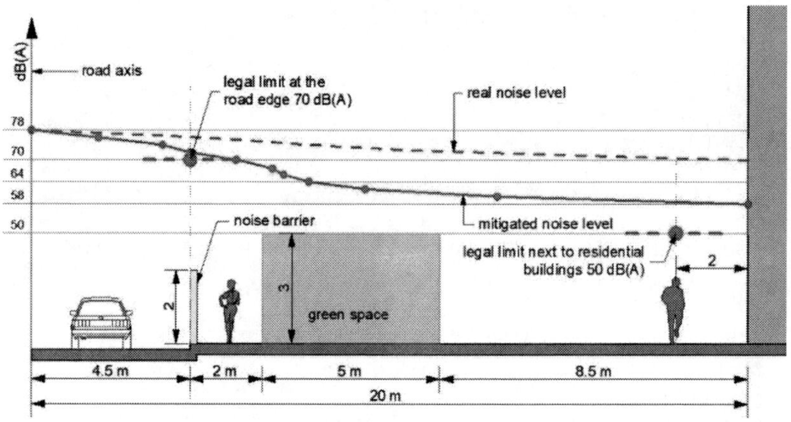

Figure 15. Highlighting the noise level - cross-sectional profile of 2 lanes street with barrier and green space (h = 3.00m).

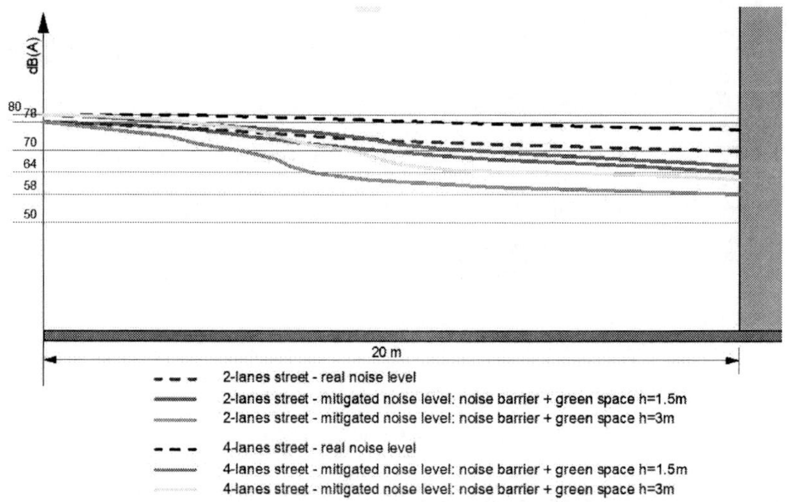

Figure 16. Highlighting the noise level - cross-sectional profile for all the simulations.

IOP Conf. Series: Materials Science and Engineering **586** (2019) 012015 doi:10.1088/1757-899X/586/1/012015

By comparing the cross sections for this street type, with the distance between the road axis and the residential buildings of 20m, it is noticed that the greatest impact on the noise level was achieved in the case of reduced traffic surface and the implementation of noise protection measures: noise absorption barrier (with a noise reduction level of 4 dB(A)) and green space with a height of 3.00 m.

4. Conclusions

Of all these tests, it can be concluded that the implementation of a green space of 3.00m height and a 2.00m noise barrier without considering a very absorbent material leads to a decrease in the level of pedestrian noise of around 12 dB(A) both for the 4 lane street and for the 2 lane street. If we were to consider the mitigation of the street capacity, from 4 lane to 2 lane street, than we could achieve a 16 dB(A) decrease in noise level, its value being 58-60 dB(A).
If the height of the green space is only 1.50m (the mean height of a hedge), the influence on the noise level is insignificant, as can be seen in figures 10 and 11. In the case of a green space height of 3.00m, the minimum noise level is 58 dB(A), while, when the green space is 1.50m high, the minimum noise level is 64dB(A). In order to have a positive impact in terms of acoustic comfort, the green space must significantly exceed the height of the acoustic barrier.
From these simulations it can be concluded that in order to reach a comfortable noise level that respects the legal limit, it is recommended to decrease traffic intensity (through various interventions on road surface capacity, promoting the public transport, etc.), to implement noise barriers of materials with a high sound absorption capacity (soft porous materials such as organic fibers or foam plastics) and a green space with a height that exceeds the height of the noise barrier and has a high density of leaves.

5. References

[1] Nieuwenhuijsen M J, Kruize H, Gidlow C, Andrusaityte S, Antó J M, Basagaña X, Cirach M, Dadvand P, Danileviciute A, Donaire-Gonzalez D, Garcia J, Jerrett M, Jones M, Julvez J, van Kempen E, van Kamp I, Maas J, Seto E, Smith G, Triguero M, Wendel-Vos W, Wright J, Zufferey J, van den Hazel P J, Lawrence R, Grazuleviciene R 2014 Positive health effects of the natural outdoor environment in typical populations in different regions in Europe (PHENOTYPE): a study programme protocol *BMJ Open* http://bmjopen.bmj.com/content/4/4/e004951.full

[2] Lebiedowska B 2005 Acoustic background and transport noise in urban areas: a note on the relative classification of the city soundscape *Transportation Research Part* ***D*** **10** p 341-45

[3] Schust M 2004 Effects of low frequency noise up to 100 Hz *Noise & Health* **6** 23 p 73-85

[4] Yamamoto K 2015 Japanese experience to reduce road traffic noise by barriers with noise reducing devices *Euro Noise* Maastricht p 33-38

[5] Vér I, Beranek L 2006 Noise and vibrations control engineering. Principles and applications *New Jersey: John Wiley&Sons, Inc. Hoboken*

[6] Amares S, Sujatmika E, Hong T, Durairaj R, Hamid H 2017 A review: Characteristics of noise absorption material *Journal of Physics*: Conference Series **908** conference 908 012005

[7] Arenas J, Crocker M 2010 Recent trends in porous absorbing materials *Sound and Vibration* **44** p 12-17

[8] Bujoreanu C, Nedeff F, Benchea M, Agop M 2017 Experimental and theoretical considerations on sound absorption performance of waste materials including the effect of backing plates *Applied Acoustics* **119** p 88-93

[9] Del Rey R, Uris A, Alba J, Candelas P 2017 Characterization of sheep wool as a sustainable material for acoustic applications *Materials* **10 (11)** 127, https://doi.org/10.3390/ma10111277

[10] Sikora J, Turkiewicz J 2010 Sound absorption coefficients of granular materials *Mechanics and Control* **29 (3)** p 149-57

[11] Gehl J 1987 Life between buildings, using public space. *New York: Van Nostrand Reinhold*

[12] Bourdeau E, Zheng X, King E, Pilla F 2015 An assessment of noise exposure along the High Line, New York City *International Congress on Noise Control Engineering, Internoise*

[13] Ştefan (Oancea) I and Budescu M 2016 An assessment of the noise level in a residential area crossed by an intensive traffic road - comparative analysis between on-site measurements and legal standards *Advanced Engineering Forum* ***21*** *Trans Tech Publications Switzerland* p 544-50
[14] STAS 10144/1-90 1990 Străzi. Profiluri transversale. Prescripţii de proiectare (Streets. Cross-section profiles. Drawing guides.)
[15] S.C. Institutul de Cercetări în Transporturi - INCERCTRANS S.A. 2014 Reactualizare hartă de zgomot pentru Municipiul Iaşi (Updating of noise map for Iasi County)
[16] STAS 10009-88 1988 Acustica în construcţii. Acustica urbană. Limite admisibile ale nivelului de zgomot (Building acoustics. Urban acoustics. Legal limits of noise level)
[17] C 125-2013 Normativ privind acustica în construcţii şi zone urbane *Monitorul Oficial al României* partea I nr. 812 bis/20.XII.2013 (Norm regarding building and urban areas acoustics)
[18] Normă tehnică din 27.01.1998 Proiectarea şi realizarea străzilor în localităţile urbane (Technical Norm. Design and construction of streets in urban areas)
[19] http://www.soundplan.com/soundplan_en-master.pdf

IOP Conf. Series: Materials Science and Engineering **586** (2019) 012016 doi:10.1088/1757-899X/586/1/012016

Lighting systems calculations for heritage buildings

C D Galațanu[1]

[1]Gheorghe Asachi" Technical University of Iași, Faculty of Civil Engineering and Building Services, 700050 Iași, Romania

E-mail: catalin.galatanu@tuiasi.ro

Abstract. The heritage buildings generate specific problems for electrical installations, with supplementary request from lighting systems. The continuous trends towards modernization are traditionally in conflict with the conservation of the architectural details. This major constraint must be considered for new circuits, but also for luminaire, when retrofit is the main approach. After the esthetical considerations, the engineering approach raises a problem in connection with the dimensioning methods. For heritage buildings, the lighting systems must cover infinity of situation, very different from classical designing methods, based on Utilization Factor or Direct Illumination Calculus for orthogonal situations. An original method of calculation is proposed, based on analytical geometry, useful for any orientation of the working plane. Even this method has important limitations, generated by the specifically architectural details, and a lighting simulation is necessary. Based on DIALuxEVO7, specific methods are presented. The importance of results interpretation is also underline.

1. Introduction

The heritage buildings are a subject of a continuous evolution, together with the society. The main parameter that imposes the rhythm of the evolution is the energy consuming, in relation with the price of the energy. All others phenomenon could be placed in the second place: comfort level and safety, water management, diversity of pollution involved and so on. Under the general research agenda of historic buildings and energy efficiency, in [1] one discover the importance of the field: "To refurbish historic buildings on a sustainable basis implies that environmental, economic as well as social and cultural aspects are taken into account. The interdisciplinary knowledge needed to handle these aspects has called for increased research activity. This demand has been reflected in several calls from the Joint Programming Initiative (JPI), as well as the EU's Seventh Frame-work Program and Horizon 2020. Several successful recent projects have focused on the refurbishment potential in historic buildings. Most of them have been characterized by topics regarding technical measures, careful passive solutions and the conversion to fossil-free energy sources." Until now, this approach was the only one. But in the same paper [1], a strong affirmation was found: "There are significant differences between the modelled (theoretic) and actual energy performance of buildings in general. The difference between the calculated energy consumption and the actual energy consumption can exceed 100% in extreme cases. User behavior is considered to be the main reason behind this significant gap between the actual and predicted energy performance of buildings." These discover are stronger than energy methodology [2], serving as a starting point for new considerations.

Another gap between calculation methods and the specificity of heritage building is available in [3], where the natural lighting was analyzed together with artificial lighting and thermal comfort: "It is not easy to select the optimal window system that provides buildings the best balance between solar energy transmission and transmission of daylight. Several calculation methods have been developed to guide engineers and architects to choose a right solution of window system, that results in good indoor climate low energy consumption and overall visual comfort. Light availability analysis characterizes every daylight system, determining several factors such as site latitude, urban context, prevailing seasonal climatic conditions (temperature, solar irradiation, solar altitude, clear sky probability), building orientation, building openings, etc."

In [3], the complexity of ancient building is again demonstrated: "the 3D model of the Hall of Two Hundred was carried out taking into account all geometric and architectural characteristics of the building, its walls, its furnishings and the outside urban context. The present furniture (consisting of 60 seats fixed at a base upright, and wooden benches along the two walls without windows) were modelled with a simplifying geometry, disregarding ornaments and inlays, the monumental marble ornaments at the south doors, but attributing the specific thermo-physical, optical, photometric and colorimetric properties to each different surface. Protrusions of the coffered ceiling were disregarded, but their surface was modelled by an equivalent surface built using the average weighted value of the reflective coefficient of multiple surfaces. These simplifications were assumed because of RAM capacity of the computer used and computational time reduction, guaranteeing the quality of the results with a system of a higher number of simulation points."

All these arguments underline the complexity of possible calculation required by the ancient buildings. The lighting systems are in a continuous modernization, due the solid state "revolution", and the calculus methodology must be reconsidered.

The ancient buildings raise also a classical problem, a conflict between the newest technology and the condition of no architecture modifications. The solution imposes the Wi-Fi technology and IoT [4] for intelligent evacuation guidance systems for improving fire safety (introducing intelligent fire safety), or photoluminescent way finding [5].

The wide range of problems rise from lighting methods in heritage buildings covers also the fresco structural problem [6] or regarding the spectral damage for lighted museum paintings [7] or [8]. The spectral composition has not only negative effect, but also a positive one, as a tool to enhance faded colors of museums artefacts [9].

Of all these obvious problems, the specificity of calculations problems for lighting systems in heritage building are the most surprising. In figure 1 one presents such example:

Figure 1. Lighting optimization for arches and vaults in Cotroceni Palace, Bucharest

The specificity of lighting configuration from Fig.1 is the surface that must be optimized: not the working plane (horizontal), but the curves one, from arches and vaults. In the same category one has dome illumination, with artificial or natural lighting. The main problem will be to find a way to represent somehow the results, not only in qualitative manner (false colors) but in quantitative manner, in association with a calculation surface.

2. New calculation method for direct illumination

The classical methodology for direct illumination from a luminaire is available in a classical monographic book [10, 196-198], but huge limitations are known, being applicable only for horizontal or vertical surfaces. A historical building has many other situations, where classical approach is not useful. A new method, presented by the author in [11], is based on analytical geometry. The geometrical considerations are presented in Fig.2, and the method could be extended for any other shapes, considering dedicated software, as MATLAB (see below).
The calculation surface is generated similar with a mesh, by the points:

X1, X2, … , XI, …, XMAX where *NX* is the number of values;
Y1, Y2, … , YJ, … , YMAX where *NY* is the number of values;
Z1, Z2, …, ZK, … , ZMAX where *NZ* is the number of values.

The ration of these vectors will be constant, but not necessary:

$$\Delta X = \frac{X_{MAX} - X_1}{N_X - 1}$$

$$\Delta Y = \frac{Y_{MAX} - Y_1}{N_Y - 1} \qquad (1)$$

$$\Delta Z = \frac{Z_{MAX} - Z_1}{N_Z - 1}$$

In this point on calculates the projection in horizontal plane of the vertical ratio increment (figure 2):

$$\Delta ZO = \Delta Z \times ctg\alpha \qquad (2)$$

The classical methodology supposes that the angles could be measured, but one propose to calculate β angle in horizontal plane, between intersection of the mesh and horizontal plane with the OX axis:

$$\beta = atg\frac{X_{MAX} - X_1}{Y_{MAX} - Y_1} \qquad (3)$$

The vertical ratio for the considered mesh:

$$ZOT = \Delta ZO \times (N_Z - 1) \qquad (4)$$

The *X, Y, Z* coordinates for current calculus point *I, J, K*:

$$X_{I,J,K} = X_1 + (I-1)\times \Delta X + \Delta ZO \times (k-1)\cos\beta$$

$$Y_{I,J,K} = Y_1 + (J-1)\times \Delta Y + \Delta ZO \times (k-1)\sin\beta \qquad (5)$$

$$Z_{I,J,K} = Z_1 + \Delta Z \times (k-1)$$

With these coordinates one calculates the current distance between every calculation point and luminaire position:

$$DPO_{I,J,K} = \sqrt{(X_{I,J,K} - XC)^2 + (Y_{I,J,K} - YC)^2} \qquad (6)$$

The angle γ is defined in vertical plane, between optical axis of the luminaire and current calculation direction, respectively on which the light intensity will be produce the direct illumination. The lighting intensity will have two components, longitudinal $I(\gamma)_{I,J,K}^{long}$ and transversal $I(\gamma)_{I,J,K}^{transv}$, except the situation when LDT files or IES files are used:

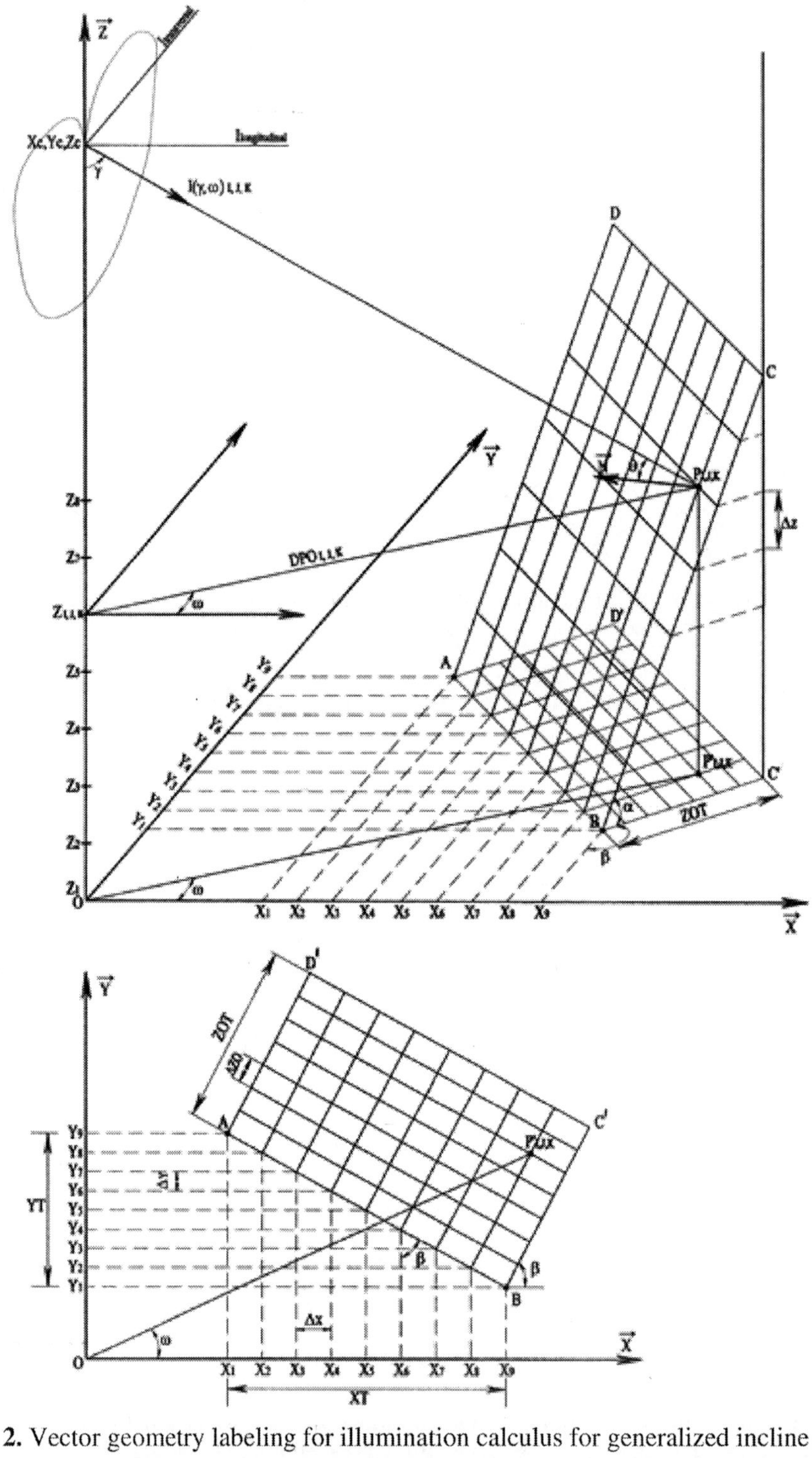

Figure 2. Vector geometry labeling for illumination calculus for generalized incline surface

IOP Conf. Series: Materials Science and Engineering **586** (2019) 012016 doi:10.1088/1757-899X/586/1/012016

$$\gamma = atg\frac{DPO_{I,J,K}}{ZC - Z_{I,J,K}} \tag{7}$$

In horizontal plane on define angle ω, between the projection of the luminous ray and the calculation point $(DPO_{I,J,K})$ and OX axis:

$$\omega = atg\frac{Y_{I,J,K} - YC}{X_{I,J,K} - XC} \tag{8}$$

The angle ω could be useful for weighted average between those two lighting intensity $I(\gamma)_{I,J,K}^{long}$ and $I(\gamma)_{I,J,K}^{transv}$, respectively:

$$I(\gamma,\omega)_{I,J,K} = \frac{I(\gamma)_{I,J,K}^{transv} \times \omega + I(\gamma)_{I,J,K}^{long} \times (90 - \omega)}{90} \tag{9}$$

For illumination calculus is necessary the angle between light intensity $I(\gamma,\omega)_{I,J,K}$ and the normal on the surface *Z*. This is the point where the classical theory wasn't capable to generate solution. Using vector algebra, the normal on the surface *Z* could be described using angles α and β:

$$\vec{N} = N_X\vec{X} + N_Y\vec{Y} + N_Z\vec{Z} \tag{10}$$

Where

$$\begin{aligned} N_X &= \cos\alpha\sin\beta \\ N_Y &= \cos\alpha\cos\beta \\ N_Z &= \sin\alpha \end{aligned} \tag{11}$$

For the current calculus point *(I, J, K)* one obtains the direction from luminaire to this point:

$$\vec{D}_{I,J,K} = X_{I,J,K}\vec{X} + Y_{I,J,K}\vec{Y} + (ZC - Z_{I,J,K})\vec{Z} \tag{12}$$

And now the angle between luminous ray and the normal on the surface *Z*:

$$\cos\,\theta = \frac{N_X X_{I,J,K} + N_Y Y_{I,J,K} + N_Z(ZC - Z_{I,J,K})}{\sqrt{N_X^2 + N_Y^2 + N_Z^2}\sqrt{X_{I,J,K}^2 + Y_{I,J,K}^2 + (ZC - Z_{I,J,K})^2}} \tag{13}$$

With all these elements, the direct illumination could be calculated for a maintenance factor M_f :

$$E_{I,J,K} = M_f\frac{I(\gamma,\omega)_{I,J,K}\cos^3\theta}{(ZC - Z_{I,J,K})^2}\; E_{I,J,K} = M_f\frac{I(\gamma,\omega)_{I,J,K}\cos^3\theta}{(ZC - Z_{I,J,K})^2} \tag{14}$$

For multiple luminaires, the effect is additive.

For the first moment, the complexity of the calculus impose to use some general programming language (as MATLAB), but DIALux are expected to give important advantages, offering a complete lighting simulation. Surprisingly, for the ancient buildings, even DIALux EVO7 has some limitations.

3. Arches Modeling In Dialux Evo7

The continuous development of DIALux EVO7 software is a professional solution, for interior and exterior lighting systems. Modeling architectural details is not a problem, grace to the importing facilities from AutoCAD and 3DS. Building the model is not the subject of this paper, but interpreting the results is a different item. In figure 3 one presents such lighting model, for a fresco, situated in a niche. This niche generates the supplementary problems, additional to those impose to the visibility of the art paining: contrast of luminance, reflections, uniformity and so.
The problems occur due the lack of the cylindrical calculation surface (or spherical) in DIALux EVO. The cylinder is realized by elements, and for every element it is possible to have any results (average

Figure 3. The Sabin Balaşa fresco, lighting solution study in Lost Steps Hall, Palace of University "A.I. Cuza" from Iaşi

illumination, minimum, maximum value, luminance also). In Fig.3 one of these cylindrical elements is highlighted. In the author's opinion, it is very difficult to analyze the results on an arch, especially when one imposes high accuracy of the model, using minimum 60 cylindrical elements for a single arch. Again, it is possible to analyze element by element, but the entire luminous situation is impossible to be obtained.

4. Arches modeling in Matlab

The author's functions for direct illumination based on previous method for generalized surfaces bring the possibility to analyze the results on any surfaces, using general graphical function from MATLAB, as surface, mesh or contour. All these functions could associate the results to a plane surface, giving

the possibility to interpret the final global configuration. Demonstrating this, the next example will give the main direction about mesh generation and result visualization.
Mesh generation refers to four main surfaces (as in figure 4), the front surface (the fresco), the superior arch, and two lateral rectangular surfaces.

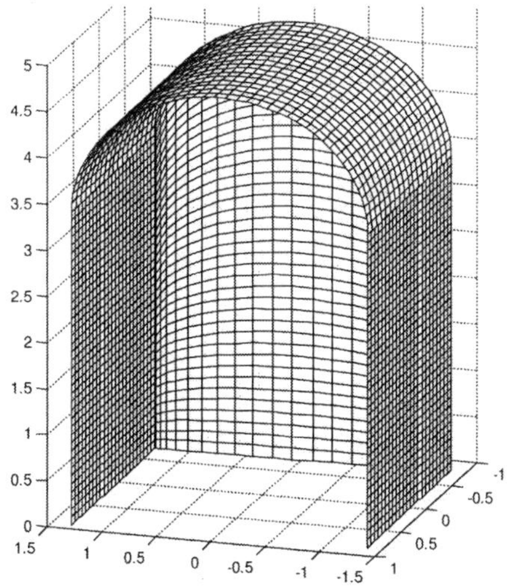

Figure 4. Mesh generated for The Sabin Balaşa fresco (Figure 3)

Mesh generation is the most important phase in building the numerical model. To understand how the superior arch mesh is generated, some lines from MATLAB coding are presented, with some comments:

```
R=1.35;      % Fresco dimension: 2,70m wide
for j=1:19 , for i=44      % elements for Y axis, elements for X axis
    Xarch(j,i)=R*cos((i-1)*pi/43);
    Yarch(j,i)=1.6/2-(j-1)*1.6/18;   % 1.6m is the deepening
    Zarch(j,i)=3.4+R*sin((i-1)*pi/43); % 3.4m is the high of lateral wall
    end, end
```

The vertical surface for the fresco is generated in similar way:

```
for j=1:39 , for i=1:24    % number of elements for vertical dimension, for horizontal ratio
    Xfresco(j,i)=R*cos((i-1)*pi/23);
    Yfresco(j,i)= -.8;
    Zfresco(j,i)=(3.4+R*sin((i-1)*pi/23))/38*(j-1);
    end , end
```

After these surfaces, two rectangular surfaces are generated directly, with mesh grid function, obtaining the mesh as in figure 4. In this point, illumination level is ready to be calculated.

5. Arch lighting modelling in Matlab

The author's function for illumination calculus, applied to the surfaces previous generated, will give the direct and reflected illuminance, as a first step in designing the lighting solution for fresco visibility. A qualitative visual effect could be observed in figure 5, in grey colour map. For the arch area, a special analysis could be realized using function *surf (Xarch, Yarch, Earch)* where illumination level *Earch* is used as a vertical parameter for visualized surface in figure 6. Other functions are available for data analysis, as *contour*.
In this moment the calculated data for the arch zone could be analyzed in a unitary matrix, as seen in figure 6, in total contrast with the situation from DIALux EVO7, as seen in figure 3.

IOP Conf. Series: Materials Science and Engineering **586** (2019) 012016 doi:10.1088/1757-899X/586/1/012016

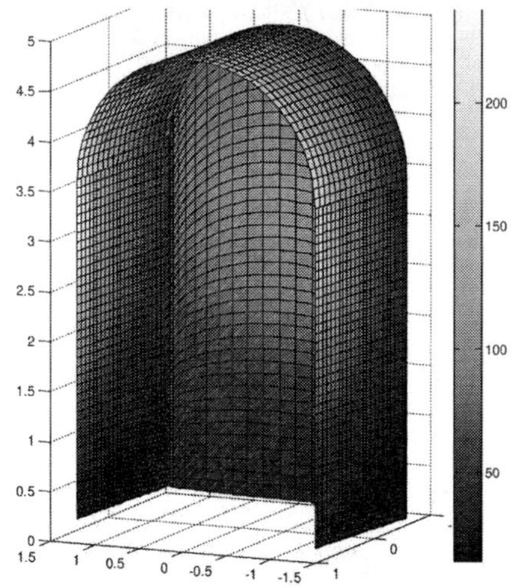

Figure 5. Illuminance level (lux) for a symmetrical lighting source with light intensity of 350 *cd* situated at (0; 1.6; 3)

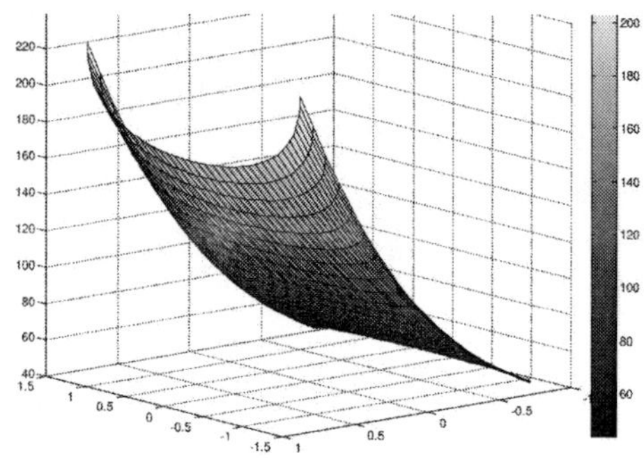

Figure 6. Illuminance level (lux) for arch zone from Fig.5, in a separate representation.

Using mesh generation for any other architectural details, the heritage building could be analyzed in details using MATLAB functions, and for the general lighting simulations the DIALux EVO will generate trustable results.

6. Conclusions

The lighting systems for heritage buildings rise specific problems in connection with preservation of architectural details, with retrofit problems and introduction of new circuits without visual aggression. Using this, other objectives could be solved, as the stability of natural dyes under light emitting diode lamps [12], the strategic objective being the enhancing the appreciation and preservation of archaeological heritage [13] and even the effect of natural light [14]. An unexpected problem, generated by the lighting design for arches, vaults and domes could be solved using vector algebra, implemented by the author in some original MATLAB function, and the results could be analyzed more efficient comparing with DIALUX EVO7. The disadvantage of DIALux is the impossibility to define calculation surfaces other than plane surfaces. The paper presents a general method for lighting design, useful for particular architectural details from heritage buildings.

7. References

[1] Berg F, Flyen Anne-Cathrine, Godbolt A L, Broström T 2017 User-driven energy efficiency in historic buildings: A review, Journal of Cultural Heritage, http://dx.doi.org/10.1016/j.culher.2017.05.009

[2] CEN, 2017 Conservation of cultural heritage – Guidelines for improving the energy performance of historic buildings TC 346, EN 16883, Comité Européen de Normalisation

[3] Balocco C, Calzolari R 2008 Natural light design for an ancient building: A case study, Journal of Cultural Heritage 9 172-178, doi:10.1016/j.culher.2007.07.007

[4] Bernardini G, Azzolini M, D'Orazio M, Quagliarini E 2016 Intelligent evacuation guidance systems for improving fire safety of Italian-style historical theatres without altering their architectural characteristics, Journal of Cultural Heritage 22, 1006–1018, http://dx.doi.org/10.1016/j.culher.2016.06.008

[5] D'Orazio M, Bernardini G, Tacconi S, Arteconi V, Quagliarini E 2016 Fire safety in Italian-style historical theatres: How photolumines-cent way finding can improve occupants' evacuation with no architect-ture modifications , Journal of Cultural Heritage 19, 492–501 http://dx.doi.org/10.1016/j.culher.2015.12.002

[6] Quagliarini E, Lenci S, Seri E 2012 On the damage of frescoes and stuccoes on the lower surface of historical flat suspended light vaults, Journal of Cultural Heritage 13 (2012) 293–303, doi:10.1016/j.culher.2011.11.008

[7] Pinilla S M, Vázquez D, Fernández-Balbuen A, Muro C, Munoz J 2016 Spectral damage model for lighted museum paintings: Oil, acrylic and gouache, Journal of Cultural Heritage 22, 931–939, http://dx.doi.org/10.1016/j.culher.2016.05.005

[8] Lowea B J, Smith C A, Fraser-Miller S J A, Paterson R, Daroux F, Ngarimu-Cameronc R, Ford B, Gordon K 2017 Light-ageing characteristics of Maori textiles: Colour, strength and molecular change, Journal of Cultural Heritage 24, 60–68 http://dx.doi.org/10.1016/j.culher.2016.11.012

[9] Viénota F, Corona G, Lavédrinea B 2011 LEDs as a tool to enhance faded colors of museums artefacts, Journal of Cultural Heritage 12, 431–440 doi:10.1016/j.culher.2011.03.007

[10] Bianchi C 1990 Luminotehnica, Editura Tehnică, Bucureşti

[11] *** GEx012 2015 Ghid de bună practică pentru proiectarea instalaţiilor de iluminat/protecţie în clădiri, Anexa la Ordinul MDRAP-833/2015

[12] Degani L, Gulmini M, Piccablotto G, Iacomussi P, Gastaldi D, Dal Bellod F, Chiantore O 2017 Stability of natural dyes under light emitting diode lamps, Journal of Cultural Heritage xxx, http://dx.doi.org/10.1016/j.culher.2017.02.002

[13] Di Salvo S 2014 Innovation in lighting for enhancing the appreciation and preservation of archaeological heritage, Journal of Cultural Heritage 15, 209–212, http://dx.doi.org/10.1016/j.culher.2013.03.001

[14] Balocco C, Frangioni E 2010 Natural lighting in the Hall of Two Hundred. A proposal for exhibition of its ancient tapestries, Journal of Cultural Heritage 11, 113–118, doi:10.1016/j.culher.2009.02.005

Acknowledgments

This research was supported by UEFISCDI grant PCAI-163CI, contract PN-III-P2-2.1-CI-2018-0965 for imaging measurement of correlated color temperature, grant SILUM-219CI, contract PN-III-P2-2.1-CI-2018-1287 for imaging measurement of luminance, grant MAXLUM-256CI, contract PN-III-P2-2.1-CI-2018-1490 for simultaneous imaging measurement of correlated color temperature and luminance.

IOP Conf. Series: Materials Science and Engineering **586** (2019) 012017 doi:10.1088/1757-899X/586/1/012017

The use of partial blanketed strands in design of prestressed precast beams

I L Stan[1], N Florea[2], P Mihai[2] and V C Tănăsucă[1]
[1] Doctoral School of the Faculty of Civil Engineering and Building Services, „Gheorghe Asachi" Technical University of Iaşi, Romania
[2] Department of Concrete, Materials, Technology and Management, Faculty of Civil Engineering and Building Services, „Gheorghe Asachi" Technical University of Iaşi, Romania

E-mail: stan.ioan.lucian@gmail.com

Abstract. Concrete is considered to be prestressed when it is subjected to a state of strain and deformation before being loaded with payloads from service stage. Efforts and deformations resulted from prestress are so established to counteract the effects produced by the transfer and service loads. Efficient design of prestressed precast beams is based on the rational use of materials: concrete and reinforcement (especially strands). Poor use of materials in design phase may result in inappropriate solutions from an economic point of view. To avoid this aspect partial blanketed strands are used. To obtain a partial blanketed strand the adhesion between it and concrete is stopped punctually, the transfer of tensions being carried out in the contact areas. In the present paper it will be illustrated the stress development on the beam resulting from the use of partial blanketed strands. To highlight these particularities the IDEA StatiCA software was used and, the results presented graphically and tabular revealed the advantages of partial blanketed strands.

1. Introduction

Partial blanketing the prestressed strands towards the ends of a beam reduces the shear strength of the sections in regions where imposed shear forces are the highest. This reduction in shear strength of the members applies to all sections, weather cracked or uncracked in flexure. Partial blanketing is effective for the design of simply supported beams because of the consequent reduction of prestress in the support area, being able to provide significant less compressive stress in the top fibres of the beams and provide the same bearing capacity.

Blanketing (debonding) can be achieved by greasing the strand, coating it with retarder, or covering it with plastic tubing. Greasing or using a retarder is risky because of the possibility of affecting other than the specified strands. Plastic tubing is preferred because it also provides an easy mean of inspection. Traditional soft sheathing is not producing the desired debonding effect due to the lack of rigidity. The rigid polymer proved to an effective way to create more clearance between steel and concrete ensuring a debonding behaviour.

2. Blanketing principles

For the proper usage of partially blanketed strands methodology is necessary to establish the number of debonded strands, the debonded length and the release pattern in correlation with the basic requirements of limiting the tensile stresses in the concrete at the top fibres at transfer and the ability to satisfy the checks imposed by the ultimate limit state and the serviceability limit state. At this point are not established specific rules for applying the partial blanketing methodology, only some general guidelines existing.

Excessive debonding may result in a negative impact upon the structural performance of prestressed precast concrete beams by producing cracking at the extremity during the strands release stage of the production process.

IOP Conf. Series: Materials Science and Engineering **586** (2019) 012017 doi:10.1088/1757-899X/586/1/012017

The general guidelines regarding using partial blanketed strands are presented next:

- the development length is measured form the end of debonded zone;
- the length of debonding zone of any strand shall satisfy limit state with consideration of total developed resistance at any section;
- the exterior strands of each horizontal row shall be fully bonded;
- not more than 25% of the total strands can be debonded;
- no more than 40% of the debonded strands, or four strands, whichever is greater, shall have the debonding terminated at any cross-section;
- not more than 40% of the strands at any one horizontal row will be debonded;
- symmetric debonding about member cross-section centreline is required;
- strands in the corners of stirrups preferably should not be debonded;
- adjoining strands form a layer of strands should not normally be debonded at the same section;
- debonded length of pairs of strands that are symmetrical positioned about the centreline of the member shall be equal;
- shear investigation shall be made in the regard to the reduce horizontal force.

3. Experimental phase

To illustrate the mechanisms that appear by using partial blanketed strands when designing precast prestressed concrete beams, a study was carried out on a number of 25 beams. The IDEA StatiCA software was used to accomplish the simulations.

Given that the design of precast prestressed concrete beams is a meticulous task it was necessary for a number of conventions to be made. By imposing a certain number of parameters to be fixed, we can focus only for the principal parameters that are correlated with the blanketing methodology.

In the Section 3.1. are detailed the fixed parameters and in Section 3.2. are detailed the variable parameters.

3.1. Fixed parameters

From this category the follow parameters are fixed:

- beams length: 25.5 m, static scheme, sections as is showed in figure 1 and figure 2;

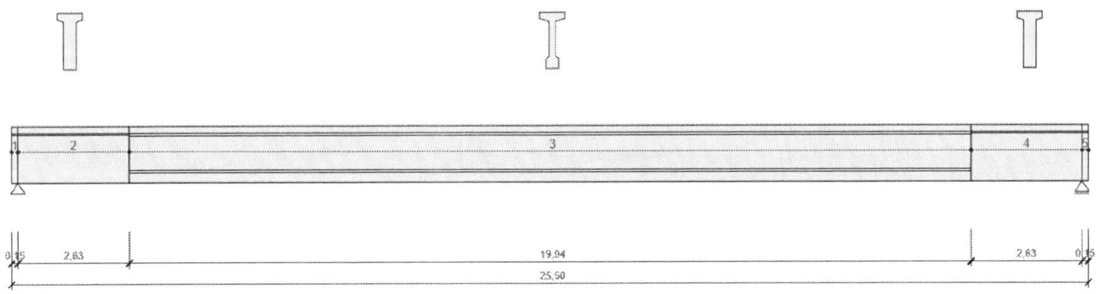

Figure 1. Static scheme

- concrete strength class (according to Eurocode 2): C50/60;
- strands used: Y1860S7-15.2;
- strands layout (a total of 16 strands being positioned in a number of 8 groups composed of 2 strands as it shown in figure 3 and figure 4);
- initial tensile stress in strands: 1100.0 MPa;
- passive reinforcement: steel grade (B500C), layout of passive reinforcement;
- load cases (self-weight, permanent load: 9.1 kN/m, variable load: 15.4 kN/m, load per construction stage G 5: 1.0 kN/m).

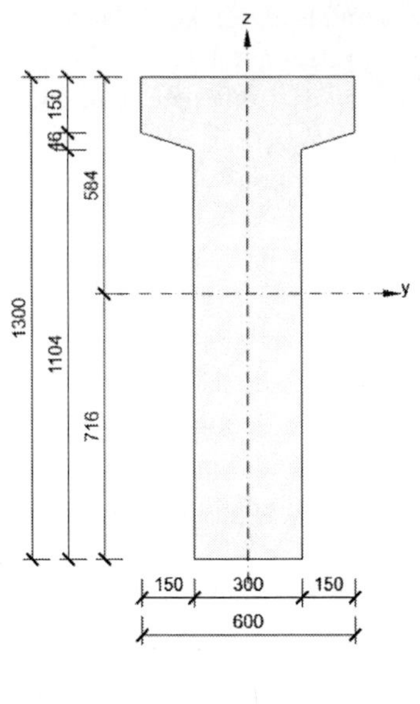

Figure 2. Section 1 as is showed in figure 1 (cm)

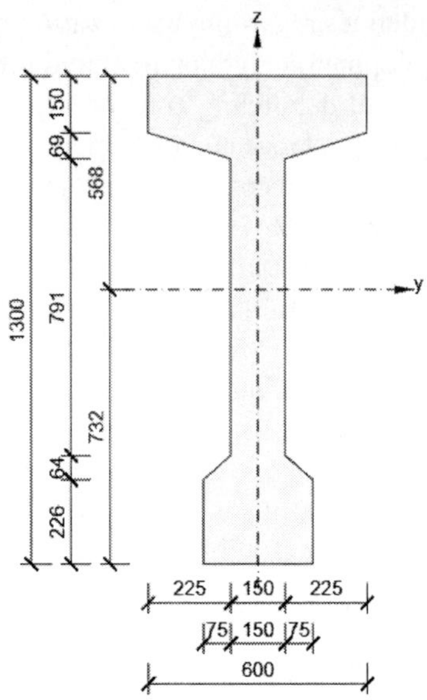

Figure 3. Section 2 as is showed in figure 1 (cm)

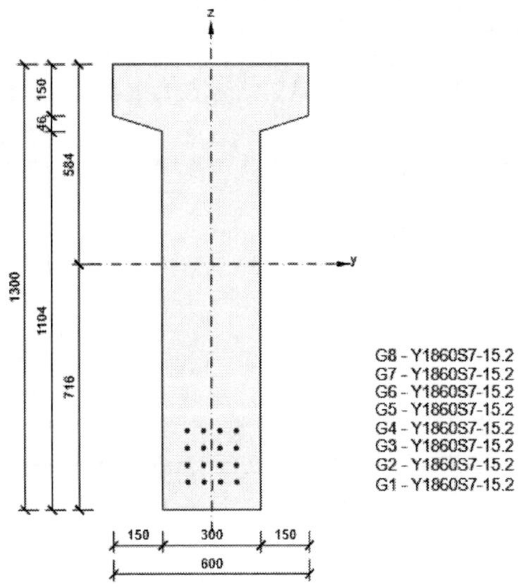

Figure 4. Strands layout in Section 1

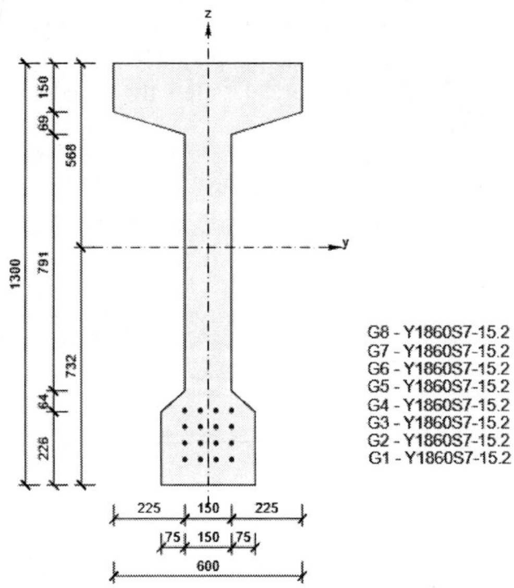

Figure 5. Strands layout in Section 2

3.2. Variable parameters

As it was stated before a study was carried out a number of 25 beams, the only parameters being alter were the positions of debonded strands and their debonded length.

To better understand how the strands groups were considered their positions were illustrated in the cross sections of the beams. The debonded length was presented in tables 1 to 7 for each beam and depending on the strands group.

Beam G 1.0.0 is the reference beam (all strands are bonded), and it will serve to highlight the difference resulted from using partial blanketed strands.

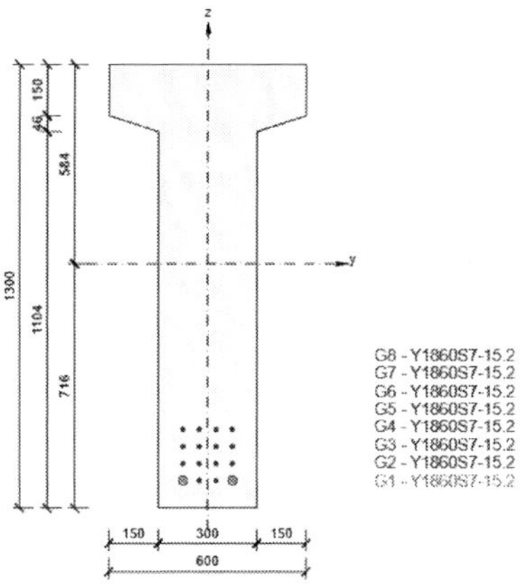

Figure 6. Debonded strands in beams G 1.1.1, G 1.1.2, G 1.1.3

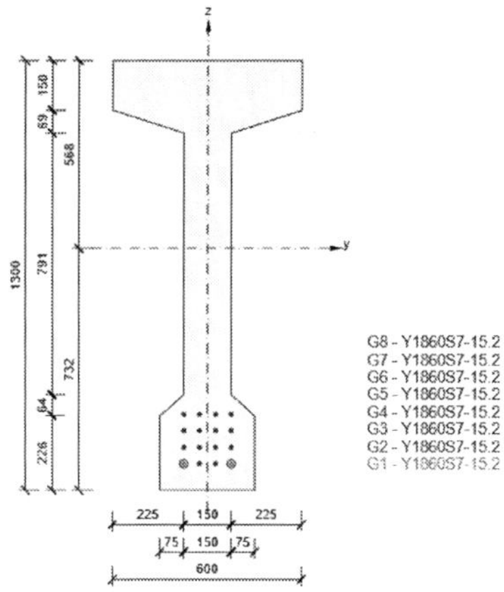

Figure 7. Debonded strands in beam G 1.1.3

Table 1. Debonding lengths (m) / strands group for beams G 1.1.1, G 1.1.2, G 1.1.3

Beam	Strands group							
	G1	G2	G3	G4	G5	G6	G7	G8
G 1.1.1	2.50	-	-	-	-	-	-	-
G 1.1.2	2.75	-	-	-	-	-	-	-
G 1.1.3	3.00	-	-	-	-	-	-	-

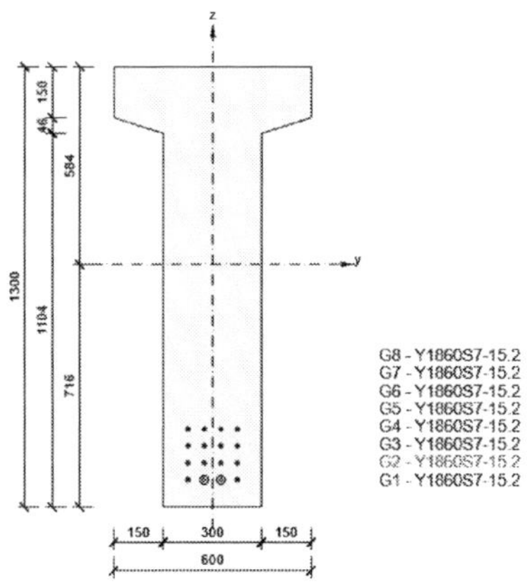

Figure 8. Debonded strands in beams G 1.2.1, G 1.2.2, G 1.2.3

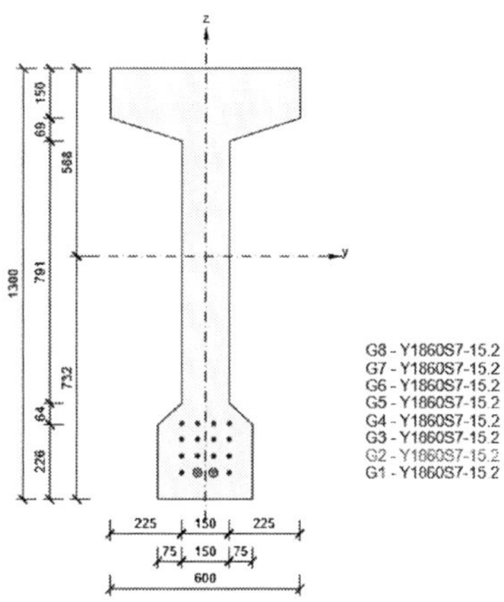

Figure 9. Debonded strands in beam G 1.2.3

Table 2. Debonding lengths (m) / strands group for beams G 1.2.1, G 1.2.2, G 1.2.3

Beam	Strands group							
	G1	G2	G3	G4	G5	G6	G7	G8
G 1.2.1	-	2.50	-	-	-	-	-	-
G 1.2.2	-	2.75	-	-	-	-	-	-
G 1.2.3	-	3.00	-	-	-	-	-	-

IOP Conf. Series: Materials Science and Engineering **586** (2019) 012017 doi:10.1088/1757-899X/586/1/012017

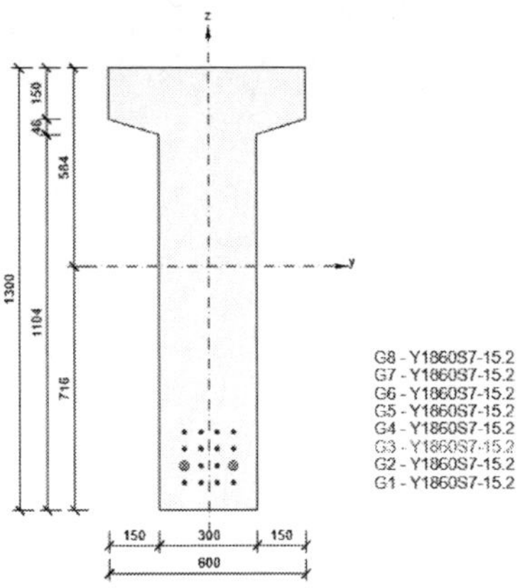

Figure 10. Debonded strands in beams G 1.3.1, G 1.3.2, G 1.3.3

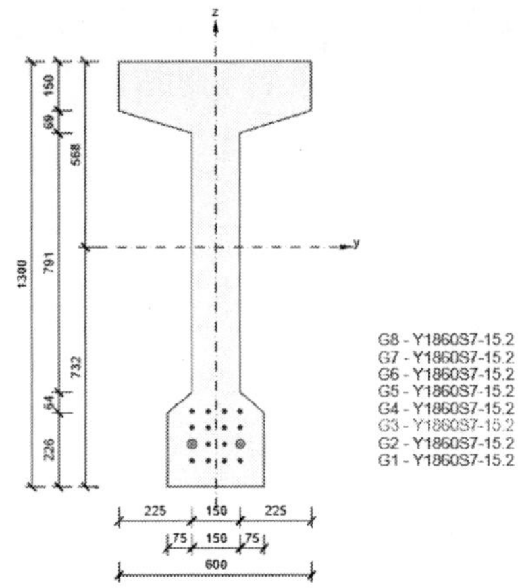

Figure 11. Debonded strands in beam G 1.3.3

Table 3. Debonding lengths (m) / strands group for beams G 1.3.1, G 1.3.2, G 1.3.3

Beam	Strands group							
	G1	G2	G3	G4	G5	G6	G7	G8
G 1.3.1	-	-	2.50	-	-	-	-	-
G 1.3.2	-	-	2.75	-	-	-	-	-
G 1.3.3	-	-	3.00	-	-	-	-	-

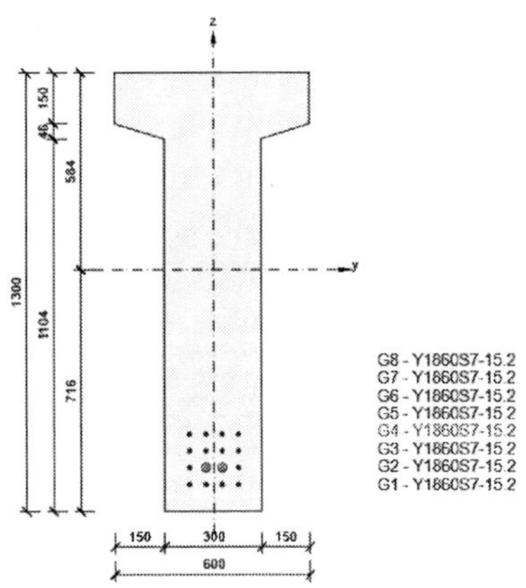

Figure 12. Debonded strands in beams G 1.4.1, G 1.4.2, G 1.4.3

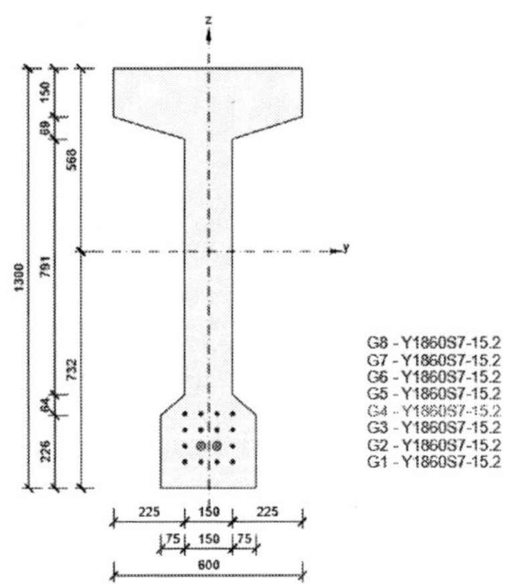

Figure 13. Debonded strands in beam G 1.4.3

Table 4. Debonding lengths (m) / strands group for beams G 1.4.1, G 1.4.2, G 1.4.3

Beam	Strands group							
	G1	G2	G3	G4	G5	G6	G7	G8
G 1.4.1	-	-	-	2.50	-	-	-	-
G 1.4.2	-	-	-	2.75	-	-	-	-
G 1.4.3	-	-	-	3.00	-	-	-	-

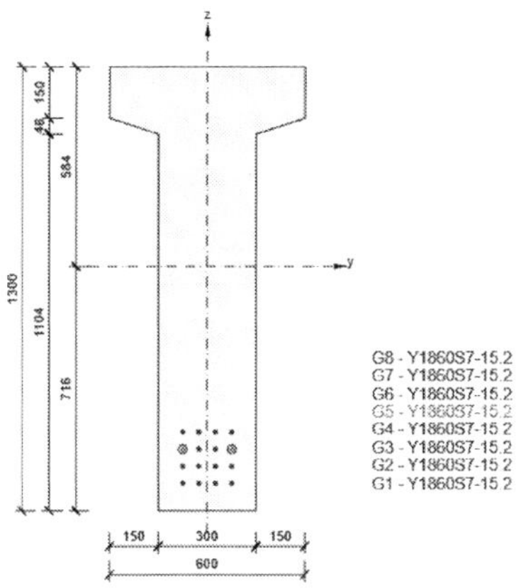

Figure 14. Debonded strands in beams G 1.5.1, G 1.5.2, G 1.5.3

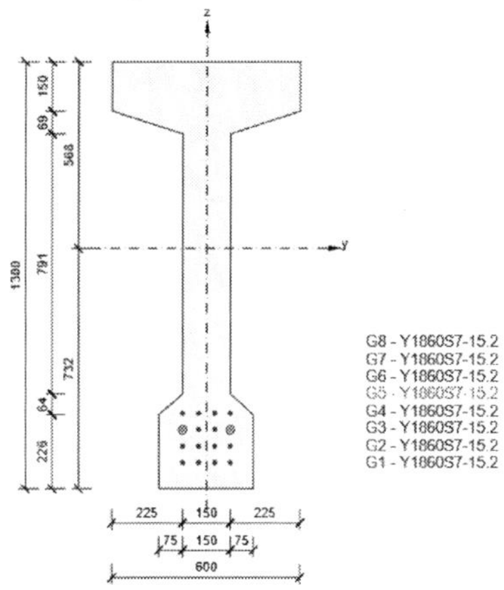

Figure 15. Debonded strands in beam G 1.5.3

Table 5. Debonding lengths (m) / strands group for beams G 1.5.1, G 1.5.2, G 1.5.3

Beam	Strands group							
	G1	G2	G3	G4	G5	G6	G7	G8
G 1.5.1	-	-	-	-	2.50	-	-	-
G 1.5.2	-	-	-	-	2.75	-	-	-
G 1.5.3	-	-	-	-	3.00	-	-	-

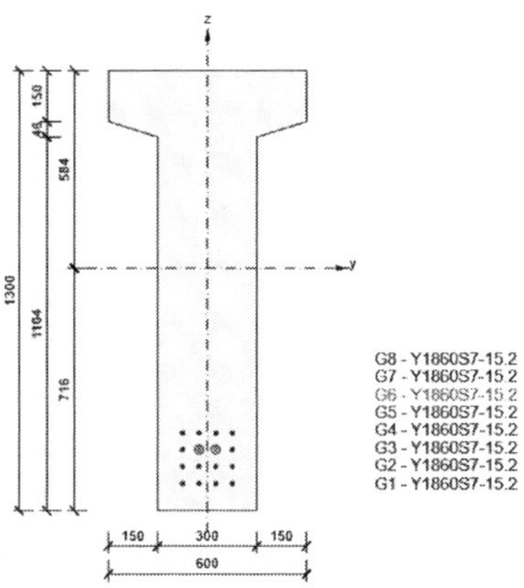

Figure 16. Debonded strands in beams G 1.6.1, G 1.6.2, G 1.6.3

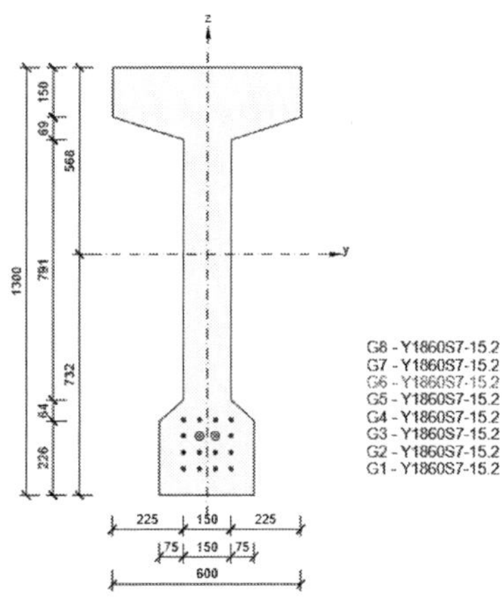

Figure 17. Debonded strands in beam G 1.6.3

Table 6. Debonding lengths (m) / strands group for beams G 1.6.1, G 1.6.2, G 1.6.3

Beam	Strands group							
	G1	G2	G3	G4	G5	G6	G7	G8
G 1.6.1	-	-	-	-	-	2.50	-	-
G 1.6.2	-	-	-	-	-	2.75	-	-
G 1.6.3	-	-	-	-	-	3.00	-	-

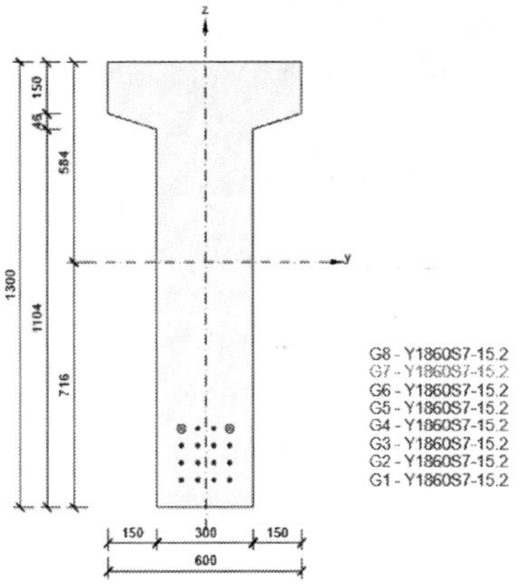

Figure 18. Debonded strands in beams G 1.7.1, G 1.7.2, G 1.7.3

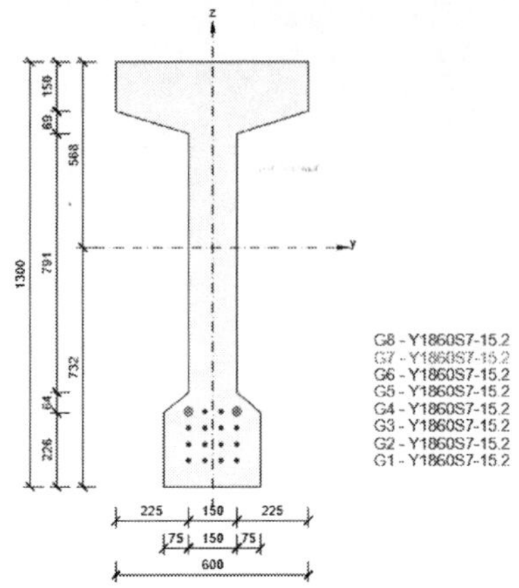

Figure 19. Debonded strands in beam G 1.7.3

Table 7. Debonding lengths (m) / strands group for beams G 1.7.1, G 1.7.2, G 1.7.3

Beam	Strands group							
	G1	G2	G3	G4	G5	G6	G7	G8
G 1.7.1	-	-	-	-	-	-	2.50	-
G 1.7.2	-	-	-	-	-	-	2.75	-
G 1.7.3	-	-	-	-	-	-	3.00	-

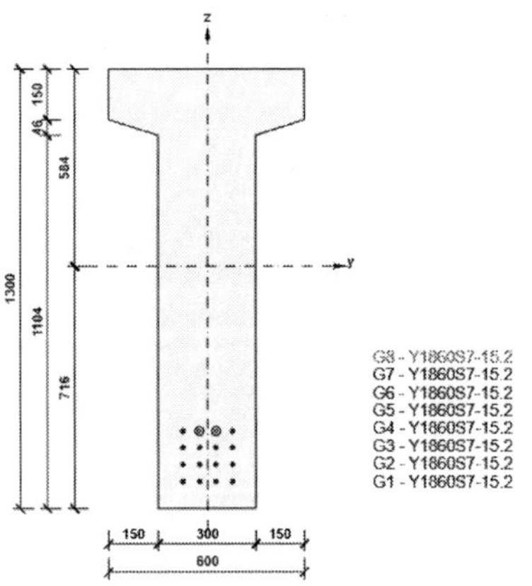

Figure 20. Debonded strands in beams G 1.8.1, G 1.8.2, G 1.8.3

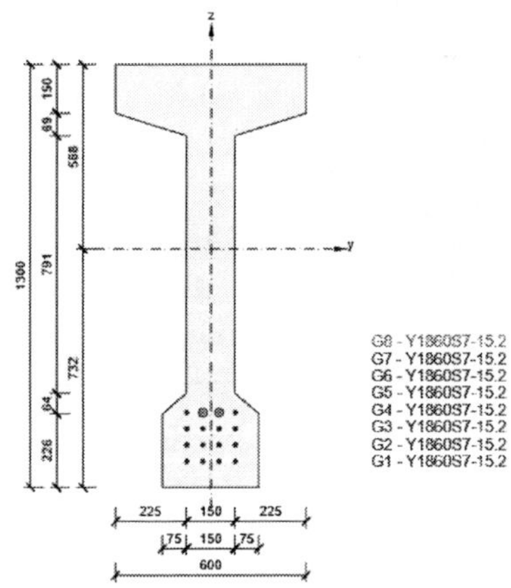

Figure 21. Debonded strands in beam G 1.8.3

Table 8. Debonding lengths (m) / strands group for beams G 1.8.1, G 1.8.2, G 1.8.3

Beam	Strands group							
	G1	G2	G3	G4	G5	G6	G7	G8
G 1.8.1	-	-	-	-	-	-	-	2.50
G 1.8.2	-	-	-	-	-	-	-	2.75
G 1.8.3	-	-	-	-	-	-	-	3.00

IOP Conf. Series: Materials Science and Engineering **586** (2019) 012017 doi:10.1088/1757-899X/586/1/012017

3.3. Results

Due to the high volume of simulations the results are presented graphically. To simplify the results presentation only the eloquent comparisons are part of this paper.

In the current section are presented the following:

- maximum tensions in the strands after stress transfer due to the different layout of debonded strands;
- maximum tensions in the strands after stress transfer due to the different debonded length;
- the bending moment (My) in the support area of the beams in the top fibers (envelope);

All 25 beam analysed meet the requirements of EUROCODE 2.

The figure 22 described the tensions development in the strands related to the reference beam (G 1.0.0). The stress loses in strands are decreasing from the strands situated near of the extreme fiber to those situated closer to the y-y axis of the beam. It can be notice that the strands situated at the same row have the same stress losses. The maximum stress after transfer is achieved in strands group G7 and G8 with a value of 1001.2 MPa, while the lowest stress after transfer is obtained in strands group G1 and G2 with a value of 992.3 MPa.

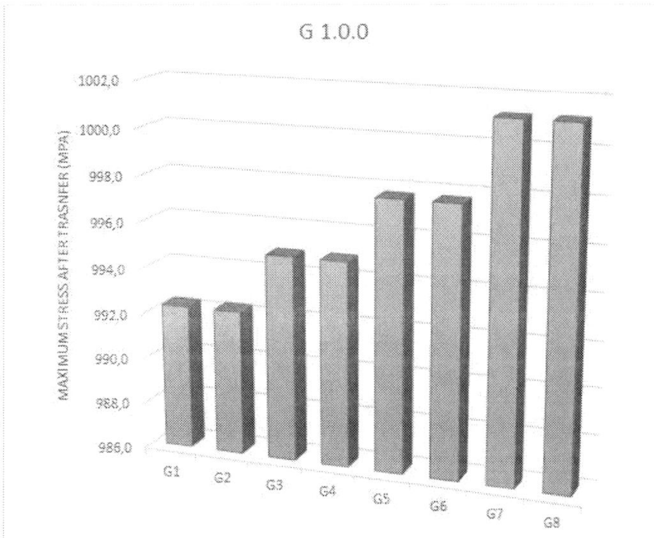

	G1	G2	G3	G4	G5	G6	G7	G8
G 1.0.0	992,3	992,3	994,9	994,9	997,7	997,7	1001,2	1001,2

Figure 22. Maximum tension stress in the strands after transfer in reference beam G 1.0.0

Figure 23 described the tensions development in the strands group related to the beams G 1.1.1, G 1.1.2, G 1.1.3. The values of maximum stress after transfer in strands group G1 are the same to that achieved to equivalent strands from the reference beam G 1.0.0. The stress obtained in the other strands group are significantly higher than that from the control beam. For instance, for strands group G8 the values of maximum stress obtained after transfer is 1009.1 MPa to 1009.8 MPa (depending of debonding length) compared to the obtain value of 1001.2 MPa on to control beam (G 1.0.0).

Similar results were obtained in the case of the beams G 1.2.1, G 1.2.2, G 1.2.3, with the only difference that the strands group with the highest stress losses are G2. Debonded length does not influence the stress losses in debonded strands for the two illustrated cases. Further investigations must be made to establish regarding the influence of debonded strands lengths situated at the extremity of the beams.

From comparing these two simulations it results that the positions of the partially debonded strands in the corner of the total strands displacement do not have a quantifiable impact.

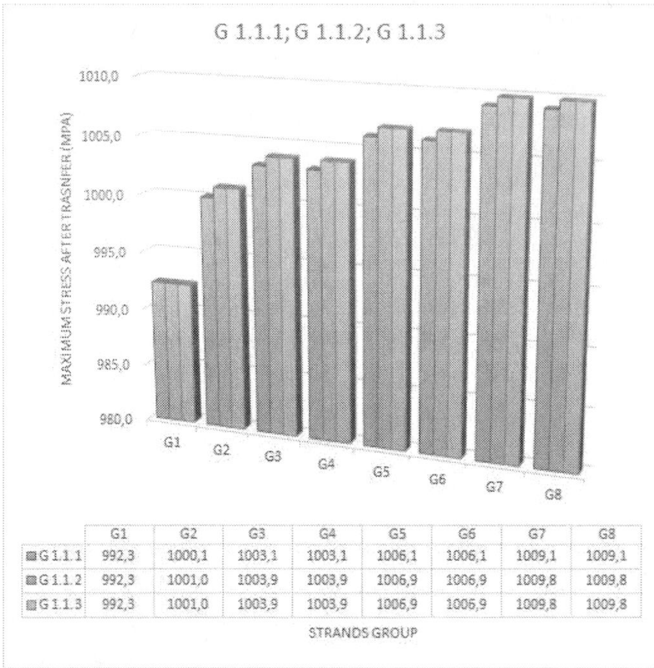

	G1	G2	G3	G4	G5	G6	G7	G8
G 1.1.1	992,3	1000,1	1003,1	1003,1	1006,1	1006,1	1009,1	1009,1
G 1.1.2	992,3	1001,0	1003,9	1003,9	1006,9	1006,9	1009,8	1009,8
G 1.1.3	992,3	1001,0	1003,9	1003,9	1006,9	1006,9	1009,8	1009,8

Figure 23. Maximum tension stress in the strands after transfer for beams G 1.1.1, G 1.1.2, G 1.1.3

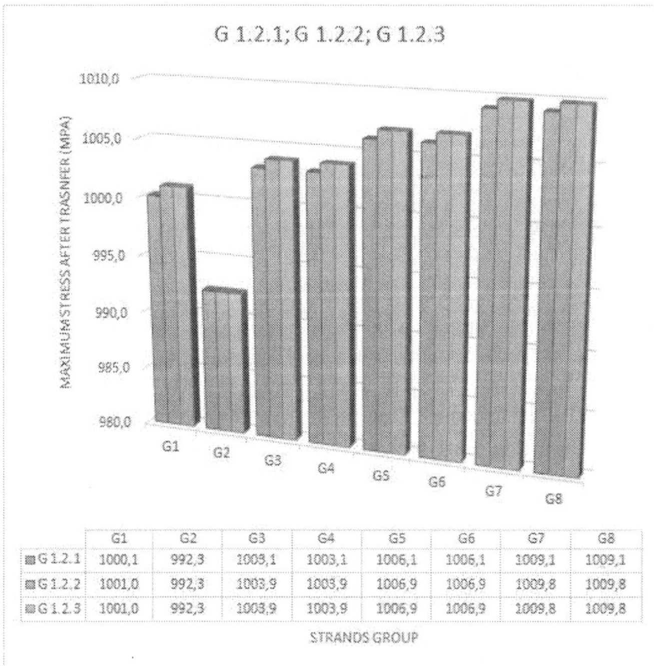

	G1	G2	G3	G4	G5	G6	G7	G8
G 1.2.1	1000,1	992,3	1003,1	1003,1	1006,1	1006,1	1009,1	1009,1
G 1.2.2	1001,0	992,3	1003,9	1003,9	1006,9	1006,9	1009,8	1009,8
G 1.2.3	1001,0	992,3	1003,9	1003,9	1006,9	1006,9	1009,8	1009,8

Figure 24. Maximum tension stress in the strands after transfer for beams G 1.2.1, G 1.2.2, G 1.2.3

In figure 25 and figure 26 we can notice similar principles as were detailed previous with the particularities resulted from the debonded strands group G3 respectively G4.

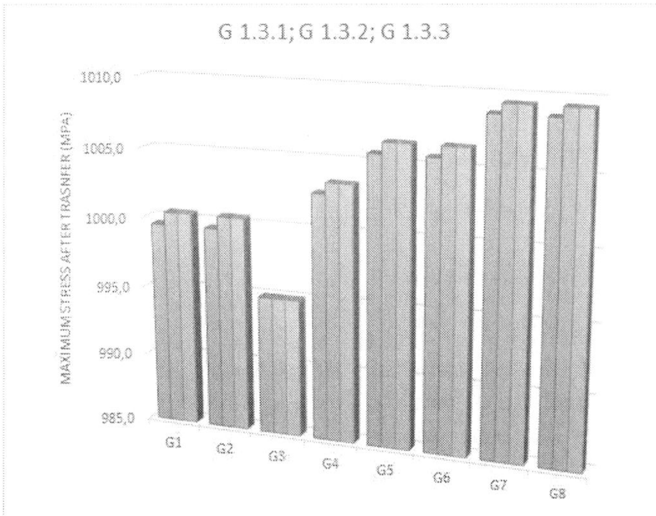

	G1	G2	G3	G4	G5	G6	G7	G8
G 1.3.1	999,5	999,5	994,9	1002,6	1005,6	1005,6	1008,7	1008,7
G 1.3.2	1000,4	1000,4	994,9	1003,4	1006,4	1006,4	1009,4	1009,4
G 1.3.3	1000,4	1000,4	994,9	1003,4	1006,4	1006,4	1009,4	1009,4

Figure 25. Maximum tension stress in the strands after transfer for beams G 1.3.1, G 1.3.2, G 1.3.3

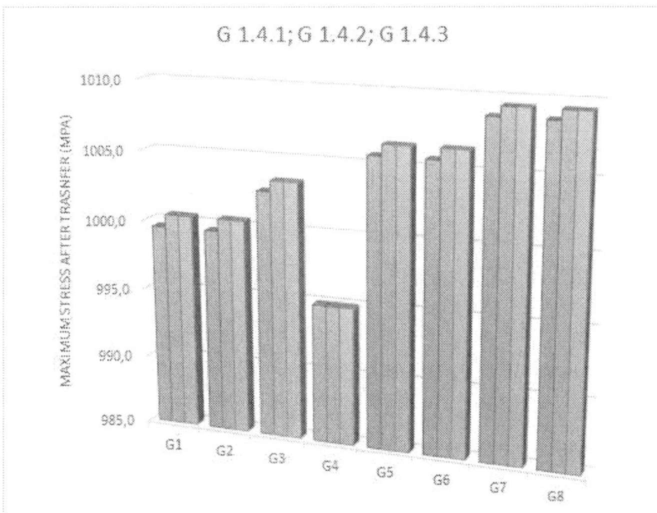

	G1	G2	G3	G4	G5	G6	G7	G8
G 1.4.1	999,5	999,5	1002,6	994,9	1005,6	1005,6	1008,7	1008,7
G 1.4.2	1000,4	1000,4	1003,4	994,9	1006,4	1006,4	1009,4	1009,4
G 1.4.3	1000,4	1000,4	1003,4	994,9	1006,4	1006,4	1009,4	1009,4

Figure 26. Maximum tension stress in the strands after transfer for beams G 1.4.1, G 1.4.2, G 1.4.3

IOP Conf. Series: Materials Science and Engineering **586** (2019) 012017 doi:10.1088/1757-899X/586/1/012017

Figure 27 and figure 28 illustrates the particularities for beams G 1.5.1, G 1.5.2, G 1.5.3 respectively for beams G 1.6.1, G 1.6.2 and G 1.6.3.

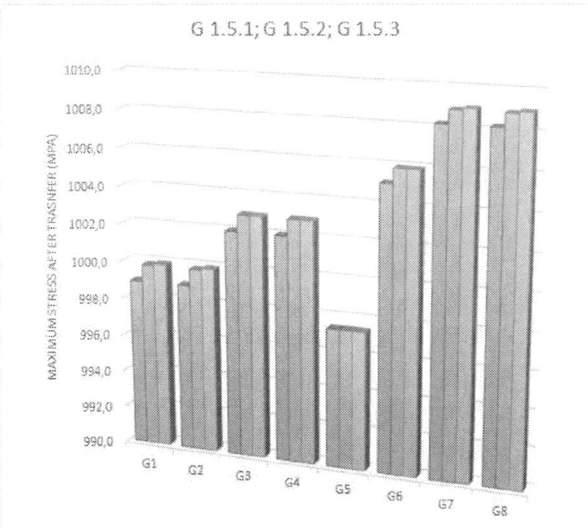

	G1	G2	G3	G4	G5	G6	G7	G8
G 1.5.1	998,9	998,9	1002,0	1002,0	997,4	1005,1	1008,2	1008,2
G 1.5.2	999,8	999,8	1002,9	1002,9	997,4	1005,9	1008,9	1008,9
G 1.5.3	999,9	999,9	1002,9	1002,9	997,4	1005,9	1009,0	1009,0

Figure 27. Maximum tension stress in the strands after transfer for beams G 1.5.1, G 1.5.2, G 1.5.3

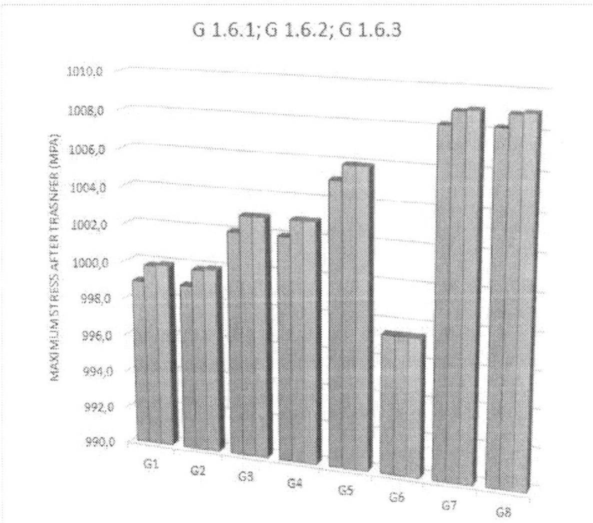

	G1	G2	G3	G4	G5	G6	G7	G8
G 1.6.1	998,9	998,9	1002,0	1002,0	1005,1	997,4	1008,2	1008,2
G 1.6.2	999,8	999,8	1002,9	1002,9	1005,9	997,4	1008,9	1008,9
G 1.6.3	999,9	999,9	1002,9	1002,9	1005,9	997,4	1009,0	1009,0

Figure 28. Maximum tension stress in the strands after transfer for beams G 1.6.1, G 1.6.2, G 1.6.3

Figure 29 and figure 23 relate the maximum stress after transfer when the partially debondend strands are situated the closest to the y-y axis.

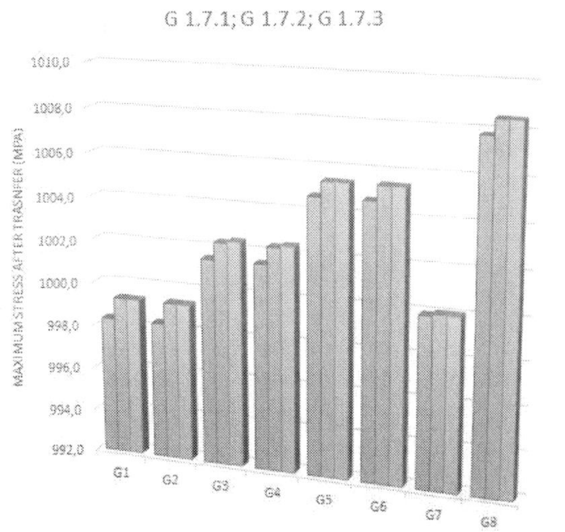

	G1	G2	G3	G4	G5	G6	G7	G8
G 1.7.1	998,3	998,3	1001,5	1001,5	1004,7	1004,7	999,9	1007,8
G 1.7.2	999,3	999,3	1002,3	1002,3	1005,4	1005,4	1000,0	1008,5
G 1.7.3	999,3	999,3	1002,4	1002,4	1005,4	1005,4	1000,0	1008,5

STRANDS GROUP

Figure 29. Maximum tension stress in the strands after transfer for beams G 1.7.1, G 1.7.2, G 1.7.3

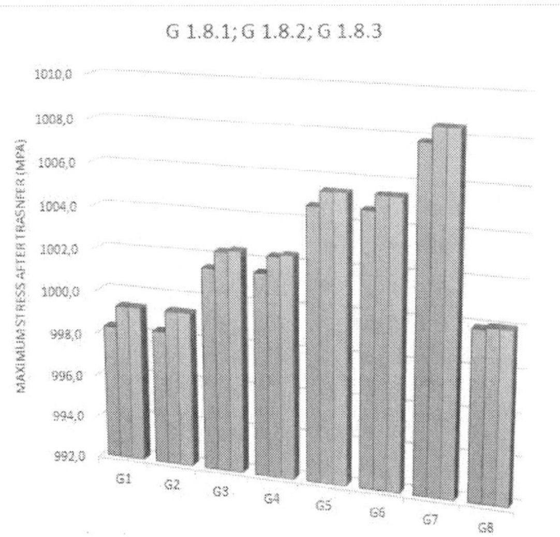

	G1	G2	G3	G4	G5	G6	G7	G8
G 1.8.1	998,3	998,3	1001,5	1001,5	1004,7	1004,7	1007,8	999,9
G 1.8.2	999,3	999,3	1002,3	1002,3	1005,4	1005,4	1008,5	1000,0
G 1.8.3	999,3	999,3	1002,4	1002,4	1005,4	1005,4	1008,5	1000,0

STRANDS GROUP

Figure 30. Maximum tension stress in the strands after transfer for beams G 1.8.1, G 1.8.2, G 1.8.3

IOP Conf. Series: Materials Science and Engineering **586** (2019) 012017 doi:10.1088/1757-899X/586/1/012017

Figure 31 illustrates the My (bending moment) development in the support area of the beams in the top fibers. The values mentioned in figure 31 are taken from the envelope. In the first column are mentioned the beams considered, in column 2 the values obtained in the case of a strands debonded length of 2.50 m, in column 3 the values obtained in the case of a strands debonded length of 2.75 m and in column 4 the values obtained in the case of a strands debonded length of 3.00 m. The biggest bending moment is obtained in the case of the reference beam (G 1.0.0) and the smallest in the case of beams G 1.1.1, G 1.1.2, G 1.1.3, G 1.2.1, G 1.2.2 and G 1.2.3. We can state that as the farther are the partially debonded strands from de y-y axis the smaller the bending moment in the support area of the beams is.

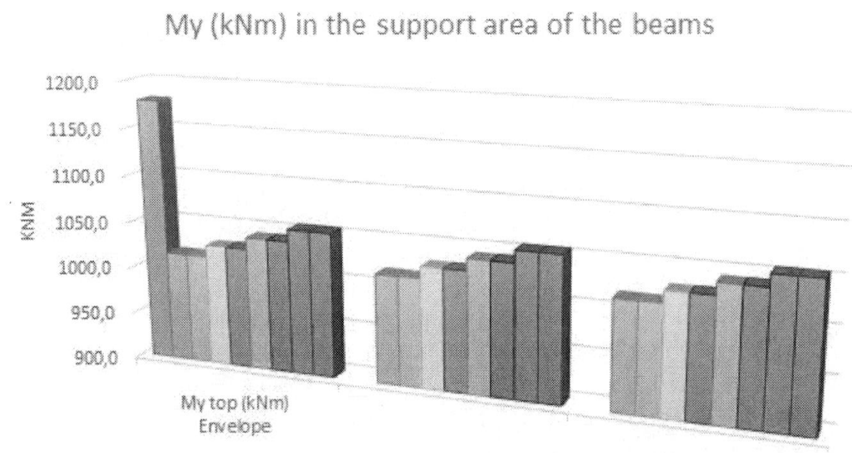

	My top (kNm) Envelope		
G 1.0.0	1179,7		
G 1.1.1 G 1.1.2 G 1.1.3	1016,1	1016,1	1016,1
G 1.2.1 G 1.2.2 G 1.2.3	1016,1	1016,1	1016,1
G 1.3.1 G 1.3.2 G 1.3.3	1028,7	1028,7	1028,7
G 1.4.1 G 1.4.2 G 1.4.3	1028,7	1028,7	1028,7
G 1.5.1 G 1.5.2 G 1.5.3	1041,4	1041,4	1041,4
G 1.6.1 G 1.6.2 G 1.6.3	1041,4	1041,4	1041,4
G 1.7.1 G 1.7.2 G 1.7.3	1054,5	1054,5	1054,5
G 1.8.1 G 1.8.2 G 1.8.3	1054,5	1054,5	1054,5

SIMULATED BEAMS

Figure 31. My (kNm) in the support area of beams in top fibers (envelope)

4. Conclusions

After interpreting the previous results, we can state the following generally applicable rules.

In the case of the beams with fully bonded strands the stress losses after transfer are decreasing from the strands situated near of the extreme fiber to those situated closer to the y-y axis of the beam. The strands situated at the same row have the same stress losses.

Regarding using partially debonded strands the stress losses are decreasing comparing to the beams with fully bonded strands excepting the debonded strands. By using partial debonded strands in designing of prestress precast concrete beams the strands are used more efficiently, the value of stress after transfer getting closer to the initial stress value.

From our research result that the shorter the length of partially debonded strand is, the higher the stress losses are. A higher number of simulations must be done to obtained a certain verdict.

Regarding My (bending moment) in the support area of the beams the highest values is obtained in the cases of the beams with fully bonded strands. For the beams with partially debonded strands the smallest values is obtained in the situation in which the debonded strands are situated further from the y-y axis. The value of the bending moment increases as the partially debonded strands are closer to the y-y axis. Regarding influence of the partially debonded strands length on to bending moment it is inconclusive. It is necesarly to run a higher number of simulations considering more debonded lengths.

The blanketing proposal must fulfil the requirements of limiting of tensile stresses in the concrete at the top fibres at transfer and the ability to sustain the imposed load tensions at the bottom fibres with the resulting reduced prestress at the debonded sections. Also, the blanketing proposal must satisfy the technical and economic criteria.

5. References

[1] SR EN 1992-1-1:2004 *Proiectarea structurilor din beton, Partea 1 - 1:Reguli generale și reguli pentru clădiri*

[2] Tertea I *Betonul precomprimat. Bazele proiectării ed Tehnică București*

[3] Rabbat B G, Kaar P H and Russell H G *Fatigue tests of prestressed girders with blanketed and draped strands*

[4] Zumbrunnen L G, Russell B W and Burns N H 1992 *Behavior of statically loaded pretensioned concrete beams with 0.5-inch diameter debonded strands*

[5] Adballa O A, Ramirez J A and Lee R H 1993 *Strand debonding in pretensioned beams – precast prestressed concrete bridges with debonded strands*

[6] Tuns I, Mihai P and Florea N 2007 *Betonul armat și precomprimat. Betonul – armătură – conlucrare*

[7] Pascu R 2008 *Betonul precomprimat. Bazele calcului*

[8] Kiss Z and Oneț T 2008 *Proiectarea structurilor de beton după SR EN 1992-1* ed Abel

[9] Mihai P 2010 *Reinforced concrete structure* ed Matei – Teiu Botez Iași

[10] Tuns I and Florea N 2010 *Betonul precomprimat* vol. 1 ed Matrix Rom București

[11] Vidales M D 2011 *Effect of partial debonding of prestressing strands on beam end cracking*

[12] Burgueno R and Sun Y 2011 *Effects of debonded strands on the production and performance of prestressed concrete beams*

[13] Pascu R and Zybaczinski A 2012 *Beton precomprimat. Calculul după Eurocode 2*

[14] Wang W 2013 *Assesement of partial strand debonding practice for AASHTO type girders*

IOP Conf. Series: Materials Science and Engineering **586** (2019) 012018 doi:10.1088/1757-899X/586/1/012018

The influence of construction and service stages on the internal forces of prestressed precast beams

I L Stan[1], N Florea[2], P Mihai[2] and V C Tănăsucă[1]
[1] Doctoral School of the Faculty of Civil Engineering and Building Services, „Gheorghe Asachi" Technical University of Iași, Romania
[2] Department of Concrete, Materials, Technology and Management, Faculty of Civil Engineering and Building Services, „Gheorghe Asachi" Technical University of Iași, Romania

E-mail: stan.ioan.lucian@gmail.com

Abstract. The accurate design of prestressed precast beams is correlated with the proper evaluation of construction and service stages. On usually are considered the following construction stages: casting, transfer of prestressing, storage time, transport and mounting stage. In the presented paper, it will be illustrated the stress development on the beams by altering the specific times between different construction and service stages as a result of using different construction technologies as well as a consequence by imposing in advance different requirements during the design phase. To highlight these particularities IDEA StatiCA software was used, and a comparison between various cases is presented.

1. Introduction

Proper design of prestressed prefabricated beams represents a difficult task, the number of parameters involved in design being correlated even with the construction stages. A construction stage represents a point, period, or step in a process of development. The shallow evaluation of the construction stages may lead to a non-conforming element, materialized through degradation, failure or irrational use of materials.

Evaluation of construction and service stages plays a significant role in the efficient design of prestressed prefabricated beams due to the fact that external forces that act on to the element vary according to them. The main construction and service stages that must be taken in consideration when designing prestressed prefabricated concrete beams are: casting (casting of fresh concrete of design member), transfer of prestressing, storage yard, transport, final support, superimposed dead load and end of design working life.

There are currently no studies that address this topic. By performing this study it is desirable to highlight the differences that are occurring when different times intervals are imposed as a result of using different construction technologies as well as a consequence by compeling in advance different requirements during the design phase. The purpose of this study is to establish if a significant differences occurs regarding the development of internal forces, and in which manner these differences influence the design of the prestressed precast beams. Another aspect of this topic is to determine if general rules regarding stress development can be established.

2. Experimental phase

To illustrate the difference occurring when design prestressed prefabricated concrete beams a study was carried out on a number of 12 beams. The IDEA StatiCA software was used to accomplish the simulations.

Given that the design of prestressed prefabricated concrete beams is a meticulous task it was necessary for a number of conventions to be made. By imposing a certain number of parameters to be fixed, it can be focused only on the principal parameters that are correlated with the stages of the construction, in this case, the imposed duration between them.

In Section 2.1. are detailed the fixed parameters and in Section 2.2. are detailed the variable parameters.

2.1. Fixed parameters

From this category the following parameters are fixed:

- beams length: 25.5 m;
- the static scheme, sections as is shown in figure 1, figure 2 and figure 3;

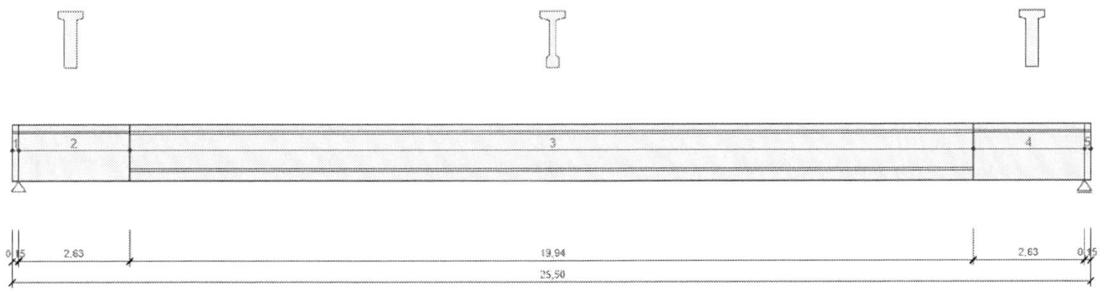

Figure 1. Static scheme

- concrete strength class (according to Eurocode 2): C50/60;
- the concrete curing was considered naturally (not the case of heat curing);
- strands used: Y1860S7-15.2;
- strands layout (a total of 16 strands as it is shown in figure 2 and figure 3);
- debonded strands (as shown in figure 2 and figure 3), debonded length (3.00 m);
- initial stress in strands: 1100.0 MPa;
- passive reinforcement: steel grade (B500C), the layout of passive reinforcement;
- load cases (self-weight, permanent load: 9.1 kN/m, variable load: 15.4 kN/m, load per construction stage G 5: 1.0 kN/m).

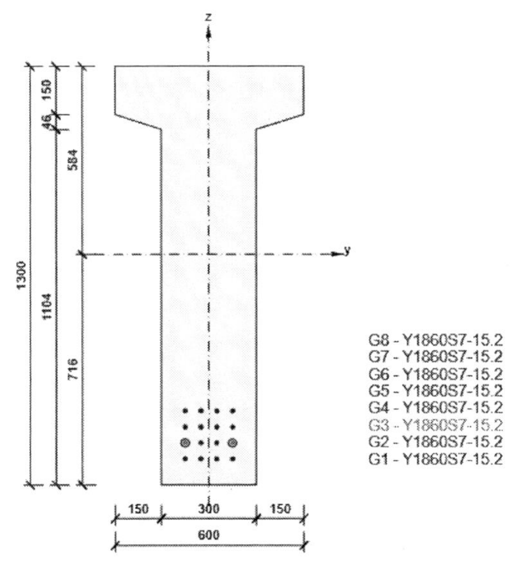

Figure 2. Section 1 as is shown in figure 1 (mm) and strands layout

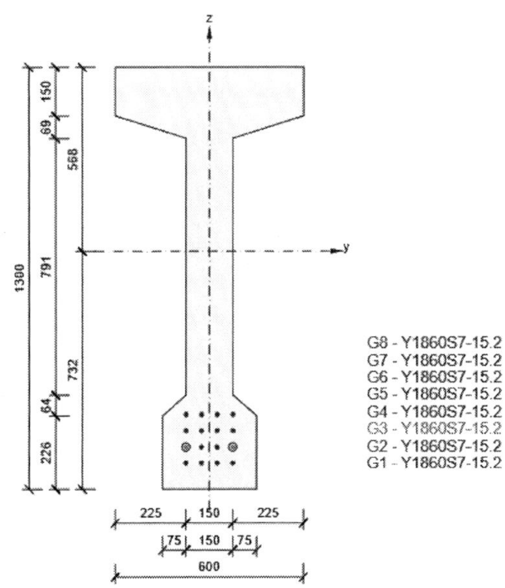

Figure 3. Section 2 as is shown in figure 1 (mm) and strands layout

IOP Conf. Series: Materials Science and Engineering **586** (2019) 012018 doi:10.1088/1757-899X/586/1/012018

2.2. Variable parameters

As it was stated before a study was carried out a number of 12 beams, the time between the construction stages being altered as a result of using different construction technologies as well as a consequence by imposing in advance different requirements during the design phase.

To better understand how the time interval was considered there are shown in the following:

Table 1. Construction and service stages for considerate beams

Beam	Stages							
	ST(1)	T1	ST(2)	ST(3)	ST(4)	ST(5)	ST(6)	ST(7)
G 1.1.1	0	25	1.92	2	29	30	60	36500
G 1.1.2	0	25	1.92	2	59	60	120	36500
G 1.2.1	0	30	2.9	3	29	30	60	36500
G 1.2.2	0	30	2.9	3	59	60	120	36500
G 1.3.1	0	35	4.5	4.6	29	30	60	36500
G 1.3.2	0	35	4.5	4.6	59	60	120	36500
G 1.4.1	0	40	7.4	7.5	29	30	60	36500
G 1.4.2	0	40	7.4	7.5	59	60	120	36500
G 1.5.1	0	45	13.3	13.4	29	30	60	36500
G 1.5.2	0	45	13.3	13.4	59	60	120	36500
G 1.6.1	0	50	28	28.1	29	30	60	36500
G 1.6.2	0	50	28	28.1	59	60	120	36500

where:

- ST(1) – casting (days);
- T1 – concrete resistance at the moment of stress transfer (MPa);
- ST(2) – transfer of prestressing (days);
- ST(3) – storage yard (days);
- ST(4) – transport (days);
- ST(5) – final support (days);
- ST(6) – superimposed dead load (days);
- ST(7) – end of design working life (days);

The structural response is presented (schematically) in figures 4 – 15, depending on time for each beam considered in the simulation.

From them it can be noticed:

- the effects of permanent load applied in construction stage „i";
- the rheological effects between construction stages „i-1" and „i";
- and the effects of prestressing applied in construction stage „i".

From the schematic representation, it can be noticed when the external forces act on the beams according to each construction stage and how the sum of forces contributes to the overall design of members. According to figures 4 – 15 it can be noticed that certain types of loads specific to a specific construction stage are rheological. The prestressed members are analyzed considering all the phenomena that are occurring in the design working life (concrete shrinkage, stress losses, etc.).

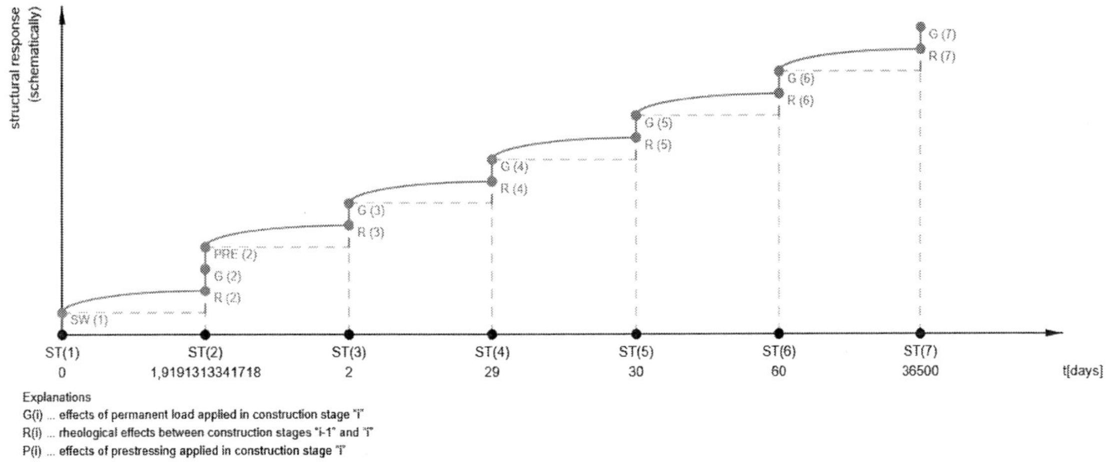

Figure 4. Constructions stages for beam G 1.1.1

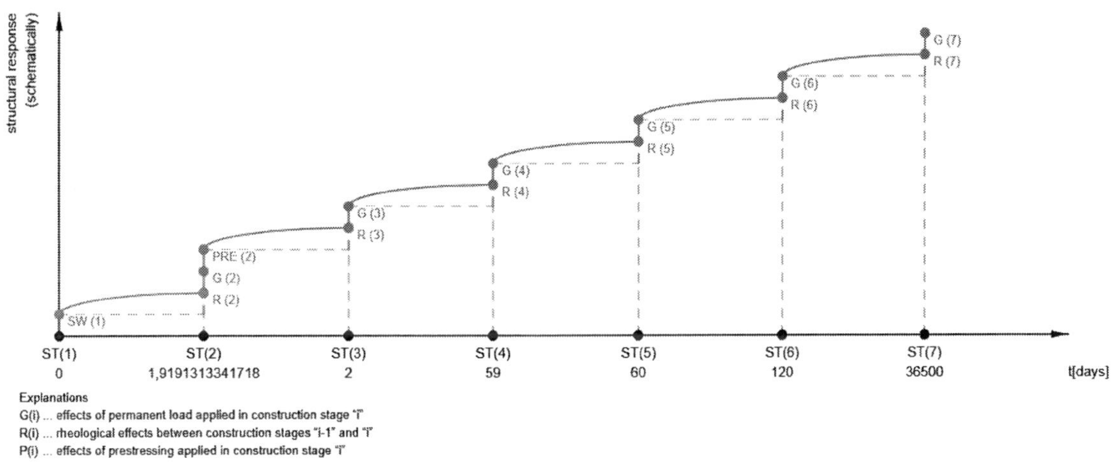

Figure 5. Constructions stages for beam G 1.1.2

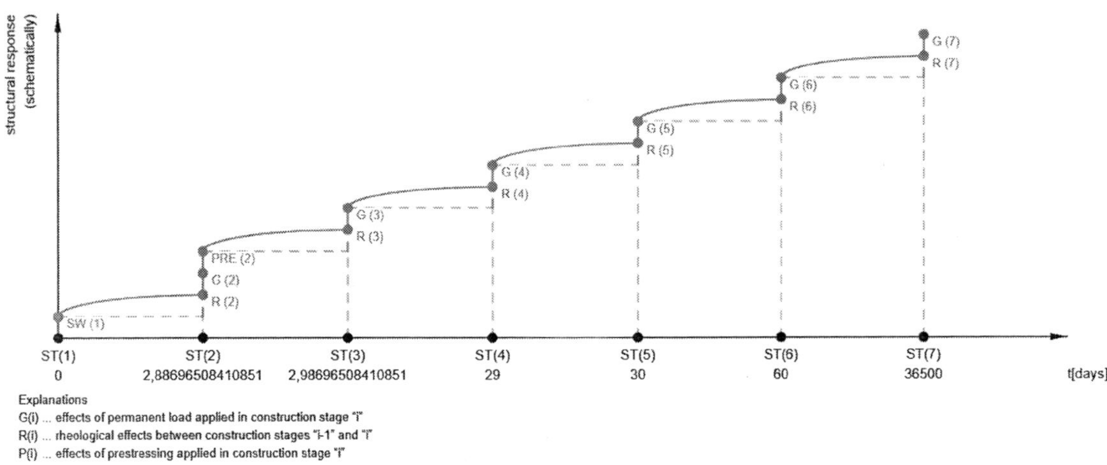

Figure 6. Constructions stages for beam G 1.2.1

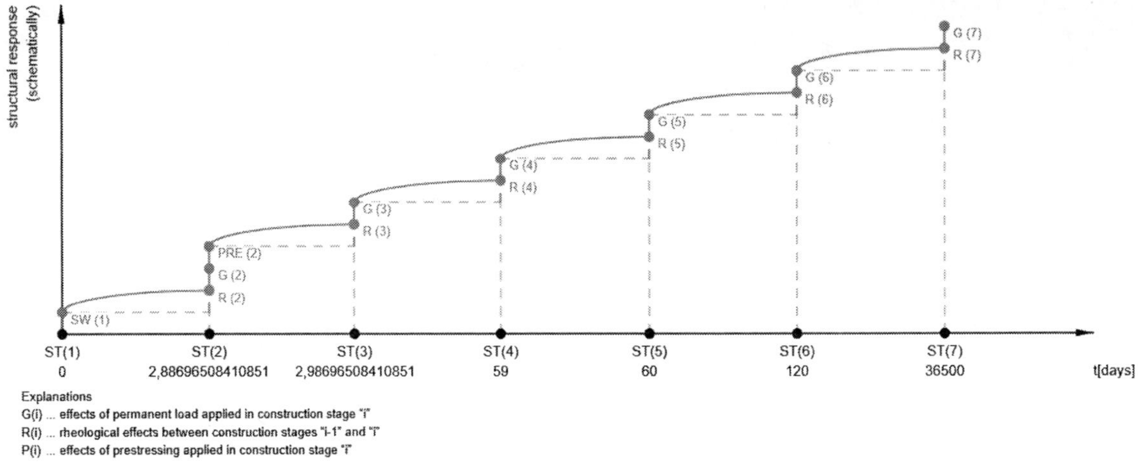

Figure 7. Constructions stages for beam G 1.2.2

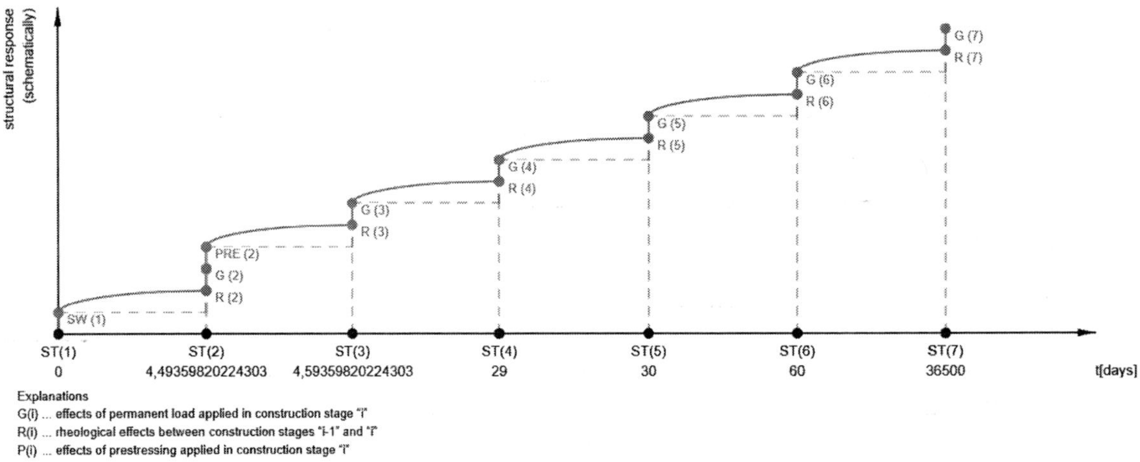

Figure 8. Constructions stages for beam G 1.3.1

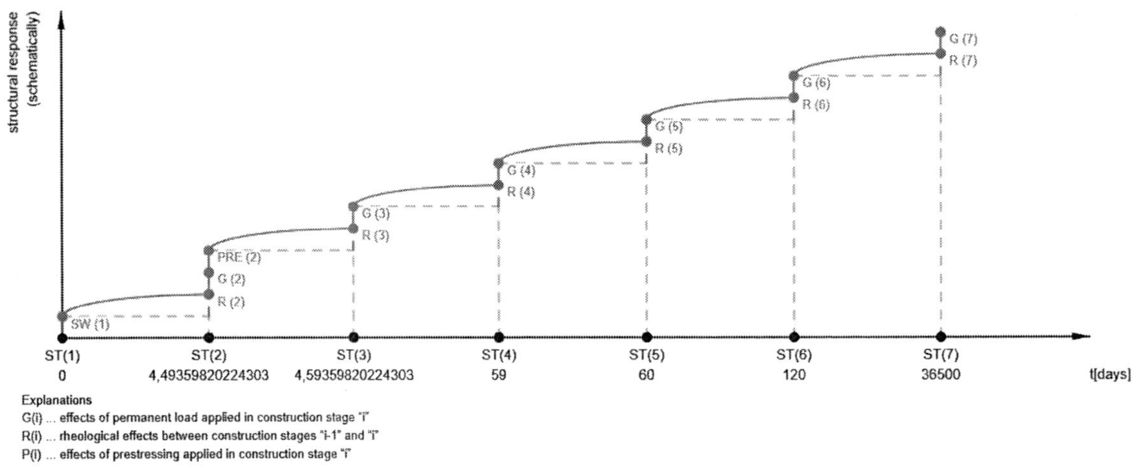

Figure 9. Constructions stages for beam G 1.3.2

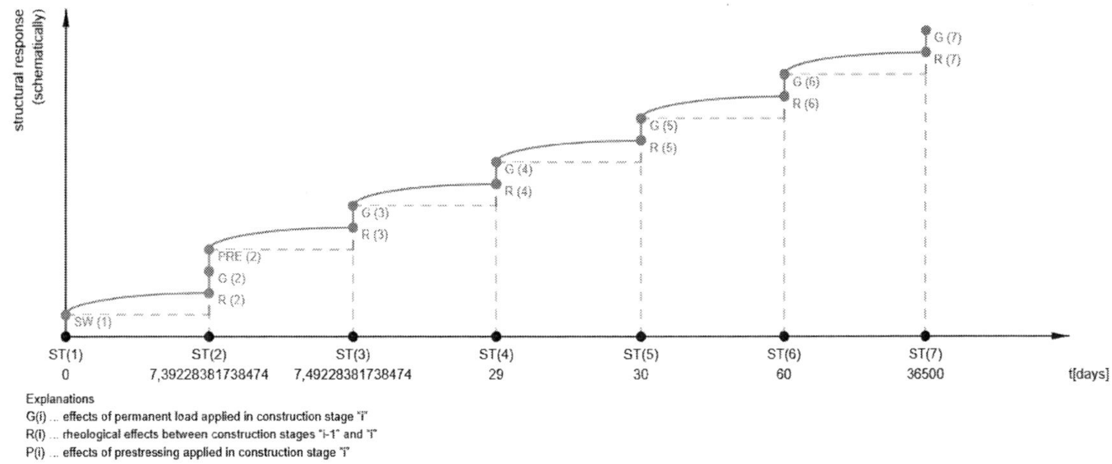

Figure 10. Constructions stages for beam G 1.4.1

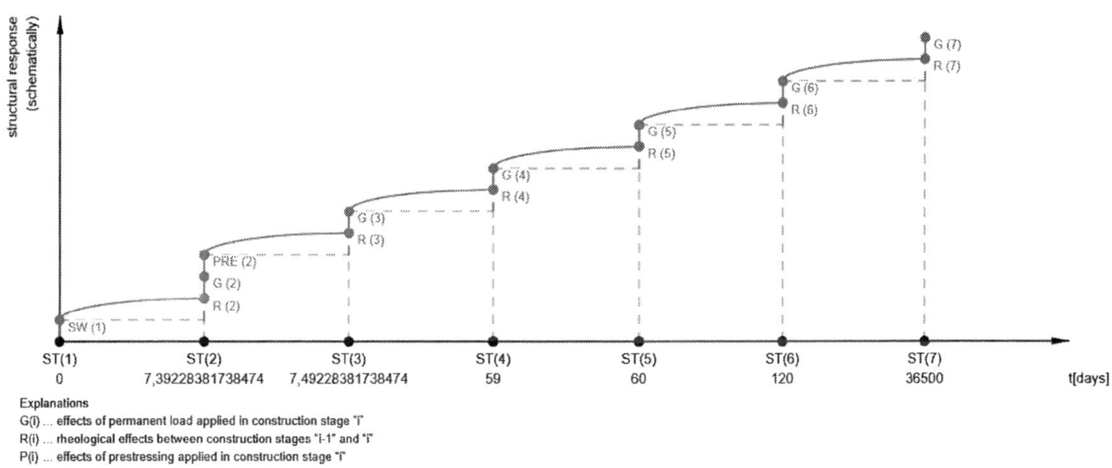

Figure 11. Constructions stages for beam G 1.4.2

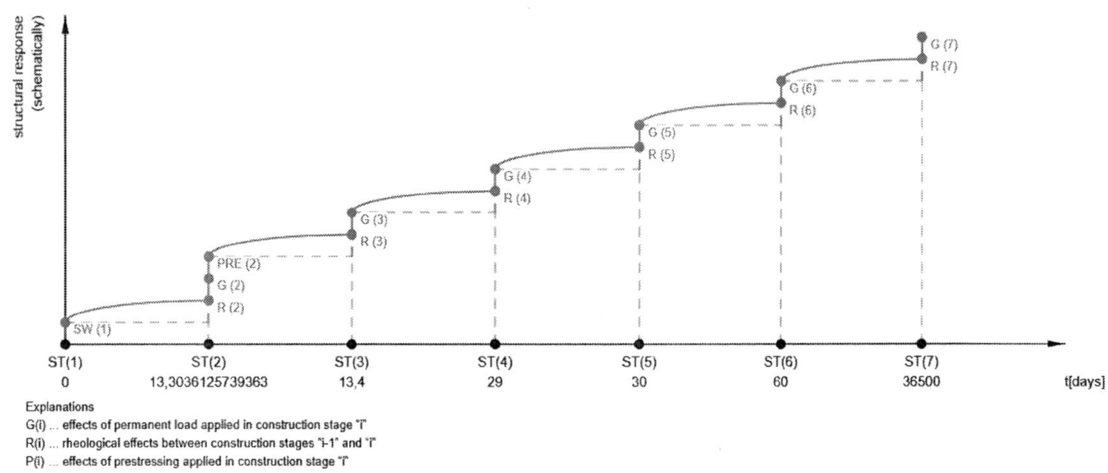

Figure 12. Constructions stages for beam G 1.5.1

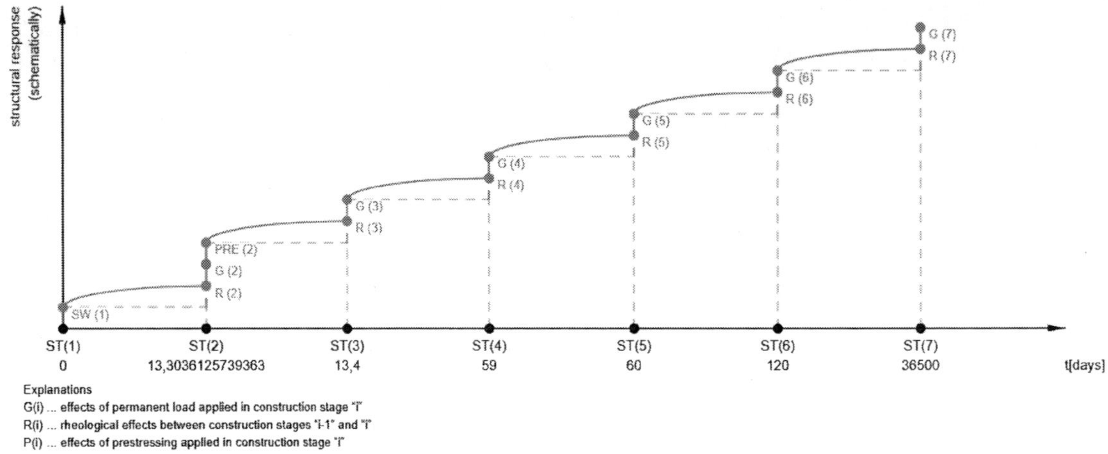

Figure 13. Constructions stages for beam G 1.5.2

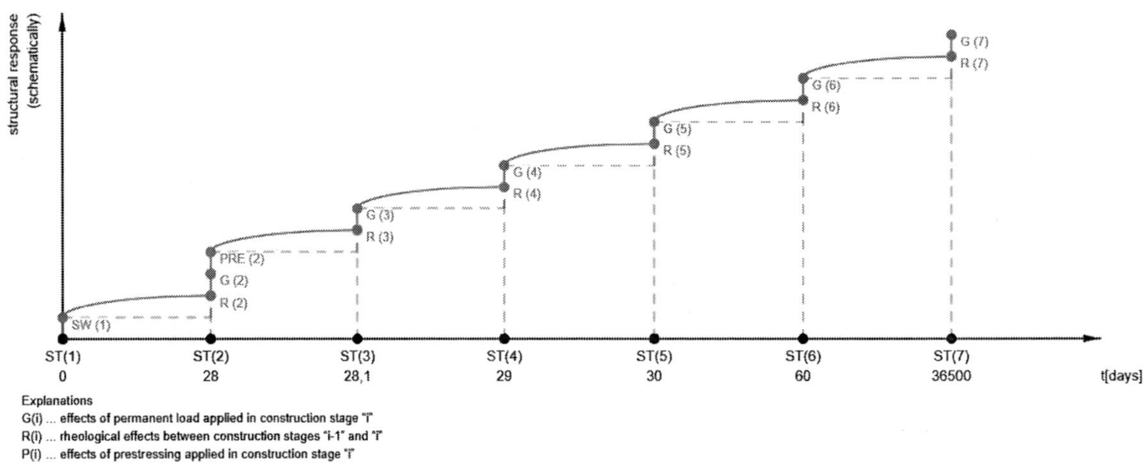

Figure 14. Constructions stages for beam G 1.6.1

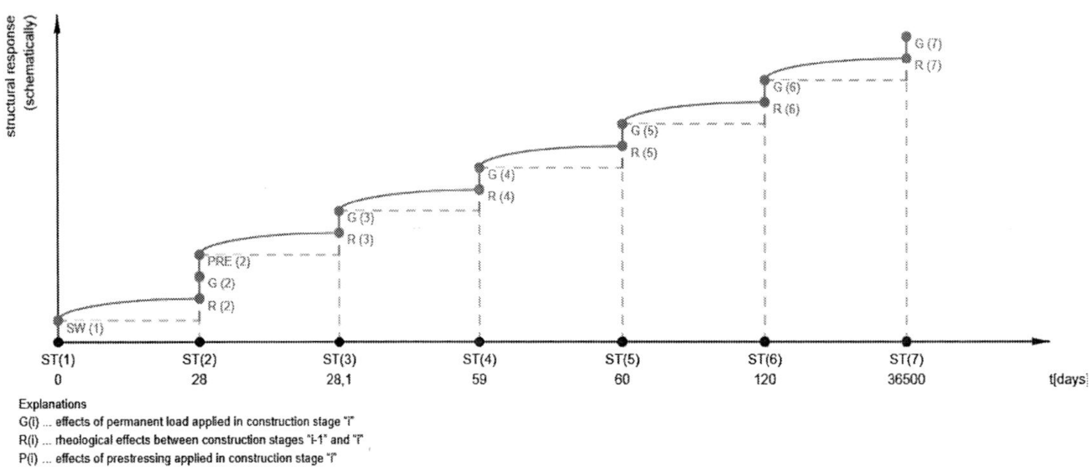

Figure 15. Constructions stages for beam G 1.6.2

2.3. Results

Due to the high volume of simulations, the results are presented graphically. To simplify the presentation of the results only the eloquent comparisons are part of this paper.

In the current section are presented the following:

- the development of axial force according to construction stages in characteristics sections;
- the development of bending moment according to construction stages in characteristic sections;
- shear force, stress limitation, crack width and deflection ratio.

Not all 12 analysed beams fulfil the requirements of EUROCODE 2.

For a proper understanding of the characteristic section consider, they are detailed below:

In the case of the axial forces, 3 characteristic sections were considered:

- section 1 (coordinates: 0.00 m);
- section 2 (coordinates: 2.75 m);
- section 3 (coordinates: 12.75 m).

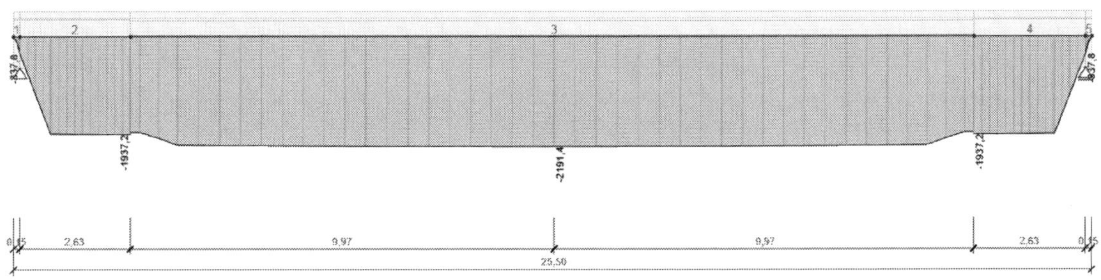

Figure 16. Axial force (envelope) for beam G 1.1.1

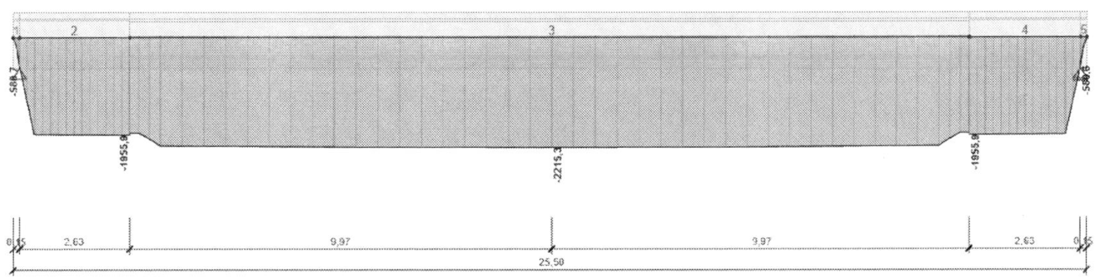

Figure 17. Axial force (envelope) for beam G 1.6.1

In the case of the shear force, there were no significant differences between the simulated beams, resulting that the shear force is not influence by the construction stages.

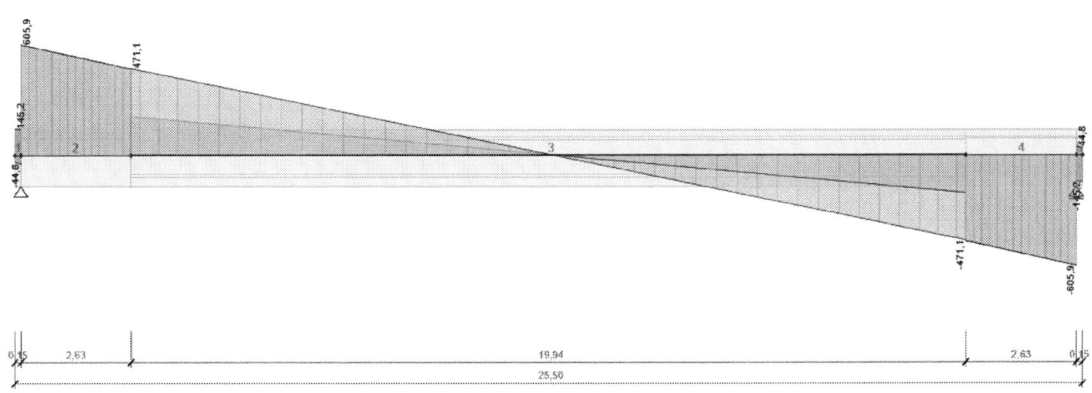

Figure 18. Shear force (envelope) for beam G 1.1.1

IOP Conf. Series: Materials Science and Engineering **586** (2019) 012018 doi:10.1088/1757-899X/586/1/012018

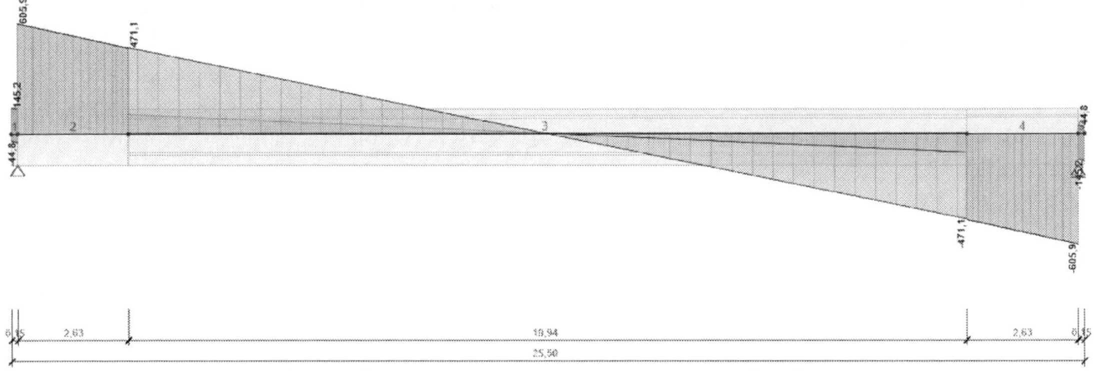

Figure 19. Shear force (envelope) for beam G 1.6.1

In the case of the bending moment 3 characteristic sections were considered:

- section 1 (coordinates: around 0.50 m);
- section 2 (coordinates: around 2.75m or 3.50 m depending on the highest values of the bending moment);
- section 3 (coordinates: 12.75 m).

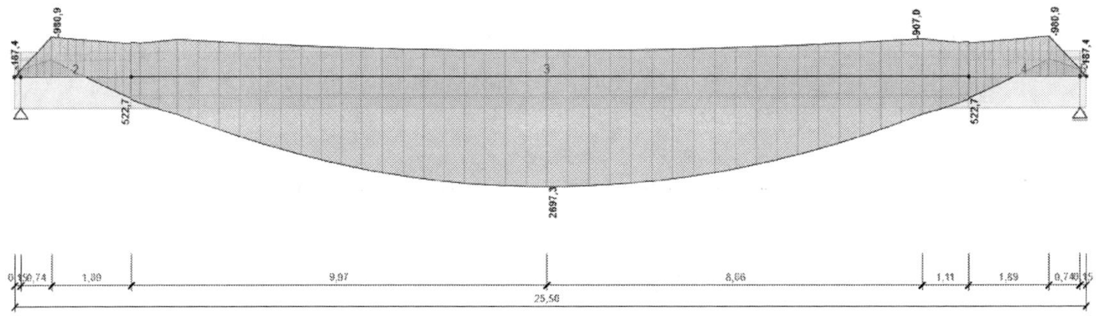

Figure 20. Bending moment (envelope) for beam G 1.1.1

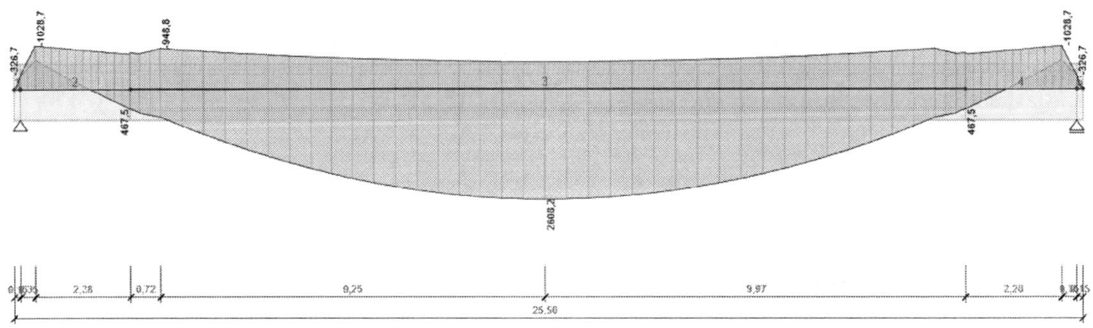

Figure 21. Bending moment (envelope) for beam G 1.6.1

Note: To simplify the presentation of the results the next conventions were made:

- internal forces developed in construction stage ST(1) are not represented;
- internal forces developed in construction stage ST(2) are presented in position 1;
- internal forces developed in construction stage ST(3) are presented in position 2;
- internal forces developed in construction stage ST(4) are presented in position 3;
- internal forces developed in construction stage ST(5) are presented in position 4;

IOP Conf. Series: Materials Science and Engineering **586** (2019) 012018 doi:10.1088/1757-899X/586/1/012018

- internal forces developed in construction stage ST(6) are presented in position 5;
- internal forces developed in construction stage ST(7) are presented in position 6.

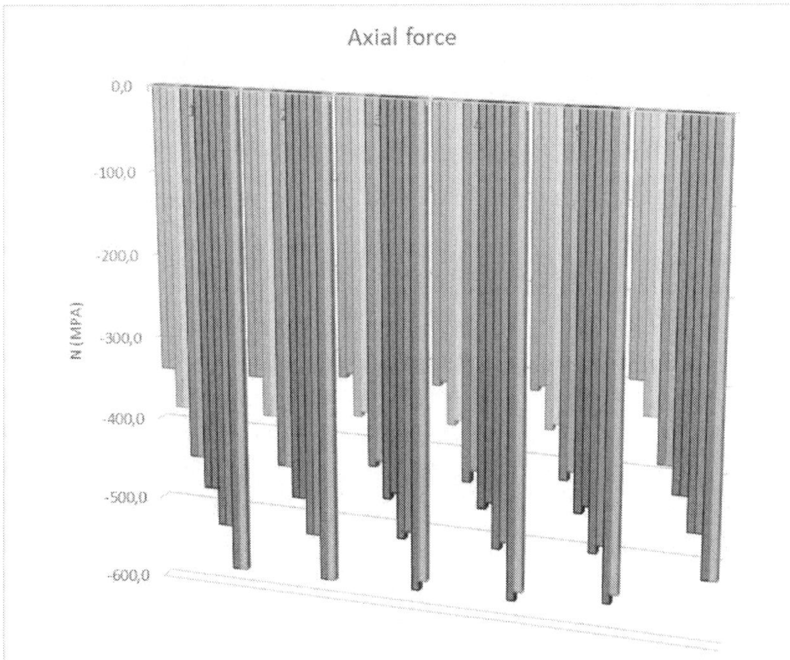

	1	2	3	4	5	6
G 1.1.1	-337,8	-337,3	-327,4	-325,9	-320,7	-297,3
G 1.1.2	-337,8	-337,3	-322,1	-320,8	-314,7	-297,4
G 1.2.1	-384,0	-383,3	-372,3	-370,9	-364,7	-337,6
G 1.2.2	-384,0	-383,3	-366,2	-364,9	-357,8	-337,7
G 1.3.1	-444,1	-443,2	-431,5	-438,0	-422,9	-391,8
G 1.3.2	-444,1	-443,2	-424,5	-423,1	-411,9	-391,8
G 1.4.1	-482,2	-481,2	-469,8	-468,3	-460,3	-425,4
G 1.4.2	-482,2	-481,2	-461,9	-460,5	-451,3	-425,4
G 1.5.1	-527,0	-525,0	-516,6	-515,0	-506,0	-467,2
G 1.5.2	-527,0	-525,0	-507,6	-506,2	-496,0	-467,3
G 1.6.1	-580,7	-579,5	-578,1	-576,0	-564,4	-521,4
G 1.6.2	-580,7	-579,5	-566,1	-564,6	-553,1	-521,5

BEAMS

Figure 22. Axial force development in characteristic section 1

In figure 21 were detailed the axial force development in characteristic section 1 (coordinates: 0.00 m) depending on the concrete strength at the moment of stress transfer and the interval between construction stages. The maximum axial force is obtained in beams G 1.6.1 and G 1.6.2 where the stress transfer was realized at 28 days (concrete reached the strength of 50 MPa). The minimum axial force is obtained in beams G 1.1.1 and G 1.1.2, where the transfer was realized at 1,92 days (concrete reached the strength of 25 MPa).
The axial force is decreasing from the moment of stress transfer to the end of design working life (as it can be noticed in figure 21) due to the effects of the applied loads and rheological effects. Related to the construction stages: *transport*, *final support* and *superimposed dead load*, it can be noticed that axial forces present in the element are slightly higher when the time between them is smaller. At the end of design working life, the axial forces have rough same values, the difference mentioned before not having a significant role for this case (section 1).

IOP Conf. Series: Materials Science and Engineering **586** (2019) 012018 doi:10.1088/1757-899X/586/1/012018

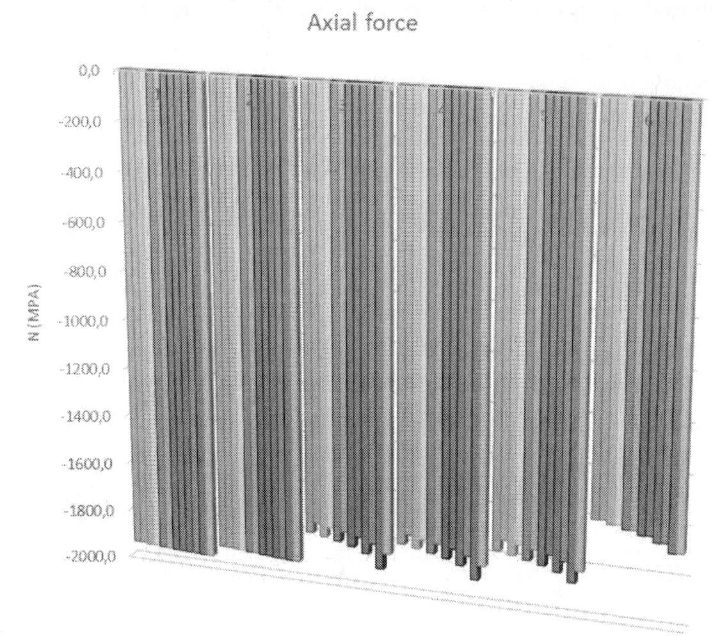

	1	2	3	4	5	6
G 1.1.1	-1937,2	-1908,7	-1799,3	-1798,6	-1777,2	-1601,6
G 1.1.2	-1937,2	-1908,7	-1757,4	-1757,4	-1730,4	-1598,9
G 1.2.1	-1942,4	-1913,0	-1810,4	-1809,6	-1788,4	-1615,3
G 1.2.2	-1942,4	-1913,0	-1768,7	-1768,7	-1742,1	-1612,7
G 1.3.1	-1946,7	-1917,8	-1822,2	-1821,5	-1800,3	-1629,4
G 1.3.2	-1946,7	-1917,8	-1780,5	-1780,5	-1754,2	-1627,7
G 1.4.1	-1950,2	-1921,9	-1836,7	-1835,9	-1814,2	-1645,5
G 1.4.2	-1950,2	-1921,9	-1794,5	-1794,5	-1768,3	-1642,8
G 1.5.1	-1953,3	-1926,0	-1858,5	-1857,5	-1834,0	-1667,3
G 1.5.2	-1953,3	-1926,0	-1814,3	-1814,3	-1787,8	-1664,7
G 1.6.1	-1955,9	-1929,5	-1914,2	-1908,3	-1868,0	-1701,1
G 1.6.2	-1955,9	-1929,5	-1848,5	-1848,3	-1819,6	-1698,4

BEAMS

Figure 23. Axial force development in characteristic section 2

In figure 23 were detailed the axial force development in characteristic section 2 (coordinates: 2.75 m) depending on the concrete strength at the moment of stress transfer and the interval between construction stages. It can be noticed that in the case of the beams with higher time interval between constructions stages the values of axial forces tends to be the same for certain stages: for example in the case of beam G 1.1.2 the value of axial force in the construction stage *transport* is the same as the value obtained in the construction stage *final support*. In the case of the beams with a shorter time interval between construction stages, this particularity cannot be noticed: for example in the case of beam G 1.1.1 the value of axial force in the construction stage *transport* is higher than the value obtained in the construction stage *final support*. It can not state that this is a particularity resulted from shortening the time intervals between construction stages, a higher number of simulations being necessary to be carried out.

Also, the values of axial forces obtained in the case of beams with shorter intervals between constructions stages (for section 2) are higher than the values of axial forces obtained in the case of beams with longer intervals between construction stages except for the construction stage *end of design working life*, as illustrated in figure 23.

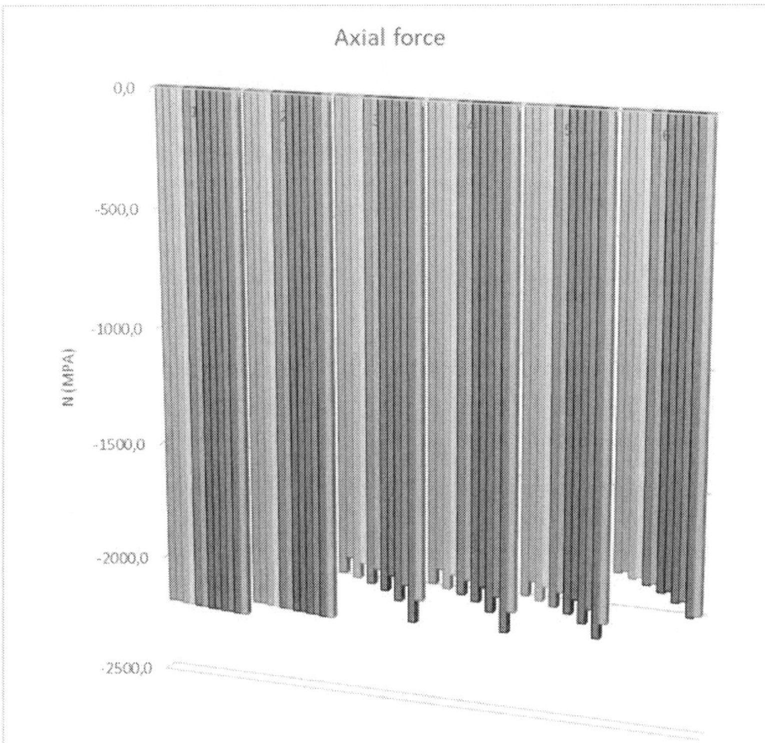

	1	2	3	4	5	6
G 1.1.1	-2191,4	-2155,5	-1978,3	-1981,4	-1986,9	-1845,1
G 1.1.2	-2191,4	-2155,5	-1912,6	-1917,1	-1923,7	-1835,8
G 1.2.1	-2197,9	-2160,9	-1994,3	-1997,4	-2003,1	-1863,8
G 1.2.2	-2197,9	-2160,9	-1928,8	-1933,2	-1940,3	-1854,6
G 1.3.1	-2203,4	-2167,0	-2012,5	-2015,6	-2021,2	-1884,1
G 1.3.2	-2203,4	-2167,0	-1946,8	-1951,3	-1958,7	-1874,8
G 1.4.1	-2207,9	-2172,4	-2036,3	-2039,3	-2044,2	-1908,8
G 1.4.2	-2207,9	-2172,4	-1969,8	-1974,3	-1981,6	-1899,6
G 1.5.1	-2211,9	-2177,8	-2073,3	-2076,1	-2078,3	-1944,2
G 1.5.2	-2211,9	-2177,8	-2004,0	-2008,5	-2015,1	-1934,9
G 1.6.1	-2215,3	-2182,6	-2161,7	-2157,9	-2136,7	-2000,0
G 1.6.2	-2215,3	-2182,6	-2062,6	-2066,9	-2070,0	-1990,7

BEAMS

Figure 24. Axial force development in characteristic section 3

In figure 24 were detailed the axial force development in characteristic section 3 (coordinates: 12.75 m) depending on the concrete resistance at the moment of stress transfer and the interval between construction stages. It can be noticed that similar behaviours are taking place as in the previous cases (section 1 and section 2). The highest value of the axial forces at the end of design working life is obtained in beam G 1.6.1. A particularity for this section consist of that axial force does not decrease linear as the construction stages are taking place, an increased being noticed for construction stages *final support* and *end of design working life*.

Some inconsistencies referring at the previous statement identified for beams G 1.6.1 and G 1.6.2. Further simulations must be carried out.

IOP Conf. Series: Materials Science and Engineering **586** (2019) 012018 doi:10.1088/1757-899X/586/1/012018

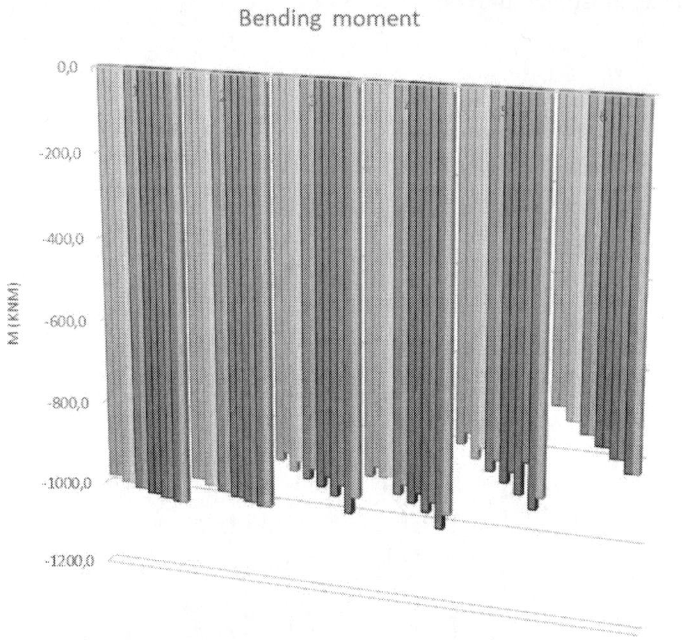

	1	2	3	4	5	6
G 1.1.1	-980,9	-964,2	-894,4	-907,8	-805,2	-693,4
G 1.1.2	-980,9	-964,2	-875,0	-883,8	-778,1	-693,0
G 1.2.1	-995,1	-977,7	-916,7	-906,8	-836,5	-725,3
G 1.2.2	-995,1	-977,7	-892,4	-903,0	-809,5	-725,0
G 1.3.1	-1006,0	-988,9	-931,9	-943,4	-862,3	-751,9
G 1.3.2	-1006,0	-988,9	-907,5	-919,5	-835,5	-751,7
G 1.4.1	-1015,0	-998,3	-947,3	-959,9	-886,0	-776,5
G 1.4.2	-1015,0	-998,3	-922,6	-935,7	-859,3	-776,2
G 1.5.1	-1022,6	-1006,4	-965,8	-979,2	-910,2	-801,5
G 1.5.2	-1022,6	-1006,4	-939,8	-953,9	-833,3	-801,3
G 1.6.1	-1028,7	-1013,2	-1003,9	-1015,1	-940,6	-831,5
G 1.6.2	-1028,7	-1013,2	-964,5	-979,2	-912,3	-831,5

BEAMS

Figure 25. Bending moment development in characteristic section 1

In figure 25 were detailed the bending moment development in characteristic section 1 (coordinates: around 0.50 m) depending on the concrete resistance at the moment of stress transfer and the interval between construction stages. The highest values are obtained for the beam G 1.6.1 and the smallest for the beam G 1.1.2. For all the beams can be noticed that bending moment does not decrease linearly as the construction stages are taking place, an increased being noticed for construction stage *final support*.

A higher bending moment is obtained in the case of beams with shorter intervals between constructions stages (for section 1) than in the case of beams longer intervals between them.

IOP Conf. Series: Materials Science and Engineering **586** (2019) 012018 doi:10.1088/1757-899X/586/1/012018

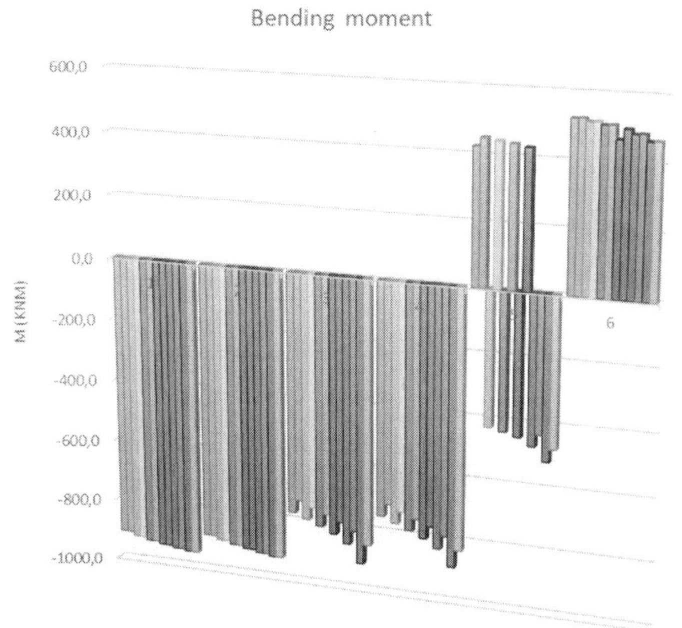

	1	2	3	4	5	6
G 1.1.1	-907,0	-884,1	-772,4	-748,1	424,7	522,7
G 1.1.2	-907,0	-884,1	-732,0	-708,5	450,9	524,2
G 1.2.1	-919,0	-895,4	-790,1	-766,7	-426,0	515,2
G 1.2.2	-919,0	-895,4	-749,8	-727,3	444,5	516,7
G 1.3.1	-928,5	-905,2	-807,3	-784,7	-435,5	507,6
G 1.3.2	-928,5	-905,2	-766,8	-745,1	438,0	509,1
G 1.4.1	-936,5	-913,8	-827,2	-805,1	-447,9	468,4
G 1.4.2	-936,5	-913,8	-786,1	-765,0	429,9	499,9
G 1.5.1	-943,3	-921,5	-854,3	-832,5	-472,1	486,2
G 1.5.2	-943,3	-921,5	-811,3	-790,6	-433,1	487,7
G 1.6.1	-948,8	-928,0	-914,2	-888,3	-515,6	467,5
G 1.6.2	-948,8	-928,0	-850,8	-830,4	-474,3	469,0

BEAMS

Figure 26. Bending moment development in characteristic section 2

In figure 26 were detailed the bending moment development in characteristic section 2 (coordinates: around 2.75m or 3.50 m depending on the highest values of the bending moment) depending on the concrete resistance at the moment of stress transfer and the interval between construction stages. The highest values are obtained for the beam G 1.6.1 and the smallest for the beam G 1.1.2. Specific for this section is the that in the construction stage *superimposed dead loads* the highest moments develop is located at the bottom fibre of the beams or ar the top fibre of the beam as the case may be as is shown in figure 25.

In figure 27 were detailed the bending moment development in characteristic section 3 (coordinates: 12.75 m) depending on the concrete strength at the moment of stress transfer and the interval between construction stages. The highest values are obtained for the beam G 1.1.1 and the smallest for the beam G 1.6.1 for the construction stage *end of design working life*. From figure 26, we can not determine a general rule that applies to all simulated beams.

IOP Conf. Series: Materials Science and Engineering **586** (2019) 012018 doi:10.1088/1757-899X/586/1/012018

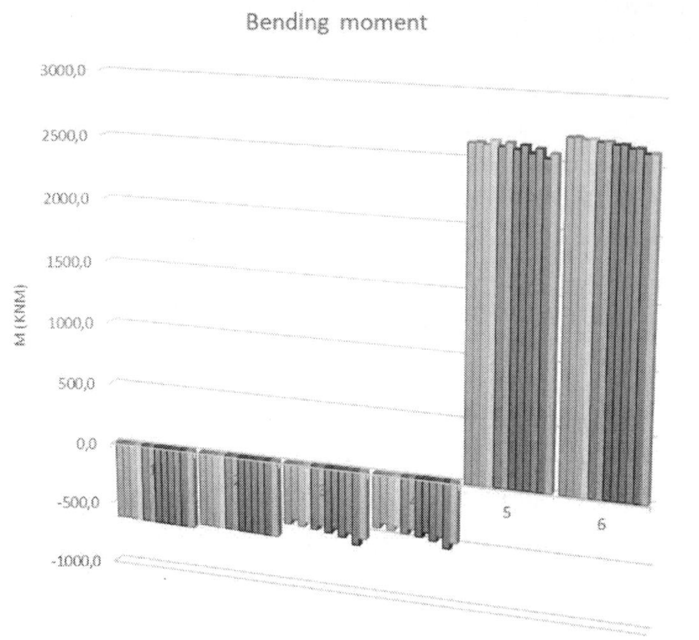

	1	2	3	4	5	6
G 1.1.1	-621,3	-600,5	-498,1	-436,7	2615,7	2697,3
G 1.1.2	-621,3	-600,5	-460,1	-399,5	2622,2	2702,7
G 1.2.1	-624,9	-603,5	-507,2	-445,8	2606,5	2686,6
G 1.2.2	-624,9	-603,5	-469,3	-408,7	2642,8	2692,0
G 1.3.1	-627,8	-606,7	-517,4	-460,0	2596,4	2675,2
G 1.3.2	-627,8	-606,7	-479,5	-418,9	2632,5	2680,6
G 1.4.1	-630,5	-610	-531,3	-469,8	2583,0	2660,8
G 1.4.2	-630,5	-610	-492,9	-432,3	2619,1	2666,2
G 1.5.1	-632,8	-613,1	-552,7	-491,1	2563,3	2640,4
G 1.5.2	-632,8	-613,1	-512,6	-452,0	2599,7	2645,8
G 1.6.1	-634,7	-615,7	-603,7	-538,3	2529,6	2608,2
G 1.6.2	-634,7	-615,7	-546,3	-485,6	2568,2	2613,7

BEAMS

Figure 27. Bending moment development in characteristic section 3

In figure 28 were detailed the ratio resulted from shear force (external loads) and the maximum shear force capable. For the beams G 1.6.1 and G 1.6.2 were obtained the smallest values while for beams G 1.1.1 and G 1.1.2 were obtained the highest values. The influence of constructions stages (*storage yard*, *transport*, *final support*, *superimposed dead load* and *end of design working life*) do not seem to have a significant influence on this ratio.

Figure 29 describes the ratio resulted from maximum allowed stress and the maximum capable stress. For beams G 1.1.1 and G 1.1.2 the ratio is higher than 100%, meaning that the elements do not satisfy the criteria of EUROCODE 2. These simulations were carried out on purpose to highlight this aspect. Also in the case of the beams G 1.2.1 and G 1.2.2 similar results were obtained with the ratio being close to 100%. The smallest ratio is obtained in the case of the beam G 1.5.2, meaning that the construction stages plays a significant role. This value was obtained when the stress transfer was done when the concrete reached a value of 45 MPa (at 13,3 days). A general rule cannot be established, further investigations were necessary to be carried out.

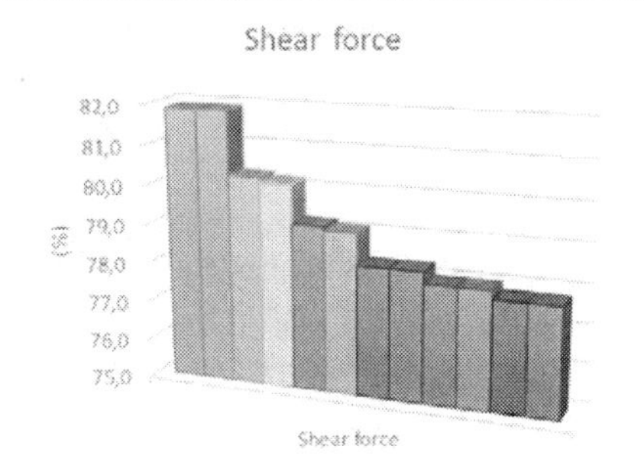

BEAMS	Shear force
G 1.1.1	81,9
G 1.1.2	81,9
G 1.2.1	80,3
G 1.2.2	80,2
G 1.3.1	79,2
G 1.3.2	79,1
G 1.4.1	78,3
G 1.4.2	78,3
G 1.5.1	78,0
G 1.5.2	78,0
G 1.6.1	77,8
G 1.6.2	77,8

Figure 28. Shear force (%)

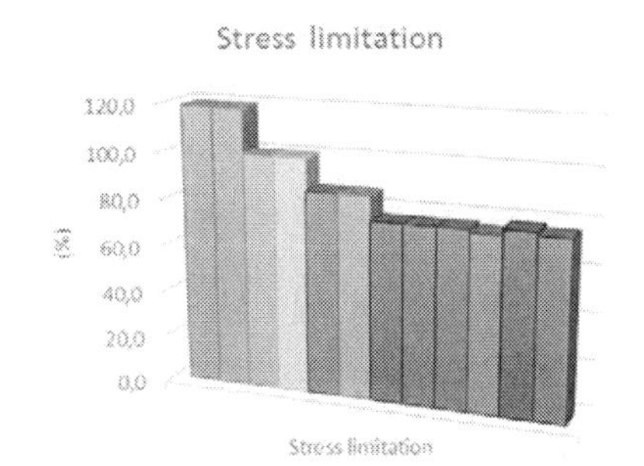

BEAMS	Stress limitation
G 1.1.1	118,8
G 1.1.2	118,8
G 1.2.1	99,9
G 1.2.2	99,9
G 1.3.1	86,3
G 1.3.2	86,3
G 1.4.1	76,1
G 1.4.2	76,1
G 1.5.1	76,1
G 1.5.2	74,7
G 1.6.1	77,3
G 1.6.2	75,9

Figure 29. Stress limitation (%)

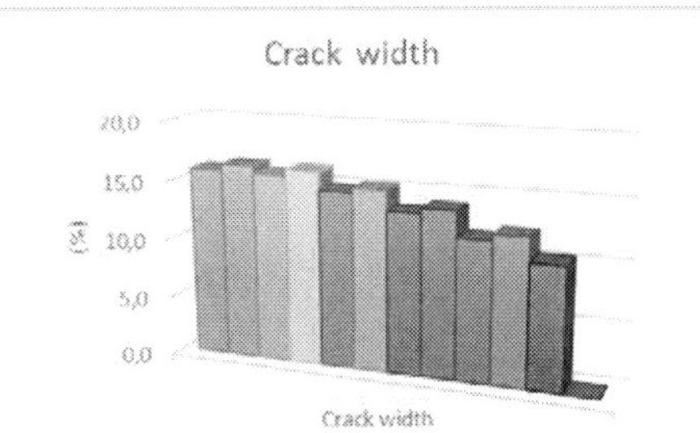

BEAMS	Crack width
G 1.1.1	15,9
G 1.1.2	16,4
G 1.2.1	15,8
G 1.2.2	16,3
G 1.3.1	14,8
G 1.3.2	15,2
G 1.4.1	13,5
G 1.4.2	14,0
G 1.5.1	11,8
G 1.5.2	12,3
G 1.6.1	10,4
G 1.6.2	0,0

Figure 30. Crack width (%)

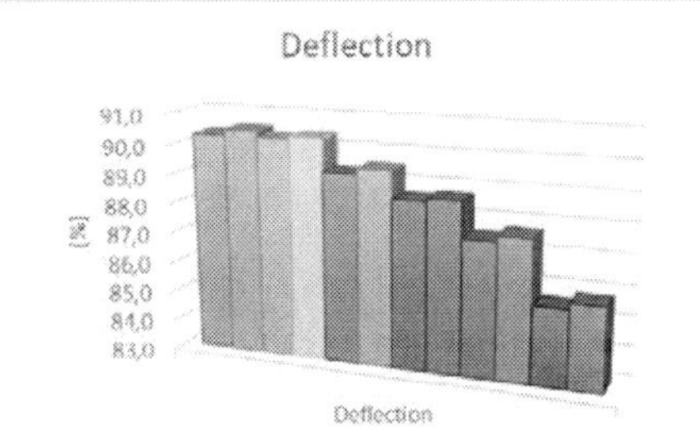

BEAMS	Deflection
G 1.1.1	90,3
G 1.1.2	90,5
G 1.2.1	90,3
G 1.2.2	90,4
G 1.3.1	89,3
G 1.3.2	89,5
G 1.4.1	88,6
G 1.4.2	88,7
G 1.5.1	87,5
G 1.5.2	87,7
G 1.6.1	85,6
G 1.6.2	85,8

Figure 31. Deflection (%)

In figure 30 were detailed the ratio resulted from crack width (existing on the beams) and the maximum accepted crack width. The highest ratio was obtained in the case of the beam G 1.1.2, meaning that construction stages play a significant role, although the maximum crack width was not reached. In the case of beam G 1.6.2 are not presented any cracks on the members.

Figure 31 describes the ratio resulted from the existing deflection and the maximum deflection accepted. Due to the fact that the values are closer, a general rule cannot be established.

3. Conclusions

After interpreting the previous results we can state the following:
While designing a prestressed prefabricated concrete member (beam) it is necessary to evaluate all the influences resulted from construction and services stages. Inadequate evaluation of these stages may lead to a faulty design. The service and construction stages (which are those and the time interval between them) are dictated by the imposing demanding. Reported to the current demanding, it is necessary that the stress transfer has to be made in a one or two days after casting in place. This procedure is not feasible due to the technological time required for the concrete to have a minimum strength unless are used special methods to speed up the curing of concrete. Such methods are very expensive involving special treatment of concrete by using special technologies (for example heated formwork). By using such methods, the member (beam) is subjected to more stress losses, resulting in elements with lower capabilities than in case of the elements who are subjected to stress transfer at an age of curing of 28 days.
The influence of construction and service stages are detailed graphically and tabular in Section 2. General rules cannot be formulated due to the complex processes that takes place. From previous results, it can be noticed that the stress transfer can be realized to an early age of curing and the beam still having similar performance with beams who had the stress transfer at the age of concrete curing of 28 days. In certain situation using special technologies with the purpose of reaching the highest concrete strength in the shortest time may lead to an element with weaker performances but manufactured in a shorter time. To clarify this aspect is necessary to conduct a technical-economic analysis.
The time between construction stages plays a significant role due to the rheological effects that are involved in the exploitation of a prestressed prefabricated concrete beam. Due to the complex process of design, it was not possible to identify a pattern.

4. References

[1] SR EN 1992-1-1:2004 *Proiectarea structurilor din beton, Partea 1 - 1:Reguli generale și reguli pentru clădiri*
[2] Tertea I *Betonul precomprimat. Bazele proiectării ed Tehnică București*
[3] Zumbrunnen L G, Russell B W and Burns N H 1992 *Behavior of statically loaded pretensioned concrete beams with 0.5-inch diameter debonded strands*
[4] Adballa O A, Ramirez J A and Lee R H 1993 *Strand debonding in pretensioned beams – precast prestressed concrete bridges with debonded strands*
[5] Tuns I, Mihai P and Florea N 2007 *Betonul armat și precomprimat. Betonul – armătură – conlucrare*
[6] Pascu R 2008 *Betonul precomprimat. Bazele calcului*
[7] Kiss Z and Oneț T 2008 *Proiectarea structurilor de beton după SR EN 1992-1* ed Abel
[8] Mihai P 2010 *Reinforced concrete structure* ed Matei – Teiu Botez Iași
[9] Tuns I and Florea N 2010 *Betonul precomprimat* vol. 1 ed Matrix Rom București
[10] Vidales M D 2011 *Effect of partial debonding of prestressing strands on beam end cracking*
[11] Burgueno R and Sun Y 2011 *Effects of debonded strands on the production and performance of prestressed concrete beams*
[12] Pascu R and Zybaczinski A 2012 *Beton precomprimat. Calculul după Eurocode 2*

Design of a high-rise building with hybrid steel-concrete structural elements

V-M Venghiac[1], M Budescu[1], S-A Băetu[1]
[1] "Gheorghe Asachi" Technical University of Iași, Faculty of Civil Engineering and Building Services, România

E-mail: mircea.venghiac@tuiasi.ro

Abstract. This paper presents the challenges encountered during the structural design of a high-rise building located in a seismic area. The construction is an office building, it has 20 stories including the basement levels and a narrow base. Some of the challenges are: providing an accurate structural system to resist the wind and earthquake actions and to design a proper foundation to avoid the overturning of the entire structure. Another important aspect is to limit the stories drift but at the same time provide an open space for the office area. Hybrid structures do not behave like reinforced concrete structures in the sense of Eurocode 2 nor like composite structures in the sense of Eurocode 4. For this reason and due to the lack of knowledge related to the problem of force transmission between steel and concrete the design of structures with hybrid steel-concrete elements can become very difficult. Several models are analysed comparing structural systems made of reinforced concrete, ones made of hybrid steel-concrete elements and several other types of systems.

1. Introduction

The strength and safety of buildings are the most important requirements that must be fulfilled and carefully designed by the structural engineer. This can be achieved by a very clear understanding of the main factors that influence the behaviour of the structure. At the same time, choosing an appropriate structural system is very important.

Since the beginning of time the main aspirations of man were to build upward and to span long distances. The Great Fire in Chicago and the restricted construction area in New York City provided ever so more the need to build upward. Influential engineers and architects teamed up and created the first and second Chicago Schools of architecture. Their work can today be seen in the skyline of these two cities. Fazlur Kahn, a Bangladeshi American structural engineer, recognized that for high-rise buildings the traditional frame system for resisting wind loads is too expensive. He proposed to view the high-rise structure as a huge upright cantilever beam. Thus the framed-tube system was created. This system was later updated to a trussed-tube for John Hancock Centre in Chicago, which has 30% less steel than similar tall buildings, and to a bundled-tube system in Chicago's Sears Tower (now the Willis Tower) which held the world's tallest building title for more than 20 years.

These structural systems are efficient for buildings over 50 stories. For 15-20 stories structures the frame system remains the best choice. Rigid frames were developed in the late 1800s for structures made of iron and reinforced concrete. However, if the base of the structure is very narrow compared to its height an inner core needs to be adopted for resisting the lateral loads. Usually, for reinforced concrete structures, the inner core is made of reinforced concrete load bearing walls and the perimeter is made of frames. Even for 10-20 stories buildings very large uplift forces appear at the base of the inner core. These forces can cause the need for very large reinforcement areas for the inner core walls and also foundation design problems. In order to distribute the inner core forces to the perimeter frames deep beams and relatively closely spaced columns on the building perimeter can be adopted. However, often times the architectural solution does not allow for these obstructions. If the buildings height-to-width ratio exceeds 6 the structural system becomes uneconomical [1].

Most modern high-rise buildings incorporate an inner core housing the staircase and elevators along with a column-free floor space between the core and perimeter columns. This system allows for greater functionality, but also effectively disconnects the two major structural elements available to resist the overturning forces. This uncoupling of the inner core from the perimeter frames greatly reduces the system's capacity of resisting lateral loads. One method for solving this problem is to incorporate outriggers. Outriggers have been incorporated in high-rise buildings within the last 35 years. An outrigger is a rigid horizontal structure designed to improve building overturning stiffness and strength by connecting the inner core to the perimeter columns. Depending on the building height range it can be placed at the top of the building, at mid-height or in pairs mid-height and at the top. In order to uniformly distribute the forces to the perimeter columns belts – such as trusses or walls encircling the building – can be implemented at the level where the outrigger is located [1].
Heavily loaded structures can be provided with steel profiles embedded in the concrete. These structures are called hybrid structures or SRC (Steel Reinforced Concrete). These structures are neither reinforced concrete structures in the sense of Eurocode 2 [2], nor composite steel-concrete structures in the sense of Eurocode 4 [3]. There is no guidance in Eurocode 4 on how to evaluate and design this type of structures. There is also no information in classical literature references [4, 5, 6, 7, 8]. Some partial design provisions are mentioned in Eurocode 8 [9] or AISC 341-10 [8], but are not very explicit. In the past decade, extensive experimental research was carried out for the study of hybrid walls [10, 11, 12, 13] and numerical models were developed, such as: multiple vertical-line-element models [14, 15], fibre beam-column models by PEER [16], and multi-layer shear element models [17, 18]. However, all this work did not lead to practical design tools [19].
The aim of this paper is to present different configurations of structural systems – some presented in this chapter – their behaviour and efficiency, for a 17 storey reinforced concrete office building located in a seismic area.

2. Modelling in Etabs

2.1. Etabs software
Etabs is an engineering software package for the structural analysis and design of buildings. The software includes the design of steel and concrete frames with automated optimization, composite beams, composite columns, steel joists and concrete and masonry walls and many other features [20].

2.2. Models description
The common features of the models taken into consideration are the following:

- height range: ground floor and 16 levels;
- ground floor height of 4.5 m and current level height of 3.5 m resulting in a total height of the building of 60.5 m;
- floor dimensions of 19.2 m x 18.0 m;
- open office area having a width of 5.4 m;
- inner core of 8.4 m x 7.2 m with 2 stair cases and 2 elevator shafts;
- inner core thickness of 60 cm for the first three levels, 50 cm for the next three levels and 40 cm for the rest having uniform 80 cm thick bulbs and the corners and at mid-span (see figure 1);
- perimeter frames made of 80 cm x 80 cm columns and 40 cm x 60 cm beams;
- 20 cm thick flat slabs;
- C40/50 concrete grade.

The models taken into consideration have the following particularities:

- model A is the base model having the particularities presented in the previous paragraphs (figure 1);
- model B has 6 HE600B profiles encased in the inner core corners (figure 2) on the first 6 levels effectively creating a hybrid steel-concrete structure. The steel grade is S355;

IOP Conf. Series: Materials Science and Engineering **586** (2019) 012019 doi:10.1088/1757-899X/586/1/012019

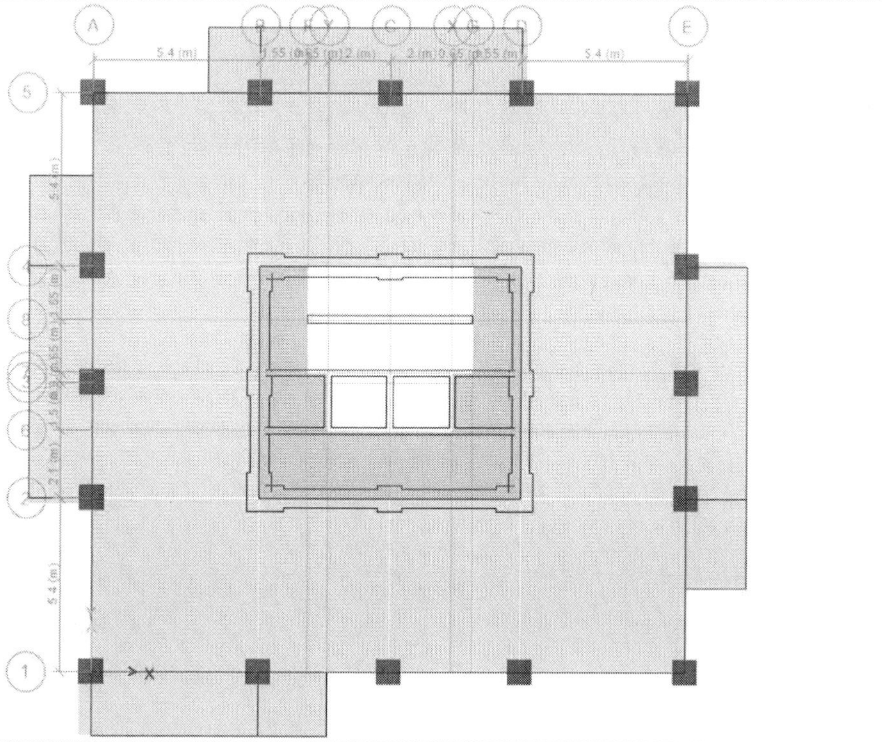

Figure 1. Model A: floor plan.

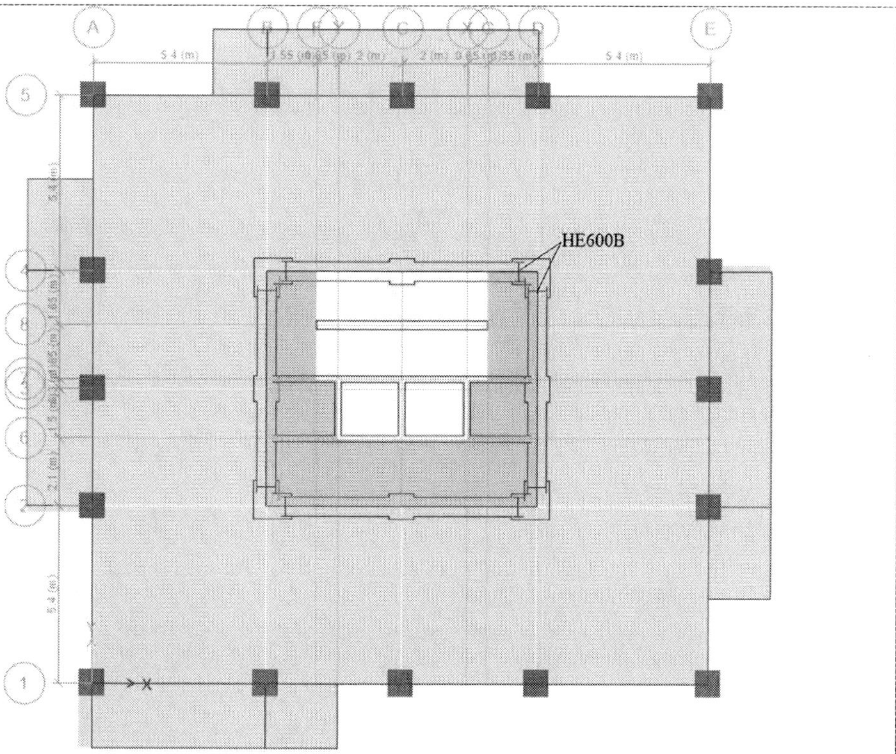

Figure 2. Model B: floor plan.

IOP Conf. Series: Materials Science and Engineering **586** (2019) 012019 doi:10.1088/1757-899X/586/1/012019

- model C has one outrigger and a truss belt on the entire perimeter placed at the top level of the building. Reinforced concrete 40 cm x 60 cm beams are used to connect the inner core to the perimeter columns (only at these levels) and 200 mm x 200 mm x 12.5 mm tube profiles are used as braces for the outrigger and belts. The steel grade is S355 (figure 3);
- model D has one outrigger and belts on the height of the 8th floor of the structure. The make-up of the outrigger and belts is the same as model C (figure 4);
- model E has two outriggers and belts: one at the top level and one on the height of the 8th floor of the structure. The make-up of the outrigger and belts is the same as model C (figure 5);
- model F has the inner core connected to the perimeter columns with 40 cm x 60 cm reinforced concrete beams (figure 6).

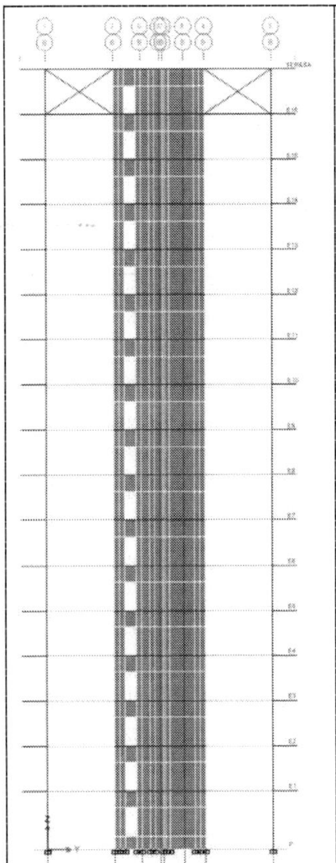

Figure 3. Model C: elevation.

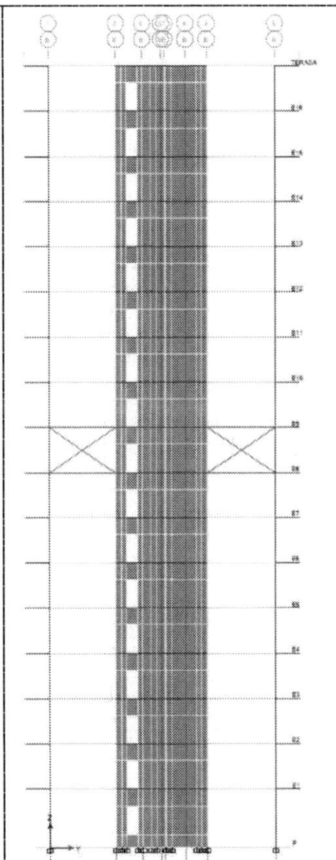

Figure 4. Model D: elevation.

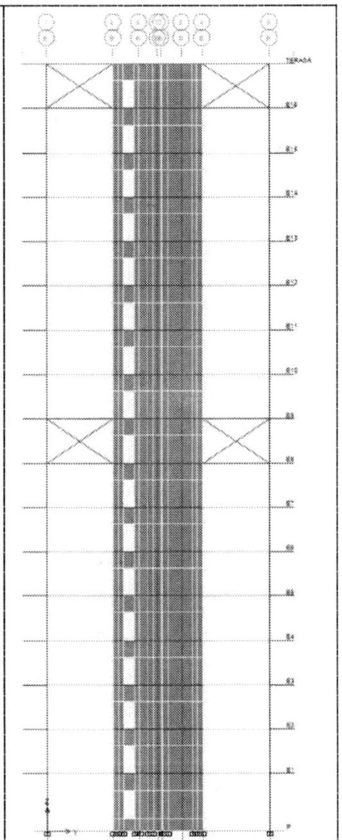

Figure 5. Model E: elevation.

The dimensions of the structural elements were established after a preliminary design according to CR 2-1-1.1/2013 [21] and SR EN 1992-1-1-2004.

The loads are the own weight of the structure estimated by the software, floor finish loads of 2.5 kN/m^2, curtain walls loads of 2.0 kN/m, live loads of 2.5 kN/m^2 in the office area and 3.0 kN/m^2 on the hallways and stairs. The horizontal loads such as wind and earthquake were evaluated according to CR 1-1-4/2012 [22] and P100-1/2013 [23], respectively. The location of the building is Iași, Romania having a peak ground acceleration a_g = 0.25g and a corner period T_C = 0.7 s.

IOP Conf. Series: Materials Science and Engineering **586** (2019) 012019 doi:10.1088/1757-899X/586/1/012019

The earthquake action is introduced in the program by using a response spectrum function. The design spectral values are computed according to P100-1/2013 by using equation (1):

$$S_d(T) = a_g \frac{\beta(T)}{q} \tag{1}$$

Where:

- $S_d(T)$ is the design/inelastic spectral value;
- a_g – peak ground acceleration;
- $\beta(T)$ – normalised elastic spectral value;
- q – behaviour factor of the structure.

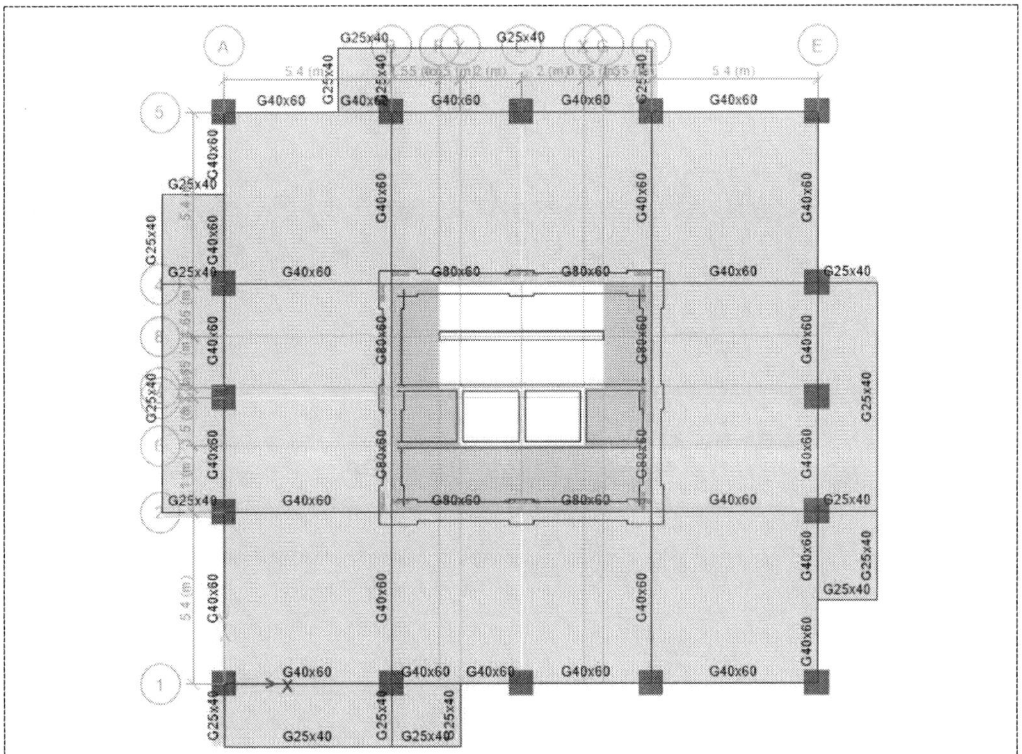

Figure 6. Model F: floor plan.

3. Results

The following results were considered for discussion:

- the required reinforcement area in the inner core of the structure;
- the story drift.

The story drifts for all models is presented in figure 7. According to P100-1/2013, annex E, the allowable storey drift in the Serviceability Limit State is 0.005 and for the Ultimate Limit State is 0.025.

The required reinforcement area for the inner core is designed using the Etabs software. The values at the base of the structure are presented in figure 8. By comparing models A and B, a reduction of reinforcement area of 11.9% was recorded. The models having outriggers and belts (models C, D and E) all present great reductions in reinforcement area (13.3%, 20.8% and 24.2%, respectively). Model F, having reinforced concrete beams linking the inner core to the perimeter columns presents a

IOP Conf. Series: Materials Science and Engineering **586** (2019) 012019 doi:10.1088/1757-899X/586/1/012019

reduction in reinforcement of 15.7% in the inner core, but reinforcement needs to be added for the additional beams.

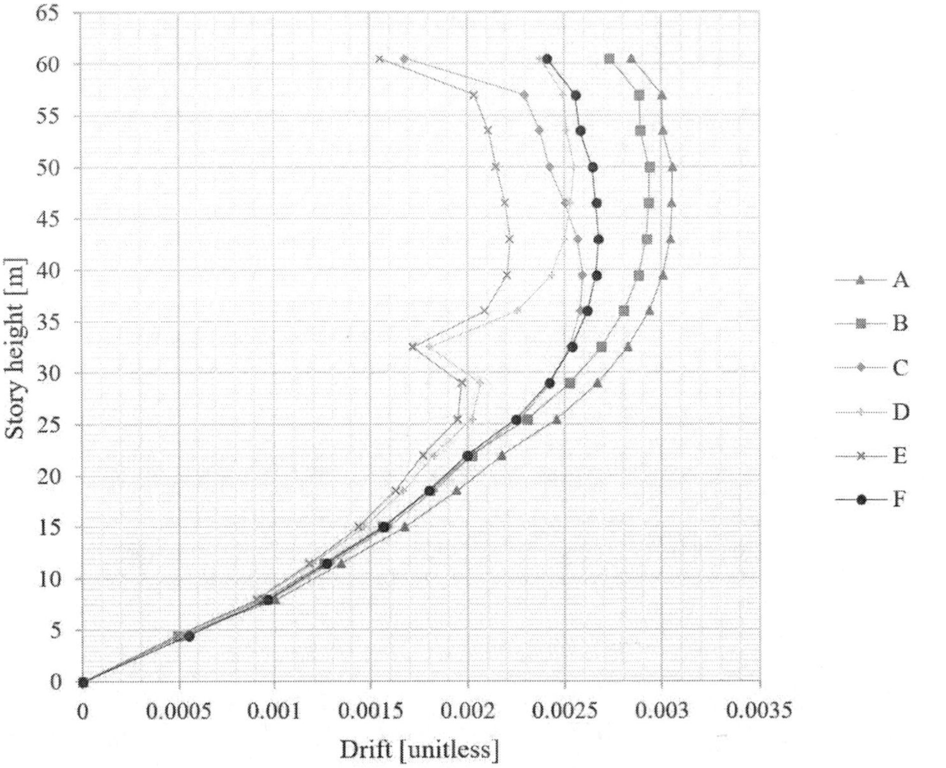

Figure 7. Story drift.

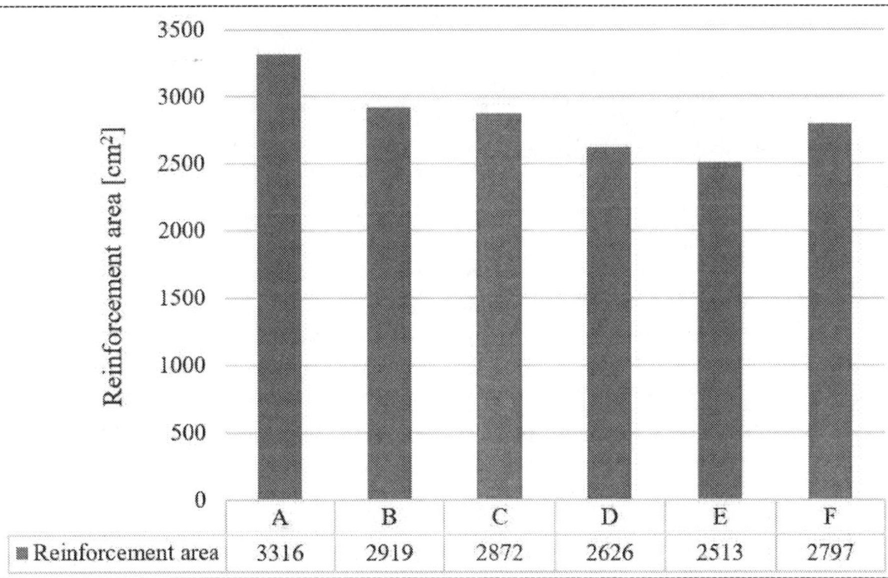

Figure 8. Required reinforcement area for the inner core: all models.

IOP Conf. Series: Materials Science and Engineering **586** (2019) 012019 doi:10.1088/1757-899X/586/1/012019

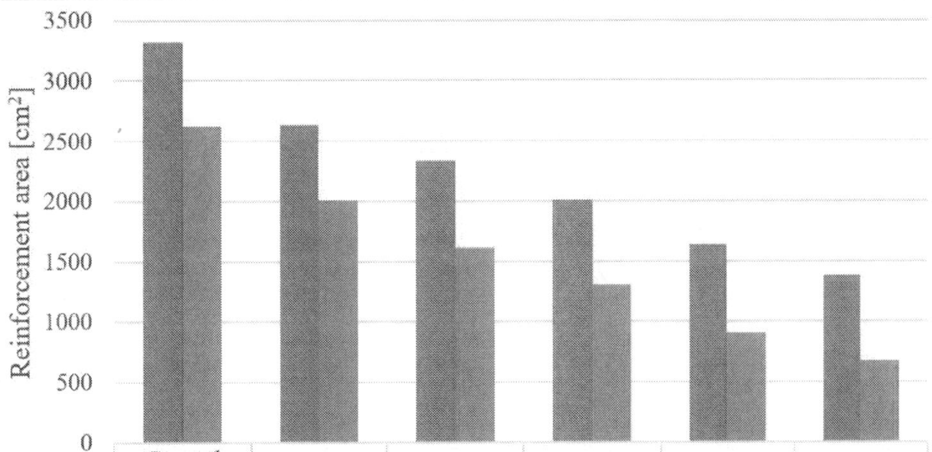

	Ground Floor	Level 1	Level 2	Level 3	Level 4	Level 5
■A	3316	2638	2332	2006	1640	1388
■D	2626	2014	1618	1308	906	675

Figure 9. Required reinforcement area for the inner core: models A and D.

The approximate quantity of steel required for one outrigger is 20 t. The reduction in reinforcement, of model D, for the first 6 levels of the structure is approximately 50.3 t. These reinforcement areas are presented in figure 9.

4. Conclusions

Structural engineers today have a great challenge to design tall buildings using cost effective solutions. At the same time, the architectural requirements became more demanding. Open office space and large windows are some of the requirements which impede the use of load bearing walls, braces, trusses or deep beams.

The structural systems analysed in this paper are usually used for buildings at least 40-50 stories tall. However, the results show that an important economy of reinforcement can be achieved by using these solutions even for buildings less than 20 stories tall. The amount of steel required for the outriggers does not exceed the amount of reinforcement saved in the inner core. The reduction of steel quantity for model D is significant (approximately 30.2 t). The main disadvantage of outriggers and truss belts is from the architectural point of view. Braces and trusses are used both in the office area and on the perimeter of the building, in the window area. However, this outcome might prove reasonable in compromising the architecture of a single level of the building with respect to the gain in material economy. Models C, D and E reflect the fact that the optimum position of one outrigger is at mid-height of the building. The efficiency of the outrigger placed at the middle of the building, in comparison to the placement at the top, is almost doubled as seen from the consumption of reinforcement (figures 8 and 9). The addition of 2 outriggers – one at the top and one at the middle of the building – show a very small improvement in comparison to the mid-height placement.

Model B shows a very small improvement. However, the use of steel profiles embedded in the inner core is beneficial for structures with very high axial forces in these areas. The main problem remains with the limited knowledge regarding the design of such structures, the load transfer and connections between steel and concrete.

The story drift does not exceed the allowable limit presented in P100-1/2013 for all models. However, a great reduction is observed for the models having outriggers (figure 7). The largest values were recorded, as expected, for the base model A and for model B having the hybrid column structural system.

5. References

[1] Taranath B S 2012 *Structural Analysis and Design of Tall Buildings* (CRC Press Taylor & Francis Group) pp 2-64

[2] Eurocode 2 2004, EN 1992-1-1 *Design of concrete structures, Part 1.1 – General rules for buildings* European Committee for Standardizations, Brussels

[3] Eurocode 4 2004, EN 1994-1-1 *Design of composite steel and concrete structures, Part 1.1 – General rules for buildings* European Committee for Standardizations, Brussels

[4] Oehlers D and Bradford M A 1995 *Composite Steel and Concrete Structural Members* (Oxford: Pergamon Press) p 549

[5] Leon R T, Kim D J and Hajjar J F 2007 Limit state response of composite columns and beam columns: Part I. Formulation of design provisions for the 2005 AISC specification *Engineering Journal* **45** 1 pp 21-46

[6] Leon R T and Hajjar J F 2008 Limit state response of composite columns and beam columns: Part II. Application of design provisions for the 2005 AISC specification *Engineering Journal* **44** 4 pp 341-358

[7] Johnson R P 2004 *Composite Structures of Steel and Concrete – Beams, Slabs, Columns and Frames for Buildings* (Oxford: Blackwell Scientific Publications) 3rd ed.

[8] AISC Standard 341-10 2010 *Seismic provisions for structural steel buildings* American Institute for Steel Construction, Chicago

[9] Eurocode 8 2004, EN 1998-1-1 *Design provisions for earthquake resistance, Part 1: General rules, seismic actions and rules for buildings* European Committee for Standardizations, Brussels

[10] Wallace J W, Orakcal K, Cherlin M and Sayre B L 2000 *Proc. of sixth ASCCS Conf. on Composite and Hybrid Structures (Los Angeles)* (University of Southern California) pp 801-808

[11] Qian J, Jiang Z and Ji X 2012 Behaviour of steel tube-reinforced concrete composite walls subjected to high axial force and cyclic loading *Engineering Journal* **36** 3 pp 173-184

[12] Ying Z, Xilin L and Yugang D 2010 *Seismic Behaviour of Composite Shear Walls with Multi Encased Steel Sections. Part 1: Experiment. The Structural Design of Tall and Special Buildings* (Wiley Online Library) pp 618-636

[13] Jim X, Sun Y, Qian J and Lu X 2015 Seismic behaviour and modelling of steel reinforced concrete (SRC) walls *Earthquake Eng. & Struct. Dyn.* **44** 6 pp 955-972

[14] Orakcal K, Wallace J W and Conte J P 2004 Flexural modelling of reinforced concrete walls-model attributes *ACI Structural Journal* **103** 2 pp 196-206

[15] Vulcano A, Bertero V V, Colotti V 1988 *Proc. of 9th World Conference on Earthquake Engineering* pp 41-46

[16] PEER/ATC 2010 *Modelling and acceptance criteria for seismic design and analysis of tall buildings* PEER/ATC 72-1 Report (Redwood City: Applied Technology Council)

[17] Miao Z, Ye L, Guan H and Lu X 2011 Evaluation of modal and traditional pushover analyses in frame-shear wall structures *Advances in Structural Engineering* **14** 5 pp 815-836

[18] Lu X, Lu X, Guan H and Ye L 2013 Collapse simulation of reinforced concrete high-rise building induced by extreme earthquakes *Earthquake Eng. & Struct. Dyn.* **42** 5 pp 705-723

[19] Plumier A 2017 Design of hybrid steel-concrete structures *Proc. of 15th National Conference on Steel Structures with International Participation* (Iasi: PIM Publishing House) pp 7-24

[20] ETABS Integrated Building Design Software 2016 *User's Guide* Computers & Structures, Inc.

[21] CR 2-1-1.1/2013 2013 *Cod de proiectare a construcţiilor cu pereţi structurali de beton armat* (in Romanian)

[22] CR 1-1-4/2012 2012 *Cod de proiectare. Evaluarea acţiunii vântului asupra construcţiilor* (in Romanian)

[23] P100-1/2013 2013 *Cod de proiectare seismică – Partea I – Prevederi de proiectare pentru clădiri* (in Romanian)

Computational Assessment of a RC Water Tank – A Comparative Static Analysis

A Filip[1], D Covatariu[2],
[1]Doctoral School of the Faculty of Civil Engineering and Building Services, "Gheorghe Asachi" Technical University of Iaşi, 700050, Romania
[2]Department of Civil and Industrial Engineering, Faculty of Civil Engineering and Building Services, "Gheorghe Asachi" Technical University of Iaşi, 700050, România

E-mail: filipalexandru91@gmail.com

Abstract. Liquids storage tanks are special structures and strategically very important. Water tanks, in particular, are important to assure continued operation of water distribution system in the event of unpredicted events. The aim of this article is to investigate the structural behaviour of a buried reinforced concrete water tank in different exploitation stages. A comparative study based on a finite element method (FEM) was considered in order to identify the most subjected areas of the structural elements (stress distribution). Numeric analysis using two advanced structural analysis and design software programs (Axis VM and SCIA Engineer) have been used for more accurate results. It was taken into consideration the extreme possibilities: case when the water tank is empty and when the water tank is at full capacity. The numerical results obtained were compared and the obtained results of both approaches has revealed minimal values differences.

Keywords: static analysis, finite element method, buried water tank;

1. Introduction

A structure must meet the exploitation, strength and stability requirements throughout the project lifetime without any significant loss of functionality and unforeseeable maintenance work [2].
In the current context, the requirements of strength, stability and safety in use must be handled with particular attention to reinforced concrete construction due to particular operating conditions.
The operating conditions of the tanks are assessed in accordance with the data collected in relation to the projected technological process taking place inside the basins, the operation of the technological equipment and associated facilities and the characteristics of the aggressive environment inside and outside the basins [1].
In general, the static actions applied to the water tanks during operation, the long-term influence of a series of climatic factors (variations in temperature, humidity) as well as the aggressive action of the corrosive agents contribute significantly to the reduction of the service life of these types of structures.
In the case of buried reinforced concrete tanks, the calculation of static actions is performed considering the liquid - structure interaction using approximate or exact calculation methods.

2. Particularities regarding the static behaviour of water tanks

In order to ensure the optimal exploitation, from a structural point of view, circular water tanks are preferred instead of rectangular ones. The radial pressure given by the stored liquid (water) is evenly distributed in the circular sections producing axial forces.
Another factor to be considered for operational safety is represented by the structural dimensions of the structure, which are limited by deformation and cracking conditions. The specific (radial or annular) deformation is independent of the tank diameter, the risk of cracking being directly proportional to the diameter of the tank [6].
The site of the structure is also a very important factor in the exploitation of the structure. It will follow

the fitting of the reservoir in the water supply scheme and the ground foundation characteristics. If the water tank is built in areas with steep slopes, unstable terrains with high compressibility or moisture sensitivity, additional measures need to be taken [8].
The site is regularly checked with regard to stability (levelling measurements on the tank and the emplacement). In case of buried water tanks, the protection against infiltrations will be represented by an effective waterproofing.
Other aspects of current operating conditions that must be considered in the structural design stage are: the use of appropriate materials for maintaining the water quality in the required quality parameters, cleaning the tank by providing a slope suitable for the foundation, ventilation, loading and unloading of the tank, protecting the pipes against corrosion [3].

3. Case study

3.1. Description of the structural system

The construction is a buried cylindrical reinforced concrete tank situated in Vişan village, Iaşi county, built during the 1980s, with the structural system presented in Figure 1.
The water tank has a circular shape, in horizontal plane, having an inner diameter of 13.40 m and a useful height of 3.80 m which is covered with earth. Within the objective under consideration, there is a pump room and a room for access to the tank

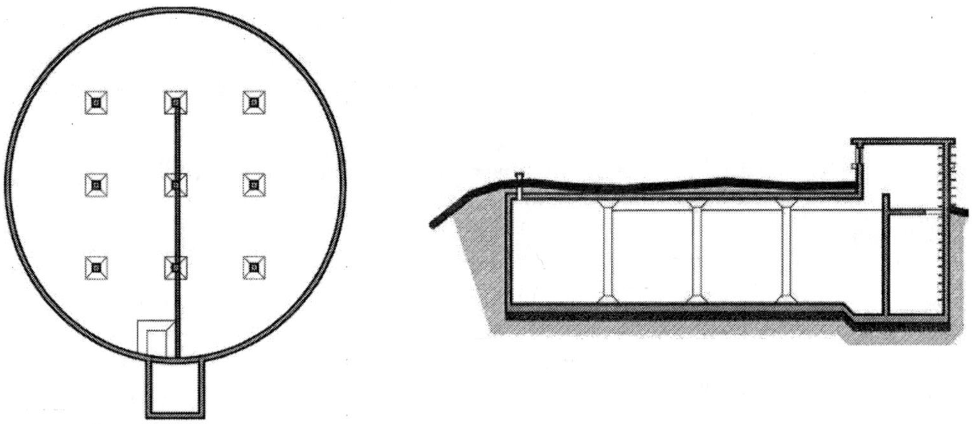

Figure 1. Buried RC water tank – Vişan village
a). Longitudinal section; b). Transversal section;

Structurally, the water tank consists of a reinforced concrete perimeter wall (20 cm thick) supported on a continuous foundation and internal reinforced concrete columns with caps at the extremities. The slab of the water tank is made of reinforced concrete and has a thickness of 20 cm, supported by the perimeter circular wall and nine interior 30 × 30 cm concrete columns.
The tank has a reinforced concrete wall inside, placed between the central columns, which partially divides the tank into two volumes and a bay next to the pump chamber.
The reinforcement of the concrete elements consists of PC52 and OB37. In order to ensure a proper rigidity and to avoid infiltrations, the water tank has been cast monolithically on site.

3.2. Comparative analysis by FEM

The analysis of constructions (static, dynamic) in general, and of water tanks in particular, has been carried out using the Finite Element Method (F.E.M.). The use of F.E.M. allows the determination of stress and displacement fields for all load categories, considering special structural shapes [4,5].
In the domain of static analysis, for water tanks, the following calculation categories are based on the finite element method:

IOP Conf. Series: Materials Science and Engineering **586** (2019) 012020 doi:10.1088/1757-899X/586/1/012020

- static linear analysis;
- static nonlinear analysis;
- identification of the failure mechanism and safety factor assessment;
- optimization of automated design.

The studied structure was modelled, discretized and analysed by using Axis VM and SCIA Engineer software programs.

The linear analysis computed in two software offer accurate results and can be efficiently and successfully used to simulate and predict the structural behaviour in different exploitation stages, in order to emphasize the stress distribution within the structural elements.

The linear static analysis was performed according to the following structural models (Figures 2, 3, 4).

AxisVM software SCIA Engineer software

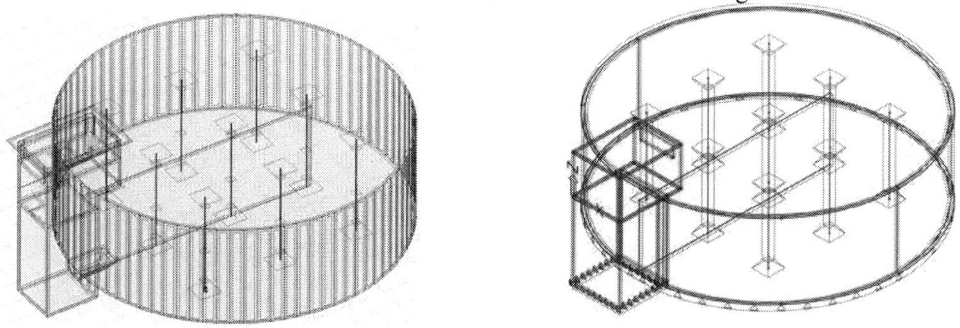

Figure 2. 3D view of the buried water tank

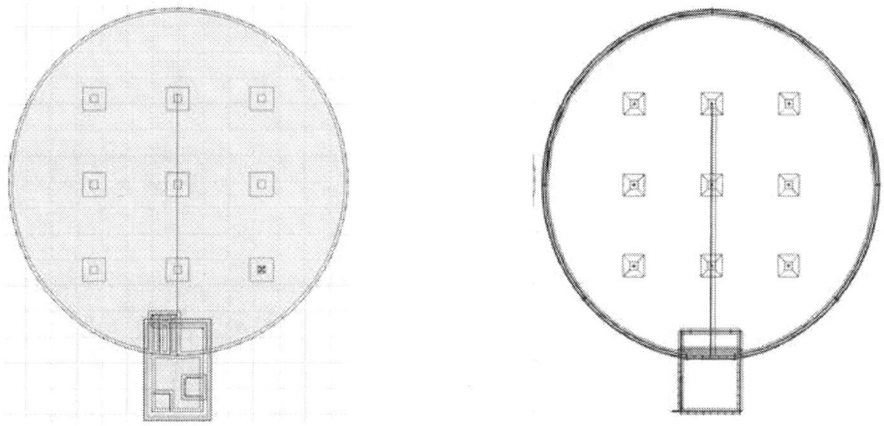

Figure 3. Longitudinal section of the buried water tank

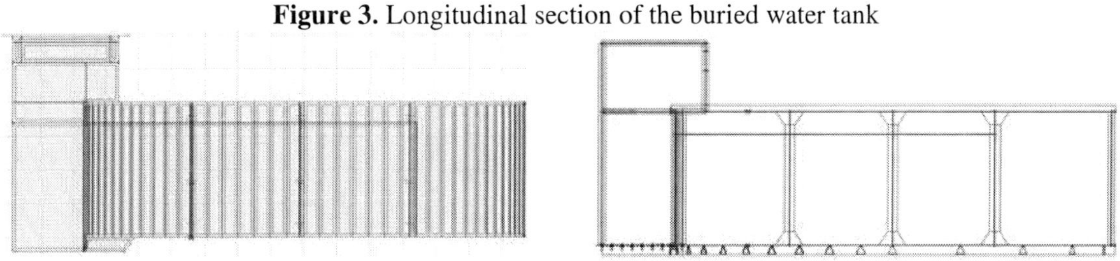

Figure 4. Transversal section of the buried water tank

The main objective of this analysis is to the assessment of structural behaviour (stress distribution) in different exploitation stages (full water tank and empty water tank).

3.3. Loading hypotheses

The accuracy of the considered analysis depends on the model's static scheme, which must be in line with the structural model (the nature of the links between the component elements, the abutments, etc.). The loads considered (Table 1) in the calculation of the water tank are of different nature:

Table 1. Loads acting on the water tank

Load	AxisVM software	SCIA Engineer software
The self-weight of the tank and its attachments (pump and tank access rooms);	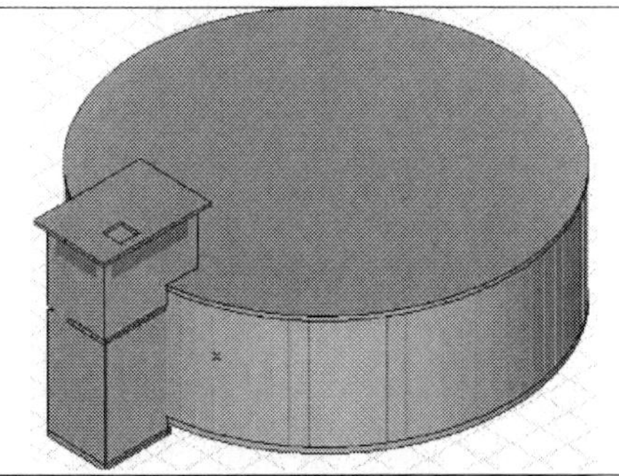	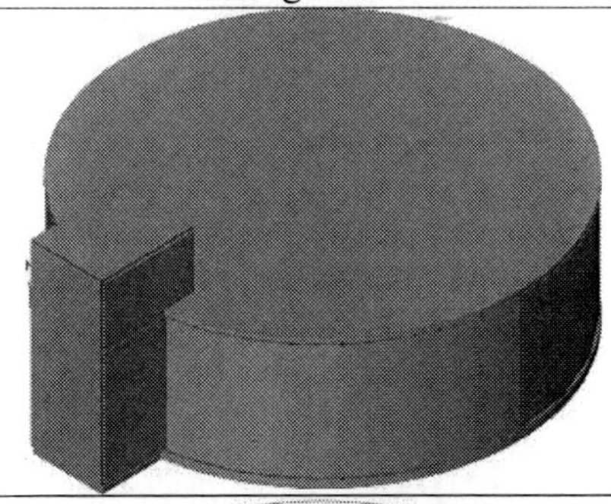
The pressure of the liquid stored in the tank (water);	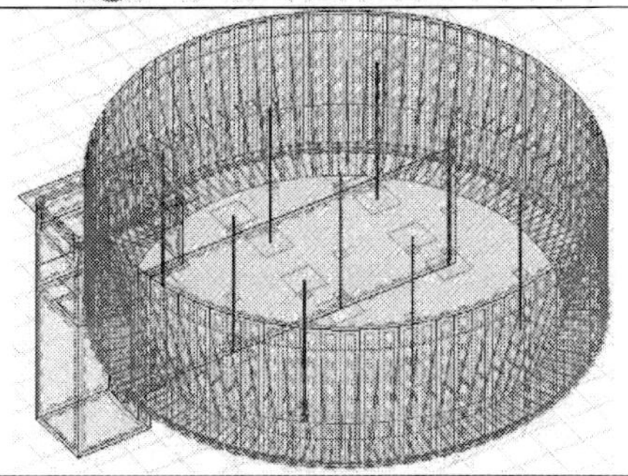	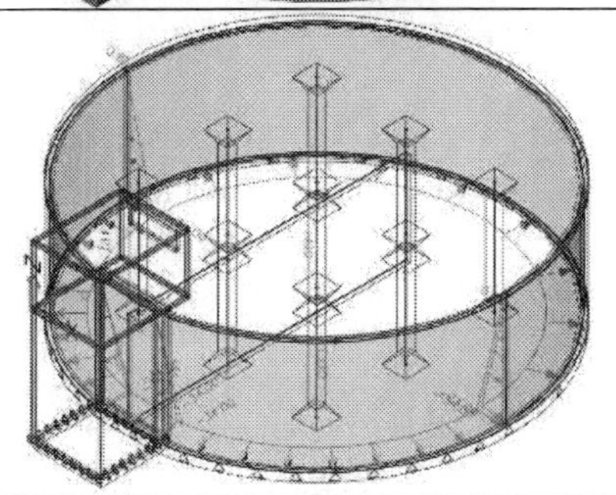
The pressure of the earth	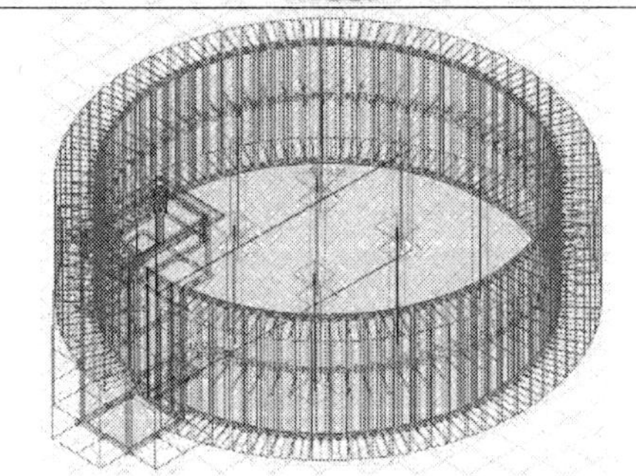	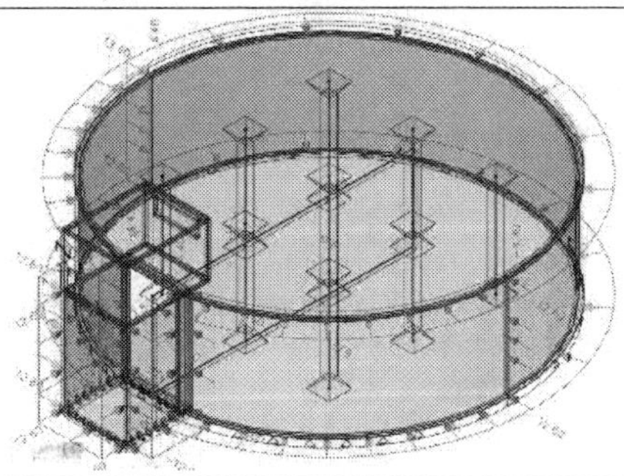
Variable actions (snow)	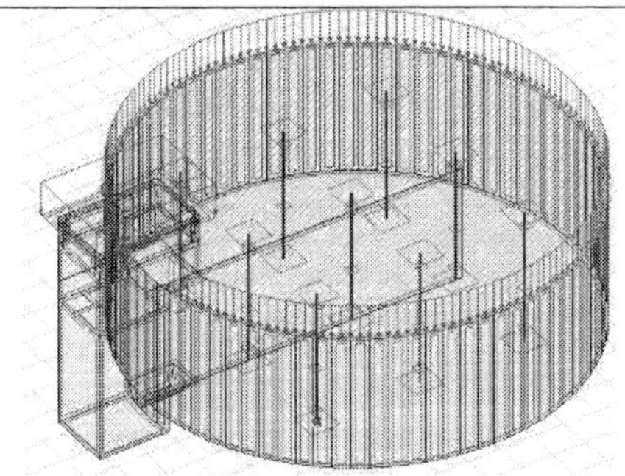	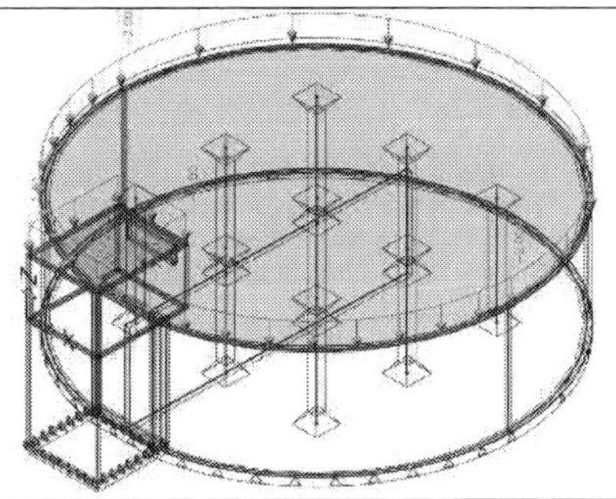

Partial safety coefficients for the combinations of actions in Service Limit State (SLS) checkings are:

- 1.00 for permanent actions;

- 0.40 for variable actions.

The load combination (quasi-permanent) for the service checking can be calculated as in equation (1) [10]:

$$\sum_{j=1}^{n} G_{k,j} + \sum_{i=1}^{m} \psi_{2,i} \cdot Q_{k,i} \quad (1);$$

where: $G_{k,j}$ – the characteristic value of the permanent action;
$Q_{k,i}$ – the characteristic value of the associated variable action;
$\psi_{2,i}$ – factor for the quasi-permanent value of the variable action.

Taking into account that the structure is considered in different exploitation stages, the static calculations are performed for the assumptions presented in Table 2.

Table 2. Loading hypothesis

	Full water tank	Empty water tank
Loads	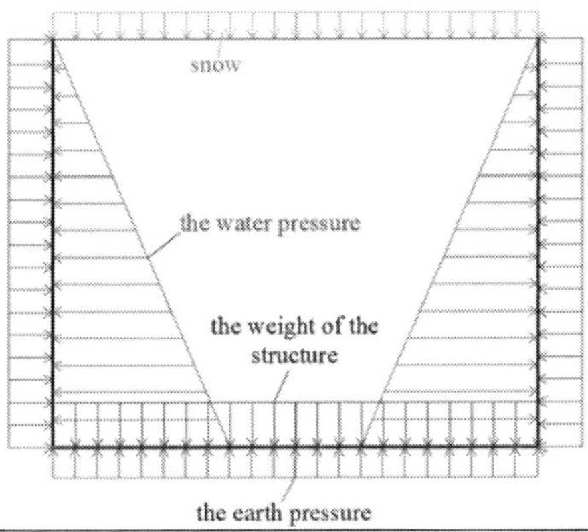	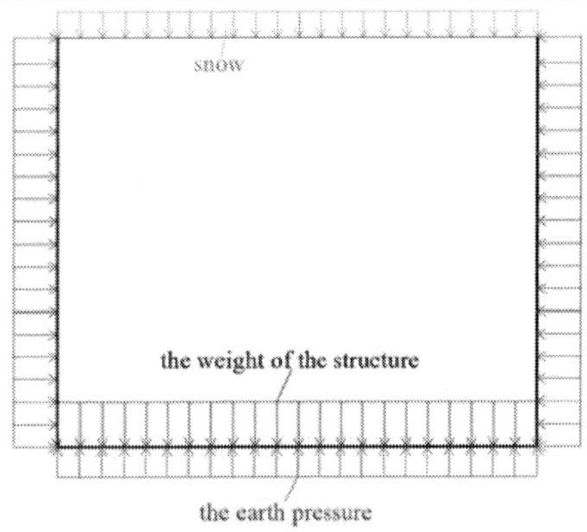
Type of combination	Service Limit State (SLS)	Service Limit State (SLS)
Load combination in respect with loading hypothesis [10]	1.00·SW+1.00·EP+0.40·S	1.00·SW+1.00·EP+0.40·S+0.40·W

where:
SW = the self-weight of the structure; EP = the earth pressure;
S = snow; W = the pressure of the stored liquid (water).

For this particular case study (buried water tank), the most important loads taken into consideration are the pressure of the stored liquid and the earth pressure.

The two loading hypotheses may be independent of one another, resulting in the need for additional iterations of the results.

3.4. Comparative analysis of the results

Meshing of a structure is a complex process of elaborating a discrete model that approximates the actual continuous structure from various points of view, such as geometric shape, load application, etc [5].

In the Axis VM program, the mesh network (Figure 5.a.), was automatically generated for the entire domain of analysis, the elements being triangularly divided, while in SCIA Engineer program (Figure 5.b.), the elements are rectangular.

Both programs automatically meshed the linear supports attached to the linear elements and the edges of the domains are considered as support elements automatically being divided into meshing.

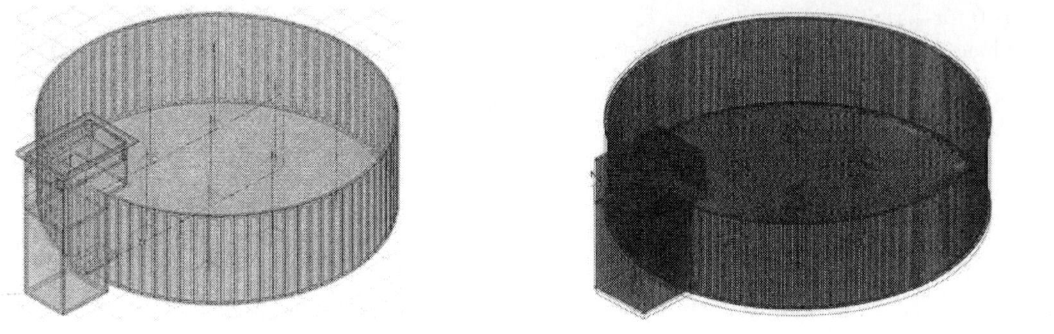

Figure 5. Meshing network
a). Axis VM software b). SCIA Engineer software

The obtained results from the linear static analysis in the software programs considering the two hypotheses (full water tank and empty water tank) are very similar, differences less than 5% for the stress distribution within the analysed structure being recorded.

3.4.1. Full water tank. In this situation, it is considering that the buried RC water tank is at its full capacity, the stress distribution being illustrated in the Figures 6, 7, 8.

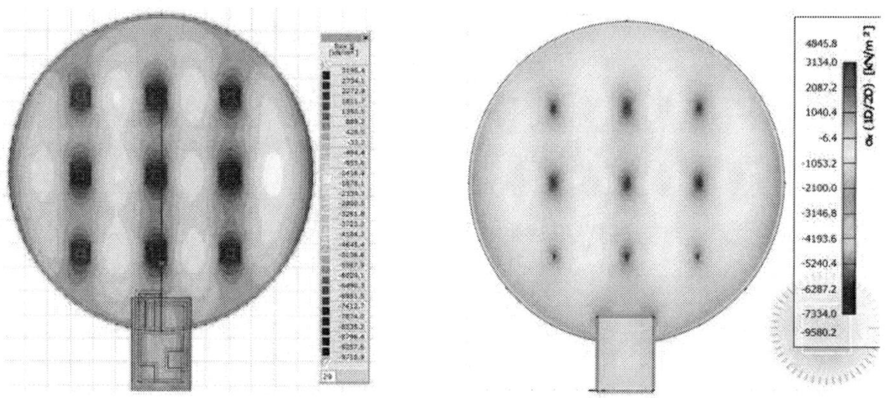

Figure 6. Stress distribution – top view of the RC buried water tank
a). Axis VM software b). SCIA Engineer software

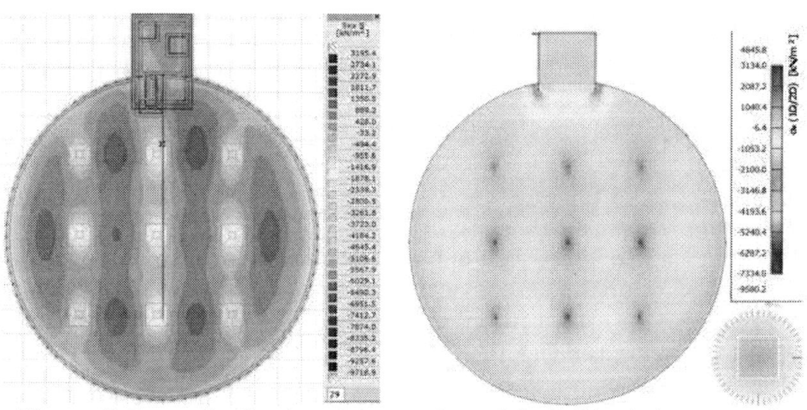

Figure 7. Stress distribution - bottom view of the RC buried water tank
a). Axis VM software b). SCIA Engineer software

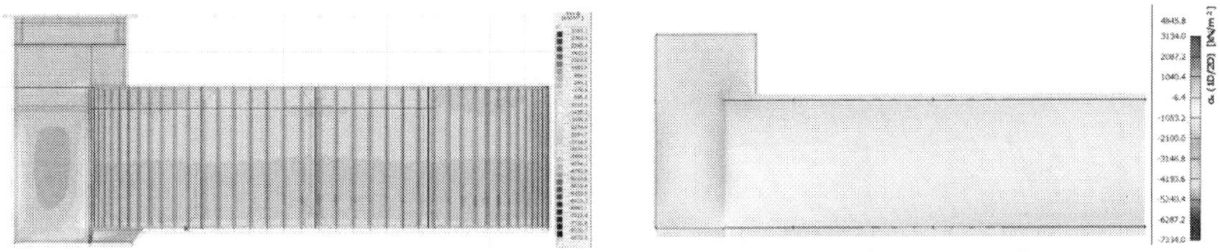

Figure 8. Stress distribution - lateral view of the RC buried water tank
a). Axis VM software b). SCIA Engineer software

In terms of stress distribution, the maximum normal stress, σ_X, computed by both programs is:
- at the top of the columns (Figure 6) - 3195.4 KN/m^2 in Axis VM and 3134.0 KN/m^2 in SCIA;
- at the base of the circular wall (Figure 8) - 1083.9 KN/m^2 in Axis VM and 1040.4 KN/m^2 in SCIA;
- at the continuous foundation level (Figure 7) - 3195.4 KN/m^2 in Axis VM and 3134.0 KN/m^2 in SCIA.

In the case of the interior wall (Figure 9), the self-weight of the element and the pressure of the water exerted on both sides, lead to mutual suppression of the effects of this action. In this situation the normal stress distribution (σ_X = 32.0 KN/m^2 in Axis VM, σ_X = 30.6 KN/m^2 in SCIA), remain constant along the whole structural element.

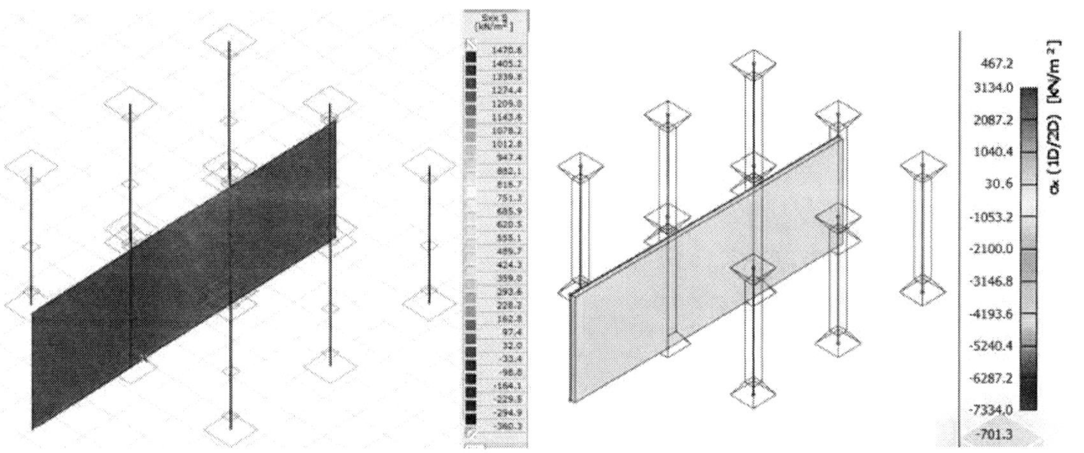

Figure 9. Perspective view of the inner wall
a). Axis VM software b). SCIA Engineer software

3.4.2. *Empty water tank.* The loading hypothesis considering that the water tank is empty, therefore not acted by the pressure of the liquid, the stress distribution is shown in Figures 11, 12, 13.

In terms of stress distribution, the maximum normal stress computed by both programs are:
- at the top of the columns (Figure 10) - 3183.1 KN/m^2 in Axis VM and 3134.0 KN/m^2 in SCIA;
- at the base of the circular wall (Figure 12) - 1081.7 KN/m^2 in Axis VM and 1040.4 KN/m^2 in SCIA;
- at the continuous foundation level (Figure 11) - 3183.1 KN/m^2 in Axis VM and 3134.0 KN/m^2 in SCIA.

In terms of stress distribution, the maximum normal stress computed by both programs are distributed symmetrically at the top of the columns (Figure 10) and in some parts of the continuous foundation (Figure 12).

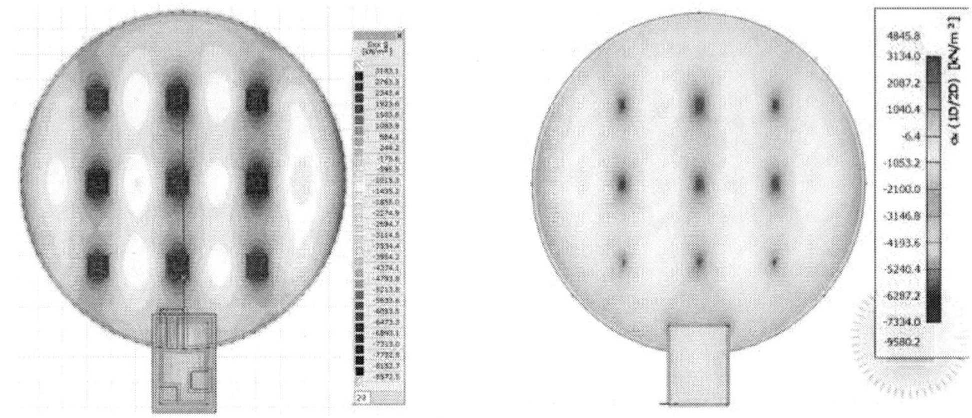

Figure 10. Stress distribution - top view of the RC buried water tank
a). Axis VM software b). SCIA Engineer software

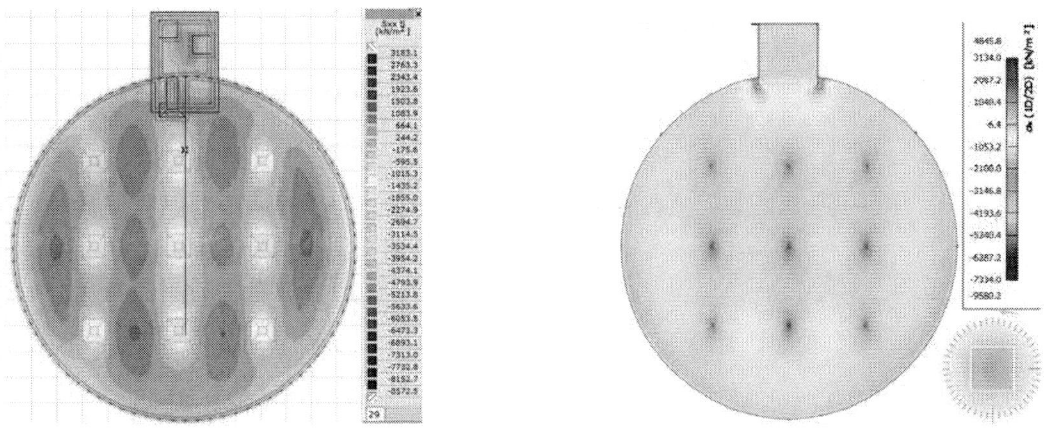

Figure 11. Stress distribution - bottom view of the RC buried water tank
a). Axis VM software b). SCIA Engineer software

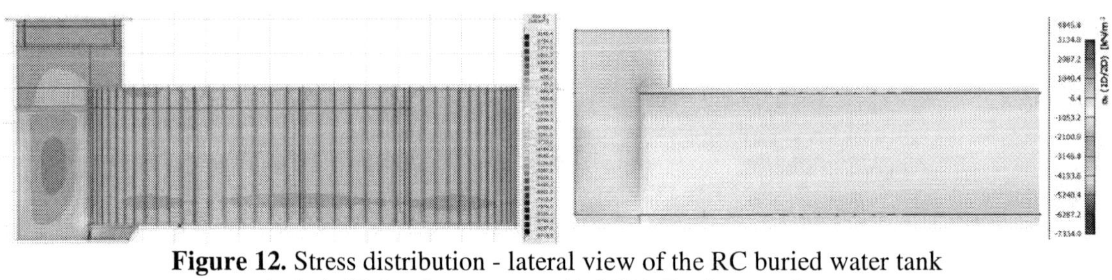

Figure 12. Stress distribution - lateral view of the RC buried water tank
a). Axis VM software b) SCIA Engineer software

In the case of the inner wall, the self-weight of the element does not influence the values of the normal stress distribution (σ_X = -6.9 KN/m^2 in Axis VM, σ_X = -6.4 KN/m^2 in SCIA), this being constant along the whole structural element (Figure 13).

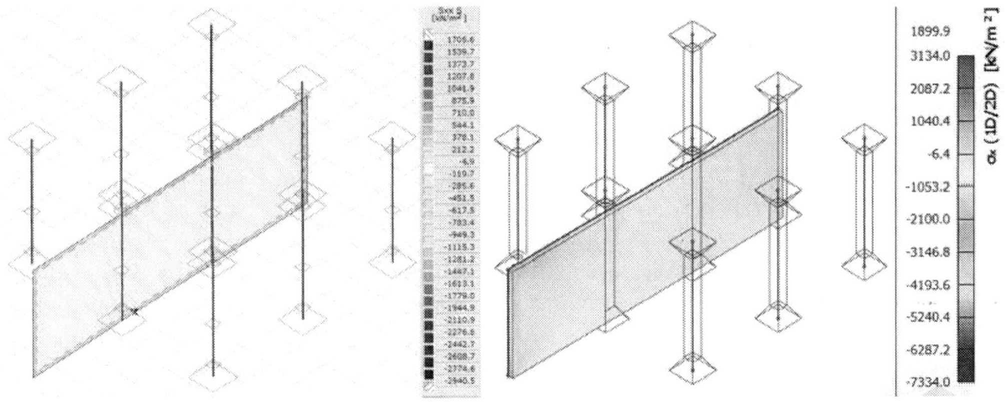

Figure 13. Perspective view of the inner wall

a) Axis VM software b) SCIA Engineer software

4. Final remarks

In the case of the full water tank the most stressed areas have been identified as follows:

- at the top of the columns (σ_X = 3195.4 KN/m^2 in Axis VM, σ_X = 3134.0 KN/m^2 in SCIA);
- at the base of the circular wall (σ_X = 1083.9 KN/m^2 in Axis VM, σ_X = 1040.4 KN/m^2 in SCIA),
- at the continuous foundation level (σ_X = 3195.4 KN/m^2 in Axis VM, σ_X = 3134.0 KN/m^2 in SCIA).

At the inner wall level, due to the liquid pressure exerted on both sides, the stress distribution has smaller values (σ_X = 32.0 KN/m^2 in Axis VM, σ_X = 30.6 KN/m^2 in SCIA).

In the situation when the water tank is empty, values of the estimated stresses are:

- at the top of the columns (σ_X = 3183.1 KN/m^2 in Axis VM, σ_X = 3134.0 KN/m^2 in SCIA);
- at the base of the circular wall (σ_X = 1081.7 KN/m^2 in Axis VM, σ_X = 1040.4 KN/m^2 in SCIA);
- at the continuous foundation level (σ_X = 3183.1 KN/m^2 in Axis VM, σ_X = 3134.0 KN/m^2 in SCIA)

At the inner wall level, the stress distribution has values (σ_X = -6.9 KN/m^2 in Axis VM, σ_X = -6.4 KN/m^2 in SCIA).

In both loading hypotheses (full water tank and empty water tank) the results obtained from the two software programs shows that the most stressed areas (red colour) are at the top of the internal columns and at the foundation level. In the same time, obtained results shows that both computer programs are compatible in order to be used for the analysis of those types of reinforced concrete structures, being considered that are realised from homogeneous material.

Taking in consideration that each FEM software program have its proper manner of introducing the structural model corroborated with the fact there are very small differences in the resulting values (less than 5%), this research can be suitable for the other types of water tanks (semi-buried, over ground).

In order to be generalised to the entire lot of the water tanks existing in Moldova county (which were realised in the same period, same conditions, same materials and same workmanships and exploited in the same conditions), it can be considered that the general structures "health" stage does not affect the proper functionality state.

References

[1] GT 052-2002 2002 Ghid privind investigarea stării tehnice a rezervoarelor, decantoarelor și metatancurilor aflate în exploatare;

[2]SR EN 1992-1-1:2004 Eurocod 2: Proiectarea structurilor din beton, - Partea 1-1: Reguli generale și reguli pentru clădiri;

[3] Hobjilă V and Luca M 2000 Complemente privind proiectarea și expertizarea unor tipuri de construcții hidrotehnice (Iași: Cermi Press) p 205, ISBN: 973-8000-38-6

[4] AxisVM X4 User Manual
[5] SCIA Engineer 18 User Manual
[6] Ajagbe W O, Ilugbo E O, Labiran J O and Ganiyu A A 2015 Analysis and Design of a Fully Submerged Undergorund Water Tank Using the Principle of Beam on Elastic Foundation (UITECH Conference) pp 1-10
[7] Dhage N P and Joshi M M 2017 Review Study on Comparison between Static and Dynamic Analysis of RCC Water Tank (IJRAT International Journal of Research in Advent Technology) pp 3-6, E-ISSN: 2321-9637;
[8] Sani J E, Nwadiogbu C P and Yisa G L 2014 Reliability of an Underground Reinforced Concrete Rectangular Water Tank vol 11 (IOSR Journal of Mechanical and Civil Engineering) pp 58-68, E- ISSN: 2278-1684;
[9] Samangany A Y, Naderi R and Shahabi M H 2013 Static and Dynamic Analysis of Storage Tanks Considering Soil-Structure Interaction vol 6 (International Research Journal of Applied and Basic Sciences) pp 515-532, ISSN: 2251-838X;
[10] CR0-2012 2012 Cod de proiectare. Bazele proiectării structurilor.

Assessment of the Time-dependent Behaviour of a Reinforced Concrete Water Tank by using the Finite Elements Method

A Filip[1], D Covatariu[2],
[1] Doctoral School of the Faculty of Civil Engineering and Building Services, "Gheorghe Asachi" Technical University of Iaşi, 700050, Romania
[2] Department of Civil and Industrial Engineering, Faculty of Civil Engineering and Building Services, "Gheorghe Asachi" Technical University of Iaşi, 700050, România

E-mail: filipalexandru91@gmail.com

Abstract. Determinants factors (design and execution errors, site location, environmental & exploitation conditions) could generate the over-time degradation of the buildings. In order to assure the safety during the service period (imposed by the norms) the time-dependent monitoring of the structure's behaviour is required. As a case study a reinforced concrete water tank was analysed. The structural analysis in two different stages was considered: initial stage (as in design project) and current stage (after approximately 35 years of exploitation). A parametric study based on a computational model (by using Finite Element Method - FEM) was carried out using SCIA Engineer software program. The computational model was designed (for both stages) by using mechanical characteristics of the material (collected from initial project) and data resulted from destructive and non-destructive tests in current stage. The objective was to estimate the level of degradations which could occur in structural elements during the exploitation period by comparing the evolution of stress distribution in the two stages. The durability factors were analysed in order to the studied structure's life-cycle assessment.

Keywords: structural behaviour, FEM method, water tank;

1. Introduction

1.1. General considerations on the durability of concrete structural elements

Buildings are usually subjected to actions from their normal exploitation, but also to exceptional actions (heavy snow, floods, fires, earthquakes).

Under these circumstances, depending on the quality of the materials used, the way in which the design assumptions made are approaching the reality, as well as the possible mistakes in execution, various degradations can occur [3].

The importance of studying the durability of concrete in water tanks is essential because the presence of water corroborated with the destructive action of the external environment and with various physico-chemical reactions favours the occurrence of degradations in the structural elements.

Due to the differences between the properties and the structure of a concrete made under laboratory conditions (according to the data included in the project) and those of a concrete put into operation (after approximately 35 years of exploitation), the appreciation of the durability of the concrete represent a complex problem.

The tracking of factors that influence the durability of a building is made throughout its existence and includes all activities of direct examination or investigation with specific observation and measurement means to maintain the quality requirements imposed by law. In addition to meeting the quality and technical standards, this cumulative action and factors (mechanical, physical, chemical) is based on the notion of durability [1,4,5].

1.2. Particularities regarding water tank's time-dependent behaviour

Site monitoring of buildings represent a systematic process of observing, examining and investigating how buildings react when they are exploited [9].
This process can be applied in the following situations:
- inadequate operation of the construction due to accidental overloads;
- producing chemical reactions between constituents of structural elements and various aggressive agents;
- corrosion of the fittings;
- the amplification of some destructive physical processes;

The main objective of tracking the behaviour of water tanks over time is to evaluate and identify changes in the current state of construction (by destructive and non-destructive methods) compared to the original one (according to the technical project data).
In this case, the operating conditions are assessed according to the nature of the technological process, the functioning of the related installations and the characteristics of the aggressive environment (the nature of the fluvial stored, the concentration of the fluid, the water solubility of the corrosive products) [8].

1.3. Time-dependent assessments

Time-depending behaviour of the water tank could be conducted considering the factors which affect the durability of concrete structures and also by knowing the mechanical characteristics of the materials taking into consideration the evolution in time [7].
In the considered case study were considered two stages:
- Initial stage;
- Current stage;

In the initial stage (design and execution process), the structure has been modelled knowing all the mechanical characteristics of the materials being collected from the technical project.
In the current stage (after approximately 35 years of exploitation) the structural system hasn't changed. By using destructive and non-destructive testing methods in order to determine the characteristic value of materials, samples of concrete have been collected.
The aim of these numerical analysis is to assess the level of degradations in the structural elements by comparing the stress distribution in the two stages and to propose different technical interventions.

2. Case study

2.1. Structural system

The analysed construction is a ground water tank made of reinforced concrete having a capacity of 5000 cubic meters. It was built in the 1980's and it is still in exploitation, providing the necessary water for the Miroslava village and the Nicolina neighbourhood in Iasi, Iasi county.
The water tank has a circular shape in the horizontal plane (Figure 1.a.) with an inner radius of 13.85 m and a useful height of the tank of 8.05 m supported by a continuous foundation (C18/22.5). Inside, there is a central reinforced concrete column with the dimensions of 70 x 70 cm sustained by an isolated footing (C18/22.5) provided at the top with a circular cap.
The wall of the tank is made of precast elements (C25/30) of 17 cm thick. At the top of the wall was made a reinforced concrete belt with the dimensions of 28x20 cm.
At the top, the tank was closed with precast T-shaped roof elements (C25/30), radially disposed, supporting one end on the top of the wall and the other on the central pillar cap.

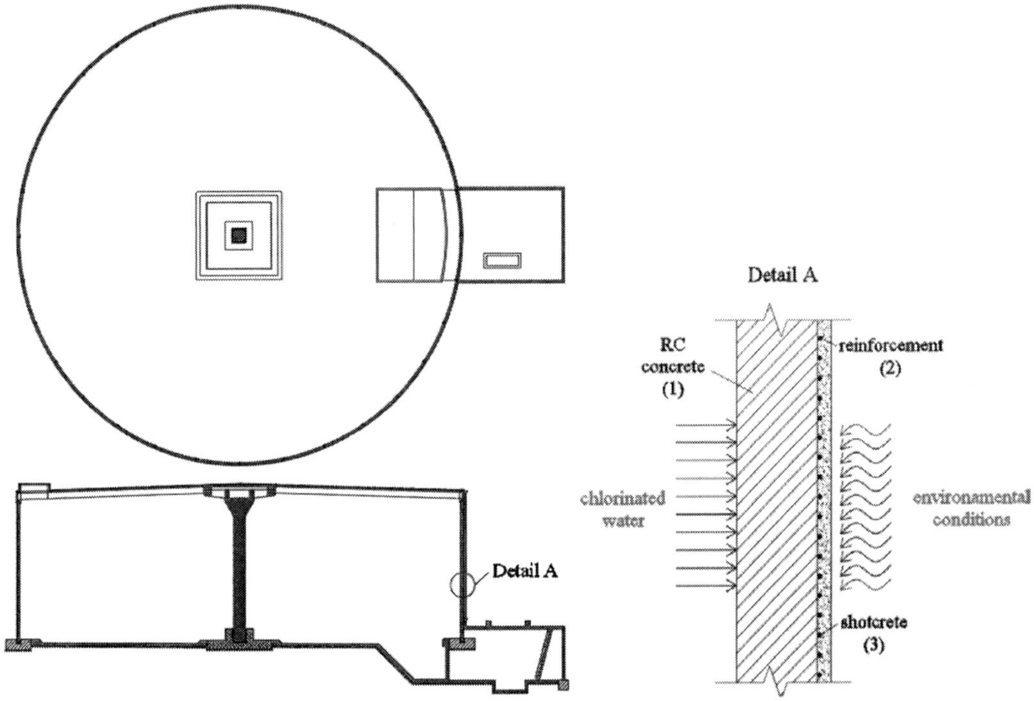

Figure 1. Water tank, Miroslava village, Iaşi county
a). Longitudinal section; b). Transversal section c). Detail A

The circular wall section was built by on three different stages (Figure 1.c.)
- in-site monolithization of precast reinforced concrete elements (1);
- post-tensioning of the circular wall with wires packed in strands SBPI Ø5 (2);
- protection at the exterior surface by applying consecutive layers of shotcrete M400 of 3 cm thick and finished with a special layer of waterproofing mortar of 1 cm thick (3).

The most common form of attack on the intimate structure of concrete is represented by the action of chloride in the chlorinated water [2].

2.2. Numerical analysis by Finite Element Method – initial stage

The analysis of constructions (static, dynamic) in general, and of water tanks in particular, has carried out by the finite element method (FEM).

In the case of water tanks, the use of the Finite Element Method (FEM) allows the determination of stress and displacement fields for all load categories, considering complicated structural shapes.

In the domain of static analysis for water tanks, the following calculation categories are based on the finite element method:
- static linear analysis;
- static nonlinear analysis;
- identification of the failure mechanism and safety factor assessment;
- optimization of automated design.

In the initial stage of the analysis, the mechanical characteristics of the materials has been collected from the technical project (Table 1).

The original values (from the project) of the control force in the strand is 2110 daN and the shotcrete layer has M400 class.

Table 1. Input data

Structural element	Concrete class
Roof element	B400 (C25/30)
Circular wall	B400 (C25/30)
Wall continuous foundation	B300 (C18/22.5)
Central column foundation	B300 (C18/22.5)

The studied structure was modelled, discretized and analysed by using SCIA Engineer software program and the linear static analysis was performed to compute the maximum efforts according to the following structural model (Figure 2).

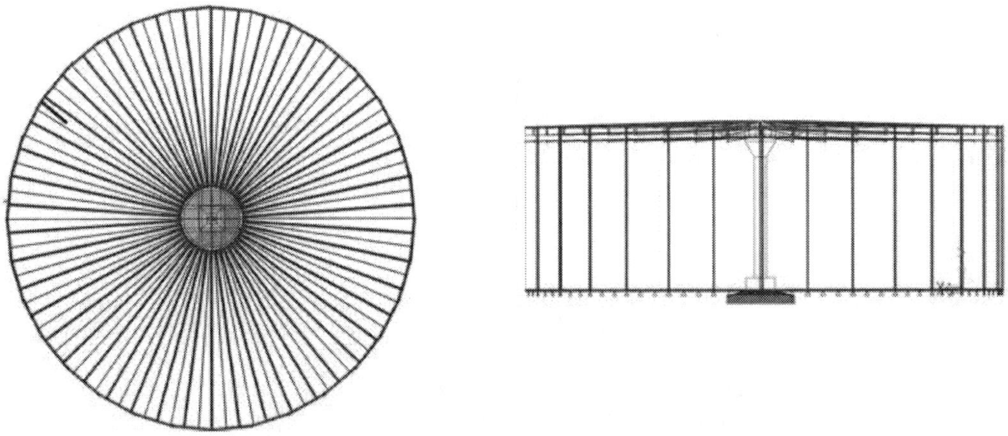

Figure 2. 2D perspective of the structure
a). Longitudinal section b). Transversal section

The accuracy of the calculation depends on the considered static scheme introduced in the software program, which must be in line with the structural model is made (the nature of the links between the component elements, the abutments, etc.). The loads involved (Table 2) in the calculation of the water tank are:

Table 2. Acting actions on the water tank

Loads	
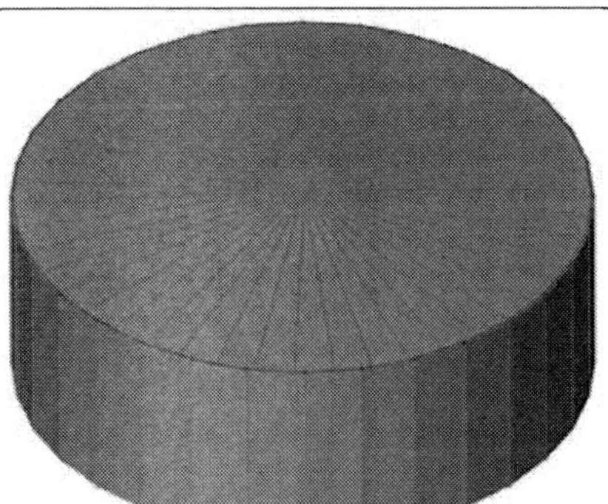 The own weight of the tank	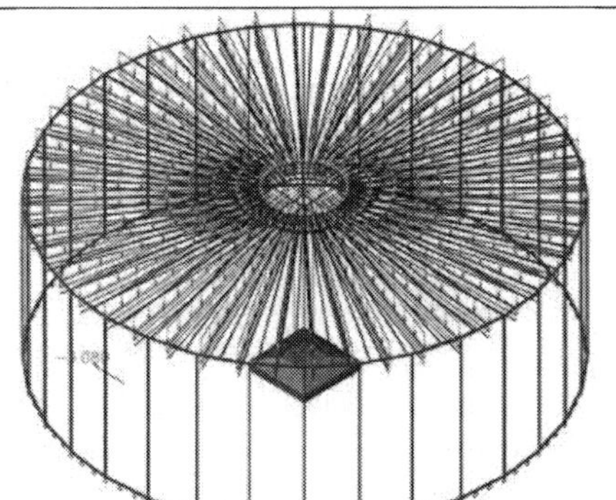 Variable actions (snow)

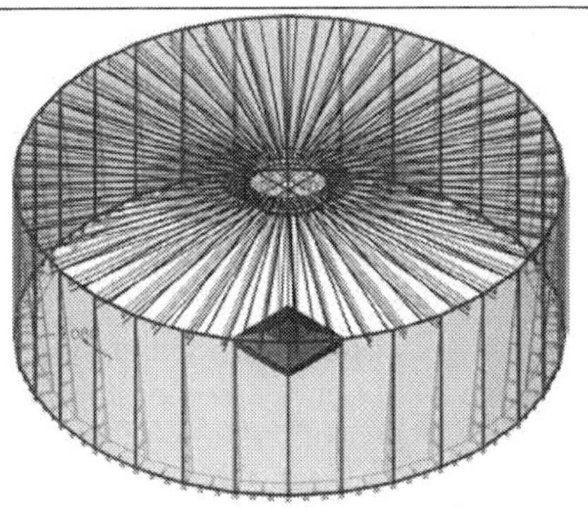	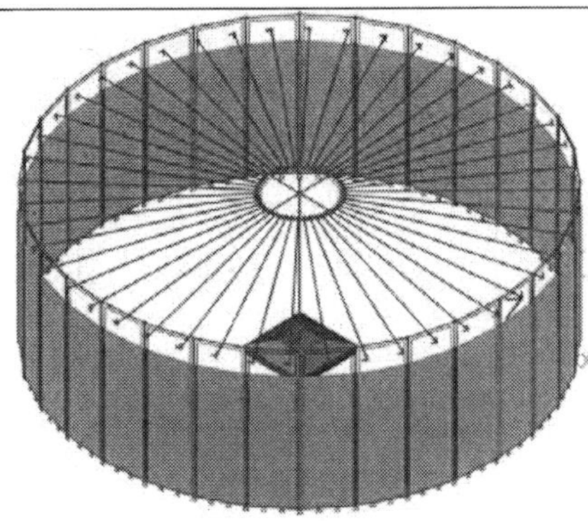
The pressure of the water stored in the tank	Post-tension force

In this case it was also taken in consideration the seismic load and the wind load, corresponding to the norms for the Iași county conditions at the moment when the structure was designed and executed. Partial safety coefficients for the combinations of actions in service limit state (SLS) checks are:

- 1.10 for permanent and accidental actions;
- 0.50 for variable actions.

The load combination (quasi-permanent) for the service check can be calculated as in equation (1) [10]:

$$\sum_{j=1}^{n} P_i + \sum_{i=1}^{m} V_i \ (1);$$

where:
P_j – the characteristic value of the permanent action;
V_i – the characteristic value of the associated variable action;

Considering that the structure is analysed in different stages (execution stage and current stage), the static calculation is performed for the following assumption (Table 3):

Table 3. Loading hypothesis

	Full water tank
Loads	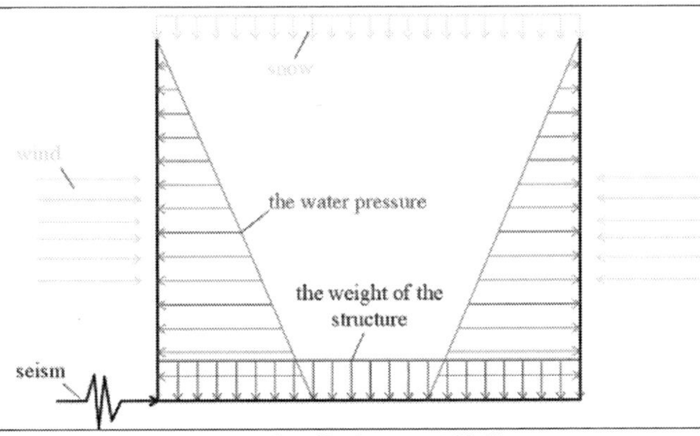
Type of combination	- service limit state (SLS)
Load combination depending on loading hypothesis [10]	1.10·SW+0.50·S + 0.50·W+1.00·SL

where:
- SW = the self-weight of the structure;
- S = snow;
- W = the pressure of the stored liquid (water);
- SL = the seismic load.

IOP Conf. Series: Materials Science and Engineering **586** (2019) 012021 doi:10.1088/1757-899X/586/1/012021

Meshing of a structure is a complex process to create a discrete model that best approximates the actual continuous structure from various points of view, such as geometric form, load application, etc.
In the SCIA Engineer program, the mesh network was generated for the entire domain of analysis, the elements being rectangular divided.

2.3. Numerical analysis by FEM – current stage

In the current stage, the mechanical characteristics of the materials has been determined by using destructive and non-destructive methods. The specimen used due to laboratory tests on concrete – from roof's elements (figure 3.a.) and shotcrete specimens – from wall's elements (figure 3.b.).

Figure 3. Test specimens
a). concrete b). shotcrete

From physical and mechanical determination, the numerical values obtained were:
- for the concrete specimens tested in the laboratory, by using destructive methods, the characteristic in-situ compressive strength is $f_{ck,\,cube}$=24 MPa, corresponding to the concrete class C20/25.
- for the shotcrete specimens, it has been used the hydraulic press (tension and compression strength determination) and physical degradation method (carbonated zones, determination of water absorption) to determine the mechanical characteristics in order to estimate the level of degradations which have occurred in the structural elements.
After the tests, the same structural model has been used in the software program by using numerical values obtained from the determinations.
In this case it was also taken into consideration the seismic load and the wind load, corresponding to norms for nowadays Iași county conditions.
Partial safety coefficients for the combinations of actions in service limit state (SLS) checks are:
- 1.00 for permanent and accidental actions;
- 0.40 for variable actions.
The load combination (quasi-permanent) for the service check can be calculated as in equation (2) [6]:

$$\sum_{j=1}^{n} G_{k,j} + \sum_{i=1}^{m} \psi_{2,i} \cdot Q_{k,i} \quad (2);$$

where:
$G_{k,j}$ – the characteristic value of the permanent action;
$Q_{k,i}$ – the characteristic value of the associated variable action;
$\psi_{2,i}$ – factor for the quasi-permanent value of the variable action.
Regarding that the structure is considered in different stages (project stage and current stage), for the second case, the static calculation is carried out according to the current norms, considering the following assumption:

$$1.00 \cdot SW + 0.40 \cdot S + 0.40 \cdot W + 1.00 \cdot SL$$

The loading hypothesis considering that the water is at its full capacity will display the maximum stress distribution (Figure 4).

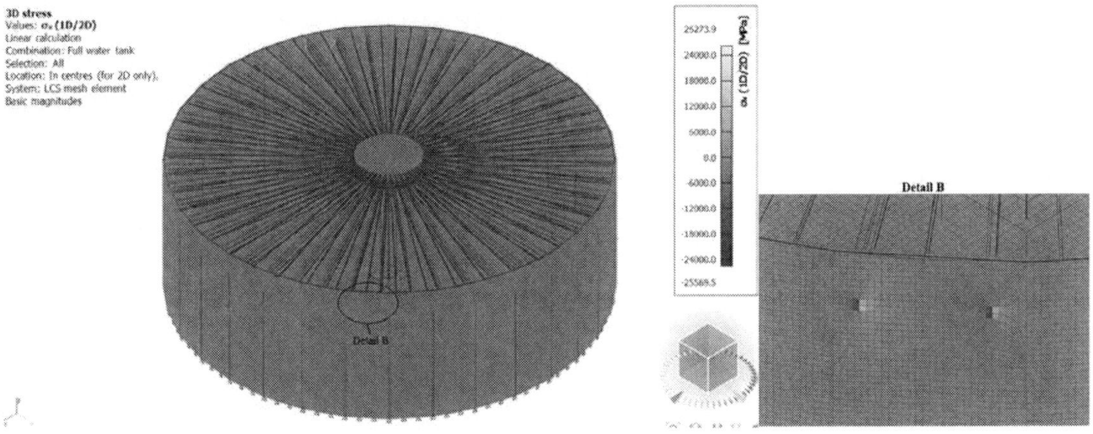

Figure 4. Stress distribution on the water tank structure – current stage

3. Final remarks

After the analysis (based on the numerical modelling) was performed, no significant discrepancies were found: the exception being the decreasing of the concrete class (from C25/30 to C20/25). Also, due to chlorinated water actions from the interior surface of the tank's walls, the concrete was degraded by carbonated occurring zones.

Internal factors (chlorinated water) has deteriorate the intimate structure of the concrete wall from the inside as well, since the structure is not waterproof (micro-cracks occurrence) and external factor (aggressive environmental agents, rainwater, snow, freeze-thaw cycles) amplifies the degradation state of the structure (due to lack of thermo-insulation).

A proper numerical analysis based on conclusive data correlated with the detailed investigation of durability factors (destructive and non-destructive test, in-situ and in specialised laboratories) represents an effective way to predict the level of degradation and to ensure a safe continuous exploitation of the structure.

References

[1] Basheer L, Kropp J and Cleland J D 2001 *Assessment of the Durability of Concrete from its Permeation Properties: A Review* vol 15 (Elsevier: Construction and Building Materials) pp 93÷103;

[2] Covatariu D and Filip A 2018 *A Comprehensive Approaching of the Water Tanks Durability's Affecting Factors in Different Stages*, vol 64 (Bulletin of the Polytechnic Institute of Jassy) pp 1÷16;

[3] Farrar C and Worden K 2006 *An Introduction to Structural Health Monitoring* 365 (Philosophical Transactions of the Royal Society) pp 303-315, doi:10.1098/rsta.2006.1928;

[4] Rozière E, Loukili A and Cussigh F 2009 *A Performance Based Approach for Durability of Concrete Exposed to Carbonation* vol 23 (Elsevier: Construction and Building Materials) pp 190÷199;

[5] Zamfirescu D and Postelnicu T 2003 *Durabilitatea betonului armat*, (Bucureşti:MatrixRom Press);

[6] CR 0-2012 2012 *Cod de proiectare. Bazele proiectării structurilor*;

[7] GT 052-02 2002 *Ghid privind investigarea stării tehnice a rezervoarelor, decantoarelor şi metatancurilor în exploatare*;

[8] NP 087-03 2003 *Normativ pentru urmărirea comportării construcţiilor hidrotehnice*;

[9] P130-1999 1999 *Normativ privind comportarea în timp a construcţiilor*;

[10] STAS 10101/0A-77 1977 *Acţiuni în construcţii. Clasificarea şi gruparea acţiunilor pentru construcţii civile şi industriale*

IOP Conf. Series: Materials Science and Engineering **586** (2019) 012022 doi:10.1088/1757-899X/586/1/012022

A simple and rapid approach for the numerical simulation of non-linear elements and examples of its application

W Loo and L Tuleasca
UNITEC Institute of Technology – Department of Civil Engineering, 139 Carrington Road, Mt Albert, Auckland, *New Zealand*

E-mail: wloo@unitec.ac.nz

Abstract. In earthquake design philosophy non-linear inelastic damage is concentrated at predetermined locations in a structure, with the aim being to keep structural members critical for collapse avoidance, elastic. Non-linear elements can include the beam column joints in steel and reinforced concrete moment resisting frames, the nail connections of timber sheathing-to-framing shear walls, and the hold-down connectors of shear walls. To numerically model the seismic behaviour of these structures, it is critical to be able to quickly and simply model their respective hysteretic behaviours. It is particularly important to identify the parameters that will allow for the accurate replication of the degrading and pinching qualities of the force-displacement relationships. The authors summarise the use of a proprietary multi-linear plastic link, available in most finite element packages, in order to achieve this. A step-by-step process to model the links, from previous research, is described. This process considers the initial elastic stiffness, the force-displacement curve up to ultimate-strength, the post ultimate-strength degrading stiffness, and the reversing stiffness. The procedure can be adopted for both linear and rotational cyclic excitation. The modelled individual elements are subjected to a series of displacement-controlled time-history schedules and validated against the results from experimental tests. These elements, representing non-linear behaviour, are in turn implemented in various numerical models of shear walls. The authors apply cyclic loading to a model wall, and then for a separately designed wall apply seismic ground accelerations associated with the 2011 Canterbury earthquakes in New Zealand, and the results are presented. While the examples provided in this paper involve timber shear walls, the numerical modelling method will be equally useful for a variety of structures of different material types and configuration.

1. Introduction

This paper provides an overview of a methodology developed by which allows for the rapid numerical modelling of a common form of non-linear relationship under seismic excitation - pinched hysteresis loops. Pinched hysteresis loops arise when a structural element, subject to cyclic displacement regime of successively increasing displacements degrades in both strength and stiffness. Pinched hysteretic behaviour can be observed at the reinforced concrete beam column joints [1], reinforced concrete shear walls [2], and are the characteristic behaviour of structures with compressive members that buckle, such as structures with conventional steel bracing [2]. In timber structures, it is the interaction of the steel connectors, primarily nail connections, with wood that provides ductility (wood in itself being brittle) [3], and in the case of timber shear walls it is the sheathing to-framing-nail connections that govern the overall force-displacement behaviour. The methodology described here was originally developed by Loo, Quenneville, and Chouw [4, 5], and in this paper the authors adopt it to investigate the behaviour of a structure subject to ground accelerations from the 2011 Canterbury earthquake. While the examples presented are timber based, the methodology can nevertheless readily be applied to any type of structure or structural element with pinched hysteretic behaviour. The paper is organised as follows:

- A mechanics-based method to determine the ultimate lateral strength of nail connections in shear walls.
- Modelling of non-linear behaviour
- Implementation of the concept in shear walls

- Implementation of the concept in shear walls subjected to seismic loading

2. Yield Model theory and the determination of lateral connection strength

European Yield Model (EYM) theory provides a mechanics-based approach to determine the lateral resistance of timber members connected by dowel elements –nails, bolts, spikes etc. [6]. EYM theory (developed in the 1940s) assumes that when a dowel connector, such as a nail or a bolt is placed under load, failure will occur either through bearing failure of the wood material, or when one or more plastic hinges develops in the dowel fastener. This gives rise to various failure modes [7]. Using equations derived from basic mechanics theory, the lateral strength associated with each of the possible modes of failure can be calculated, for dowels in both single shear and double shear. The lowest calculated lateral strength of the possible modes of failure will be that connection's ultimate lateral strength.

The EYM theory identifies six possible failure modes for connections in single shear and four possible failure modes for connections in double shear. These modes are designated as Im, Is, II, IIIm, IIIs, and IV. All six of these modes can apply to single shear connections, but only Im, Is, IIIs, and IV, are possible for double shear connections. The reader is referred to [6, 7] for detailed descriptions of these failure modes. Using the formulas associated with each of the modes, ultimate loads can be calculated and plotted for various sheathing thicknesses and for standard nail penetrations. The example shown below is for nails in double shear connecting pine plywood sheathing to pine framing (Figure 1). The procedure is described in detail in [4, 8].

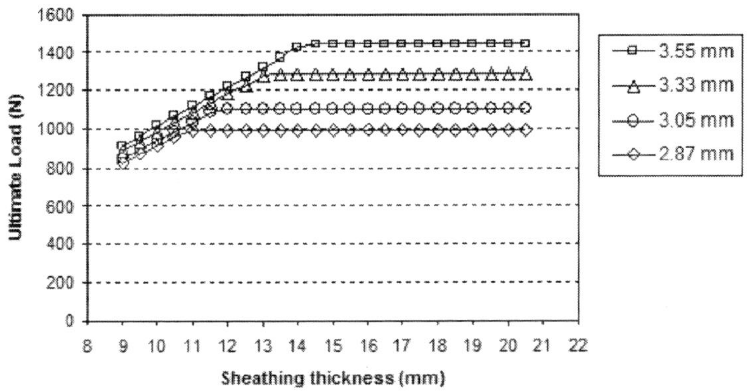

Figure 1. Ultimate nail loads from General dowel formulae – nails in double shear (for OSB sheathing SG=0.64, and for SPF framing, SG=0.42).

The strengths calculated from EYM, while useful from a design perspective, appear to be conservative. From a review of the actual data from lateral strength tests, the EYM consistently underestimates the actual ultimate lateral load. This is possibly because the EYM ignores end fixity of the dowel, inter-member friction between sheathing and framing, and string resistance which develops from nail tension. A discussion of the influence of each of these factors is provided in [4]. While the EYM values would be suitable for design, they would not be realistic enough for numerical modelling, where the aim of modelling is to replicate as closely as possible actual wall behaviour. Following a review of the literature on actual testing [9-11], an equation relating the measured values of ultimate lateral strength to the theoretical EYM values was proposed for single shear [4]:

$$F_{ult} = 994\ln\left(F_{ult(EYM)}\right) - 5536 \tag{1}$$

For double shear, the Loo, Quenneville, and Chouw [4], proposed:

$$F_{ult} = \frac{F_{ult(EYM)} + 3976\ln\left(0.5F_{ult(EYM)}\right) - 22144}{3} \tag{2}$$

IOP Conf. Series: Materials Science and Engineering **586** (2019) 012022 doi:10.1088/1757-899X/586/1/012022

In summary, to determine the lateral strength, the EYM is first used to obtain an initial value, and then adjusted using equations (1) single shear or (2) double shear.

3. Modelling the non-linear behaviour

This discussion is based on detailed one found in [4, 8]. Non-linear springs (multi-linear plastic link elements) can be used to model the sheathing to the nailed connections. The multi-linear plastic link allows definition of a nonlinear load slip curve for shear deformation in two perpendicular directions. Strength and stiffness degradation can be modelled and hysteretic behaviour observed. In SAP2000 [12] link elements connecting two nodes consists of six springs. Three springs relate to translation (U1, U2, and U3), and three springs relate to rotation (R1, R2, and R3) - refer Figure 2.

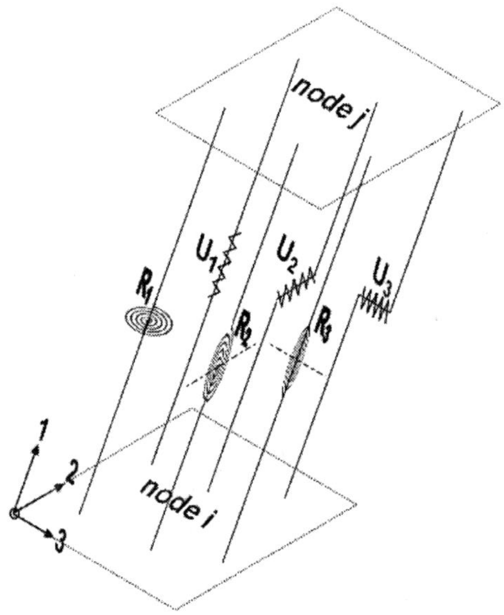

Figure 2. Non-linear element connecting two nodes.

For modelling of nail connections, only the U2 and U3 directions which relate to shear (Figure 3.16) are considered when defining the force displacement relationship. U2 is the vertical shear direction, and U3 the horizontal shear direction. So, for vertical stud framing members, U2 is parallel to the grain and U3 perpendicular to the grain. For top and bottom plates, these switch around – U3 is parallel to the grain and U2 is perpendicular to grain. The load displacement relationship for both U2 and U3 are assumed to be the same (Dolan and Madsen [9] found the force-displacement behaviour of nails in sheathing to be largely directionally independent). All the nail connections will be oriented horizontally along their length i.e. the orientation of U1 will lie in the horizontal direction. Note that any nail withdrawal effect is considered represented by the U2 and U3 defined relationships – the force displacement behaviour (as with actual testing), simply records laterally applied force against the observed displacement – the displacement itself could be the sum of various parts – nail withdrawal included. The load displacement relationship for the hysteretic envelope curve (or simply the monotonic load displacement curve) is input for the U2 and U3 directions (assumed same for both directions). This can be carried out rapidly – SAP2000 allows the upload of the data sets from an Excel Spreadsheet. The Multilinear plastic element has three available hysteresis types for selection; Kinematic, Takeda, and Pivot. For hysteretic behaviour with pinching, the pivot type is selected, and there are parameters which can be adjusted to account for the shape of the envelope curve, the degrading stiffness, and the pinching strength. The steps described in [4, 8] are used to model the element. An example is an element modelling the behaviour through wood of a 3 mm nail. It is set up as shown in Figure 3(a), subject to the displacement schedule of Figure 3(b), to produce the hysteretic relationship of Figure 3(c). It is shown in [4] that this result closely matches the result from actual testing carried out by Dinehart et al. [13].

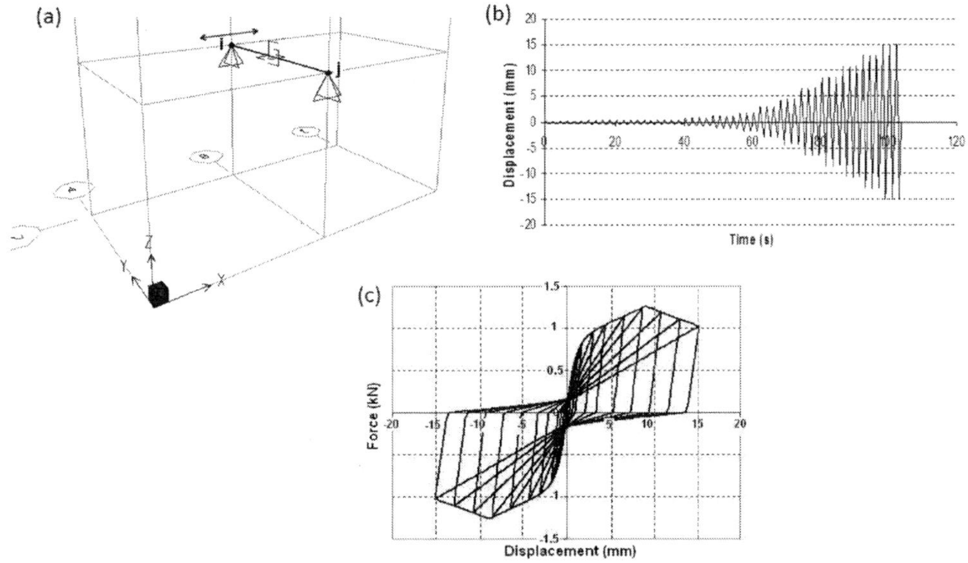

Figure 3. (a) Model nail connection in SAP2000, (b) displacement protocol, and (c) resulting hysteretic behaviour.

4. Implementation in a timber shear wall

In order to explore the force-displacement characteristics of sheathing-to-framing timber shear walls, a 2.44 x 2.44 m sheathing to framing shear wall was numerically modelled in SAP2000 [12], with the nail connections represented by multi-linear plastic links modelled in the manner described in the preceding sections. The sheathing was modelled as 12 mm thick F11 grade plywood, and the framing studs were of 38 mm x 89 mm Spruce Pine Fir with an MOE of 9500 MPa and density of 420 kg/m3. The studs were spaced at 406 mm, and the top beam was represented by a steel beam (to replicate the experimental conditions in tests carried out on shear walls by Varoglu et al. [14]). One nail element was used to represent the behaviour of four actual nails, each of ultimate lateral strength 1500 N, and an assumed yield strength of 750 N (half of the ultimate strength). The model wall was subjected to a cyclic load protocol with a maximum horizontal displacement applied at the top corner of 125 mm. The wall was structurally designed for a yield strength of 13 kN, and an ultimate strength of 26 kN (twice the yield strength). Figure 4(a) shows the model wall in SAP2000, while Fig 4(b) shows the hysteretic behaviour after application of the displacement cycles. The numerically obtained hysteretic behaviour (Figure 4(b)) replicates the form of the typical pinched behaviour of timber shear walls from experimental studies [14, 15].

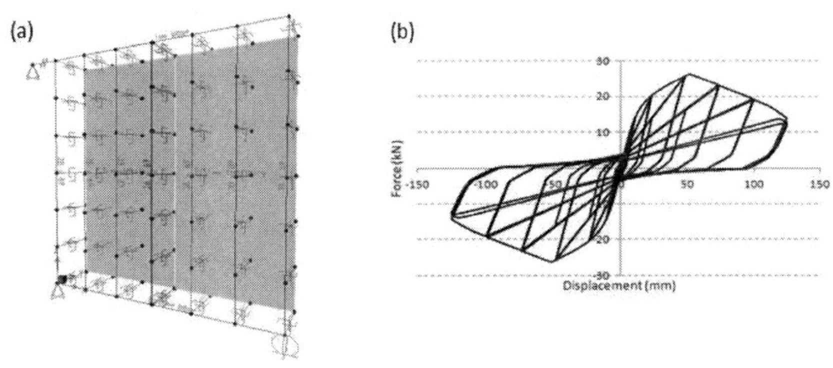

Figure 4. (a) Model wall with sheathing to framing nail connections in SAP2000, and (b) simulated hysteretic behaviour.

5. Earthquake loading

The authors modelled a shear wall using the methodology of the preceding section and subjected it to a sequence of simulated earthquake loadings. Earthquake motions from the destructive 22nd February 2011 Christchurch earthquake (magnitude 6.3 M_L) were applied. The data used were from three different sites in the Christchurch central business district: Christchurch Cathedral College, Christchurch Hospital, and Christchurch Resthaven. From the response spectra shown in Figure 5, it is seen that the seismic demands of this earthquake significantly exceeded the code required level of design at the time.

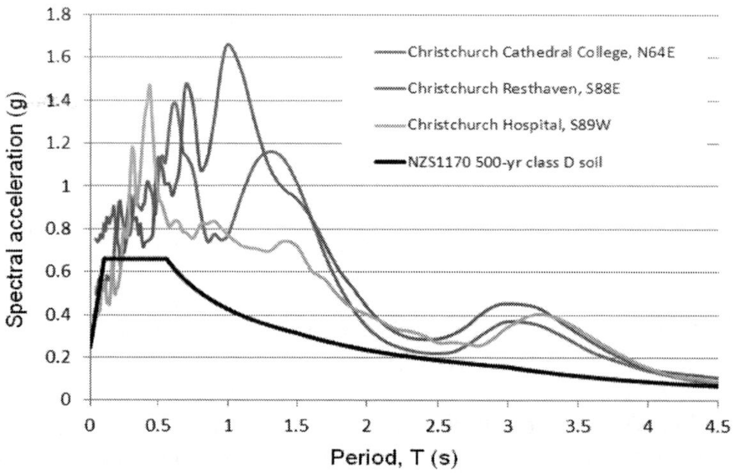

Figure 5. Spectral accelerations for the February 2011 Christchurch earthquake at three sites, and the ULS design spectrum (500 yr return period) for NZS1170.5 [16], Type D (soft) soils (Spectra produced from data provided by GeoNet NZ [17])

The first set of simulations involved scaled down accelerations corresponding to a design level event. The second set of simulations applied the actual, un-scaled acceleration records. Details are shown in Table 1.

Table 1. Earthquake events used for numerical simulation[1,2]

Site	Componen t	PGA (g)	ULS scale factor[3]	Scaled PGA (g)
Christchurch Cathedral College	N96E	0.483	0.7	0.338
Christchurch Hospital	S89W	0.359	0.6	0.215
Christchurch Resthaven	S88E	0.718	0.7	0.503

[1]Data from GeoNet NZ [17]
[2]Event 10 km south-east of Christchurch 2011 Feb 21, 23:51:42 UT Dist 8km Depth 5Rkm Ml 6.30
[3]500 year return period, NZS1170.5 soil type 'D' (deep and soft soils) [16]

The model wall was of a 2.44 x 2.44 m midply wall (design ductility, μ=3) designed for an ultimate limit state event in Christchurch (for a description of midply walls refer [14]). It consisted of two vertical 12 mm thick plywood sheathing panels, sandwiched between 90 x 45 mm laminated-veneer-lumber studs. Finite elements were used to model 4 mm diameter nails in double shear, at required spacings of 500 mm. The wall was assigned a seismic mass of 4550 kg at its top, which corresponded to a design base shear from NZS1170.5 [16] of about 9600 kN (assuming a fundamental period of vibration of less than 0.4 seconds).
The hysteretic response of the walls to the scaled ULS events, and then the actual, un-scaled events are shown in Figure 6.

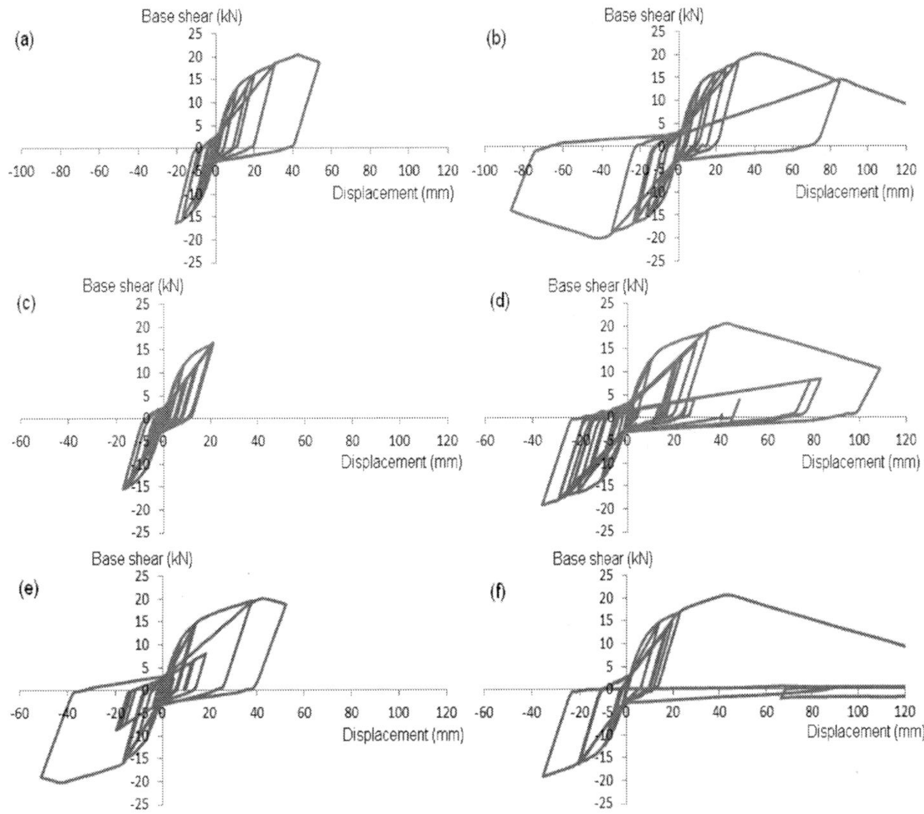

Figure 6. Response of model wall to (a) Christchurch Cathedral ULS event, and (b) actual event, and (c) Christchurch Hospital ULS event, and (d) actual event, and (e) Resthaven ULS event, and (f) actual event.

From Figure 6, it is seen that the wall survived the three design level events, maintaining its lateral strength in each case, while incurring some inelastic damage. This response was in line with the intention of the ductile design - which allows the possibility of yielding, but without collapse of the structure during an earthquake.

The ultimate lateral strength of the wall was found to be a little over 20 kN. From inspection of the force-displacement records, the yield strength is approximately 10 kN, which corresponds closely to the intended design lateral strength of 9.6 kN. All three of the walls completely failed when placed under the actual, un-scaled earthquake loadings. This was as expected because the demands of this earthquake, as previously mentioned, significantly exceeded the relevant code design requirements.

6. Conclusions

A numerical method to model the behaviour of non-linear elements with pinching hysteretic behaviour is introduced. The method requires little if any programming but rather simply the configuration of proprietary elements that are available in most finite element software packages. While pinched hysteretic behaviour can be associated with braced frames, concrete shear walls, and the beam-column connection of reinforced concrete structures, it almost invariably occurs in timber structures with slender steel connectors (most typically nails). Therefore, the example of nail connections is presented. The first step is to apply a mechanics-based approach to determine the theoretical lateral ultimate strength of a nail connection. This method appears to apply a value that is overly conservative, which is acceptable for design, but not for numerical simulation, in which the actual behaviour should be approximated as closely as possible. The relationship between theoretical and actual strengths (collated from

IOP Conf. Series: Materials Science and Engineering **586** (2019) 012022 doi:10.1088/1757-899X/586/1/012022

experimental results carried out by other researchers) is presented. An example of a model nail connection is subjected to cyclic loading, with the resulting hysteresis behaviour providing an excellent representation of an experimentally obtained result. These elements are then implemented in a timber shear wall and subject to simulations of cyclic loading. The pinched hysteretic behaviour reflects that of actual timber shear walls, in which the force-displacement behaviour is largely governed by the behaviour of the connectors. Finally, the method is used to test a wall subject to ground accelerations from the 2011 Canterbury earthquakes. The ground accelerations are first scaled to a ULS level earthquake and the model walls behave as expected, with non-linear damage but no collapse. The actual unscaled accelerations are next applied, and the walls experience complete collapse after a sequence of cycles of pinched hysteretic shape. The method presented to model pinched hysteretic behaviour has potential not only in the research of timber structures but will also find applicability in concrete and steel structures, as well as rigid timber structures which are held down by nail brackets.

7. References

[1] Sengupta P, Li B 2011 *Hysteresis Behaviour of Reinforced Concrete Non-Ductile Beam-Column Joints*, in *Proceedings of the ninth Pacific Conference on Earthquake Engineering* Auckland, New Zealand

[2] Narayanan R, Roberts T M 2005 *Structures Subjected to Repeated Loading: Stability and strength*, London and New York: Taylor & Francis

[3] Buchanan A 2007 *Timber Design Guide 3rd edn.* 2nd ed, Wellington, New Zealand: Timber Industry Federation Inc

[4] Loo W Y, Quenneville P, Chouw N 2012 *A numerical approach for simulating the behaviour of timber shear walls.* Structural Engineering and Mechanics, **42**(3): p. 383-407

[5] Loo WY, Quenneville P, Chouw N 2012 *A numerical study of the seismic behaviour of timber shear walls with slip-friction connectors.* Engineering Structures, **34**(22): p. 233-243

[6] American-Forest-and-Paper-Association 1999 *General Dowel Equations for Calculating Lateral Connection Values, Technical Report 12.* 2nd ed, Washington DC

[7] Aghayere A, Vigil J 2007 *Structural Wood Design: A Practice-oriented Approach Using the ASD Method.* 2nd ed, New York: John Wiley and Sons Inc

[8] Loo WY 2010 *The behaviour of timber shear walls with slip-friction connectors*, in *Department of Civil and Environmental Engineering*, University of Auckland: Auckland

[9] Dolan JD, Madsen B 1992 *Monotonic and cyclic nail connection tests.* Canadian Journal of Civil Engineering, **19**(1): p. 97-104

[10] Coyne T 2007 *NEES Wood: Framing to sheathing connection tests*, in *Proceedings of Earthquake Engineering Symposium for Young Researchers,* University at Buffalo, Seattle.

[11] Ekiert C, Hong J 2006 *Framing-to-sheathing connection tests in support of NEESWood Project.* NEESWood Project, State University of New York at University at Buffalo, Buffalo, NY

[12] Computers-and-Structures-Inc 2009 *SAP2000 v14, Integrated solution for structural analysis and design*, California, Berkeley

[13] Dinehart D W, Hoffman R M, Blasetti A S 2006 *Finite element modeling of wood shear walls with viscoelastic polymers*, in *Proceedings of the 9th World Conference on Timber Engineering*, Oregon, Portland.

[14] Varoglu E, Karacabeyli E, Stiemer S, Ni C 2006 *Midply wood shear wall system: Concept and performance in static and cyclic testing,* Journal of Structural Engineering, **132**(9): p. 1417-1425

[15] Popovski M, Karacabeyli E 2012 *Seismic behaviour of cross-laminated timber structures*, in *Proceedings of the World Conference on Timber Engineering*, Auckland, New Zealand

[16] NZS1170.5 2004 *Structural design actions part 5: Earthquake actions.* Wellington, New Zealand, Standards New Zealand

[17] GeoNetNZ 2018 *Geological hazard information for New Zealand.* 17/08/18]; Available from: http://ftp.geonet.org.nz/strong/processed/Proc/2011/02_Christchurch_mainshock_extended_pass_band/.

IOP Conf. Series: Materials Science and Engineering **586** (2019) 012023 doi:10.1088/1757-899X/586/1/012023

Finite element modelling of different strengthening strategies for reinforced concrete deep beams

V Gergely[1], M Pop[1], C Campian[1], N Chira[1]
[1] Technical University of Cluj Napoca, Baritiu Street no 25, ZIP 400027, Romania

E-mail: maria.pop@dst.utcluj.ro

Abstract. The paper is focused on the behavior of deep beams retrofitted with FRP sheets and layers In order to understand the behavior of these beams, and an advanced nonlinear finite element model is used. The nonlinear analyses were performed based on the Modified Compression Field Theory which is a smeared rotating crack approach implemented in program VecTor2. The predictions of the finite element models were compared to the test results in order to validate the models and to further understand the behavior of the test specimens. Finally, the validated model was used to perform parametrical studies to investigate the effect of FRP sheets and UHPFRC layers on the behavior of true-scale deep beams. The variables investigated in a systematic manner included the aspect ratio of the member, the layout of the FRP sheets, the thickness and fiber content of the UHPFRC layers. The results were used to compare the effectiveness of the two retrofitting techniques.

1. Introduction

Reinforced concrete deep beams are important structural elements, which transmit loads from the upper structure to widely spaced lower supporting elements. By definition, such members have small shear-span-to-depth ratios and their behavior is governed by shear. Due to the fact that these beams are used in lower parts of the building for architectural purposes or for shopping areas, their size has to be big enough to support the whole upper structure. In some cases these transfer beams can reach up to 2-3 floors height. In the case of bridges, cap beams create the connection between the deck and the columns. Thus their main role is to concentrate the distributed forces into point ones. Their span can vary from 25 to 150m, depending on the traffic on the bridge, the environmental conditions, etc.

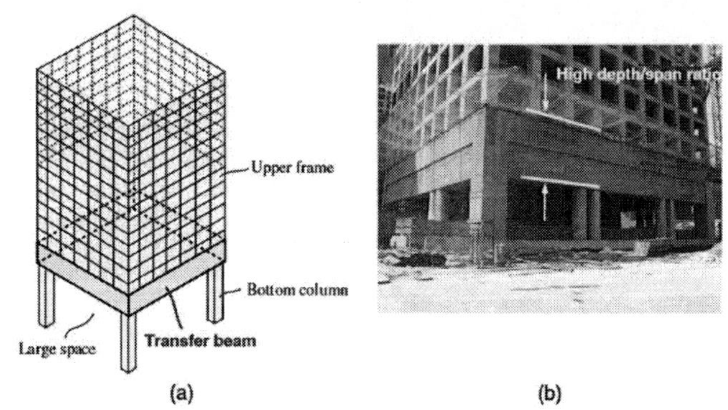

Figure 1. Deep transfer girders in high-rise buildings (Nie et al. 2017) [1].

Because some deep beams have suffered degradation in existing buildings and bridges, Fiber Reinforced Polymer (FRP) or Ultra-High Performance Fiber Reinforced Concrete (UHPFRC) can be used for their retrofit.

In this article we used an advanced nonlinear finite element model in order to understand the behavior of deep beams retrofitted with FRP sheets and UHPFRC layers and to try to make a comparison between these two solutions. For this purpose, general information about the behavior of FRP and UHPFRC is gathered and summarized. The selected approach for modelling is a smeared rotating crack as formulated in the Modified Compression Field Theory for elements subjected to shear.

2. Strengthening with FRP sheets

FRP strengthening systems have three main components, respectively adhesives, matrices and fibers. Adhesives bind together the concrete and FRP material in order to get a composite action (fib Bulletin no. 14 [8]). Because the load transfer is passing through the adhesive layer also, it is important to have necessary shear and bond strengths. One of the most used adhesives is epoxy, due to the fact that it has great strength characteristics and is commonly found. The key to a successful composite action lies on several preparation aspects. Such preparations are namely adherent materials, mixing temperatures, surface preparations, etc.

Firstly, when the material is transformed into fibrous form, its stiffness and strength is enhanced compared to the bulk form. Also, due to the high length/diameter ratio, the load transfer from the matrix to the fiber is an effective one [8]. This gives FRP composites a high tensile strength. As seen in Figure 2. Carbon FRP-s can have as high as 2000 MPa ultimate strength in tension. It can also be seen, that unlike steel, which has a bilinear response, FRP sheets behave linear elastically. This phenomenon can be attributed to the fibrous nature.

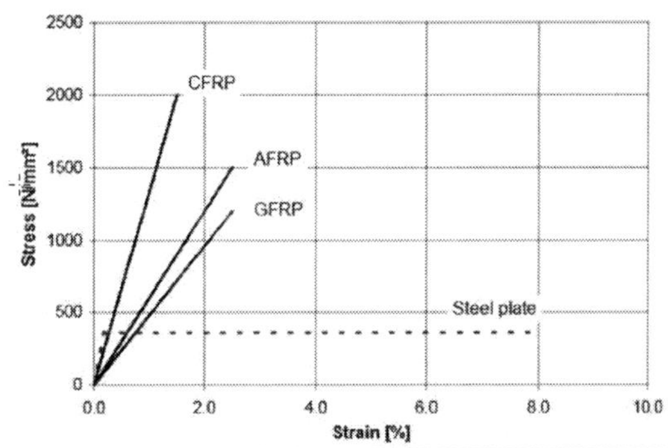

Figure 2. Stress-strain behavior of reinforcement for external bonding (Matthys 2000).

CFRP = carbon FRP, AFRP = aramid FRP, GFRP = glass FRP.

Secondly, due to FRP sheets superior characteristics, the thicknesses of the repairing layers are in the order of few mm. This combination of high strength and low cross sectional area greatly reduce the extra weight added to the structure, and also helps in maintaining aspects which are close to the initial architectural design. On the other hand, due to the thin layers, the material behaves only in tension, having almost zero compressive strength.

The title is set 17 point Times Bold, flush left, unjustified. The first letter of the title should be capitalized with the rest in lower case. It should not be indented, as shown above. Leave 28 mm of space above the title and 10 mm after the title.

2.1. Advantages and disadvantages of FRP sheets

There are several criteria, when selecting a proper strengthening system, namely the condition in which the structure is, environmental conditions, cost effectiveness being the most important one.

Mathys (2000) compared FRP sheets rehabilitation system with the steel slab bonding one. In the following list are some of the advantages of the FRP sheets:
- Unlike steel, FRP sheets don't need protection against corrosion.
- FRP elements are light weight rehabilitation tools and can be purchased in any kind of dimensions. These factors make them flexible and of easy use. On the contrary, steel plates are heavier, come in predefined dimensions, and when applied in large projects, a proper connection system has to be developed between the plates.
- Since FRP sheets are thinner than steel plates and have a large bond width/thickness ratio, stresses which are produced at the interface of materials are low. Hence in many cases there is no need of mechanical anchorage.

FRP elements have excellent fatigue behavior.
- Small thickness and lack of mechanical anchorages make FRP sheets aesthetically pleasing. Also different types of finishing layers can be applied.

When comparison of two retrofit systems is made, it is also important to highlight the disadvantages, which for FRP are:
- In comparison to steel, FRP has higher initial cost. CFRP can be at least 6 times more expensive than steel plates. This difference can be reduced by the labour cost and future reparation expenses.
- FRP elements have low transverse strength. This makes FRP sheets more vulnerable to direct impact action. Hence, in some cases, a protective layer has to be added.
- FRP retrofit systems are exposed to high temperatures (e.g. in case of fire), some epoxy resins starting to soften even at low temperatures (45-70°C) [8].

All in all, this shows that there are several factors which have to be taken into consideration, when using FRP sheets as retrofitting systems. As future research is made in order to make the material more accessible for the market, so will its use be seen in more structural retrofitting works.

2.2. Modelling of deep beams strengthened with CFRP sheets

In recent times retrofit actions on old buildings being past their service life became imminent. Especially deep beams suffered damage due to constant loading, degradations due to hars environmental conditions, aging etc. For this, use of novel materials such as Fiber-Reinforced Polymers (FRP) and Ultra High Performance Fiber-Reinforced Concrete (UHPFRC) as possible retrofitting options will be tested.

In order to study the behavior of deep beams strengthened with FRP sheets, several articles concerning the topic were reviewed and test data were collected. One major criteria was to find experimental studies, which focused on deep beams that is beams with shear-span-to-depth ratios a/d smaller or equal to 2. This way a proper shear failure could develop in the samples prior to flexural failure. Another criteria was to find tests, which had samples large enough for the size of them not the affect the outcome of the results. An experimental study was selected, namely Bukhari et al. (2013) [2].

In this experimental program the main focus was on the behavior of different CFRP configurations applied to short beams, which were designed to fail in shear. For this cause the beams were reinforced only with longitudinal bars, adding no shear reinforcement. The same type of CFRP sheets were used for all the test specimens (there were to groups, A and B, where group A had CFRP sheets over the whole height of the beam, while group B was covered only half of the height). Each beam was tested under a symmetrical three-point loading and the deflection was measured at mid-span and at supports. The geometrical properties are presented in Figure 3.

Figure 4 shows the finite element model (FEM) of beam C-3 built in program VecTor2. As one can see, the beam is defined as a simply supported one, having a pin support on the left side and a roller on the right. The load was applied as displacements uniformly distributed on the top steel plate in order to capture the post-peak behavior of the beam. Steel plates (material C) were placed in both support and load points.

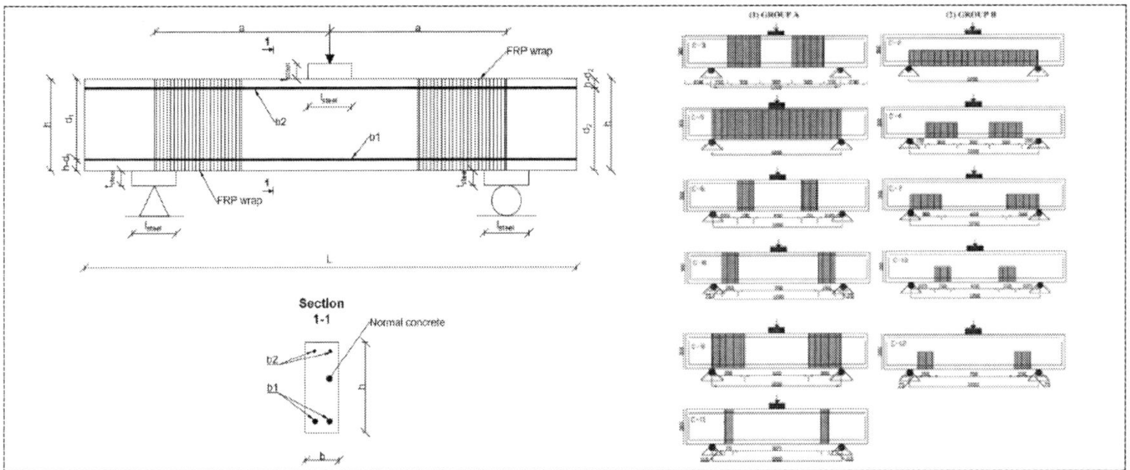

Figure 3. Test setup of beams retrofitted with UHPFRC (Bukhari et al series) [2].

The FEM consists of two types of finite elements: quadrilateral elements, used for the modelling of concrete and steel plates, and truss elements used for the longitudinal reinforcement. A perfect bond is defined between the longitudinal reinforcement and the surrounding concrete, as well as between the concrete and retrofitting material. The mesh size used in the model for the quadrilateral elements is of 15x15 mm, to avoid unrealistically stiff behavior. Because the FE program works in a 2D, with principal axes X and Y, the thickness of the element is given as property.

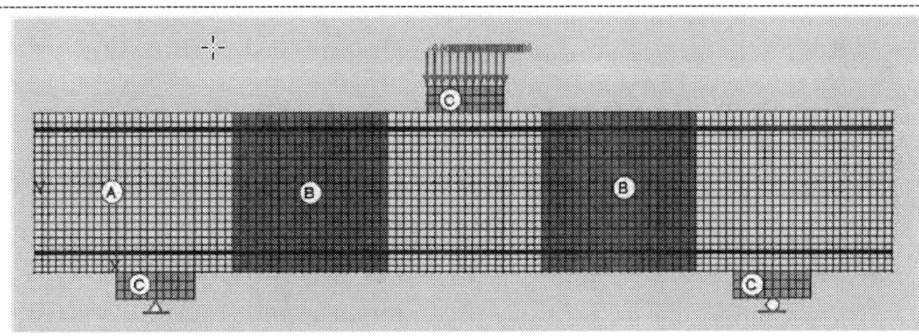

Figure 4. Finite element model of Bukhari et al. test series (model C-3) [2].

The test specimens from the two distinct groups showed different behavior which was captured with different accuracy by the finite element models. Group A, where the FRP sheets was applied on the whole height of the beam, showed better agreement between experiment and prediction results.

3. Strengthening with Ultra High Performance Fiber-Reinforced Concrete (UHPFRC)

When talking about ultra-high performance materials, UHPFRC is one of the major subjects. This novel material is composed of two main components, namely matrix and fibers. In the term matrix are included all the components of the matrix, such as cement, water, aggregates, additives, etc. Fibers are mixed in discontinuous and randomly oriented ways inside the matrix. They are mainly made of steel, carbon or glass. These two components are connected by bonding, hence one major characteristic of UHPFRC is its bond strength.

In Figure 5 a test beam retrofitted with UHPFRC transversal section is presented.

Figure 5. Test beam retrofitted with UHPFRC jacket (Martinola et al. 2010) [3].

3.1. UHPFRC jacket applications in retrofitting

Due to its superior strength and durability, UHPFRC is used more and more in retrofitting old RC and steel structures. Since the early 2000s it was a preferential choice in projects such as rehabilitation of bridges, buildings, maritime signalization structures etc.

An example is a building from GENEVA (Moreillon et al. 2013), erected in the 1920s, and has a RC slab supported by masonry wall type of setup. Due to requirements for higher load bearing capacity, its retrofit was imminent. UHPFRC layer of 30 mm with passive reinforcement was proposed as a retrofit design. This approach was selected due to reduced construction time, only small increase in dead load and the ability to neglect foundation strengthening. The connection between the old slab and the new UHPFRC layer is realized through Gritblasting, as seen in figure 6, thus the roughened surface connects the two layers.

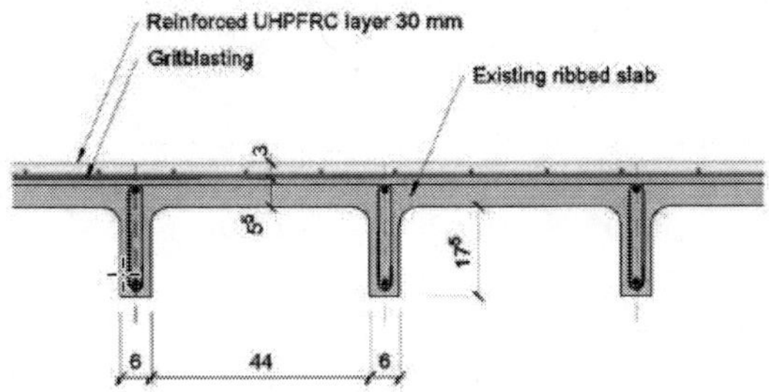

Figure 6. Cross section of the strengthening procedure of the ribbed slab (Moreillon et al. 2013) [5]

3.2. Modelling of deep beams strengthened with UHPCFRC

Another experimental program (Martinola et al. 2010) [3] was taken from the literature to investigate the effect of UHPFRC strengthening on full scale reinforced concrete beams.

One beam with reinforcement was used as reference, while the other two were retrofitted with 40mm thick UHPFRC layers on bottom and side faces. In order to create an adequate bond strength between reinforced concrete and strengthening material, the beams were sandblasted prior to application of UHPFRC jackets. The jackets were applied directly to the beams, without vibration. Curing of test specimens was carried out at ambient temperature and humidity.

Also, as in the first case, a finite element model was made, also built in program VecTor2.

It can be observed that in the FEM the critical crack is located in the middle of the beam, while in the in the experiment it was located near the left load point. This can happen due to material imperfections or localization of stresses in the reinforcement.

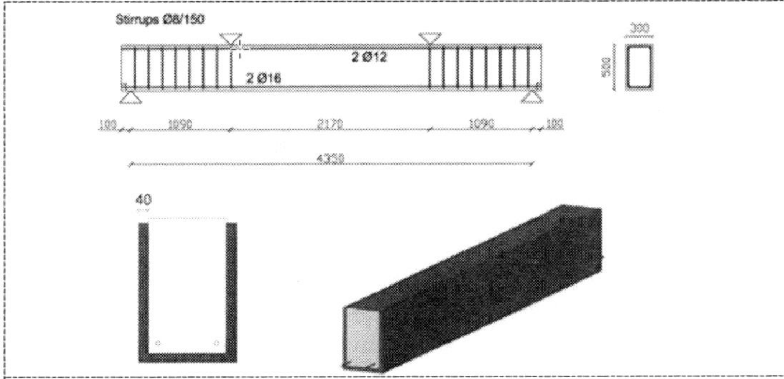

Figure 7. Geometry of specimens and strengthening scheme (Martinola et al. 2010)[3].

Nevertheless several similarities can also be noticed, such as the multiple vertical cracks between the two loading points, or the existence of cracks at 45° between loading plates and supports. One can identify the horizontal cracks at the level of longitudinal reinforcements, which indicate shear failure. In Figure 8 the crack patterns of FEM and experiment are presented and analyzed. For FEM the thin red lines represent the cracks in the finite elements, where the line thickness is indicative of the crack width. In FEM the FRP sheets prevented shear cracks and made the reinforcement yield, thus flexural failure was achieved. In case of experiment though flexural and shear cracks appeared at the same time, due to the cause of delamination of FRP sheets. Still the behavior of the prediction and experiment were in good accordance.

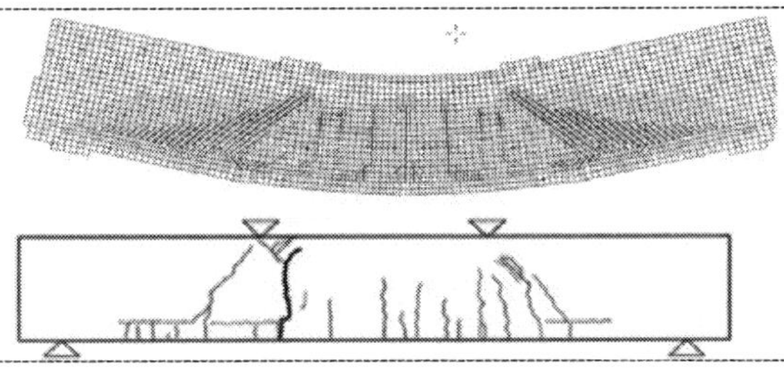

Figure 8. Comparison of crack patterns at failure load between FE model and experiment.

As shown in Figure 9, the predicted crack patterns of the beam are compared to the measured pattern at failure.

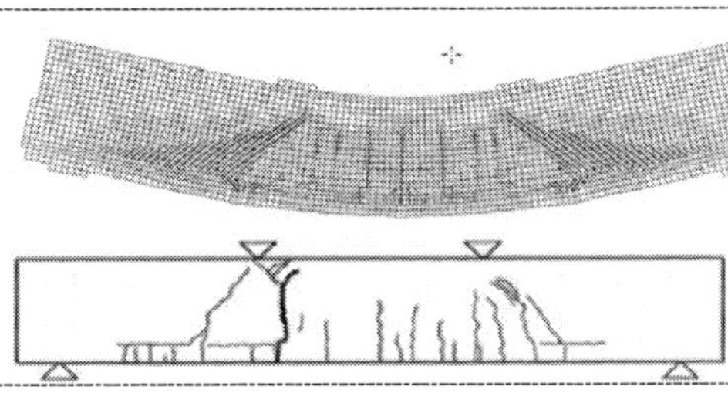

Figure 9. Comparison of crack patterns at failure load between FE model and experiment.

The FEM captures very well the behavior of the experimental beam at failure load. In both cases the critical crack appears next to the left loading point. It indicates flexural failure with localization of stresses. This observation is attested by the FEM, where the longitudinal reinforcement yields at the given moment and the fibers are pulled out of the UHPFRC layer. Also the smaller cracks appearing between the loading points in the experiment are well predicted by the FEM.

4. Effect of FRP and UHPFRC retrofit beams

In order to study the effect of FRP and UHPFRC retrofit on deep beams, a parametric study was carried out. Two parameters were studied in the case of FRP retrofit, namely span-to-depth ratio and FRP wrap configuration. For UHPFRC three variables were defined, namely span to-depth ratio, fiber volume ratio and layer thickness.

Figure 10 shows the geometrical properties of the beams from the parametric study. As one can see, some parameters are defined as variable, as follows: shear span a= 1900mm, 3800mm or 5700mm; L= 4800mm, 8600mm or 12400mm; and n x FRP wrap which represents the number of FRP wraps on a beam. This varies from beam to beam, having configurations of width/spacing of 100mm/100mm, 100mm/200mm, and 100mm/300mm or fully wrapped for each span-to-depth ratio case.

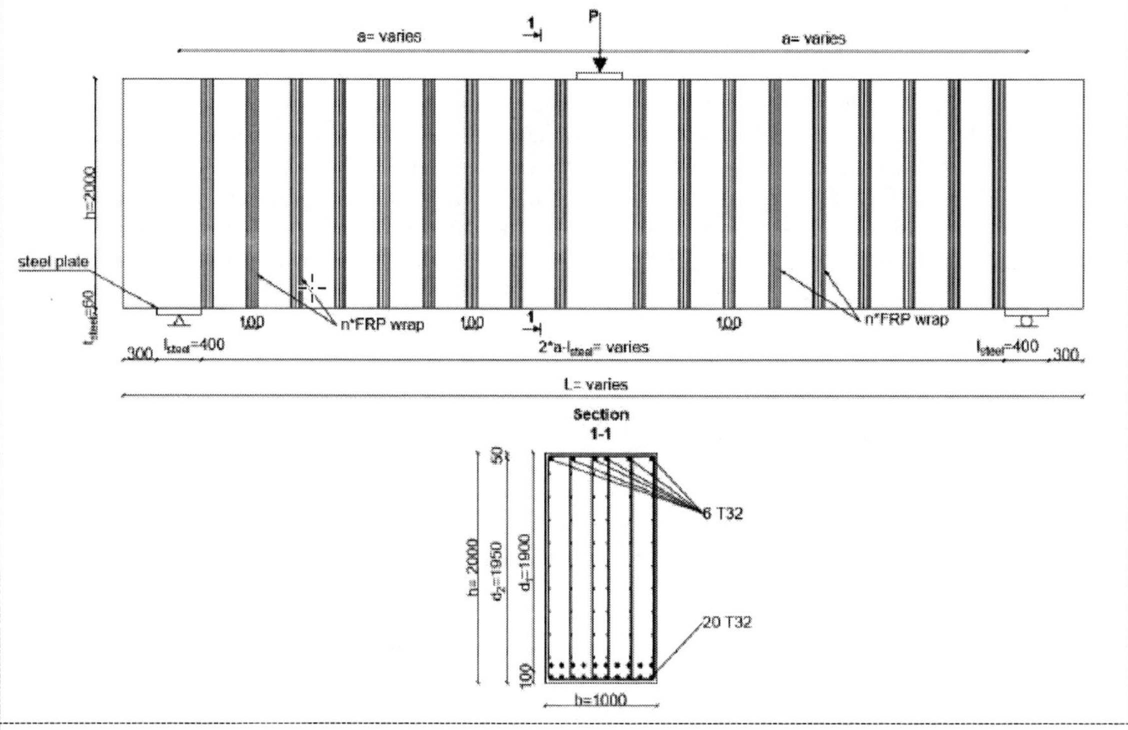

Figure 10. Geometrical parameters of beams retrofitted with FRP wraps.

4.1. UHPFRC retrofitted beam

Figure 11 shows the geometrical properties of beams from the parametric study. As one can see, some parameters are defined as variable. For these the values are as follows: shear span a= 1900mm, 3800mm or 5700mm, L= 4800mm, 8600mm or 12400mm,UHPFRC layer thickness UHPFRC= 50mm or 100mm; A final parameter, namely the fiber volume ratio has the given values Vf= 1, 2 or 3%.

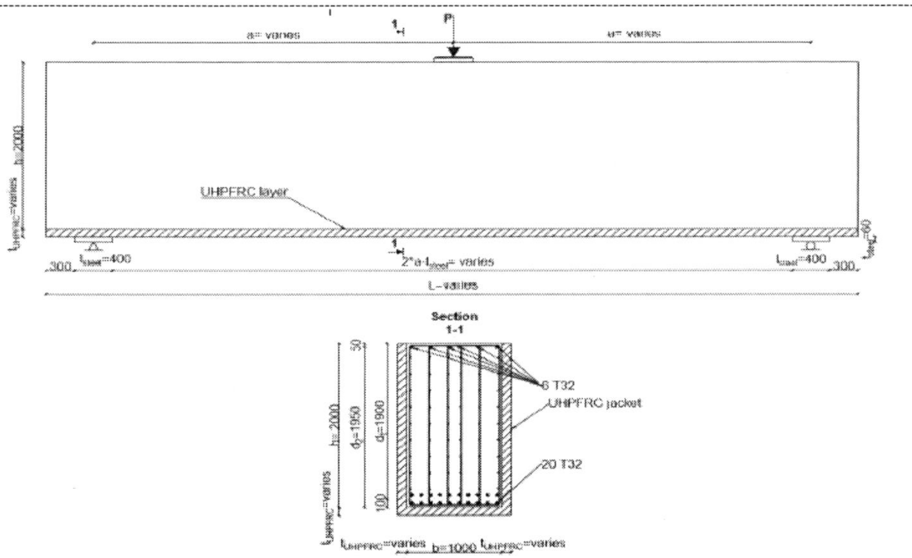

Figure 11. Geometrical parameters of beams retrofitted with UHPFRC layers.

5. Finite element models for retrofitted beams

In Figure 12 one can see the finite element model of beam FRPB-2-pw-100-200 built in program VecTor2 as Figure 13 shows the finite element model of beam UHPFRCB-2-100-1 built in the same program.

The first one has a span-to-depth ratio equal to two, and is wrapped with FRP strips of 100mm width with spacing of 200mm between them. The experimental specimens were modelled as half-beams, thus compilation time was greatly reduced. One roller was positioned at the end of the beam, while several others were installed in the place where the middle of the beam is in reality. Thus the whole system acts like a simple supported beam, and continuity of beam in the middle is not compromised. Several displacements of 100 mm were placed on the loading plate at the top of the beam in order to capture the post-peak behavior of the beam. Steel plates were placed over the support and under the load to prevent concrete crushing.

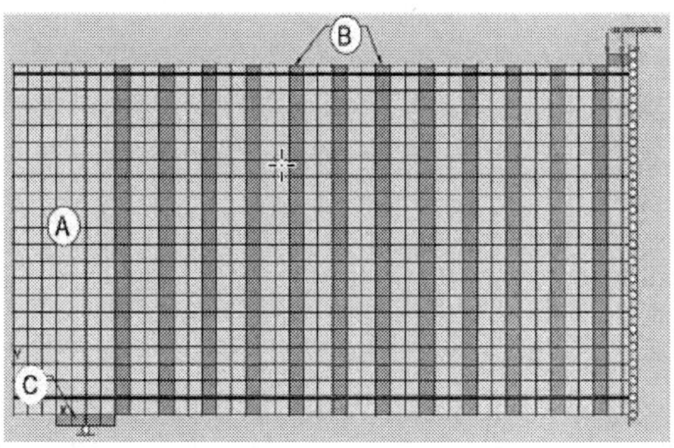

Figure 12. Finite element model of beam FRPB-2.

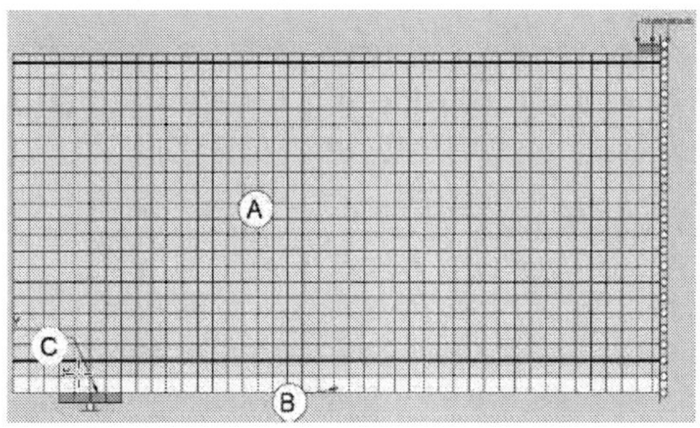

Figure 13. Finite element model of beam UHPFRCB-2.

6. Results

Best retrofit methods from both FRP and UHPFRC are taken and compared. Nevertheless what can be noticed is a loss of ductility of beams, as retrofit methods are applied. All retrofitted specimens reach

IOP Conf. Series: Materials Science and Engineering **586** (2019) 012023 doi:10.1088/1757-899X/586/1/012023

failure load at less deflection than the control specimen. Same remarks can be made about FRP whole wrap retrofit and 50mm thick UHPFRC jacket retrofit as before, with the difference being the not so effective increase of strength with respect to control beam.This is caused by longitudinal reinforcement yielding prior to full exploitation of FRP and UHPFRC retrofits.

Figure 14 shows the load-deflection curves of FRP and UHPFRC retrofit for deep beams (a/d=1). For this analysis the best retrofit methods from both FRP and UHPFRC are taken and compared. As mentioned before, both retrofit methods increase stiffness of beams up to failure load. It is very important to notice how FRP whole wrap retrofit and 50mm thick UHPFRC jacket retrofit have almost identical curves prior to load failure. This shows that on one hand for the presented parametric study UHPFRC retrofitting does not cause loss of ductility of element, and on second hand FRP retrofit can be as effective as UHPFRC if applied on whole surface of the beam. As it can be seen, 100mm thick UHPFRC jacket retrofit has the greatest impact on both stiffness and load capacity of beams. Neither in this case can a loss of ductility be observed.

In Figure 15 the load-deflection curves of FRP and UHPFRC retrofit for medium beams (a/d=2) are presented. As before, best retrofit methods from both FRP and UHPFRC are taken and compared. In case of beams with a/d=2 the same remarks can be said as before. Nevertheless what can be noticed is a loss of ductility of beams, as retrofit methods are applied. All retrofitted specimens reach failure load at less deflection than the control specimen. Same remarks can be made about FRP whole wrap retrofit and 50mm thick UHPFRC jacket retrofit as before, with the difference being the not so effective increase of strength with respect to control beam. As mentioned before, this is caused by longitudinal reinforcement yielding prior to full exploitation of FRP and UHPFRC retrofits.

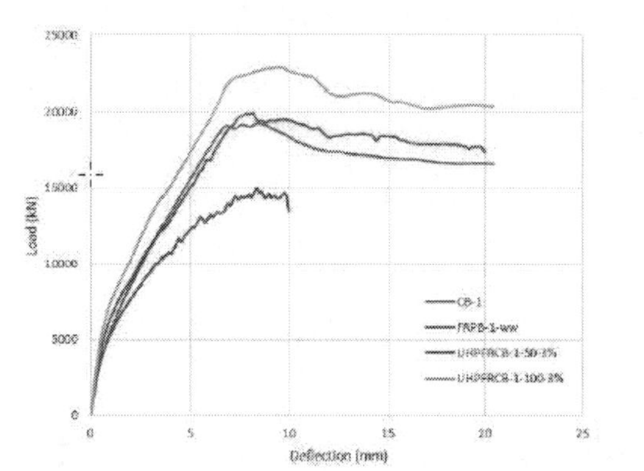

Figure 14. Load-deflection curves of FRP and UHPFRC Retrofit for deep beams (a/d=1).

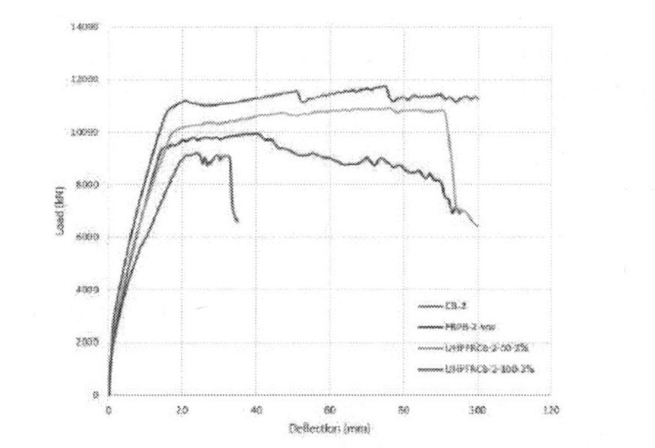

Figure 15. Load-deflection curves of FRP and UHPFRC retrofit for medium beams (a/d=2).

7. Conclusions

A number of tests were collected and studied in order to understand the effect of the test variables. The predictions of the finite element models were compared to the test results in order to validate the models and to further understand the behavior of the test specimens.

Finally, the validated model was used to perform parametrical studies to investigate the effect of FRP sheets and UHPFRC layers on the behavior of true-scale deep beams. The variables investigated in a systematic manner included the aspect ratio of the member, the layout of the FRP sheets, the thickness and fiber content of the UHPFRC layers, and the results were used to compare the effectiveness of the two retrofitting techniques. Based on these investigations we can assume that the finite element model based on the Modified Compression Field Theory captures well the complete behavior of FRP-retrofitted deep beams tested in the structural laboratory. Exceptions to this observation were the beams in which the FRP sheets were not applied to the entire depth of the section and exhibited early deboning. It was shown that early deboning can be estimated reasonably well by imposing a limit on the strains in FRP sheets.

The nonlinear analyses were terminated when the strain reached the limit value recommended in American design code ACI 440.2R-08. As UHPFRC-retrofitted test specimens did not exhibit delamination of the UHPFRC layers, the nonlinear finite element model captured well the behavior of all studied deep beams. This included agreement in terms of load-deflection response, crack patterns and failure modes. It was however noted the models produced slightly higher stiffness that observed. The parametric study demonstrated that both FRP and UHPFRC retrofitting strategies have beneficial effect on the shear strengths of deep beams. Strength increase of up to 30% in case of FRP-retrofitting strategy was observed, while for UHPFRC-retrofitting strategy the value was 53%. A favourable effect in terms of ductility was observed when the UHPFRC layers changed the failure mode from shear to flexural.

8. References

[1] Nie J G, Pan W H, Tao M X, Zhu Y Z 2017 Experimental and numerical investigations of composite frames with innovative composite transfer beams, *J. Struct. Eng.* 143

[2] Bukhari I A, Vollum R, Ahmad S, Sagaseta J 2013 Shear Strengthening of Short Span Reinforced Concrete Beams with CFRP Sheets, *Arab J Sci Eng*

[3] Martinola G, Meda A, Plizzari G A, Rinaldi Z 2010 Strengthening and repair of RC beams with fiber reinforced concrete, *Journal Cement & Concrete Composites*

[4] ACI Committee 440 2008 Guide for the Design and Construction of Externally Bonded FRP Systems for Strengthening Concrete Structures, American Concrete Institute, Farmington Hills, MI

[5] Moreillon L, Menetrey P 2013 Rehabilitation and strengthening of existing RC structures with UHPFRC: various applications", AFGC-ACI-fib-RILEM Int. Symposium on Ultra-High-Performance Fibre-Reinforced Concrete, Marseille, France

[6] Gergely V 2017 Finite Element Modelling of Different Strengthening Strategies for Reinforced Concrete Deep Beams, Disertation thesis, Universite de Liege, Faculty of Applied Sciences

[7] Da Silva W R L, Svec O, Thrane L N, Pade C 2017 Effect of fibre orientation on the tensile strength of ultra-high performance steel fibre-reinforced self-compacting concrete, AFGC-ACI-fib-RILEM Int. Symposium on Ultra-High-Performance Fibre-Reinforced Concrete, Montpellier, France

[8] fib Task-group 9.3 2001 Externally Bonded FRP Reinforcement for RC Structures, International Federation for Structural Concrete, fib Bulletin 14, pp 1-138

[9] European Committee for Standardization EN 1992-1-1 Eurocode 2 2004 Design of Concrete Structures - Part 1-1: General Rules and Rules for Buildings", CEN, Brussels

[10] Constantinescu H, Gherman O, Negrutiu C, Sosa I P 2014 Mechanical Properties of Hardened High Strength Concrete, 8th International Conference Interdisciplinarity in Engineering, INTER-ENG 2014, 9-10 October 2014, Tirgu-Mures, Romania

[11] Negrutiu C, Sosa I P, Constantinescu H, Heghes B 2014 Crack Analysis of Reinforced High Strength Concrete Elements in Simulated Aggressive Environments, 8th International Conference Interdisciplinarity in Engineering, INTER-ENG 2014, 9-10 October 2014, Tirgu-Mures, Romania (ISI Proceedings)

Acknowledgments

Authors wish to acknowledge the assistance from the professors, Boyan Mihaylov, Luc Courard, Jean-Marc Franssen and Frédéric Cerfontaine, of the Universty of Liege, Faculty of Applied Sciences.

IOP Conf. Series: Materials Science and Engineering **586** (2019) 012024 doi:10.1088/1757-899X/586/1/012024

Seismic vulnerability of historical orthodox churches in Romania: numerical modelling and retrofit solutions

M Bârnaure[1], M Coman[2], D Marcu[2]
[1] Faculty of Civil, Industrial and Agricultural Buildings, Technical University of Civil Engineering, Bucharest, 122 – 124 Lacul Tei Blv, 020396, Bucharest, *Romania*
[2] Popp & Asociații SRL, 136 Calea Griviței 136, 010737, Bucharest, *Romania*

E-mail: mircea.barnaure@utcb.ro

Abstract. Historical orthodox churches in Romania are generally masonry structures, built without a clear understanding of earthquake demands. Past earthquakes have shown specific damage patterns for these buildings: damages in domes and bell towers, damages of the church nave (particularly a general longitudinal dislocation), damages of the infrastructure. As some of the existing buildings have high architectural or historical value, their structural strengthening is important. Typical failure mechanisms for churches under earthquake loads are identified. Based on these mechanisms, the methods for assessing the seismic vulnerability of existing churches are discussed. A historical listed church is considered as case study. The considered numerical model is presented. Qualitative and quantitative results obtained from the numerical model for the un-retrofitted building are analysed. A retrofit solution is proposed. Comparisons are made between the expected structural behaviour of the church in its present state and after structural strengthening. Conclusions are drawn regarding the seismic vulnerability of this type of buildings.

1. Earthquake damage patterns for orthodox churches

An important part of the cultural heritage buildings in Romania is represented by masonry churches. The preservation of these monumental buildings with cultural and artistic value is a stringent issue, considering their earthquake vulnerability. This vulnerability is both related to the particularity of the construction materials as to the specific structural and architectonic configuration.

Masonry typically displays limited compression and low tensile and shear strength. For the old buildings, materials often deteriorated in time, when lacking proper maintenance and repairing works, due to mechanical stresses (in particular during earthquake events) and environmental actions, which lead to a further decrease of their initial strength.

In terms of geometry, orthodox churches have few transverse walls to connect the longitudinal facades. The walls are high and slender. The roofs are vaulted and most of them have the specific shapes of Byzantine vaults and domes with indirect support by the pillars through the pendentives. The bell towers are often slender. The combination of slender walls, flexible floors, inadequate connection between the different parts and eccentric position of bell towers make the churches particularly vulnerable to earthquakes.

Due to their specific characteristics, churches exhibit typical earthquake damage, as identified in [1] and [2]. The general deterioration pattern consists in three types of damages: damages of bell towers, damages of the church nave, damages of the foundations.

The bell towers represent some of the most vulnerable elements, in many cases partial or even full collapse being observed [1]. As the towers are slender and can suffer a high dynamic amplification during earthquakes, important tensile and shear stresses can occur. Depending on their geometry and the shape of the window openings, failure areas are typically observed at the lower level of the windows or at the base of the tower (figure 1).

Damage of the church nave consists in longitudinal and transverse dislocations (figure 1). The longitudinal dislocation almost always extends from the altar to the porch, dividing the church in two halves. It is due to the lack of transverse walls, the lack of a rigid connection at the roof level and the out of plane forces induced on the façade walls by the vaulted roofs.
The transversal fractures generally appear in the axes of the naos, narthex and porch [1] and are systematically observed in the weaker parts of the nave walls, more precisely near the narrow window openings [1], [2].

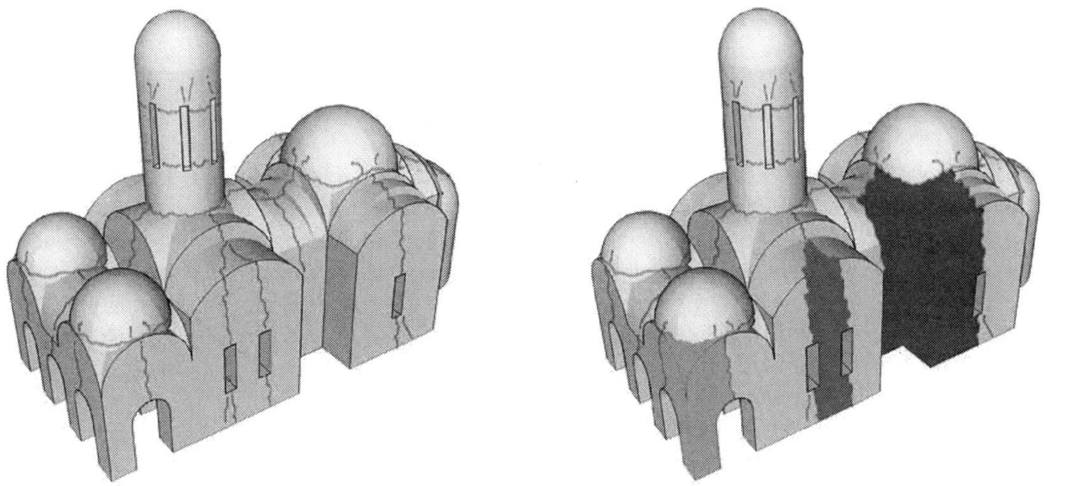

Figure 1. Typical earthquake damage patterns for churches and identification of some of the blocks in which the church is separated

The combined longitudinal and transversal dislocations lead to the fragmentation of the church into multiple independent blocks and failure mostly occurs out of the plane of the facades. Examples of such blocks are highlighted in figure 1 (right). The probable block geometry for different geometrical typical configurations of orthodox churches is studied in [3].

Sometimes the longitudinal and transversal fractures of the nave also propagate to the foundations. This is most often the case for the longitudinal crack at the base of the window in the axis of the altar [1], but it is also observed for the foundations of the longitudinal façade walls [1], [2].

2. Numerical modelling for the assessment of earthquake vulnerability of churches

The numerical modelling of historical masonry structures is an intricate issue. This is mainly due to the degrading nonlinear anisotropic cyclic behaviour of masonry and to the complex interaction between in plane and out of plane failure patterns. With the increase in computational power in recent years, it is possible to create refined finite element models that can reproduce this behaviour. As shown in [4], accurate results can be obtained by using numerical models where the non-linear behaviour of masonry is described by a continuum plasticity-based damage model. In these models, different strength, stiffness degradation and recovery effects are defined for tension and compression and the main failure mechanisms are tensile cracking and compressive crushing.
The difficulties related to this approach, besides the significant computational efforts, are related to correctly define the input data. In fact, for old complex monumental buildings, there are numerous uncertainties regarding both the material properties and the geometrical or stiffness characteristics. Regarding the materials, old buildings generally suffered multiple changes through time (local degradations, reparations or modifications using different materials, etc.) so that the material characteristics are not uniform as would be for new constructions. Even more, there is a difficulty related to measuring the material characteristics, as only limited destructive tests are possible. In terms of global parameters, the presence of structural elements with curved geometry (arches, vaults,

IOP Conf. Series: Materials Science and Engineering **586** (2019) 012024 doi:10.1088/1757-899X/586/1/012024

domes), the stiffness of the floors and the connection between orthogonal walls or between walls and floors highly influence the dynamic behaviour of the structure. Therefore, it is difficult to calibrate the FEM model as to correctly represent the actual building behaviour and there are uncertainties related to the obtained results.
For this reason, in many cases a simplified approach is adopted. As shown in [5], churches can be subdivided into macro-elements - architectonic elements characterised by a proper seismic behaviour, almost independently from the rest of the structure (facade, dome, bell tower etc.). For each macro-element, the damage modes and the possible kinematic mechanism of collapse can be identified. Complete numerical models can be obtained by defining the laws governing the connections between the various macro-elements. The studies described in [6], [7] and [8] conclude that macro-modelling are adequate solutions for the seismic assessment of unreinforced masonry buildings, as they require lower computational resources, allow easier interpretation of results and provide satisfactory accuracy when compared to the more complex FEM approach.
Still, for practicing engineers even more simplified methods might be necessary. Many commercially available structural modelling software, to which they are accustomed, often only allow considering the in-plane behaviour of masonry walls. Also, for the current structural design, linear models are often used. In the case of churches, where the seismic behaviour is dominated by the nonlinear out-of-plane response, it is obvious that such numerical models are not accurate for the evaluation of the structural safety of the existing buildings. Yet, a combined method can be easily used in practice. For the assessment of the out-of-plane failure modes for the initial building, the maximum forces for each macro-element can be computed as described in [9], without making use of complex nonlinear numerical models. The strengthening solution generally consists in introducing elements that stiffen the roof and tie together the walls such that out-of-plane failure is avoided, combined with new elements that strengthen the walls in their plane. Having this in mind, a numerical model that only accounts for in-plane failure can estimate with a good degree of accuracy the necessary strength and stiffness of the newly introduced structural elements. Such an approach is further described.

3. Case study: historical listed church in Buzău county

3.1. The existing building: Gârlaşi Church (St. Constantine and Helena)

The construction of the analysed church (as written in the carved stone above the door of the porch) began in 1780. During its existence, it suffered numerous damages due to earthquakes and fires. Following the earthquake of 1802, the vaults of the church broke down. In 1806 the church burned and was later repaired. In 1835, a new tower of the church was built, but during the 1838 earthquake several new cracks appeared in the walls. In 1932, extensive restoration works began, which lasted until 1935. During the 1977 earthquake, the church suffered further damage, mainly cracks in the two pillars inside the nave, cracking of interior and outer walls, serious damage at the ceiling of the porch and at the two vaults of the nave and the altar. To prevent further damage, the circular columns were clad in reinforced concrete and horizontal steel ties were installed at the base of the vaulted roofs in 1980.

3.2. Strengthening solution

In order to enhance the seismic strength of the existing building, new reinforced concrete elements connected to the existing masonry are proposed. Their position and dimensions consider the initial geometry, such that they do not affect the volume and the interior decorations of the listed building.
A general connection between the walls at the roof level is ensured by creating a continuous tie-beam at the base of the vaults on the perimeter of the building connected by low-height transversal beams cast on top of the existing vaults (figure 2 - left and figure 3).
Vertical tie-columns are embedded in the existing masonry walls and clad on the exterior with brick masonry so that the exterior aspect of the building is unchanged. These tie columns (figure 2) are

continuous from the foundation to the top of the walls where they are connected by a tie-beam (figure 3).

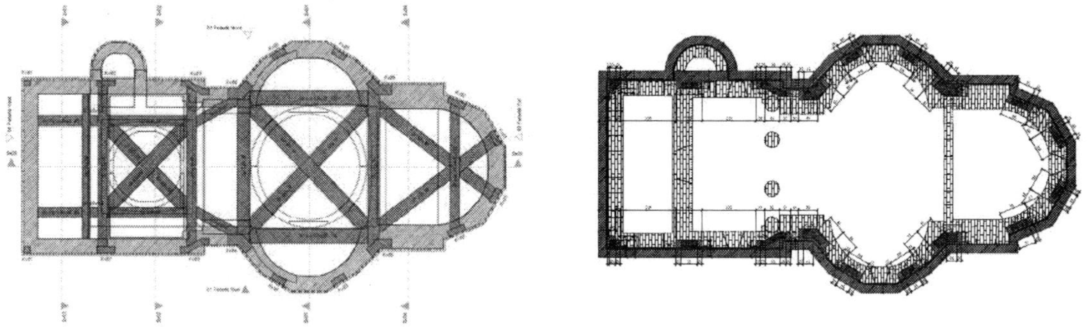

Figure 2. Vertical embedded tie-columns and tie-beams at the roof level (left) and foundations (right)

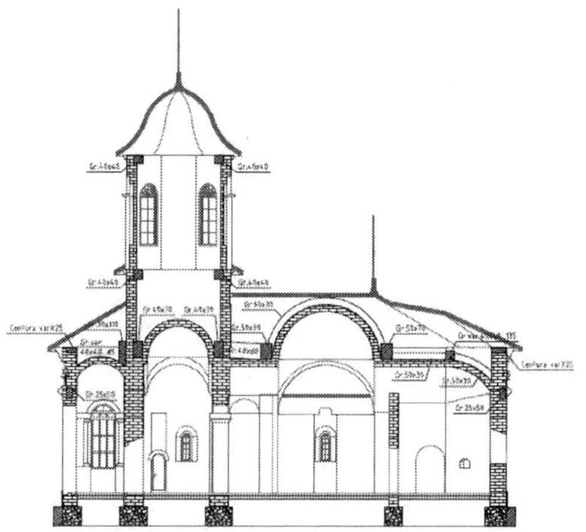

Figure 3. Position of tie-beams proposed in the strengthening solution

For the bell tower, 8 vertical tie-columns on the interior of the existing walls are connected by horizontal tie-beams at the top of the wall, above the windows and at the base of the tower. These tie-columns rest at their base on transverse beams that transmit the charges to the exterior walls (figure 3). The existing foundations are strengthened by a new RC foundation beam placed on their outer side (figure 2 - right).

3.3. Numerical modelling

The numerical model was implemented using the Etabs software [10]. The masonry walls as well as the foundations were defined using shell elements. The vaults were defined using linear elements in order to account for the lack of bending connection at their base to the walls. The geometry of the wooden roof elements was not included in the model, an equivalent weight was defined on the vaults and walls. In figure 4 the geometry of the existing building (left) and the strengthening RC elements, highlighted in purple (right) are shown, as they were defined in the model.

In the strengthened configuration model, several construction stages were defined in order to account for the stress levels that exist in the masonry elements during the consolidation works. The analysis assumes elastic behaviour (considering the ductility factor). Nonlinear behaviour was only defined at the level of the foundations, as to account for compression only in the soil.

The dynamic characteristics and the distribution of lateral forces were based on modal analysis. The load cases were defined using equivalent lateral forces along the height of the building and considering different predominant direction for the earthquake.

IOP Conf. Series: Materials Science and Engineering **586** (2019) 012024 doi:10.1088/1757-899X/586/1/012024

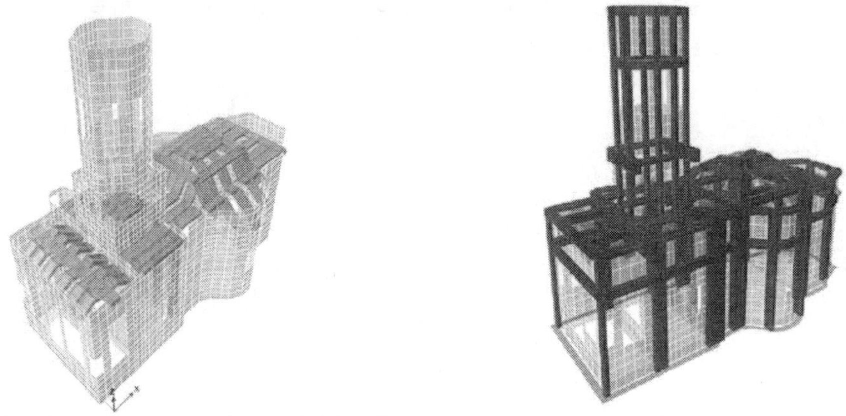

Figure 4. Numerical model for the initial (left) and strengthened (right) situation

4. Results and discussion

Comparisons were done between the initial and strengthened situation in terms of stresses and displacements. Regarding the initial configuration, a qualitative analysis was done as to identify the most vulnerable elements. As shown in [11], the identification of the most vulnerable elements can be done, with a good degree of accuracy, based on the normalised virtual work. In figure 5 the potential failure areas for the walls and vaults, as determined based on this method, are highlighted.

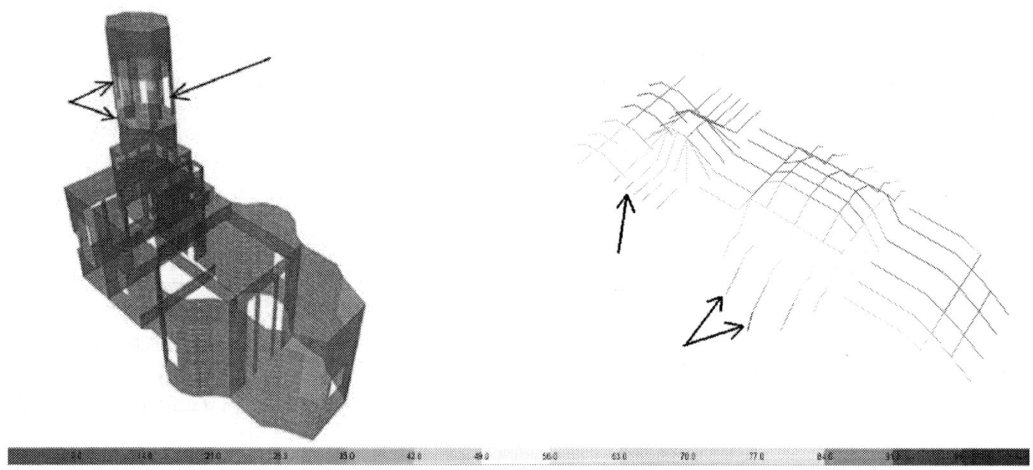

Figure 5. Identification of the potential failure areas based on normalised virtual work for masonry walls (left) and vaults (right) for one of the analysed earthquake load scenarios

Regarding the analysis in terms of stresses, both principal compression and tensile stresses were computed, for each load case, and the zones with high stresses were identified and analysed. The computed absolute maximum values of stresses are diminished by only a few percent after consolidation, but the extents of areas with high stresses is reduced.

In figure 6, the principal compression stresses diagrams are traced. The highlighted (purple) areas correspond to values higher than 2 MPa. The principal tensile stresses diagrams are traced in figure 7, where the highlighted (blue) areas correspond to values higher than 0.6 MPa. It can be observed that the elements where an exceedance of these chosen values can occur are substantially reduced after consolidation and therefore the risk of damage occurring in the existing walls is diminished.

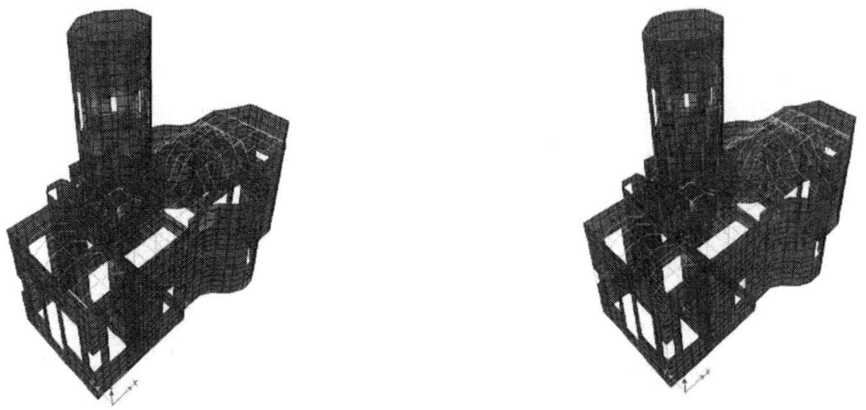

Figure 6. Principal compression stresses in the masonry walls for the initial (left) and the strengthened (right) configuration.

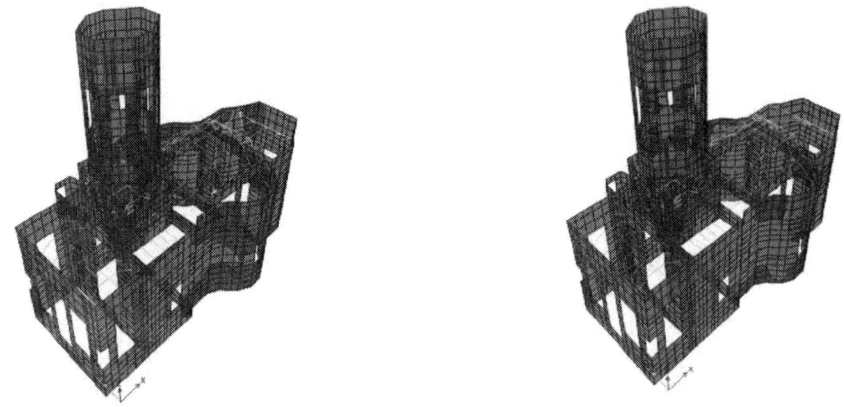

Figure 7. Principal tensile stresses in the masonry walls for the initial (left) and the strengthened (right) configuration.

The axial and flexion forces in the masonry vaults were also analysed. In terms of pure axial forces (figure 8), the maximum efforts are practically equal before and after consolidation (the maximum computed compression stress value is equal to 0.65 MPa). In figure 8, the compression forces are shown, not the stresses.

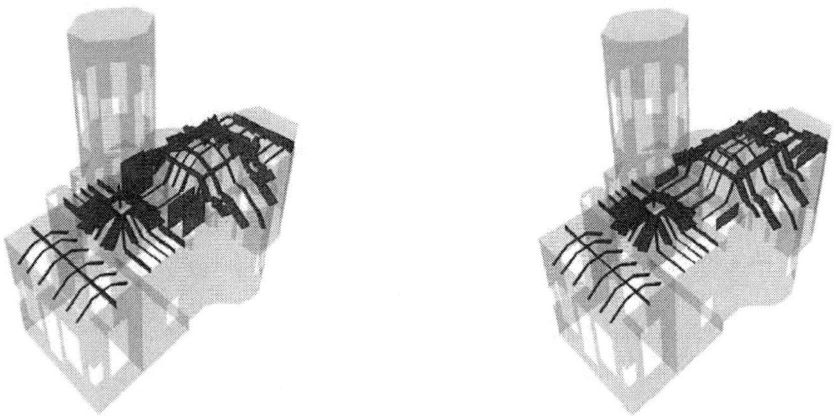

Figure 8. Compression in the masonry vaults under earthquake loads for the initial (left) and the strengthened (right) configuration

In terms of bending forces (figure 9), the strengthening solution leads to a diminishment of almost 50% of the maximum computed values in the initial situation. It is obvious that the risk of cracking or breaking of the vaults is greatly reduced after the intervention. In figure 9, the bending moments on the elements are shown, not the stresses. The maximum computed stress is 1.44 MPa.

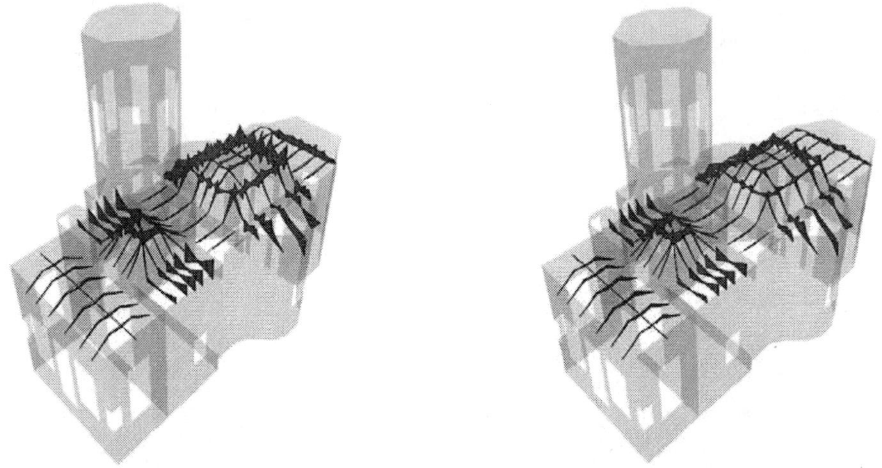

Figure 9. Flexion in the masonry vaults under earthquake loads for the initial (left) and the strengthened (right) configuration

For the check of the foundations, a Winkler model (linear force-deflection relationship) was considered. No tensile forces in the soil are allowed in the numerical model. The strengthening solution for the foundations was chosen such that the maximum pressure on soil during earthquake is similar to the maximum pressure under vertical loads for the existing building. As previously stated, the numerical model took into account the stresses in the existing elements during the construction works. In figure 10, left, the computed soil pressure after completion of the intervention works is traced. It can be noted that a significant pressure exists under the old foundations, while the new foundations are less charged, due to limited additional deflection. A similar situation appears under earthquake loads (figure 10, right), where the maximum soil pressure values correspond to areas under the initial foundation beams.

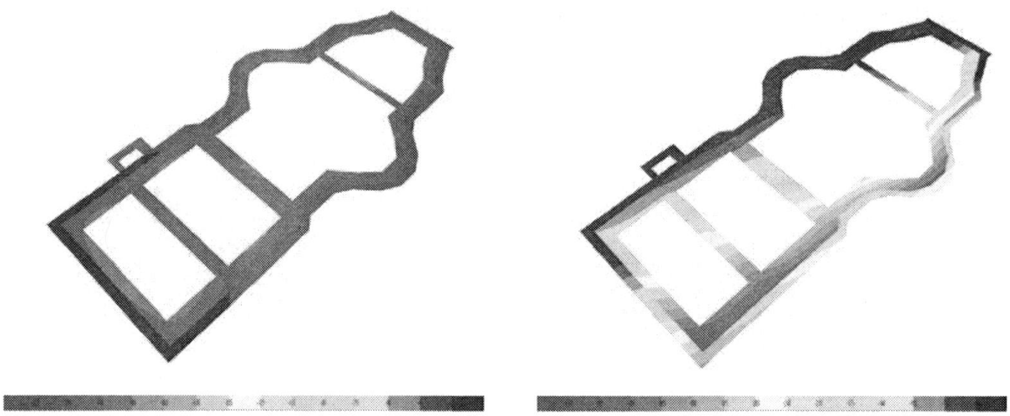

Figure 10. Reaction on foundations after completion of strengthening works (left) and for one of the analysed earthquake load scenarios (right)

IOP Conf. Series: Materials Science and Engineering **586** (2019) 012024 doi:10.1088/1757-899X/586/1/012024

5. Conclusions

The historic masonry churches exhibit specific damage patterns under earthquake loads. Their seismic behaviour is very complex, being governed by nonlinear out-of-plane response. As shown in the literature, the vulnerability of this type of buildings can be precisely established using refined FEM models or more simplified macro-models. There are two significant problems related to implementing these methods in practical cases. Firstly, these methods are very sensitive to input data (geometric, stiffness or material properties) that is difficult to assess for historical buildings. Secondly, practicing engineers are more accustomed to using commercial software that does not always allow for defining complex material behaviour.

A simplified analysis method is conducted, that is based on the fact that strengthening solutions generally consists in introducing elements that stiffen the roof and tie together the walls such that out-of-plane failure is avoided, combined with new elements that strengthen the walls in their plane. Analysis is made on linear numerical models, considering code-based ductility factor and material properties. Construction stages are defined in order to account to stress levels in the existing elements. Compression-only behaviour in the foundation soil is considered. This solution can be used for assessing the strengthened situation, while for the initial evaluation it must be combined with a separate analysis of the out-of-plane strength for each macro-element.

Results show that the strengthening solution is appropriate for reducing stress levels for vaults, masonry walls and foundations alike. The numerical models allow to conduct code-base deflection and strength checks for the masonry and RC elements. As this method is easy to implement it can be used in current practice for the design of rehabilitation works.

6. References

[1] Cismigiu A and Cismigiu M 1996 The Seismic Pathology of Religion Related Buildings in Romania and their Treatment According to Article 10 of Venice Chart Paper No. 582 *Proc. 11th World Conf. on Earthquake Engineering, Acapulco Mexico June 23-28 1996*

[2] Crisan M 2003 *Structural Restoration of Orthodox Churches - in Romanian* (Bucharest: "Ion Mincu" University Publishing House)

[3] Cosma O 2010 Risk evaluation of orthodox churches, using collapse plastic mechanism *Proc. Int. Conf. on Risk Management, Assessment and Mitigation RIMA 10 Bucharest Romania 20-22 April 2010*

[4] Valente M and Milani G 2018 Damage assessment and partial failure mechanisms activation of historical masonry churches under seismic actions: Three case studies in Mantua *Engineering Failure Analysis* **92** 495

[5] Lagomarsino S 1998 A new methodology for the post-earthquake investigation of ancient churches *Proc. 11th European Conference on Earthquake Engineering Paris France 6-11 September 1998* pp 6-11

[6] Pantò B, Cannizzaro F, Caddemi S and Caliò I 2016 3D macro-element modelling approach for seismic assessment of historical masonry churches *Advances in Engineering Software* **97** 40

[7] Clementi F, Gazzani V, Poiani M and Lenci S. 2016 Assessment of seismic behaviour of heritage masonry buildings using numerical modelling *Journal of Building Engineering* **8** 29

[8] Marques R and Lourenço PB 2011 Possibilities and comparison of structural component models for the seismic assessment of modern unreinforced masonry buildings *Computers & Structures* **89** 2079

[9] Beolchini et al. 2002 *Repertoire of damage mechanisms, intervention techniques and related costs in masonry buildings – in Italian* (Marche, ed. ITC CNR L'Aquila)

[10] CSI ETABS 15.2.2 2015 Computers & Structures Berkeley USA

[11] Stoica ND and Barnaure M 2015 Energy dissipation-an indicator of potential structural damages for masonry structures *Romanian journal of materials* **45** 255

IOP Conf. Series: Materials Science and Engineering **586** (2019) 012025 doi:10.1088/1757-899X/586/1/012025

Parametric FEM analysis of steel beam-to-column connection with extended end-plate and diagonal stiffeners

M Stașcov[1] and V M Venghiac[1]
[1] The Faculty of Civil Engineering and Building Services, "Gheorghe Asachi" Technical University of Iași, *Romania*,

E-mail: stascov.mihail@gmail.com

Abstract. Connection behaviour is the most important factor in the steel structures behaviour subjected to cyclic actions. To avoid formation of plastic hinges in the column, it is stiffened with additional diagonal stiffeners. This paper presents an FEM analysis of extended end-plate beam-to-column connections with diagonal stiffeners subjected to the cycling loading proposed by FEMA350/SAC protocol. Models were analysed with the ANSYS finite element analysis program. The study analyses the energy dissipation capacity, the stress and deformation state. The parameters are the direction of singular diagonal stiffeners, influence of crossing diagonal stiffeners and lack of stiffeners. The design of the connections is based on the strong column-weak beam principle.

1. Introduction

Earthquake are natural, complex geological phenomena, which are exposed by disordered earth crust movements, resulting from the sudden release of energy in Earth's crust. This movements, are with relatively short duration, and have rapid variations of directions, speed and acceleration of soil. Vibrations result either from tectonic plate movements, or from geological layers failure.

Earthquakes can act on the environment by altering the state of equilibrium of superficial ground structures; subsidence and liquefying of saturated sand deposits. Earthquakes can produce negative effects on constructions and assemblies, such as: the total or partial destruction of vulnerable buildings, the destruction of structural or non-structural elements.

Seismic hazard is the imminent probability of an earthquake occurring in a certain area and with an intensity exceeding a certain threshold. The Romanian design rules of the structures provide for the probabilistic determination of seismic hazard using the mean of recurrence period (IMR). In Romania, the risk of a major earthquake is very high, so greater attention should be paid to the behaviour of structures subject to seismic action.

Steel structures are very widespread due to their proper behaviour in seismic actions, the relatively simple process of execution and maintenance. In the case of degradations, they can be brought back into operation in a shorter time, simpler and cheaper workmanship than the case of reinforced concrete or masonry structures [1].

Thus, beam-to-column joints are regarded as a critical area of the structure because in the joint or the associated area, the plastic hinges which are formed can lead to collapse of the entire structure. To improve the behaviour of the structure, it is necessary to improve the performance of the joint. In actual practice, the proper design of the joint is a problem in ensuring the strength and stability of the structure.

There is a dependency relationship between the stiffness, the strength, the ductility of the joint and the response of the structure to the seismic action [2].

Steel structures are very common in areas with high seismic risk due to ductile behaviour, efficient energy dissipation, and the relatively low complexity of execution [3]. Stress concentrations, weld defects and material imperfections are some of the reasons why fragile breaking of welded joints occurred [4].

To improve the behaviour of steel beam-to-column joints, a process of research and innovation has begun with the main purpose of preventing the failure of steel structures subjected to seismic actions. Considering the requirement of ductile yield and energy dissipation of steel structures, namely the possibility of plastic deformation without major loss of strength. The design codes require the use of seismic force reduction factors, which means the significant reduction of seismic design forces, at the same time, it was intended to create new types of joints with an increased degree of ductility and energy dissipation capacity, respecting the principle of "strong column – weak beam" [5] , according to which the plastic joint develops exclusively in the beam end and at the column base.

The major degradations that occurred in structures following the (1994) Northridge and (1995) Kobe earthquakes have shown that the beam-to-column joints fail brittle [6] [7].

Numerous studies have been performed to analyze the behavior of the joint under cycling loading, determined by the Northridge and Kobe earthquakes [8, 9, 10, 11, 12, 13, 14, 15, 16, 17, 18, 19]

2. Analysed models

The ductile behaviour of the joint is provided by directing the formation of the yielding mechanism in the end plate or beam end [20] The main objective of the presented analyses is to study the state of stress and deformation of the components of beam-to-column joints with end plate and stiffeners on the column flanges. End plate joints where the rigidities of the column and the beam are approximately the same, show some yielding phenomena at the column flanges, phenomena which should be avoided.

The considered nodes have as a common element the column – HEA300, the IPE360 beam and the bolts M22, group 10.9. The steel grade in beam, column, end plate, and stiffeners is S235. The profiles chosen for the beam and column have similar rigidities.

Node 1 has 20mm thick end plate, M22 bolts, 12mm stiffeners on the column at the beam flanges and 10mm diagonal upward stiffeners. Node 2 is similar to node 1 except the direction of diagonal stiffeners, that are applied in downward direction. Node 3 is similar to nodes 1 and 2, but diagonal stiffeners are intersected. The nodes are shown in figure 3. For an analysis of the connections under cycling loading, it has been developed a model using the ANSYS finite element software.

The static scheme of the analysed models is presented in figure 1. The load applied to the models is the alternate displacement, according to FEMA/SAC loading protocol (FEMA 350, 2005) [21] is shown in figure 2, it should be noted that the maximum displacement applied corresponds to the rotation θ of 0.1 radians.

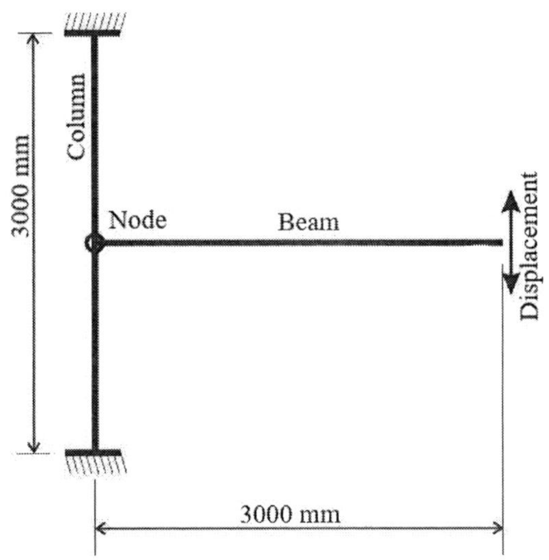

Figure 1. The static scheme.

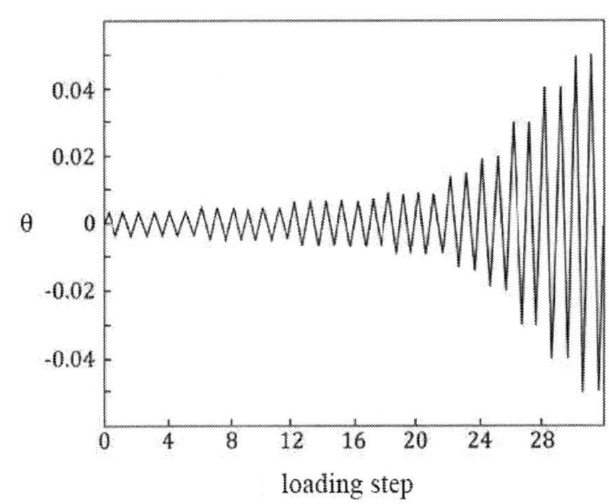

Figure 2. Loading protocol FEMA/SAC.

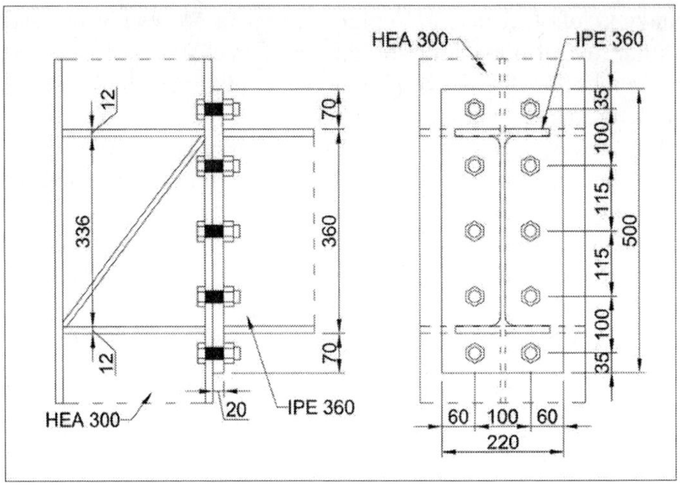

Node 1

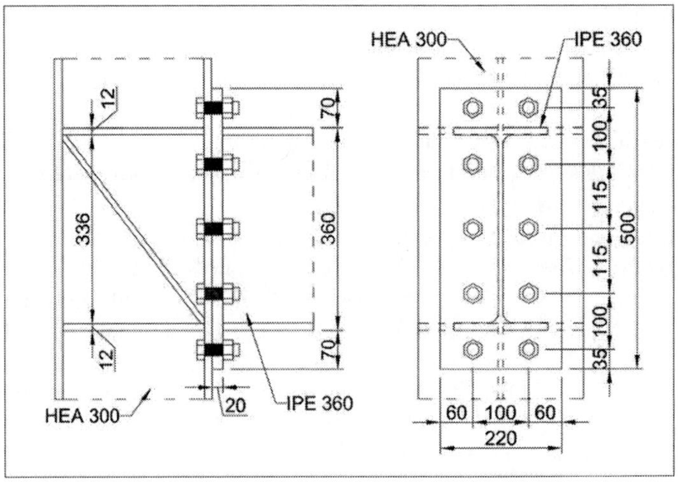

Node 2

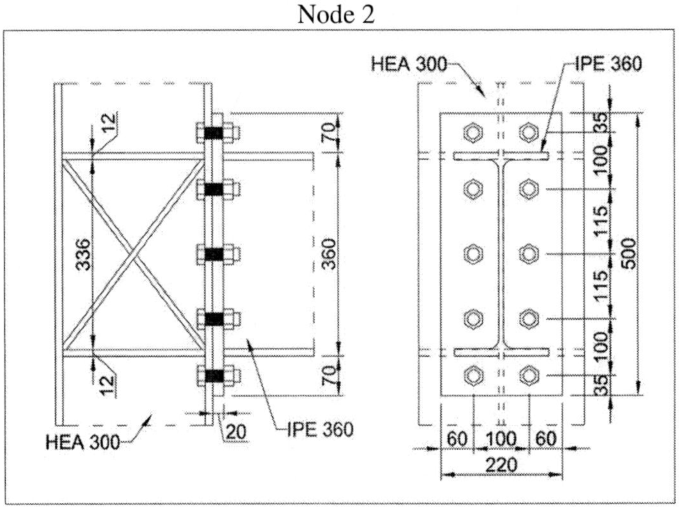

Node 3

Figure 3. Analysed nodes.

The FEMA/SAC loading protocol assumes that lateral displacements of the top and bottom of the column are prevented (figure 4).

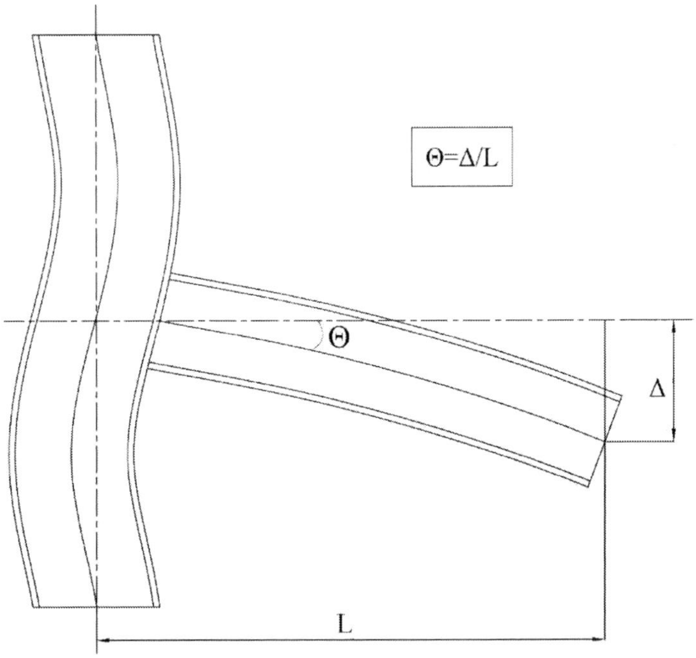

Figure 4. Rotation θ of the joint in terms of the beam end displacement.

The nonlinear behaviour of steel was modelled using the simple bilinear characteristic curve is shown in figure 5.

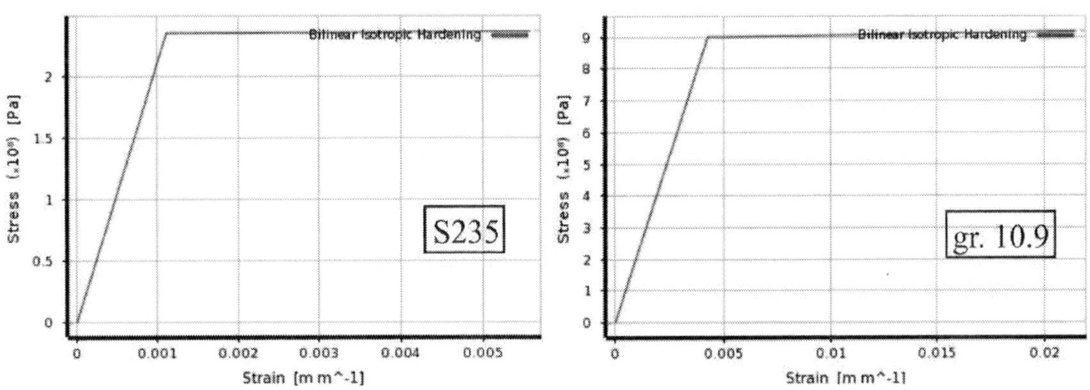

Figure 5. Rotation θ of the joint in terms of the beam end displacement.

Analyses were performed on half of the structure using the symmetry option in order to reduce the number of nodes and finite elements, and to reduce the actual time of computation.

IOP Conf. Series: Materials Science and Engineering **586** (2019) 012025 doi:10.1088/1757-899X/586/1/012025

3. Results

The hysteretic curves are presented in figure 6, figure 8, figure 10, and figure 12.

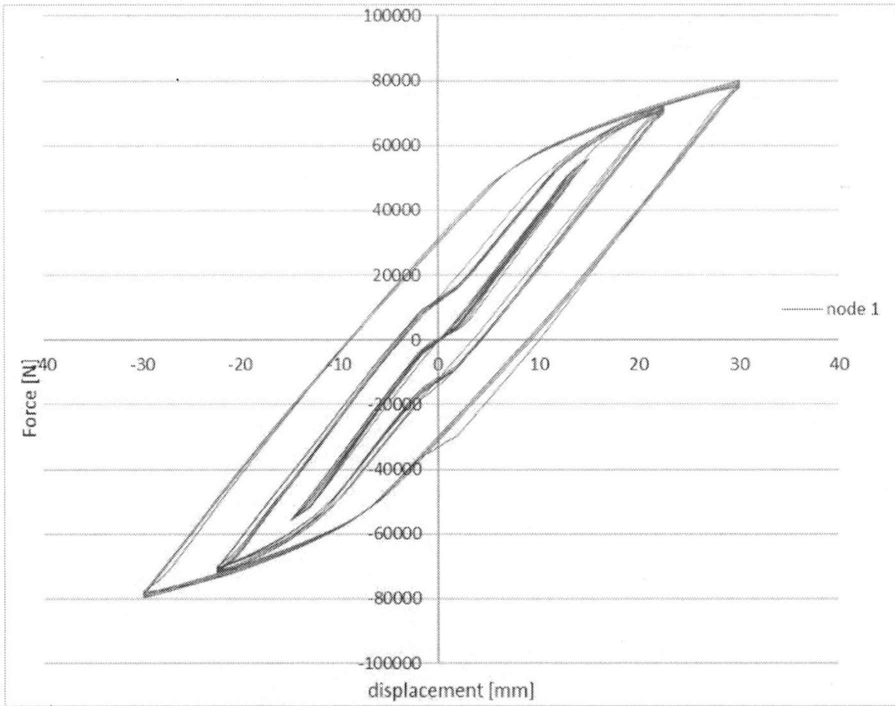

Figure 6. Force-displacement curve for node 1.

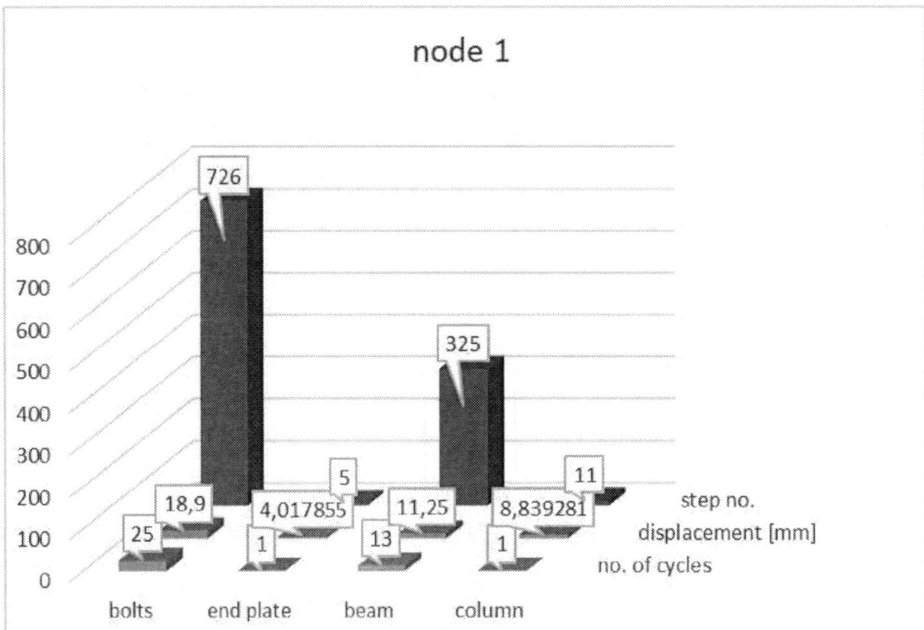

Figure 7. Moment of yielding of joint elements for node 1.

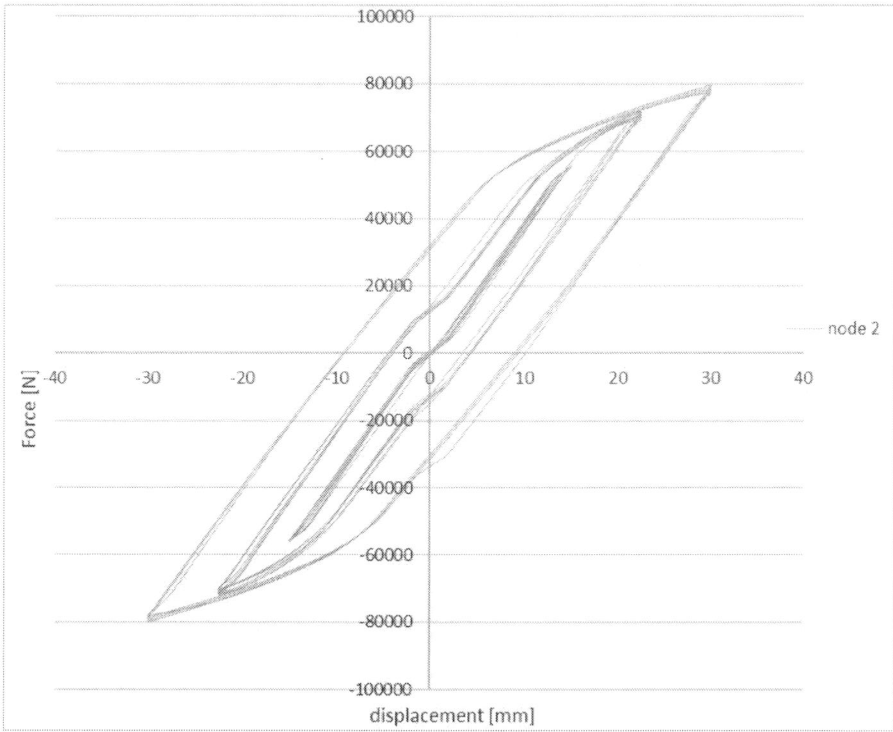

Figure 8. Force-displacement curve for node 2.

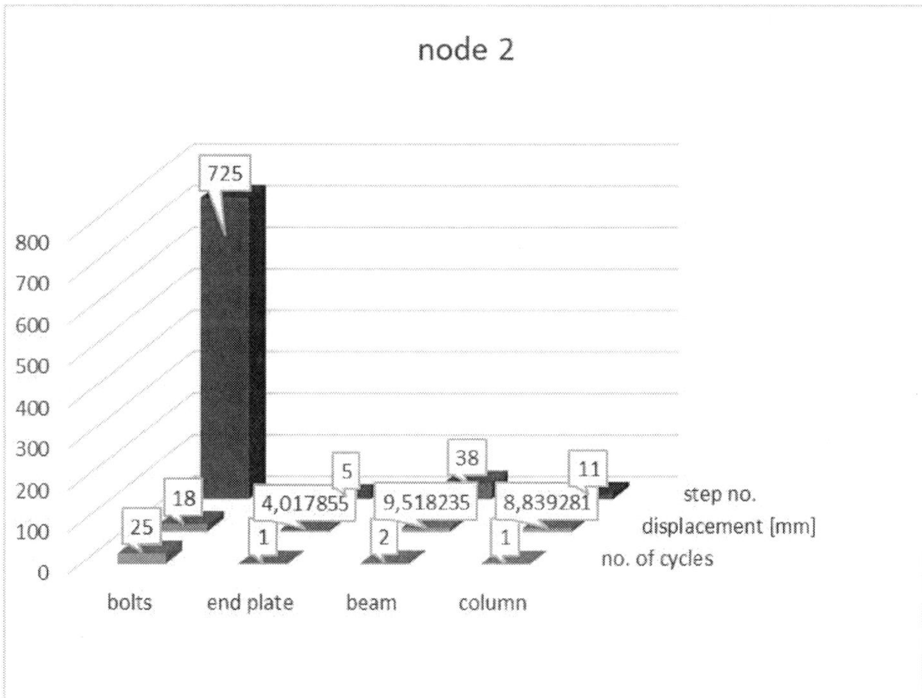

Figure 9. Moment of yielding of joint elements for node 2.

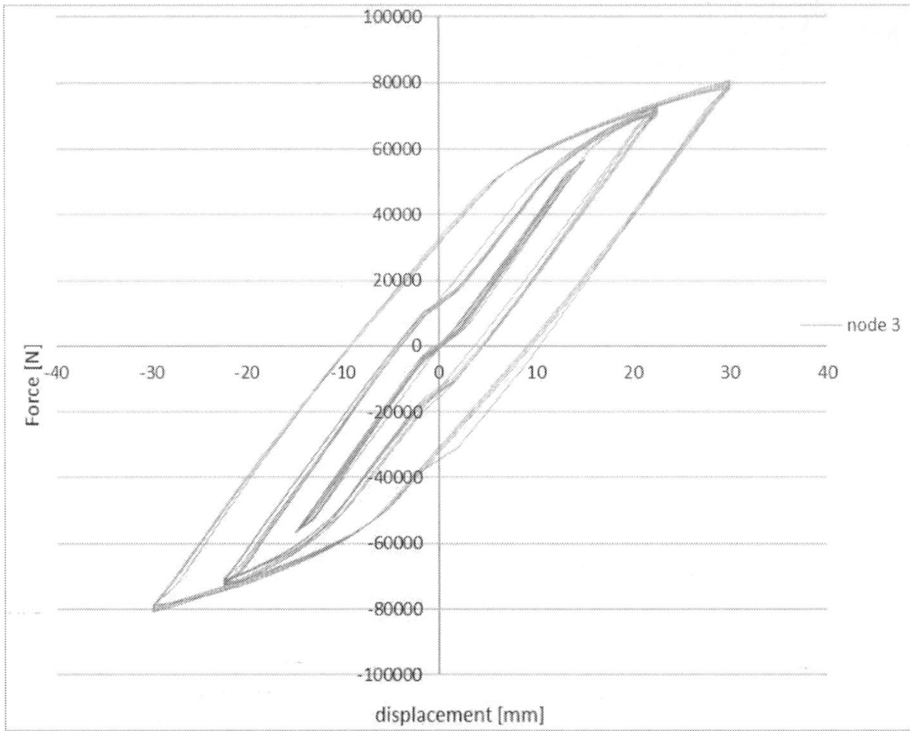

Figure 10. Force-displacement curve for node 3.

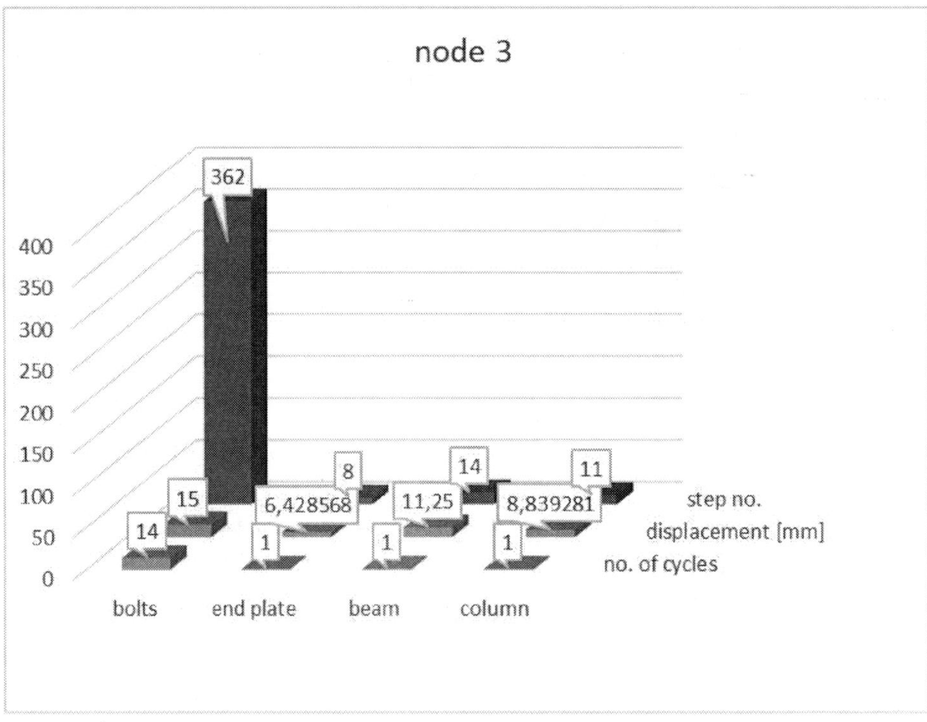

Figure 11. Moment of yielding of joint elements for node 3.

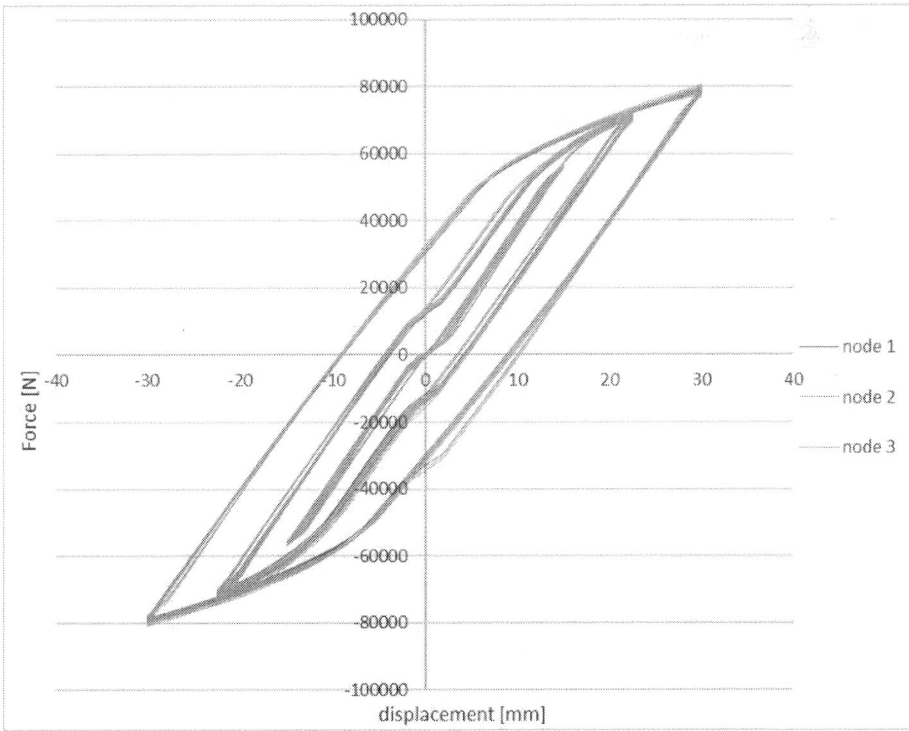

Figure 12. Force-displacement curve for nodes 1, 2, and 3.

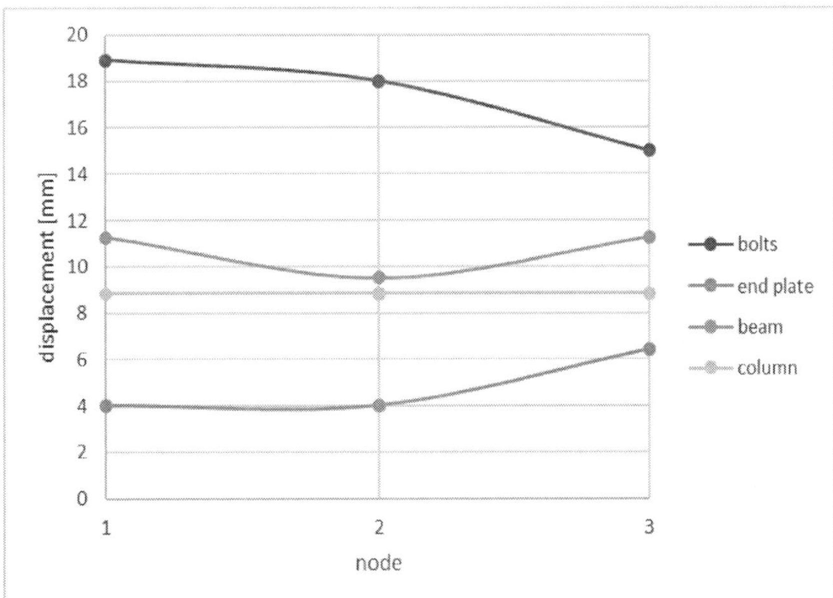

Figure 13. Displacement at which joint elements are yielded.

The stresses that develop in the end plates are shown in figure 14. In figure 15 and figure 16 are presented the stresses appearing in the flanges of the column. These stresses are recorded at the maximum values of the load.

IOP Conf. Series: Materials Science and Engineering **586** (2019) 012025 doi:10.1088/1757-899X/586/1/012025

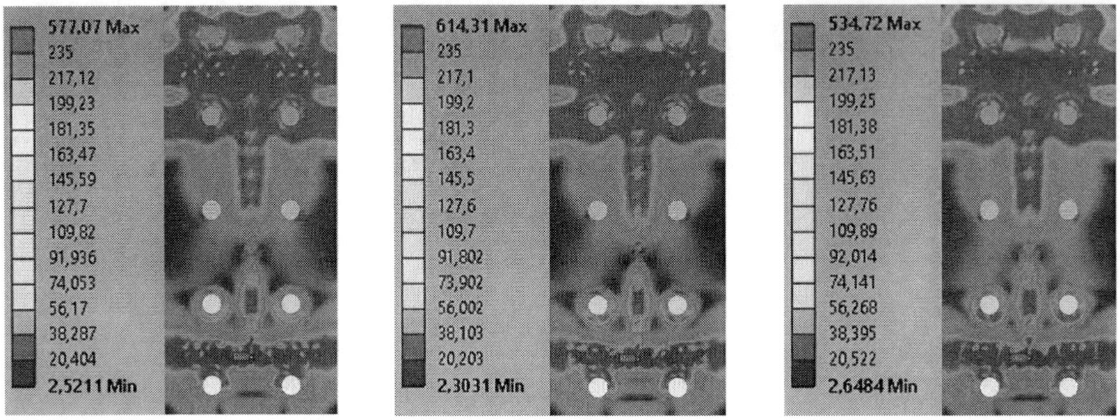

Node 1 Node 2 Node 3

Figure 14. Von-Mises stress in end plate at maximum load imposed according to the FEMA/SAC loading protocol

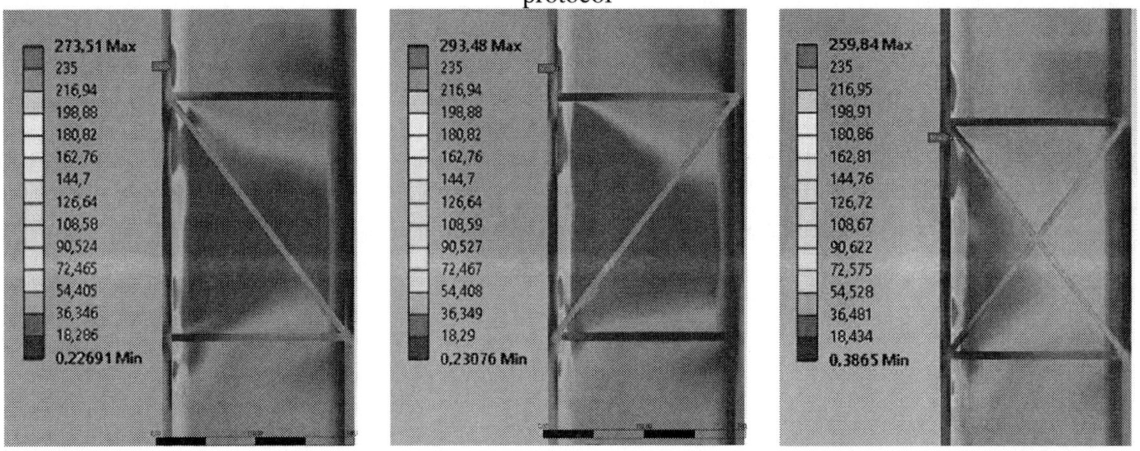

Node 1 Node 2 Node 3

Figure 15. Von-Mises stress in midle column flange at maximum load imposed according to the FEMA/SAC loading protocol

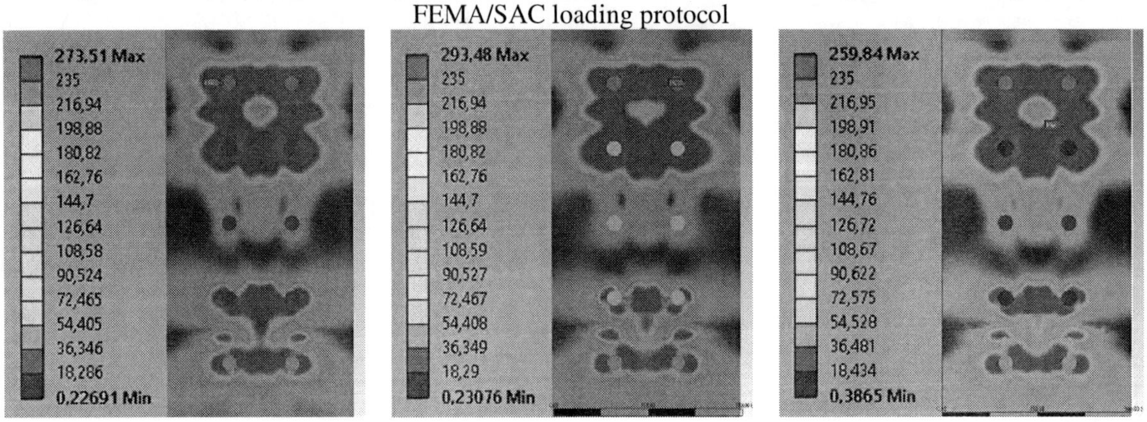

Node 1 Node 2 Node 3

Figure 16. Von-Mises stress in column flange at maximum load imposed according to the FEMA/SAC loading protocol

In figure 12 it is observed that the joints have almost identical behaviour.
In figure 14 is observed a stress concentration in the end plate, at the level of beam flanges for all analysed models. At nodes 1 and 2 intensity of the stresses is lower than those at node 3, but differences are insignificant.
In figure 15 it is observed that the distribution of stresses on the column flanges is uneven. In the case of node 3, the stress distribution is lower than in case of node 1 and 2.
In figure 16 is a rather uniform distribution of the stress at the level of the column. Node 3 shows a higher intensity of tension in the column flange; here plasticizing phenomena are present.

4. Conclusions

According to the results presented in figure 12, it is observed that the direction of application of the diagonal stiffeners does not have a significant influence on the energy dissipation capacity, or the increase of the stiffness of the joint. Instead, the presence of these stiffeners significantly reduces the energy dissipation capacity compared to nodes without diagonal stiffeners. In figure 17 it can be noticed that the node without diagonal stiffeners represents a hysteretic curve significantly larger than the nodes with diagonal stiffeners, but the resistance is lower.

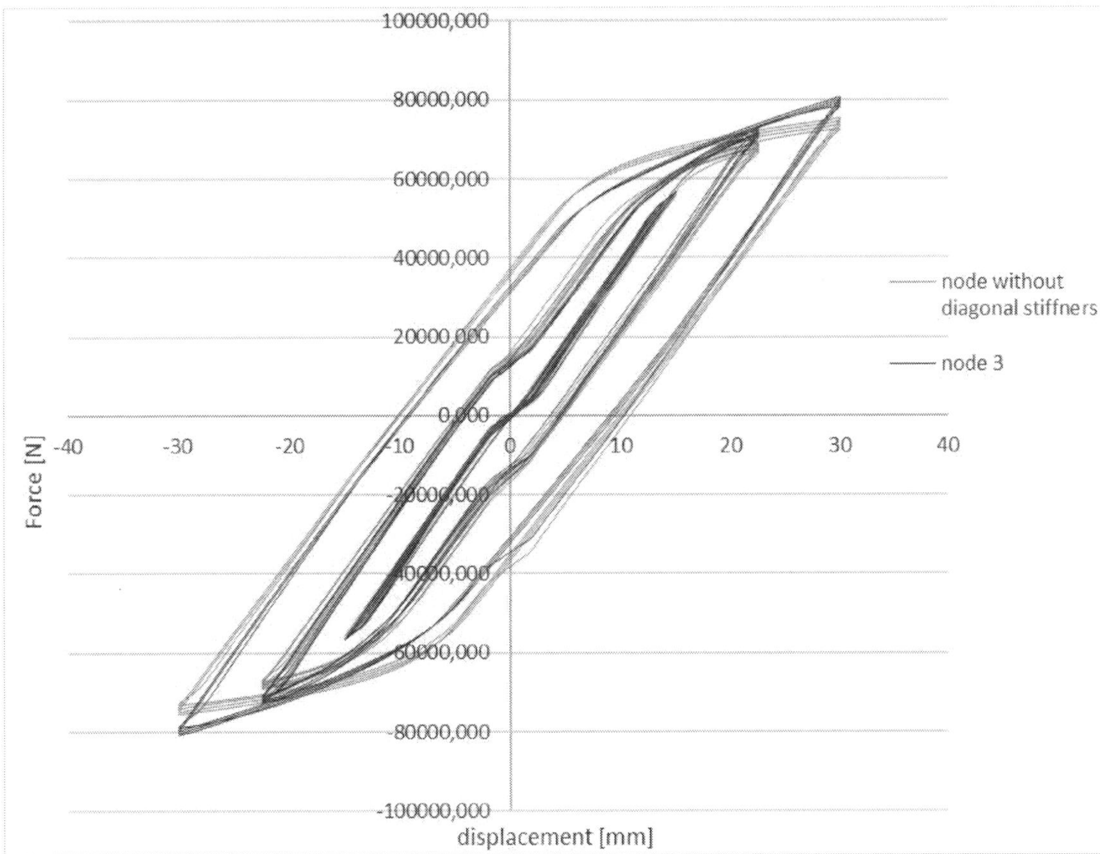

Figure 17. Force-displacement curve for node 3 and node without diagonal stiffeners.

The stresses distribution in the end plate is uniform but at the level of the column flanges the stresses distribution is not as uniform. In all three nodes dissipation capacity is reduced compared to the node without diagonal stiffeners, but the strength of these types of joints is significantly higher.

IOP Conf. Series: Materials Science and Engineering **586** (2019) 012025 doi:10.1088/1757-899X/586/1/012025

The similar distribution of stresses in the end plate and almost identical hysteretic behavior of nodes 1, 2, and 3 are due to the almost identical make-up of the nodes, and the position of the single diagonal direction is insignificant in the case of bidirectional cycling loading.
Supplying of additional diagonal stiffeners or even stiffeners as conformal to the node 3 lead in the increased end plate stiffness and a reduction of stresses in the column.
The similar stiffness between the beam and the column is avoided, but the phenomenon of yielding in the column are not avoided, the column being much more stressed, the end plate being highly demanded and susceptible to failure.

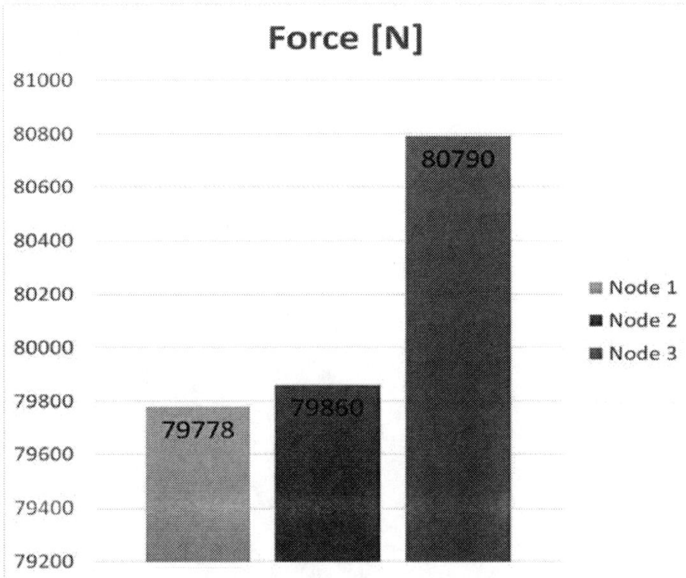

Figure 18. The force corresponding to the maximum displacement.

5. References

[1] Venghiac V M, Staşcov M, Budescu 2017 Analize privind îmbinările grindă-stâlp cu şuruburi. *Proceedings of the 15th National Conference on Steel Structures with International Participation – 15 CONMET, ISSN 2559-0812,* pp. 227-237

[2] Diaz C, Victoria M, Marti P and Querin O M 2011 FE model of beam-to-column extended end-plate joints. *Journal of Constructional Steel Research, 67,* pp. 1578-1590

[3] Calado L, Proença J, Espinha M, Castiglioni C 2013 Hysteretic behaviour of dissipative bolted fuses for earthquake resistant steel frames. *Journal of Constructional Steel Research, 85,* p. 151–162

[4] Staşcov M, Veghiac M V 2017 Beam-to-column connections with histeretic dampers. *Buletinul Institutului Politehnic din Iaşi, 63(67),* pp. 75-88

[5] Eurocod 3. SR EN 1993-1 2006 *Proiectarea structurilor din oţel, ASOCIAŢIA DE STANDARTIZARE DIN ROMÂNIA.* Bucureşti

[6] Engelhardt M D, Winnebeger T, Zekany A J, Ptyaraj T J 1996 The dog bone connections: part 2. *Modern Steel Construction*

[7] Engelhardt M D, Husain A S 1993 Cyclic loading performance of welded flange-bolted web connections. *Journal of Structural Engineering, 119(12),* pp. 3537-3550

IOP Conf. Series: Materials Science and Engineering **586** (2019) 012025 doi:10.1088/1757-899X/586/1/012025

[8] Bravo M, Herrera R 2014 Performance under cyclic load of built-up T-stubs for Double T moment connections. *Journal of Constructional Steel Research, 103,* pp. 117-130
[9] Brunesi E, Nascimbene R and Rassati G A 2014 Response of partially-restrained bolted beam-to-column connections under cyclic loads. *Journal of Constructional Steel Research, 97,* pp. 24-38
[10] Hu F, Shi G, Bai Y, Shi Y 2014 Seismic performance of prefabricated steel beam-to-column connections. *Journal of Constructional Steel Research, 102,* pp. 204-216
[11] Moshaly E, El-Heweity M, Abou-Elfath H and Osman M 2011 Finite element analysis of beam-to-column joints in steel frames under cyclic loading. *Alexandria Engineering Journal, 50,* pp. 91-104.
[12] Naughton D T, Tsavdaridis K D, Maraveas C and Nicolaou A 2017 Pushover Analysis of Steel Seismic Resistant Frames with Reduced Web Section and Reduced Beam Section Connections. *Frontiers in Built Environment 3:59*
[13] Rahnavard R, Hassanipour A, Siahpolo N 2015 Analytical study on new types of reduced beam sections moment connections affecting cycling behavior. *Case Studies in Structural Engineering, 3,* pp. 33-51
[14] Roeder CW, Shneider S P, Carpenter J E 1993 Seismic behavoir of moment-resisting steel frames: Analytical study. *Journal of Structural Engineering, 119(6),* pp. 1866-1884
[15] Rui L, Bijan S, Zhong T, Md Kamrul H 2017 Cyclic behaviour of composite joints with reduced beam sections.. *Engineering Structures, 136,* pp. 329-344
[16] Shneider S P, Roeder C W, Carpenter J E 1993 Seismic behavoir of moment resisting steel frames: Experimental study. *Journal of Structural Engineering, 119(6),* pp. 1885-1902
[17] Sofias C E, Kalfas C N and Pachoumis D T 2014 Experimental and FEM analysis of reduced beam section moment endplate connections under cyclic loading. *Engineering Structures, 59,* pp. 320-329
[18] Tsavdaridis K D, Papadopoulos T 2016 A FE parametric study of RWS beam-to-column bolted connections with cellular beams. *Journal of Constructional Steel Research, 116,* pp. 92-113
[19] Valente M, Castiglioni C, Kanyilmaz A 2017 A Numerical investigations of repairable dissipative bolted fuses for earthquake resistant composite steel frames. *Engineering Structures, 131,* p. 275–292
[20] Maggi Y I, Gonçalves R M, Leon R T, Ribeiro L F L 2005 Parametric analysis of steel bolted end plate connections using finite element modelling. *Journal of Constructional Research, 61,* pp. 689-708
[21] FEMA 350 2005. *Recomended seismic design criteria for new steel moment-frame buildings..* Washington DC, USA: Federal Emergency Management Agency

Nonlinear FEM analysis of steel beam-to-column connections with extended end plate

M Staşcov[1], V M Venghiac[1] and S G Luca[1]
[1] The Faculty of Civil Engineering and Building Services, "Gheorghe Asachi" Technical University of Iaşi, *Romania.*

E-mail: stascov.mihail@gmail.com

Abstract. Even if it has been shown that steel beam-to-column connections with extended end plate has a semi-continuous behaviour, many structural analyses consider these joints either fixed or pinned. For advanced structural analysis of connections and frames it is necessary to know the behaviour of the joint. The characteristics of the joint can be identified on the moment-rotation curve. This paper presents a comparative study of the moment-rotation curve for a standardized node calculated according Eurocode 3, and obtained by FEM analysis using IDEASTATICA and ANSYS finite element analysis program. The analyses were performed for the models with non pre-tensioned bolts. Also, an additional study was performed in ANSYS to identify the influence of the thickness of the extended end plate.

1. Introduction

The distribution of forces acting on the structure are influenced by the strength and rigidity of the entire structure. The characteristics of the structure are influenced by the characteristics of the joints. Thus, the characteristics of the joints, such as strength, stiffness and plastic deformability (rotation) directly influences the behaviour of the entire structure, figure 1.

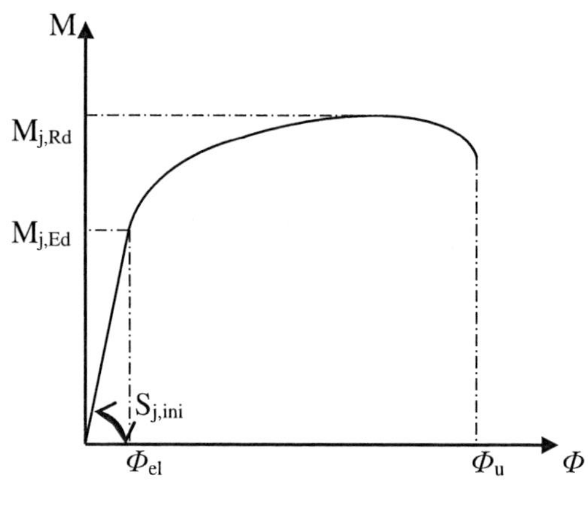

Figure 1. The interdependent relation bending moment-rotation.

The failure mode of steel structures subjected to cyclic actions such as earthquakes, creates the most problems. Conventional Pre-Northridge joints, which involve direct welding of the beam to the column flange, have proved to be ineffective due to the brittle failure before the plastic hinge can be formed. In that case, a small amount of energy is absorbed, which leads to a reduced plastic deformation capacity [1].

The traditional approach in designing of steel frames structures implies that joints were either pinned of perfectly rigid. Research performed after the Northridge (1994), and Kobe (1995) earthquakes showed that joints considered perfectly rigid had a semi-rigid behaviour, and some joints considered pinned had capacity to transfer the bending moment, thus was introduced the notion of semi-rigid joint, shown in figure 2c.

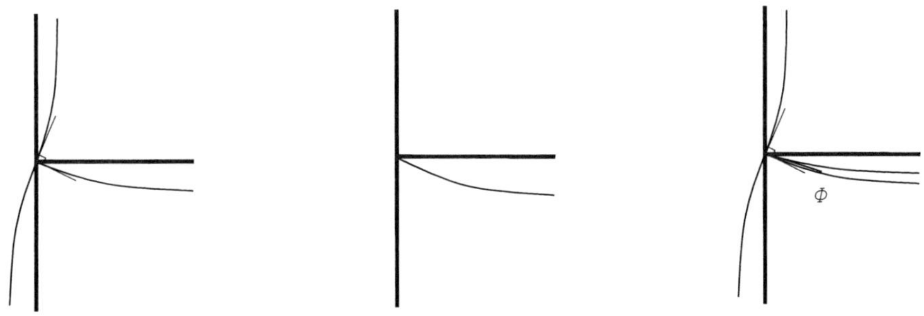

a. Rigid node b. Articulated node c. Semi-rigid node

Figure 2. Types of nodes depending on the beam and column relationship.

At the same time, the design rules for steel structures design, such as EC3 [2, 3, 4] require respecting the "strong column – weak beam" principle in designing of steel frame structures (figure 3).

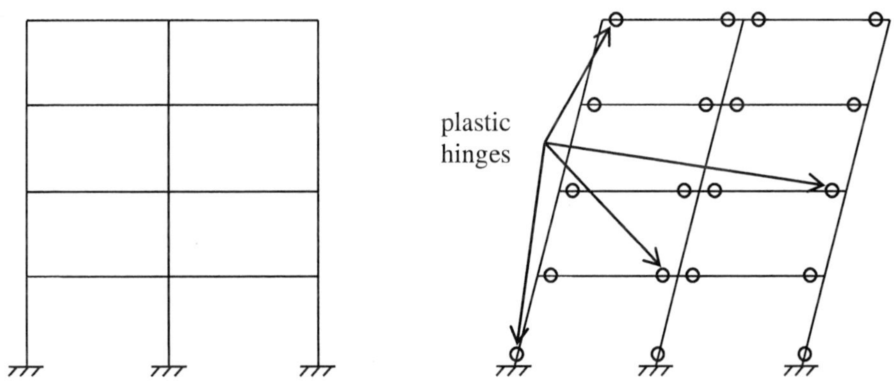

a. Steel frame structure b. Steel frame structure after plastic hinges are formed

Figure 3. "Strong column – weak beam" principle.

In order to satisfy this requirement joints were proposed with reduced section beams, reduced beam flanges sections RBS [5, 6], and reduced web sections RWS [7]. Additional stiffening of the end of the beam or additional stiffeners on the column flanges were also proposed.

Design of the joints implies the assessment of the capability of each element (component method) [2]. The evaluation of all components is quite complicated due to the geometrical and material nonlinearity and due to the effects of local imperfections and residual forces [8, 9]. Despite these difficulties in the past two decades, a large number of advanced FE studies have been performed for the study of end plate joints with a large variety of geometries, to obtain values for resistance and ductility [8, 10, 11, 12, 13, 14, 15]. Generally, these studies have attempted to provide the response to the behavior of the joints such as interdependent relation bending moment-rotation curves, or hysteretic curves.

Due to the possibility of varying the number of the bolts, their diameter, the dimensions and thickness of the end plate, this type of joint show a wide variation of the characteristics. The purpose of this study is to develop the comparison of the result for a type joint, obtained according to the European designing rules [2], obtained from numerical modeling with FEM program specialized in the design and check of the steel joints IDEA StatiCa, and FEM multiphysics program Ansys Workbench, at the same time, to develop a FEM model for end plate joints, and obtaining of the interdependent relation bending moment – rotation curves, to study the influence of the thickness of the end plate.

2. Analysed models

By directing the formation of the plastic hinge in the beam end or the end plate, the ductile behaviour of the joint can be ensured [15]. Three configurations of the end plate joints were analyzed. The considered nodes have common elements: HEB300 column, IPE360 beam, and M22, 10.9 grade high strength bolts, with normal clamping. The steel grade in the beam, column, end plate and stiffeners is S235. Figure 4 shows the analyzed nodes.

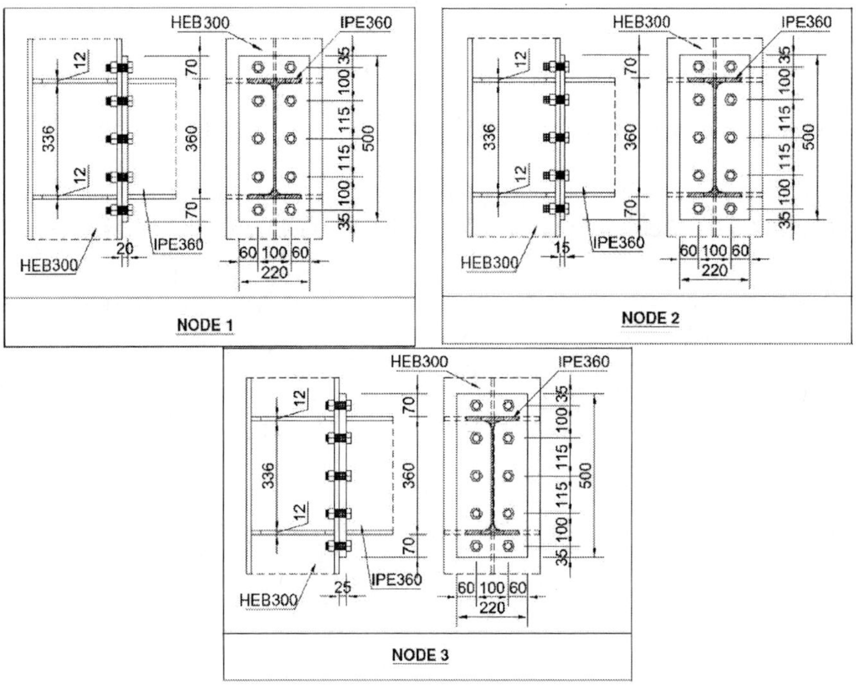

Figure 4. Analyzed nodes.

3. EC3 analysis

The design standard for steel structures [2] (Eurocod 3. SR EN 1993-1, 2006) provide for element verification by the component method. Only the dimensions and characteristics of the node elements are required. Typically, the design of the joints according to EC3 is very labour intensive, manual computation requires a lot of time [16]. Following the calculation based on the component method according to (Eurocod 3. SR EN 1993-1, 2006), the following values were obtained for:

- $M_{j,Rd} = 231,96 Knm$
- $S_{j,ini} = 6,691*10^{10}$ Nmm/rad

With these values, the simplified, bi-linear interdependent bending moment-rotation curve were created, for the joint with 20mm thick end plate, figure 5.

IOP Conf. Series: Materials Science and Engineering **586** (2019) 012026 doi:10.1088/1757-899X/586/1/012026

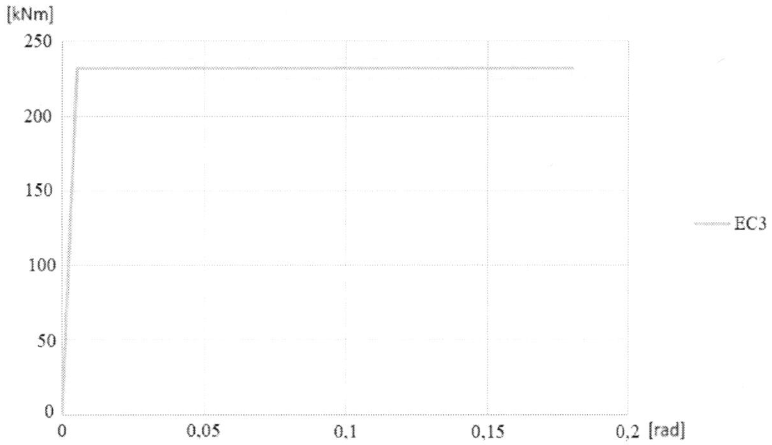

Figure 5. The interdependent relation bending moment-rotation bending moment-rotation according EC3.

4. IDEA StatiCa Analysis

The IDEA StatiCa Connections program was used to analyse the joint. IDEA StatiCa Connections is a program based on Component Based Finite Element Model (CBFEM). The CBFEM method keeps the key elements of the component method and combine them with the FEM calculation capabilities.

IDEA StatiCa Connections can perform two types of analysis:

- Static analysis with material and contact nonlinearities to analyse the state of stress and deformation.
- Second grade analysis for buckling analysis.

The materials of component for the FEM analysis are identical to the materials considered for the EC3 analysis.

Shel elements are used for modelling of the elements of the joint. Bi-dimensional elements with four nodes in the corners of the elements are used, each node have six degrees of freedom: three translations and three rotations.

Contacts between plates have a major impact in the redistribution of forces in the joint. Contacts between plates are modelled according to the "Penalty" method. The advantage of the "penalty" method is the automatically assemble of the model. This method is controlled by an iterative algorithm to achieve better numerical performance.

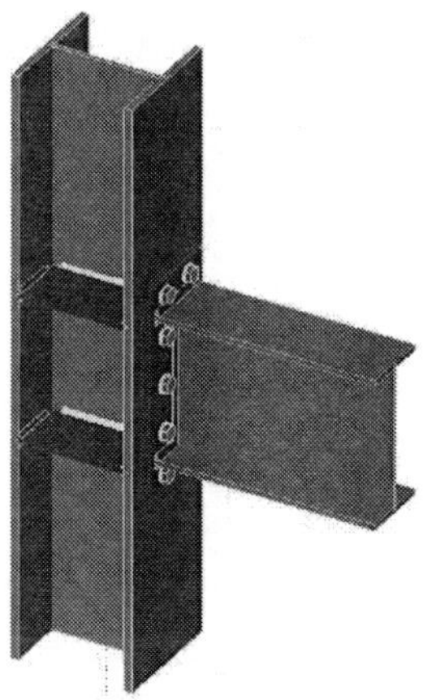

Figure 6. Analyzed node configuration (IDEA StatiCa).

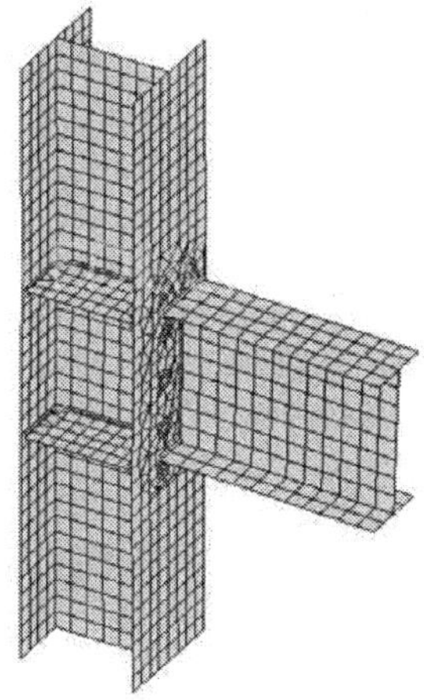

Figure 7. IDEA StatiCa mesh elements.

IOP Conf. Series: Materials Science and Engineering **586** (2019) 012026 doi:10.1088/1757-899X/586/1/012026

The Multi-Point Constraint (MPC) contact method is used for welding contact modelling. This method allows connecting elements with different meshing densities.
Following the analysis in the IDEA StatiCa Connections program the following values were obtained for:

- $M_{j,Rd}$ = *237,3Knm*
- $S_{j,ini}$ = *10,81*10*^{10} *Nmm/rad*

Also, was resulting the interdependent relation bending moment-rotation curve, figure 8.

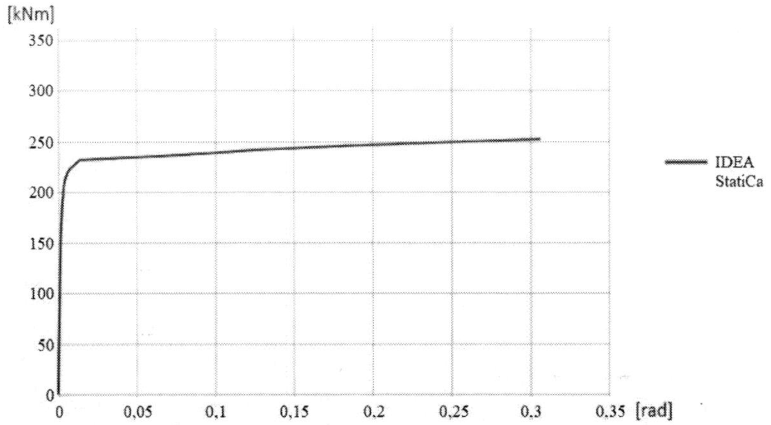

Figure 8. The interdependent relation bending moment-rotation according IDEA StatiCa.

5. Ansys analysis

Analyses were performed with the FEM program Ansys Workbench. Three-dimensional elements "Solid 185" were used to model the joint elements. The "Solid 185" element have 8 nodes, and each node have 3 degrees of freedom, orthogonal translation.
The contacts were modelled with elements "Conta185" and "Targe170", to represent the contact and sliding of the elements. This element is located on the surfaces of solid 3D element and has the same geometric characteristics as the solid element with which it is connected. For modelling the contacts, the same methods were used as for modelling in IDEA StatiCa, namely "Pure penalty" for plate contact and "MPC" for welding contact.

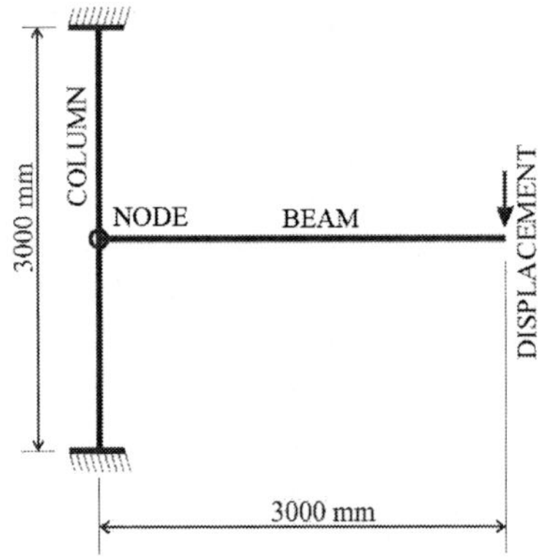

Figure 9. Static scheme

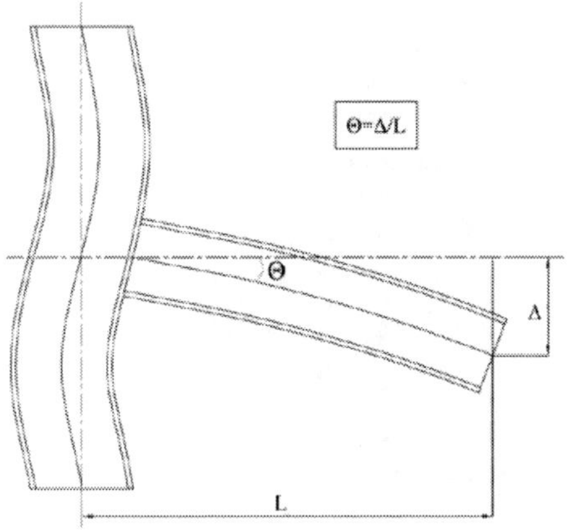

Figure 10. Rotation θ of the joint in terms of the beam end displacement.

IOP Conf. Series: Materials Science and Engineering **586** (2019) 012026 doi:10.1088/1757-899X/586/1/012026

The materials used in the modelling are described by a bilinear stress-strain curve, simplified according to Von-Mises criteria to define plasticity. Due to the fact that Ansys Workbench is not a program specifically designed for calculating joints, some manipulations were required. It is considered a column of 3m height and a 3m beam, according to the static scheme, figure 9.
The loading type is displacement mode. For the evaluation of the rotation, the geometric transformation was used, according to figure 10. The meshed model is presented in figure 11.

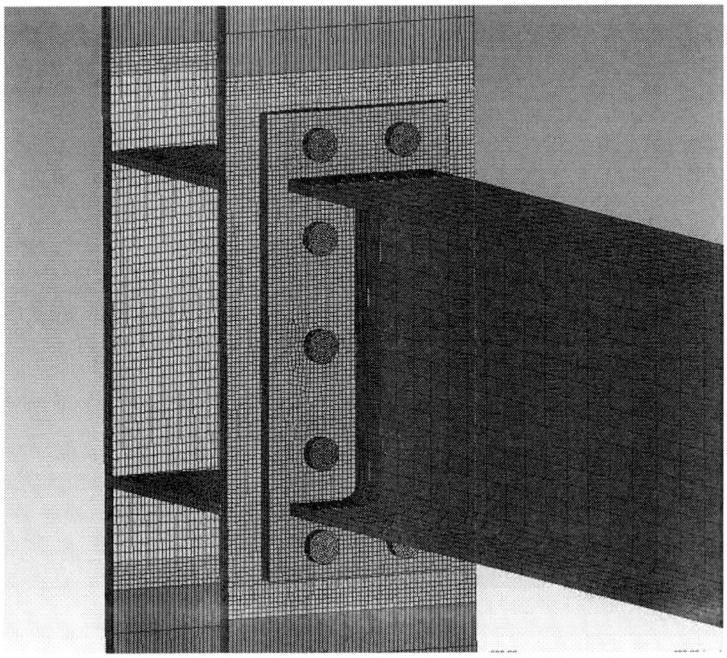

Figure 11. Meshed model in Ansys Workbench.

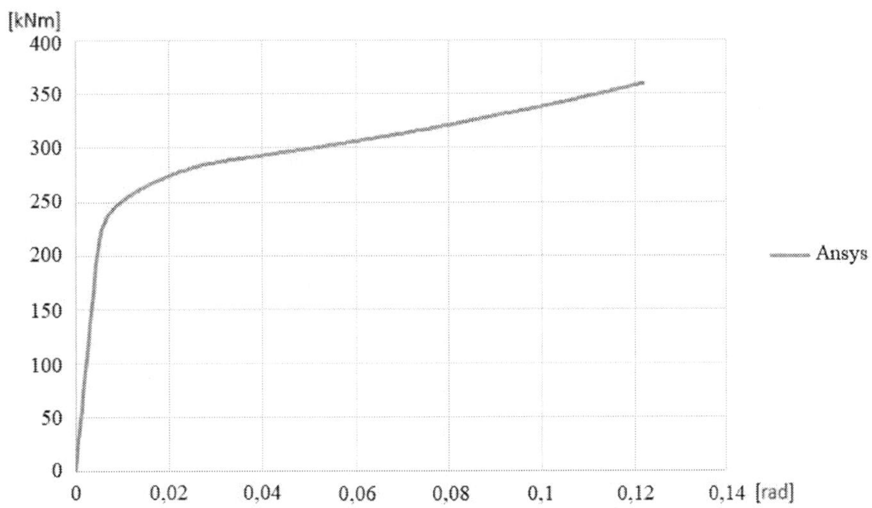

Figure 12. The interdependent relation bending moment-rotation according Ansys Workbench (node 1, t=20mm).

Figure 12 shows the interdependent relation bending moment-rotation curve for node 1, with a 20mm end plate thickness resulting from FEM modelling in Ansys Workbench.

IOP Conf. Series: Materials Science and Engineering **586** (2019) 012026 doi:10.1088/1757-899X/586/1/012026

In Ansys the joints with end plate thickness t=15mm and t=25mm were analysed, to study the influence of the end plate on the node behaviour. For these joints the characteristic curves were obtained for interdependent relation bending moment- rotation (Figure 13, Figure 14).

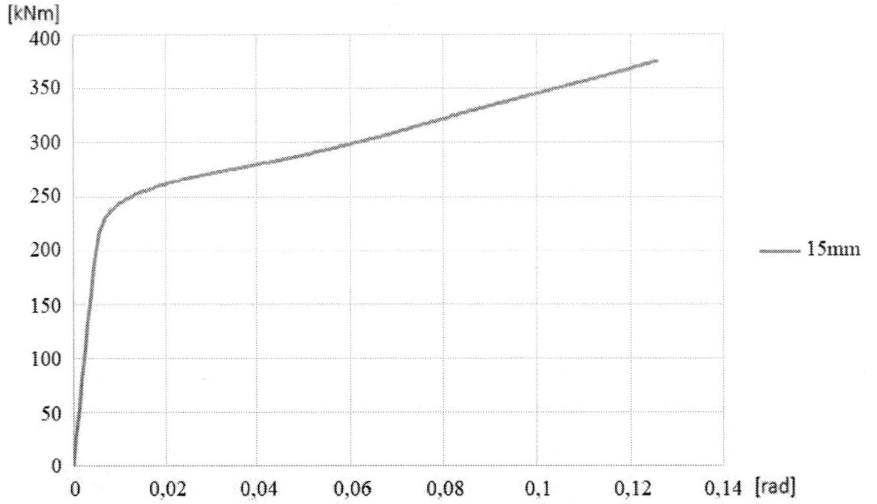

Figure 13. The interdependent relation bending moment-rotation according Ansys Workbench (node 2, t=15mm).

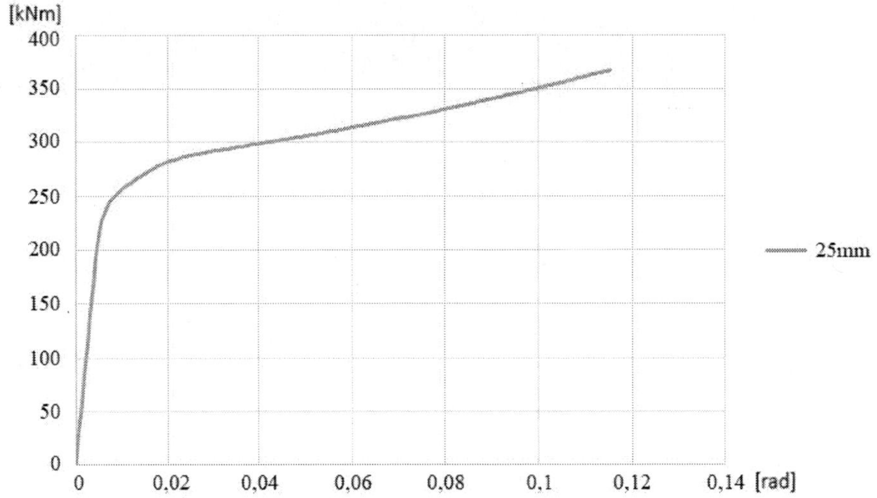

Figure 14. The interdependent relation bending moment-rotation according Ansys Workbench (node 3, t=25mm).

6. Observations

Figure 15 shows the interdependent relation bending moment-rotation curves obtained by three different methods: calculated manually according to the component method in EC3; obtained from finite element modelling using the IDEA StatiCa program and obtained by finite element modelling with Ansys Workbench.

In figure 15 it can be noticed that the interdependent relation bending moment-rotation, obtained from EC3 manual calculation and following modelling in the IDEA StatiCa provide approximately similar values. These similarities are due to the fact that both calculations are based on the method of

components for the calculation and designing of the joints. At the same time differences between the slopes of the initial stiffness and the "consolidation" slopes are observed after the plastic hinges are formed. These differences result from the linear and simplified mode in manual computation and complex phenomena that are considered in the FEM computation.
The results from Ansys provide superior values for the moment, but the initial stiffness slope is similar to those in EC3. The differences of moment values are due to the fact that the bending moment is added to the beam stiffness and other elements of the joint and the plastic hinge is distributed to other element due to the isotropic hardening properties.
Figure 16 shows the interdependent relation bending moment-rotation for nodes 1, 2 and 3 obtained using the Ansys Workbench.

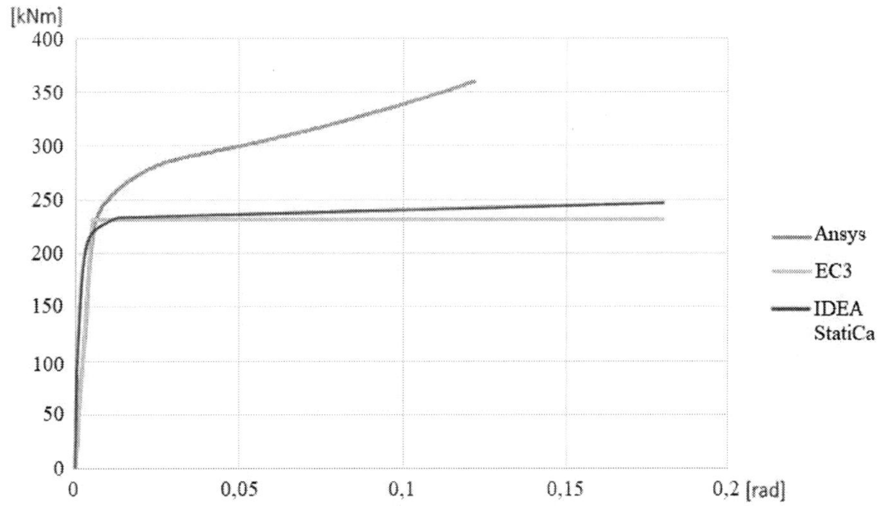

Figure 15. The interdependent relation bending moment-rotation according EC3; IDEA StatiCa; Ansys Workbench (t=20mm).

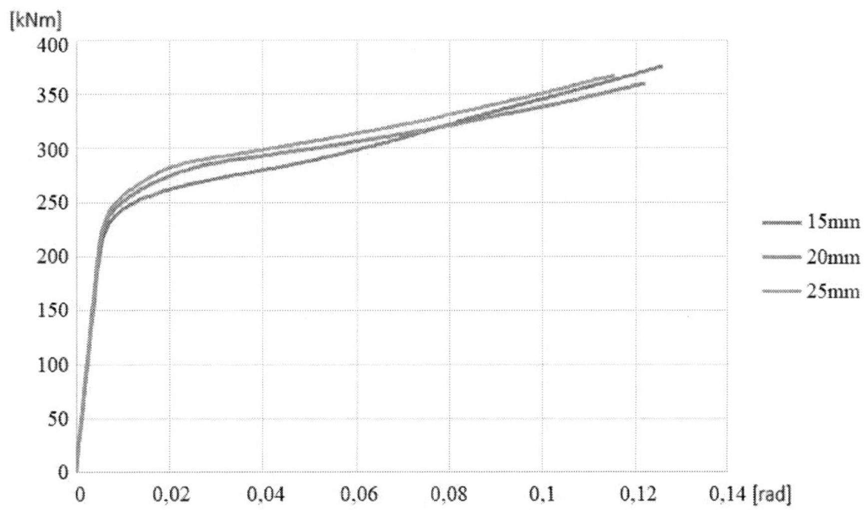

Figure 16. The interdependent relation bending moment-rotation according EC3; IDEA StatiCa; Ansys Workbench (t=20mm).

IOP Conf. Series: Materials Science and Engineering **586** (2019) 012026 doi:10.1088/1757-899X/586/1/012026

In figure 16 it is observed that the slope of the initial stiffness is similar in the elastic domain. It is noted that the bending moment capacity is dependent on the thickness of the end plate, with increasing thickness of the end plate increases the resistance to the bending moment. Figure 16 confirms that the behaviour of the joint is governed by the geometry of its components, especially the thickness of the end plate. Increasing the thickness of the end plate can lead to the failure of the column flange, which is inadmissible in the design of the joints, because that contradicts the "strong column – weak beam" principle. It was also confirmed that the thickness of the end plate influences the diameter of the bolts, and in order to obtain an efficient design, the ratio between the thickness of the end plate and the diameter of the bolts should be approximately equal to 1.2 (Venghiac, V.M., Stașcov, M., Budescu, M., 2017).

7. Conclusions

The results were focused on the study of the influence of variation of the thickness of the end plate on the behaviour of the joints as well as on the differences between the calculation methods. The FEM models developed with the Ansys program have shown that the thickness of the end plate is one of the most important parameters in terms of joint behaviour and moment capacity. According to the component method from EC3, resistance and rotation capacity of the connection is directly influenced by the resistance and ductility of weaker components in the joint. This component must act ductile and leave the possibility of redistribution of the internal forces to the other components. Bolts and welds are fragile, in those areas stress concentrations must be avoided, and these elements should be strengthened, to ensure the plastic behaviour and deformations of the ductile components.

It has been observed that the strength is proportional to the thickness of the end plate. At the same time, these results have demonstrated that FEM models are powerful tool for improving knowledge of joint design. The substantial differences between the results obtained by two different calculation programs indicate that the IDEA StatiCa Connection program is a program designed exclusively for the calculation of the joint, namely it has all safety and load reduction coefficients set according to the EC3 provisions, which substantially reduces the moment capacity of the joint.

The Ansys Workbench program is a multiphysics program that is not only designed for joint designing, and has no pre-set safety coefficients, it provides results that are provided only of element geometry and material properties. For this reason, the Ansys models/results must be adjusted to EC3 computing conditions.

8. References

[1] Stașcov M, Veghiac M V 2017 Beam-to-column connections with histeretic dampers. *Buletinul Institutului Politehnic din Iași, 63(67),* pp. 75-88

[2] Eurocod 3. SR EN 1993-1 2006 *Proiectarea structurilor din oțel, ASOCIAȚIA DE STANDARTIZARE DIN ROMÂNIA.* București

[3] FEMA 350 2005 *Recomended seismic design criteria for new steel moment-frame buildings..* Washington DC, USA: Federal Emergency Management Agency

[4] ANSI/AISC341 2016 *Seismic Provisions for Structural Steel Buildings.* Chicago (IL): American Institute of Steel Construction

[5] Engelhardt M D, Winnebeger T, Zekany A J, Ptyaraj T J 1996 The dog bone connections: part 2. *Modern Steel Construction*

[6] Engelhardt M D, Husain A S 1993 Cyclic loading performance of welded flange-bolted web connections. *Journal of Structural Engineering, 119(12),* pp. 3537-3550

[7] Tsavdaridis K D, Papadopoulos T 2016 A FE parametric study of RWS beam-to-column bolted connections with cellular beams. *Journal of Constructional Steel Research, 116,* pp. 92-113

[8] Maggi Y I, Gonçalves R M, Leon R T, Ribeiro L F L 2005 Parametric analysis of steel bolted end plate connections using finite element modelling. *Journal of Constructional Research, 61,* pp. 689-708

[9] Stankiewicz B 2013 Parametric analysis of stiffness of bolted end-plate connections of I beams using finite element method.. *Journal of Civil Engineering, Environment, Vol. 30, No 60 (2/13),* pp. 231-242
[10] Abolmaali A, Matthys J H, Farooqi M, Choi Y 2005. Development of moment – rotation model equations for flush end-plate connections. *Journal of Constructional Steel Research, Vol. 61,* pp. 1595-1612
[11] Chen S, Du G 2007 Influence of initial imperfection on the behaviour of extended bolted end-plate connections for portal frames. *Journal of Constructional Steel Research, Vol. 63,* pp. 211-220
[12] Brunesi E, Nascimbene R and Rassati G A 2014 Response of partially-restrained bolted beam-to-column connections under cyclic loads. *Journal of Constructional Steel Research, 97,* pp. 24-38
[13] Jablonska-Krysiewicz A 2015 Finite element modeling of the behaviour of steel end-plate connections. *Czasopismo Inżynierii Lądowej, Środowiska i Architektury,* pp. 173-178
[14] Moshaly E, El-Heweity M, Abou-Elfath H and Osman M 2011. Finite element analysis of beam-to-column joints in steel frames under cyclic loading. *Alexandria Engineering Journal, 50,* pp. 91-104
[15] Venghiac V M, Stașcov M, Budescu M 2017 Analize privind îmbinările grindă-stâlp cu șuruburi. *Proceedings of the 15th National Conference on Steel Structures with International Participation – 15 CONMET, ISSN 2559-0812,* pp. 227-237
[16] Dubina D, Grecea D, Ciutina A, Danku G, Vulcu C 2010 *Calculul și proiectarea îmbinărilor structurale din oțel în conformitate cu SR-EN 1993-1-8 Recomandări, comentarii și exemple de aplicare Redactarea II.* Timișoara

IOP Conf. Series: Materials Science and Engineering **586** (2019) 012027 doi:10.1088/1757-899X/586/1/012027

Study of the interface behaviour between fictile tubules bricks and mortar: numerical and experimental analysis

C Scuro[1], S Tiberti[2], S Porzio[1], R S Olivito[1] and G Milani[2]
[1]Department of Civil Engineering, University of Calabria, Via §Ponte Pietro Bucci, Rende 87036, *Italy.*
[2]Department of Architecture, Built Environment and Construction Engineering, Technical University of Milan, Piazzale Leonardo da Vinci, Milan 20133, *Italy.*

E-mail: carmelo.scuro@unical.it

Abstract. This paper presents an investigation on the interaction between hollow clay bricks (*caroselli*) and mortar, aiming at obtaining frictional properties needed for a full description of the composite masonry material derived from these two constituents. A series of three-point bending tests is carried out on prism-shaped specimens representing typical arrangements of *caroselli* within mortar. Since no building code or guideline exists regulating this kind of tests, the specimens are built respecting the 1÷4 ratio between the transversal dimension and the length of the prism requested for mortar specimens subjected to three-point bending tests. The frictional properties of the interaction between the two constitutive materials are then derived through a series of back analyses concerning four different hypotheses on the actual degree of transmission of tangential stresses between *caroselli* and mortar. The numerical analyses are performed with the commercial finite element software ABAQUS. The obtained results sustain the experimental findings.

1. Introduction
The cultural heritage of different areas located around the Mediterranean Sea consists of several unique features and structural elements, often conceived to solve practical issues in the construction of buildings. A shining example is represented by a peculiar type of hollow clay bricks known as "fictile tubules". Three distinct categories can be generally identified with this definition - plain tubules, amphorae, and *caroselli* - depending on their geometrical characteristics and structural use. In particular, *caroselli* present a squat cylindrical shape, closed by a dome-like upper part, possibly provided with a small opening at the top [1]. They arguably originated during the 2nd Century A.D. in the Roman provinces of Northern Africa to bypass issues related to the shortage of wood needed to erect centrings [2]. In fact, they were used to build light-weight, self-supporting vaults and domes in several constructions, thus effectively overcoming the necessity of centrings during construction. The rough external surface of *caroselli* is characterised by an unevenness spiralling towards the upper part, which was created by the potter during the moulding phase to help each clay element adhering to the surrounding mortar.

2. Classification of hollow clay elements
Historically, a vast variety of hollow clay elements have been used to build domed and vaulted structures. For this reason, it is possible to create a classification based on their geometrical characteristics and different use in building techniques [3-4]. The larger category of hollow clay bricks consists of three general elements: tubules, amphorae and *caroselli.* The tubules are hollow, syringe-like elements manufactured so that they slot into each other (Figure 1). To guarantee the adhesion between these elements and the mortar, the external surface of the former is left rough. Over time, these tubules have taken different shapes, bulging in the central part or tapered on the end section. The most significant differences are observed in the length of the tip (upper end section) that allows alignment of the joint between multiple elements. The amphorae are ceramic elements strongly resembling the vessels

of the same name, whose shape is characterised by a bulging central section and a spindle-shaped bottom part. They present two handles tying the neck to the central part (Figure 2). Like the previous case, during the manufacturing phase the external surfaces are left rough to improve the adhesion with the mortar. Eventually, the fictile elements called *caroselli* are unusual structural elements made with clay (Figure 3) [5]. It is important to underline the fact that the name of these elements derives from a dialect term that identifies the moneybox, due to their geometry. This name is different and changes according to the geographical location of production. The term mostly used is *caroselli*, but there is trace of other names such as "pigniatelli", "caccavelli" or "bubbole" [1-6].

Figure 1. Example of tubule [5].

Figure 2. Example of amphora [7].

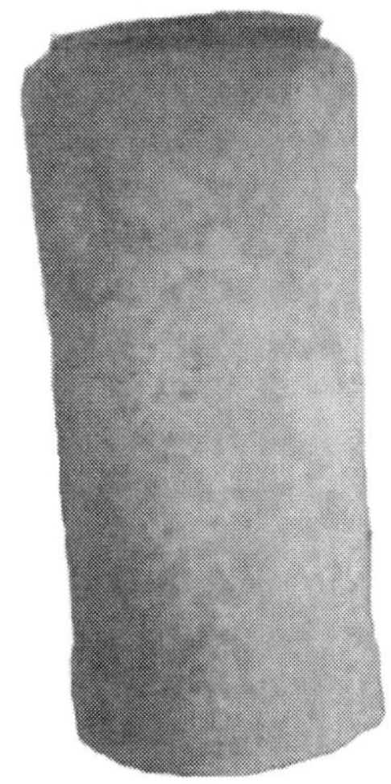

Figure 3. Example of *carosello* [2].

2.1. Construction with fictile tubules
During the construction phase, the tip of one tubule is inserted into the next element – characterised by a perforated base – thus ensuring the interlocking between adjoining elements (Figure 4) [8]. This technique allows the creation of arches joined side by side by lime mortar, so that a barrel vault can be erected without the use of traditional centring. The actual centring consists of a series of cordons compound of tubules inserted one inside another and following a generating curve, all converging in a keystone tubule that is uniquely produced without an apex (Figures 4-5) [9]. The vaults and domes constructed in this manner consist entirely of fictile tubules without the use of Roman concrete (*opus caementicium*).

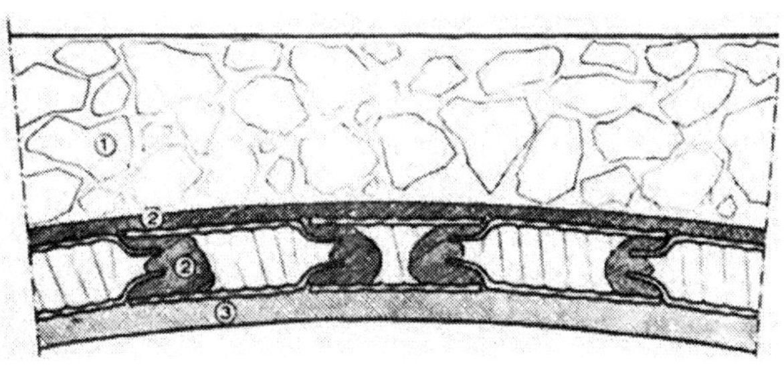

Figure 4. Draw of a keystone [10].

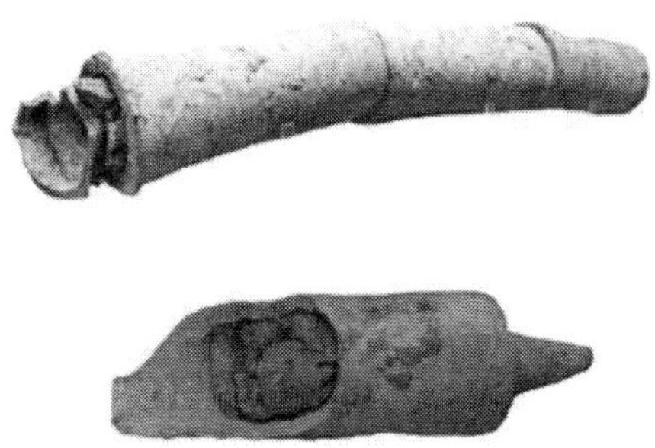

Figure 5. Detail of the joints of the fictile tubules [9].

IOP Conf. Series: Materials Science and Engineering **586** (2019) 012027 doi:10.1088/1757-899X/586/1/012027

2.2. Construction with amphorae

In historical buildings, amphorae are positioned inside the structure before casting the cement jet to ensure the formation of voids and consequently decrease the stress due to the dead loads [2].

These elements are found in different locations of historic buildings, typically included in masonry walls: for example, at the bottom of the wall, in the reins of the arches, in the filling of old concrete structures, and in the roofing [5]. They are used to lighten the weight of the roofs and the horizontal thrust on the perimeter walls. In the case of cavity walls, amphorae are also used to ensure internal ventilation. Famous examples are the mausoleum of Helena in Rome (Figure 6) where the amphorae are arranged in two lines, one running parallel above the other, and the bleachers of the Circus of Maxentius in Rome where the number of amphorae employed is almost equal to 9000 [11].

The positioning of the amphorae appears more effective in the case of the Temple of Minerva Medica (81-96, Rome), where they are arranged on top of the windows, easing the springing load [12]. In the case of architectural findings dating back to the early Christian era, especially in the Mediterranean area, the amphorae are placed on the thickness of vault with their opening visible on the intrados. In this case, they were used to improve the acoustics of the rooms [13].

Figure 6. Mausoleum of Helena, Rome [14].

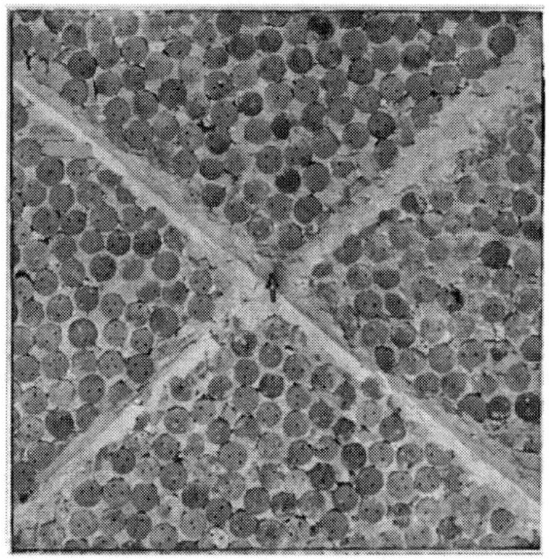

Figure 7. Cross vault built with *caroselli* [5].

2.3. Construction with caroselli

Caroselli are used to build barrel vaults, cross vaults, domes, floors, partitions, also serving decorative purposes [4]. Due to their unique shape, *caroselli* cannot not be inserted one inside another to build curved structures.

Conversely, they are placed in a vertical position on the centring or on a bed of bricks and then kept together with gypsum or lime mortar. In the case of barrel vaults, cross vaults or domes, *caroselli* are radially assembled following overlapping lines; their disposition over two lines is staggered in such a way that each *carosello* of the upper line is placed midway between two *caroselli* of the bottom line. In general, their lower base points downwards to form the intrados surface of the structure.

To complete the main structure, the completion jet is cast in two steps: the first consists in the actual casting of mortar, the second in the insertion of stone chips to better fill the voids between adjoining *caroselli*.

In the Italian southern region of Calabria there are several examples of barrel and cross vaults built with *caroselli*, especially in the small village of Bisignano (close to Cosenza) and the cities of Crotone, Soverato, and Monterosso Calabro (Figure 7). In the same region, *caroselli* are also part of a patented technique used to build a specific type of aseismic structures [15].

3. Problem of adhesion observed in experimental tests

In a previous work, the structural behaviour of an arch built with *caroselli* was investigated [16]. This arch presents an internal span of 1.5 m, a width equal to 0.5 m and a thickness of 15 cm. The *caroselli* are here arranged in a staggered manner, with their circular base tangent to the intrados of the arch. Mortar joints of approximately 2 cm separate each row of elements. The arch is built using a wooden centring. After the completion of the arch, its upper side is covered with a final cast of mortar with a thickness of about 3 cm [6;16].

A possible equivalent isostatic scheme for the analytical model is the three-hinged arch. Two steel beams are fixed with a single bolt to two blocks of concrete at the base of the arch to prevent relative displacements during the test. Initially, only two hinges are supposed to be present at each base, leading to supposing a two-hinged arch as a feasible equivalent isostatic scheme. However, the keystone section rotates in a way that is not consistent with the previous hypothesis, which may be caused by a non-perfect adhesion between mortar and fictile tubules, as shown in the Figure 8. This figure shows how the cracks develop inside the mortar without affecting the fictile tubules, suggesting the analysis of this problem in the present work.

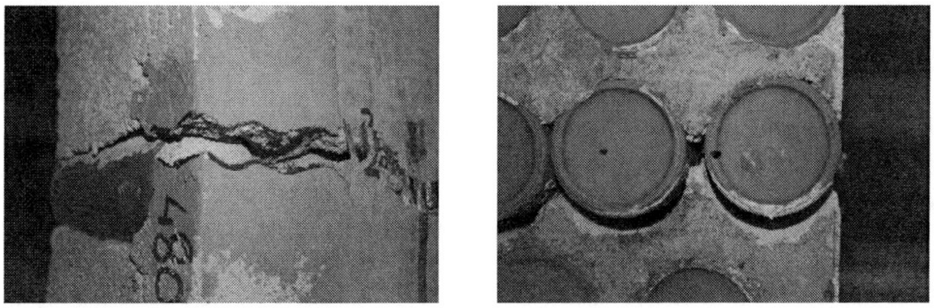

Figure 8. Developed of the cracks in the arch during the experimental test [13].

4. Experimental study/research

The aim of the experimental campaign is to investigate the interface behaviour between fictile tubules bricks and mortar. The experimental tests are carried out at the Materials and Structures Engineering Laboratory of Civil Department of University of Calabria.

4.1 Preparation of the specimens

The dimension of the investigated prism-shaped specimens is 48×16×15 cm^3, respecting the 1:4 ratio between the transversal dimension and the length of the prism requested for mortar specimens subjected to three-point bending tests. The formworks are built in a specialised carpentry workshop and consist of multilayer wood panels. The connection between adjacent wood panels is ensured by iron screws that facilitate disassembly after 28 days of maturation of the mortar. Before casting the mortar in the prism-shaped specimens, the frameworks have been coated with a material that facilitates the form stripping.

The fictile tubules here used are labelled NFTs (New Fictile Tubules) since they are produced in a potter factory with innovative and faster techniques, and then refined on the potter's wheel. They are 13 cm high, their diameter is about 6 cm, and they present a thickness of 6 mm [3-5]. The mortar is characterised by a compressive strength of about 1.8 MPa and consists of 9 parts of sand, 2 of lime mortar and 1 of cement.

A total number of 11 fictile tubules is used for each prism-shaped specimen. The elements are arranged in a staggered manner as shown in Figure 9. Mortar joints of approximately 1.2 cm separate each element from the nearest ones (Figure 10). The beam is completed by casting a 2 cm thick layer of mortar to cover the tips of the fictile tubules (Figure 11).

IOP Conf. Series: Materials Science and Engineering **586** (2019) 012027 doi:10.1088/1757-899X/586/1/012027

Figure 9. Placement of the fictile tubules.

Figure 10. Mortar joints.

Figure 11. Final phase of the mortar casting.

4.2 Three-points bending tests

A series of three-point bending tests is carried out on three prism-shaped specimens to assess the actual degree of adhesion between mortar and fictile tubules. Given the similarities between these prism-shaped specimens and mortar specimens, the tests are performed in accordance with EN 1015/11 [17]. The three-point bending testing machine consists of three steel rollers: two are placed at the base of the specimen for support, the other is located on the upper side for the application of the displacement (Figure 12). This is applied at a uniform rate with a speed of 0.2 mm/min.

Figure 12. Three-point bending testing machine.

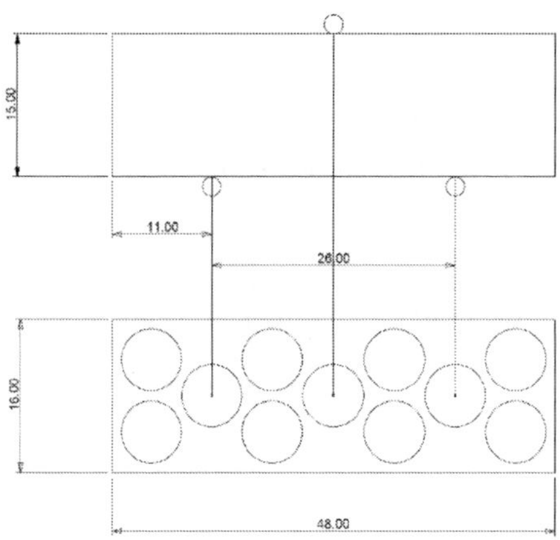

Figure 13. Scheme of three-point bending test.

Figure 14 shows the load-displacement diagrams for the three considered specimens. In all three cases, the failure of the specimen occurs due to a single crack forming centrally and developing all along the height of the prism, which are shown in Figure 15, Figure 16, and Figure 17 for Specimens 1, 2, and 3, respectively. This crack develops inside the mortar, extending along the contact surface between the mortar and the fictile tubules, up to the top where the force is applied.

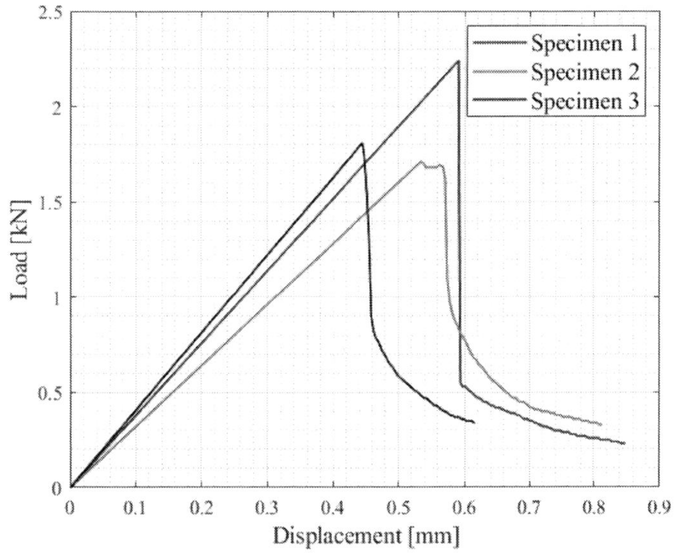

Figure 14. Load-displacement diagrams for the three specimens.

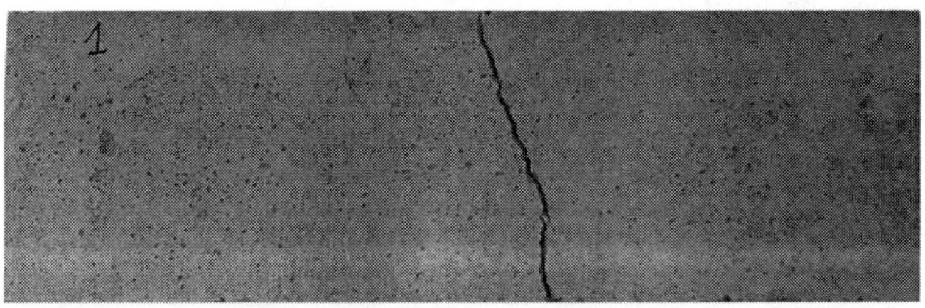

Figure 15. Development of the crack during the three-point bending test in the Specimen 1.

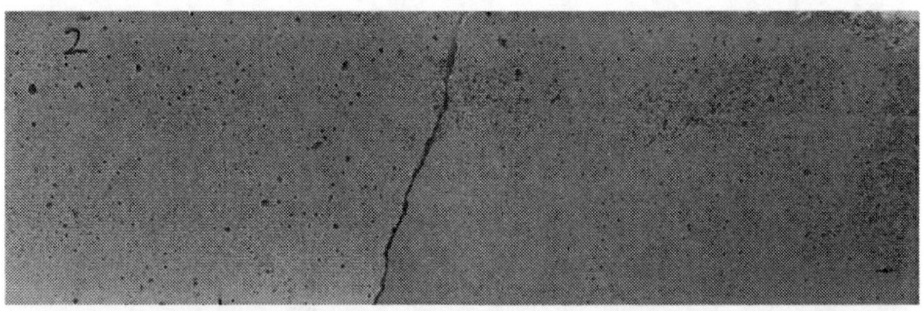

Figure 16. Development of the crack during the three-point bending test in the Specimen 2.

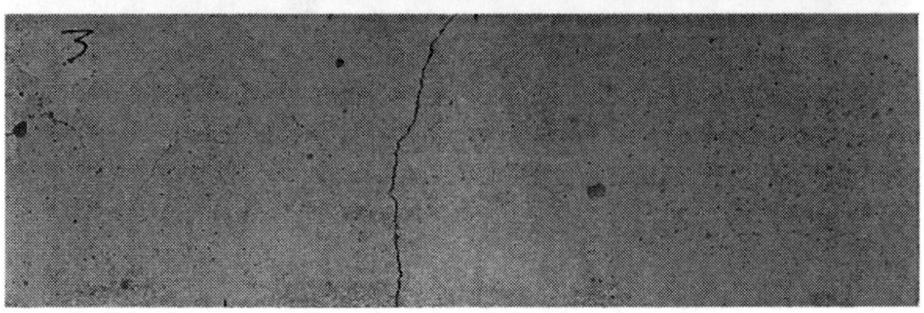

Figure 17. Development of the crack during the three-point bending test in the Specimen 3.

5. Numerical modelling of interface behaviour

The numerical analyses are carried out in the commercial finite element software Abaqus. The analytical model of one prism-shaped specimen is created as follows: first, the models of a single NFT and of the mortar *caldana* are separately generated, then 11 instances of the NFT model are assembled into the *caldana*.

IOP Conf. Series: Materials Science and Engineering **586** (2019) 012027 doi:10.1088/1757-899X/586/1/012027

The complete model is eventually meshed using tetrahedral finite elements. The finite element for the model of the prism-shaped specimen is shown in Figure 18, consisting of 8593 nodes and 28425 elements.

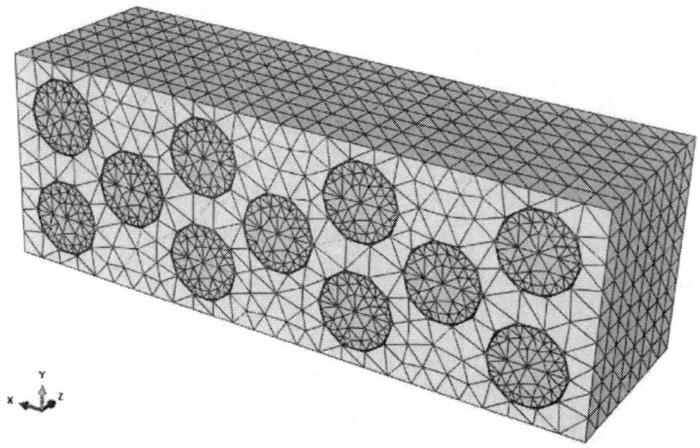

Figure 18. Finite element mesh for the analytical model of the prism-shaped specimen.

Two distinct material models are used to simulate the behaviour of the constituents. NFTs are modelled with the "brittle cracking model" available in the Abaqus library, which is suitable for brittle materials such as clay. It is based on the assumption of infinitely resistant compressive behaviour, whereas the tensile behaviour can be formulated as function of the tensile strength and mode I fracture energy. The elastic parameters and mechanical properties used for modelling the NFTs are listed in Table 1, and they are based on values whose reliability in representing the behaviour of NFTs has been assessed in a previous work [5].

Table 1. Elastic parameters and mechanical properties for modelling the NFTs.

Parameter	Value
Young's modulus	3900 MPa
Poisson's ratio	0.15
Tensile strength	3 MPa
Mode I fracture energy	0.087 N/mm

Conversely, mortar is modelled with the "concrete damaged plasticity model" also available in the Abaqus library, which is conceived for quasi-brittle materials such as concrete and mortar as well. It assumes distinct behaviours in compression and tension, and the latter can still be formulated as function of the tensile strength and mode I fracture energy. Also, it allows the modelling of damage in the material, both in compression and tension. The elastic parameters and mechanical properties used for modelling mortar are listed in Table 2, and they are again based on values whose reliability in representing the behaviour of mortar has been assessed in a previous work [5].

IOP Conf. Series: Materials Science and Engineering **586** (2019) 012027 doi:10.1088/1757-899X/586/1/012027

Table 2. Elastic parameters and mechanical properties for modelling mortar.

Parameter	Value
Young's modulus	190 MPa
Poisson's ratio	0.2
Tensile strength	0.6 MPa
Mode I fracture energy	0.018 N/mm
Strain corresponding to damaged material	0.5

The interface behaviour between NFTs and mortar is assessed by creating 11 interactions in Abaqus, one for each NFT. These interactions connect the contact surfaces between one NFT and the surrounding mortar. The interface behaviour is modelled considering four cases, each representing a possible configuration in terms of transmission of tangential stresses. All the cases are created by setting a specific "tangential behaviour" in the formulation of the interaction property, and they are listed in Table 3.

Table 3. Description of the four considered cases for modelling the interface behaviour between NFT and mortar.

Case	Description
Frictionless	No transmission of tangential stresses
Friction 0.3	Transmission of 30% of tangential stresses
Friction 0.6	Transmission of 60% of tangential stresses
Rough	Complete transmission of tangential stresses

The three-point bending test is simulated with an analysis in Abaqus/Explicit. The three steel rollers are not modelled for sake of simplicity. The rollers at the base are substituted by one hinge and one roller (to represent an isostatic scheme), while the velocity applied on the upper roller is directly applied to the mid upper section of the model.
The load-displacement diagrams for the four cases are shown in Figure 19, where they are compared against the envelope of the experimental data. It is evident how only the case labelled "Friction 0.6" is

IOP Conf. Series: Materials Science and Engineering **586** (2019) 012027 doi:10.1088/1757-899X/586/1/012027

able to grasp the actual interaction between NFTs and mortar. The first two cases are not capable of reaching the lower bound of the experimental envelope, whereas the last case overestimates the stiffness of the specimen. Figure 20 displays the analytical crack development for the case "Friction 0.6", which is consistent with that observed in reality for the three specimens.

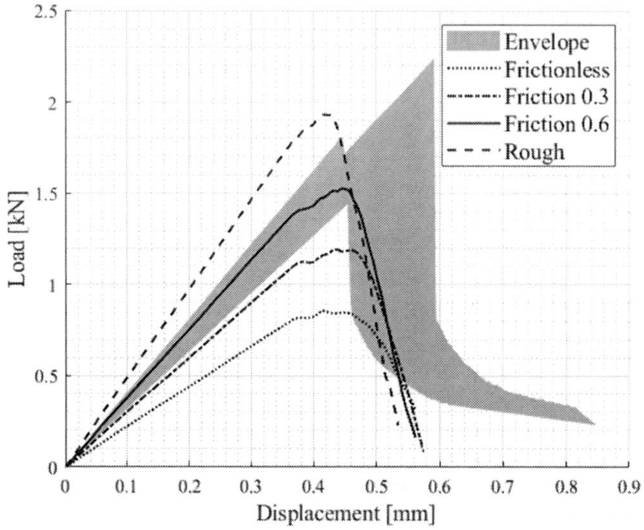

Figure 19. Load-displacement diagrams of the four interaction cases plotted against the envelope of the experimental data.

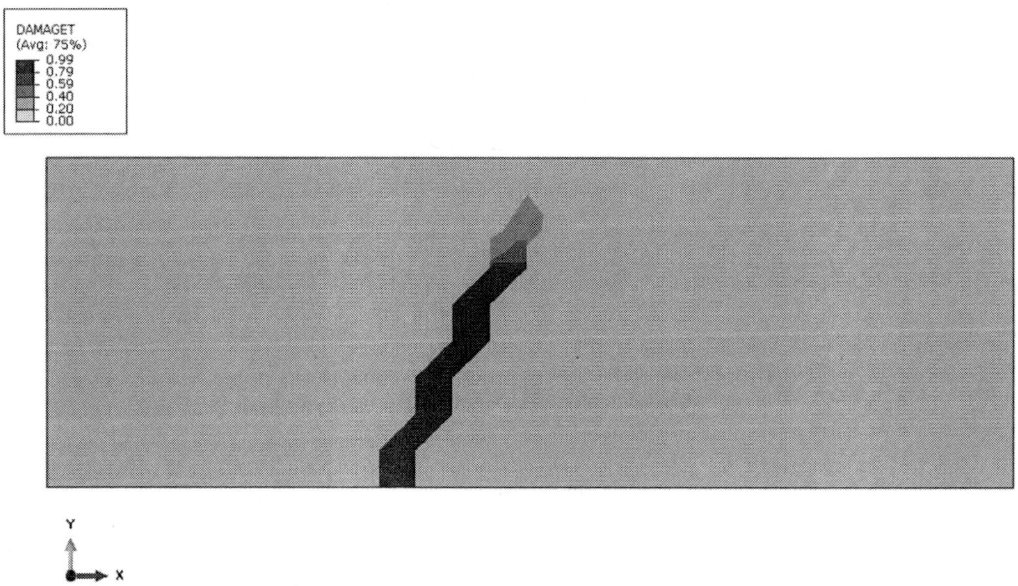

Figure 20. Development of the crack during the analytical three-point bending test for the interaction case "Friction 0.6".

6. Conclusions

In this work, the interface behaviour between hollow clay bricks called fictile tubules and mortar has been investigated. The investigation has been performed through a series of three-point bending tests on

IOP Conf. Series: Materials Science and Engineering **586** (2019) 012027 doi:10.1088/1757-899X/586/1/012027

three specimens, which consist of 11 fictile tubules embedded in mortar. The specimens are prepared assuring consistency with the 1:4 ratio between the transversal and longitudinal dimensions of the specimens used in three-point bending tests of mortar. The experimental results are then used as the basis for a series of back analyses carried out in the finite element code Abaqus, where four different cases of frictional properties are investigated. It is observed that the working hypothesis of non-perfect adhesion between tubules and mortar is correct, and that the actual interaction behaviour is well replicated by setting a 60% of transmission of tangential stresses between mortar and tubules.
In the follow-up of this paper, the newfound frictional properties will be applied to the analytical model of the arch to further corroborate the validity of these parameters assessing their suitability to simulate the outcome of the static vertical test on the arch itself.

7. References

[1] Olivito RS, Scuro C, Codispoti R 2016 Static analysis of masonry kilns built with fictile tubules bricks. In: Si-mos TE, Kalogiratou Z, Monovasilis T (eds) 12th International Conference of Computational Methods in Sciences and Engineering, Athens, Greece, 2016. AIP Conference Proceedings, vol 1790, art n 130006

[2] Olivito RS, Scuro C, Codispoti R 2016 Experimental analysis of typical hollow clay fictile tubules of cultural heritage of Mediterranean area. In: Van Balen K, Verstrynge E (eds) SAHC 2016 – Structural Analysis Historical Constructions – Anamnesis, diagnosis, therapy, controls, Leuven, Belgium, 2016. CRC Press, pp 1362–1369

[3] Tiberti S, Scuro C, Codispoti R, et al 2017 Traditional Masonry Arches and Domes with Fictile Tubules in Mediterranean Seismic Areas: Advanced Numerical Models and Experimentation. In: Papadrakakis M, Fragiadakis M (eds) 6th ECCOMAS Thematic Conference on Computational Methods in Structural Dynamics and Earthquake Engineering, Rhodes Island, Greece, June 2017. Eccomas Proceedia COMPDYN, vol 1, pp 2431–2448

[4] Tiberti S, Scuro C, Codispoti R, et al 2017 Post-Cracking FRCM Strengthening of an Arch Composed by Hollow Clay Elements Embedded in Mortar: Experimental Investigations and Numerical Analyses. In: Di Tommaso A, Gentilini C, Castellazzi G (eds) 5th International Conference on Mechanics of Masonry Structures Strengthened with Composite Materials, Bologna, Italy, 2017. Key Engineering Materials, vol 747, pp 142–149

[5] Scuro C, Tiberti S, Codispoti R, Milani G, Olivito R S 2018. Fictile tubules: A traditional Mediterranean construction technique for masonry vaulted systems. Construction and Building Materials, 193, 84-96.

[6] Tiberti S, Scuro C, Codispoti R, et al 2017 Masonry structures built with fictile tubules: Experimental and numerical analyses. In AIP Conference Proceedings (Vol. 1906, No. 1, p. 090010). AIP Publishing.

[7] Tsantini E, Jiménez-Piqué E, Montana G, Randazzo L 2015 Strength of pre-Roman amphorae: Comparison of the different types, Journal of Archaeological Science: Reports 2 (2015) 405-417.

[8] Storz S 1997, "La tecnica edilizia romana e paleocristiana delle volte e cupole a tubi fittili", p. 28 (il saggio alle pp. 23-41) in Claudia Conforti (a cura di),Lo specchio del cielo, Milano, Electa, 1997, pp. 309.

[9] Acocella A 2014 Anfore e tubuli fittili per volte, L'architettura di pietra, 2014 (accessed February 7th, 2018) (in Italian) http://www.architetturadipietra.it/wp/?p=6509.

[10] Arslan E A 1967 "Il significato spaziale delle volte sottili romane e paleocristiane", pp. 185-193 in Mesopotamia II, Torino, Giappichelli Editore

[11] Lancaster L C 2005 Concrete vaulted construction in imperial Rome, New York: Cambridge University Press

[12] Choisy A 1873, L'art de batir chez es Romains, Paris

[13] Scuro C 2017 Caratterizzazione meccanica e fisico chimica dei tubuli fittili: prove statiche su modelli in scala e modellazione numerica (Ph.D. Diss.), University of Calabria (in Italian)

[14] Lézine A 1954 Les voûtes romaines à tubes emboîtés et les croisées d'ogives de Bulla-Regia, Karthago 5 168–181 (in French).

[15] Tiberti S, Scuro C, Codispoti R, Olivito R S, Milani G 2019 Experimental and numerical analysis of historical aseismic construction system. In Structural Analysis of Historical Constructions (pp. 910-918). Springer, Cham

[16] Scuro C, Lamonaca F, Codispoti R, Carnì D L, & Olivito R S 2018 Experimental and numerical analysis on masonry arch built with fictile tubules bricks. Measurement, 130, 246-254

[17] 2007 Ente Italiano di Normazione UNI, UNI EN 1015-11:2007, Methods of Test for Mortar for Masonry – Part 11: Determination of Flexural and Compressive Strength of Hardened Mortar, Rome, Italy

IOP Conf. Series: Materials Science and Engineering **586** (2019) 012028 doi:10.1088/1757-899X/586/1/012028

Interlaminar damage numerical modelling based on delamination growth of multi-layered composites

I Hudişteanu, N Ţăranu, D N Isopescu, V Munteanu, O Neculai and D Ungureanu

"Gheorghe Asachi" Technical University of Iasi, Faculty of Civil Engineering and Building Services, Department of Civil and Industrial Engineering, Blvd. Mangeron, No. 1, 700050, Iasi, Romania

E-mail: iulianahudisteanu@ce.tuiasi.ro

Abstract. A growing interest is nowadays directed towards the numerical modelling and monitoring of the interlaminar damage evolution of composite laminates, but also to the numerical mathematical models and techniques for simulation of the interface delamination. Modelling and investigation of the mechanical behaviour and fracture modes of the multi-layered composites is a complex issue, since other important parameters, such as fibre orientation angles, stacking sequences and configuration of the composite laminates are added to the anisotropic character of the composite materials. Different failure modes may occur on multi-layered composites, while the delamination type fracture may lead to a considerable decrease of the load carrying capacity of the composite structures. Interlaminar stresses, which occur in composite laminates as an effect of the presence of the free edges, have a major influence on the delamination onset and the delamination growth. The paper summarizes the numerical methods and the approaches considered in the simulation of the delamination growth of composite laminates. The obtained results are figured in terms of the opening displacement at the interface crack tip and stress distributions on the plies of the multi-layered composites.

1. Introduction

The multi-layered composites are composed of unidirectional layers of fibre reinforced composite materials, stacked together to ensure the required mechanical properties, such as strength, and the in plane and bending stiffness [1-3]. The prediction of damage occurrence in composite laminates is a present concern, since various configurations are increasingly used in engineering applications.

Depending on the loading, specimen defects or mechanical properties of the material, various failure modes, such as fibre breakage, matrix cracking or interface debonding and delamination, may occur in multi-layered composites [4-7].

Nowadays, a growing interest is directed towards the analysis of the mechanical behaviour and damage evolution of composite laminates, especially of the delamination; an increasing number of mathematical models and theories, as well as experimental tests have been proposed, [8-13].

Delamination failure occurs at the interface between two adjacent layers and might lead to a considerable decrease of the load carrying capacity of the laminated composites. The interlaminar stresses have a major influence on the interlaminar damage onset and delamination growth.

2. Numerical modelling description

A balanced symmetric specially orthotropic composite laminated structure having the configuration $[0/15/30/45/90/-45]_s$ is investigated in this paper. The numerical analysis is performed in ANSYS Workbench and ANSYS Composite Prep/Post [14], the multi-layered composite having the total length of 85 mm, the width equal to 25 mm and an initial crack length of 10 mm. The fibre orientations of the

IOP Conf. Series: Materials Science and Engineering **586** (2019) 012028 doi:10.1088/1757-899X/586/1/012028

corresponding layers of the analysed configuration of a multi-layered composite are defined with respect to the local system of axes. It forms an angle θ with respect to the x axis, which is considered a positive angle in the (x, y) plane and negative in the (x, -y). The definition of the model for each elementary layer is represented in Fig. 1.

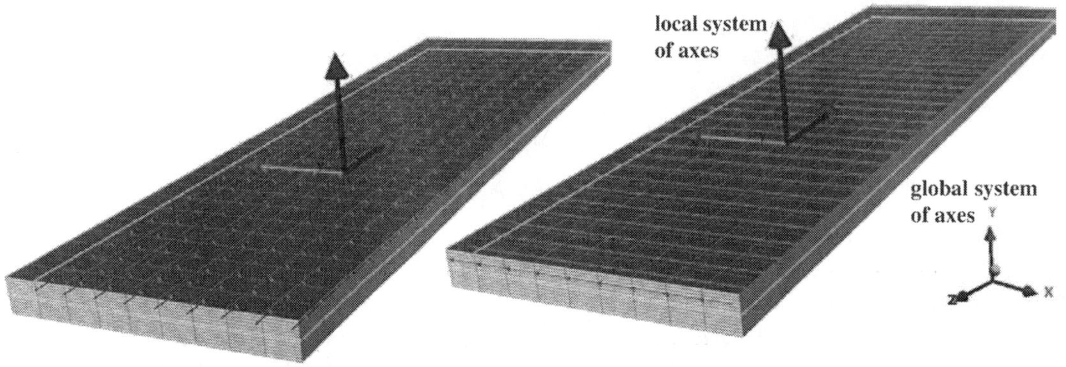

a) 0° fibre orientation angle b) 90° fibre orientation angle

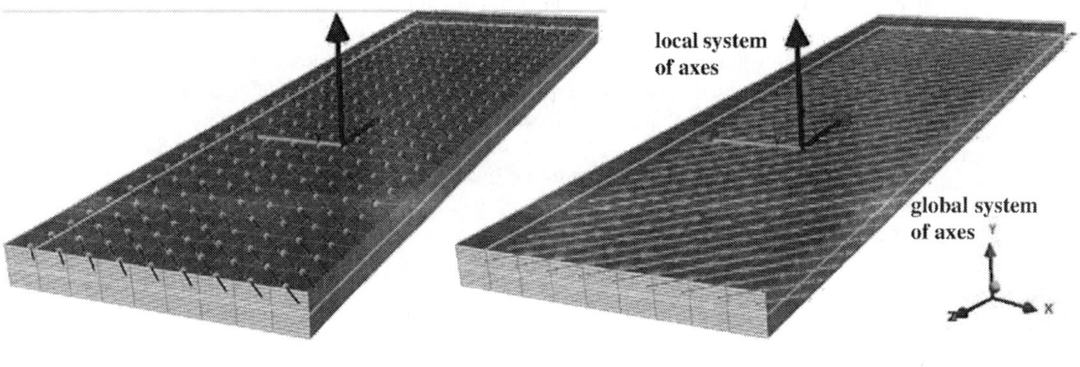

c) 45° fibre orientation angle d) -45° fibre orientation angle

Figure 1. Several fibre orientation angles of the balanced laminate (according to ANSYS)

Depending on the fibre orientations and the adopted configuration, the stacking sequence is defined in ANSYS Composite Pre-processing. Moreover, an interface is considered between the adjacent layers, located at the middle plane, for tailoring a delamination failure towards a specified trajectory. The interface is essential to establish a delamination numerical modelling analysis type.

The initial crack separates the open interface from the free edge (caused by a pre-existent delamination or due to some manufacturing defects) to the undamaged interface. The investigation of the delamination evolution is presented in Fig. 2. Several details, such as the local axes and the pre-meshed crack definition, as well as the specified trajectory of the delamination evolution are presented.

Since the interlaminar failure is investigated and the stress state of interest is a three-dimensional one, the numerical model is also conceived using 3D finite elements. Therefore, a through-thickness mesh is realized on the plies of the composite laminates, as represented in Fig. 3.

The boundary conditions of the numerical model are illustrated in Fig. 4. The composite laminates are fixed at one end, while at the free edge, two opposite and equal remote displacements, acting perpendicular to the exterior layers, are imposed.

The balanced composite laminate is made of unidirectional layers consisting of S glass fibers embedded in an epoxy resin, with the mechanical properties of the laminae presented in Table 1.

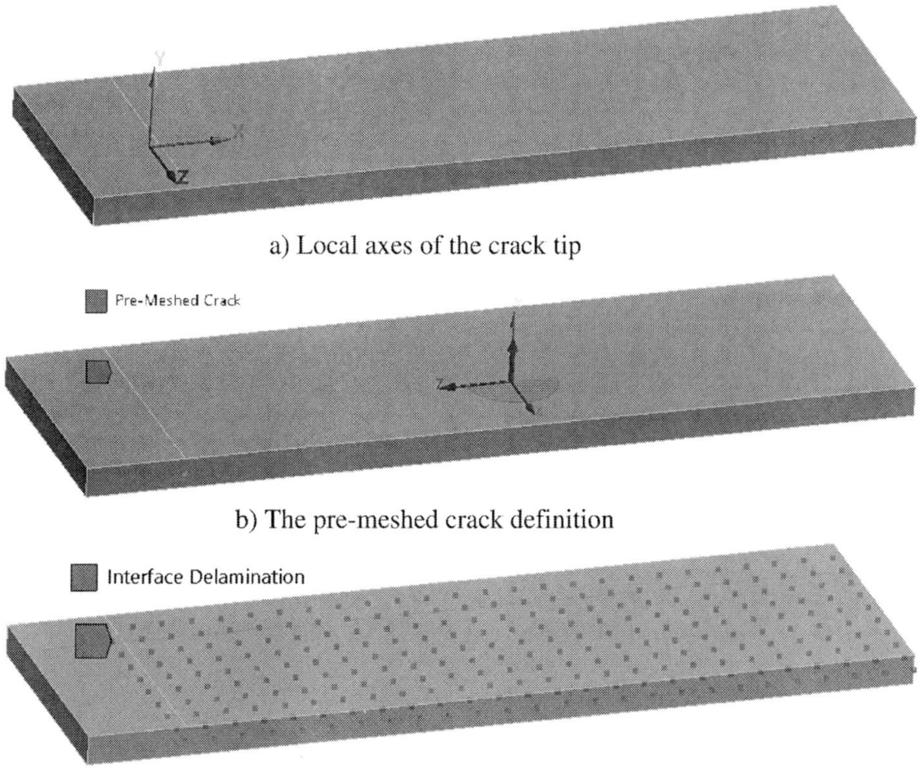

a) Local axes of the crack tip

b) The pre-meshed crack definition

c) Definition of the delamination propagation interface

Figure 2. Initial crack and direction of propagation definition of interlaminar damages at the interface

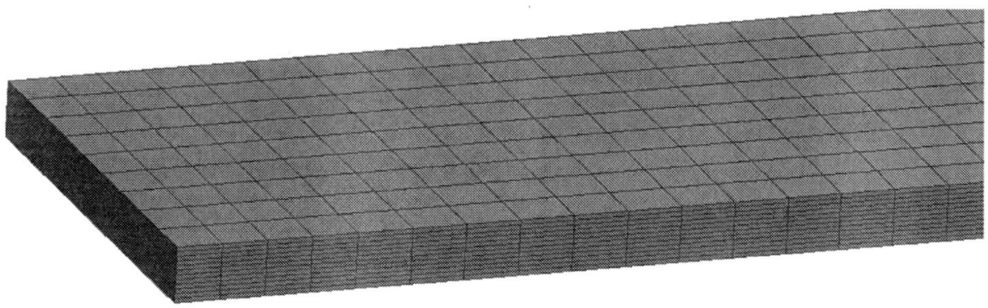

Figure 3. Through-thickness mesh on the plies of the composite laminate

Table 1. Mechanical properties of *S glass-epoxy* composite laminae* [15]

E_1 [GPa]	E_2 [GPa]	G_{12} [GPa]	υ_{12}	f_{1t} [MPa]	f_{1c} [MPa]	f_{2t} [MPa]	f_{2c} [MPa]	f_{12s} [MPa]
52.94	13.93	5.07	0.292	2836	1122	62.53	125.1	58.29

*where: E_1 and E_2 represent the longitudinal and transverse modulus, respectively; G_{12} – the in-plane shear modulus; υ_{12} – the in-plane major Poisson's ratio; f_{1t} and f_{1c} are the longitudinal tensile and compressive strength; f_{2t} and f_{2c} represent the transverse tensile and compressive strength, respectively; f_{12s} represent the in-plane shear strength.

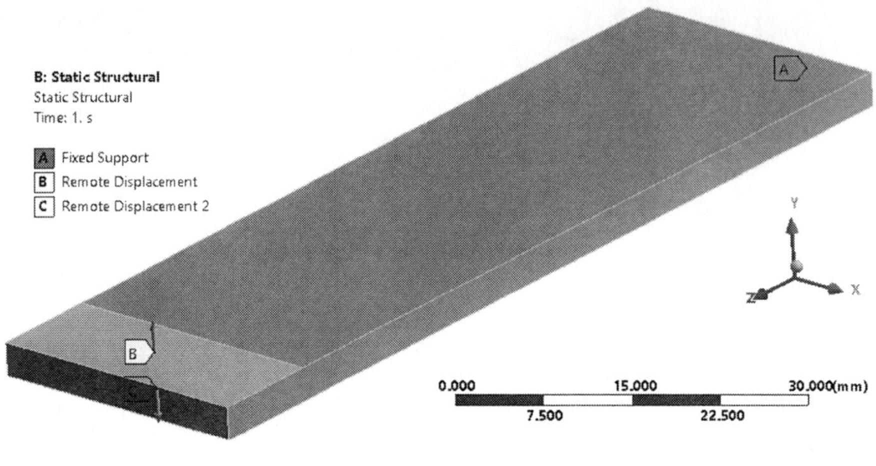

Figure 4. Boundary conditions corresponding to the numerical model

3. Results and Discussion on Interlaminar Delamination Growth

In order to perform a delamination behaviour analysis, an initial displacement equal to 40 mm, in an opening mode fracture type, established after a number of trials, is imposed to the balanced laminate. The initial crack opening and the contour plot of equivalent stresses as well as the state of stresses on the close vicinity of the crack tip delamination growth are presented in Figs. 5-7.

The separation of the layers adjacent to the interface and the propagation of the delamination is initiated from the loaded free edge towards the fixed end.

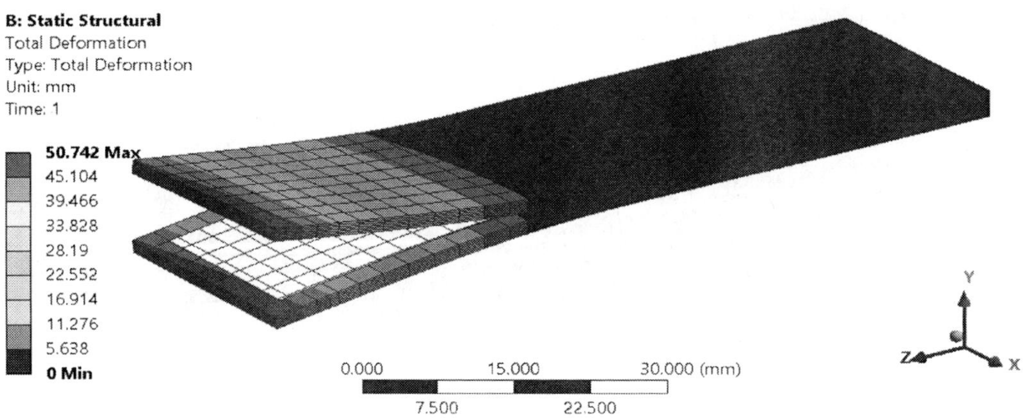

Figure 5. Opening displacement at the crack tip

The obtained results show that the opening displacement at the crack tip corresponding to the initial imposed displacement is 50.742 mm, while the maximum equivalent stresses occur on the exterior layers of the balanced laminate. For an improved visualization of the stress distributions on the laminae of the balanced laminate, a cross-section and a longitudinal section of the multi-layered structure is represented in the vicinity of the crack tip. The displacement jump and the stress distributions are symmetric with respect to the middle plane.

Subsequently, the interlaminar damage evolution and the final delamination are investigated. This ultimate displacement is described by the stage when the reactive force can no longer resist the delamination process.

IOP Conf. Series: Materials Science and Engineering **586** (2019) 012028 doi:10.1088/1757-899X/586/1/012028

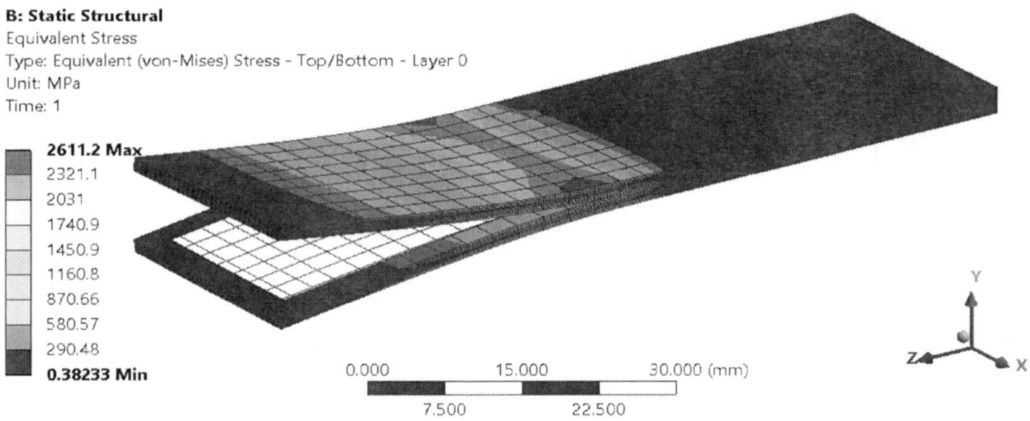

Figure 6. Stress distributions on the balanced laminate for the initial imposed displacement

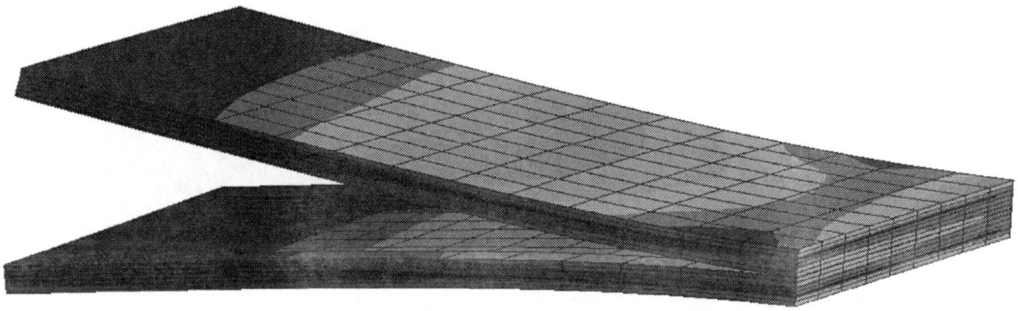

a) Cross-section

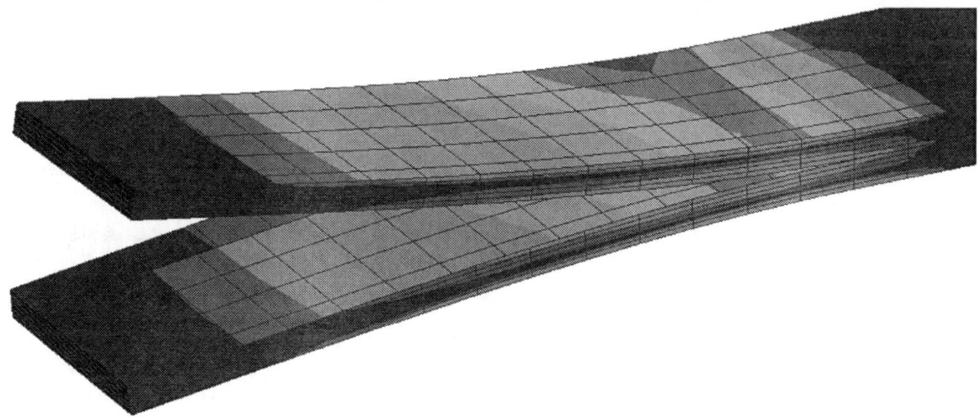

b) Longitudinal section

Figure 7. Stress state in the close vicinity of the crack tip for the balanced laminate

The traction-separation curve for the considered laminate is illustrated in Fig. 8, for an ultimate imposed displacement of 200 mm. This diagram shows the relation between the displacement jumps and the reactive forces that resist against the separation of the cohesive surfaces.

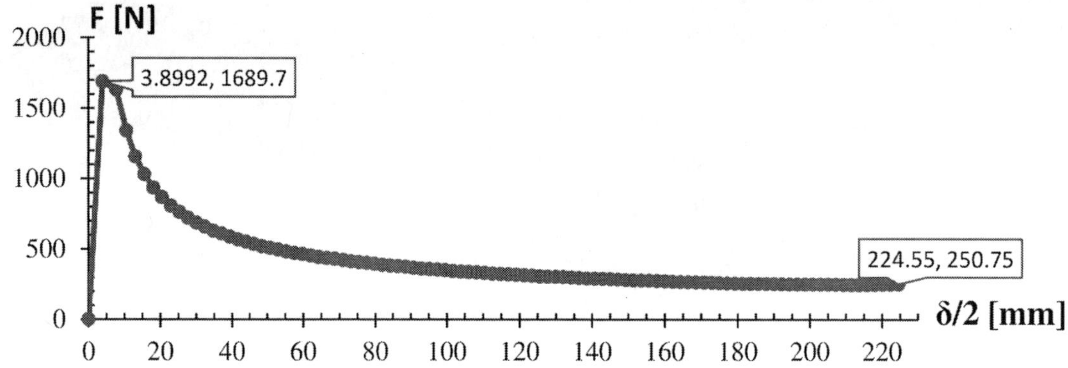

Figure 8. Traction-separation curve of balanced laminate for ultimate displacement Δ = 200 mm

The results obtained are in good agreement with those available in the literature [16], the traction-separation curve having the same variation.

The maximum displacement jumps corresponding to the last failure stages and the stress distributions are shown for the ultimate imposed displacements of 200 mm, Figs. 9-10.

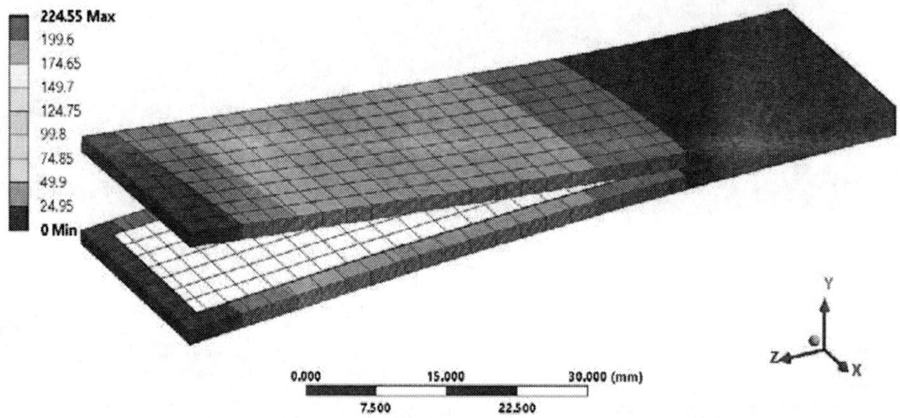

Figure 9. Ultimate crack tip opening displacement, [mm], for balanced laminate, Δ=200 mm

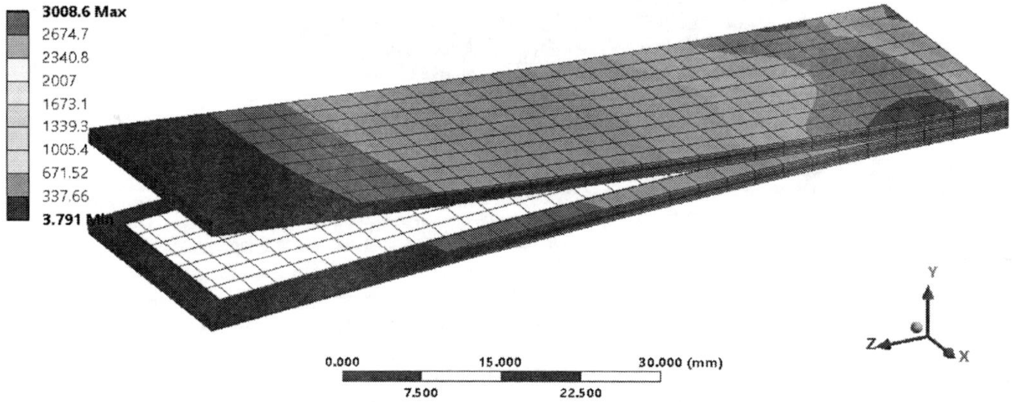

Figure 10. Equivalent stress distribution for the balanced laminate for ultimate imposed Δ=200 mm

IOP Conf. Series: Materials Science and Engineering **586** (2019) 012028 doi:10.1088/1757-899X/586/1/012028

4. Conclusions

The numerical modelling of the interlaminar damage onset and the damage evolution for a balanced composite laminate is presented in the paper, as an effect to the imposed displacements in an opening mode fracture type, in order to analyse the delamination propagation.

The results are figured in terms of displacement jumps and distribution of equivalent stresses on the plies of the laminate for an initial imposed displacement up to the final delamination of the considered adjacent laminas.

The simulation of the delamination behaviour of the composite laminates is often required, since the interlaminar failure occurs frequently in multi-layered composite elements.

References

[1] Daniel I M, Ishai O 2006 *Engineering mechanics of composite materials*, Second Edition, Oxford University Press, Oxford, New York.

[2] Herakovich C T 1998 *Mechanics of Fibrous Composites*, John Wiley & Sons, Inc., Univ. of Virginia, United States of America.

[3] Girão Coelho A M 2016 *Finite Element Guidelines for Simulation of Delamination Dominated Failures in Composite Materials Validated by Case Studies*, Arch Computat Methods Eng, 23, 363-388.

[4] Dogan F, Hadavinia H, Donchev T, Bhonge P S 2012 *Delamination of impacted composite structures by cohesive zone interface elements and tiebreak contact*, Cent Eur J Eng, 2 (4), 612-626.

[5] Amiri-Rad A, Mashayekhi M 2017 *A Cohesive Zone Approach for Fatigue-Driven Delamination Analysis in Composite Materials*, Applied Composite Materials, 24, 751-769.

[6] La Saponara V, Muliana H, Haj-Ali R, Kardomateas G A 2002 *Experimental and numerical analysis of delamination growth in double cantilever laminated beams*, Engineering Fracture Mechanics, 69 (6), 687–699.

[7] Park K, Paulino G H 2011 *Cohesive Zone Models: A Critical Review of Traction-Separation Relationships Across Fracture Surfaces*, Applied Mechanics Reviews, 64 (6), 06802.

[8] Bienias J, Dadej K, Surowska B 2017 *Interlaminar fracture toughness of glass and carbon reinforced multidirectional fiber metal laminates*, Engineering Fracture Mechanics, 175, 127–145.

[9] Long S, Yao X, Zhang X 2015 *Delamination prediction in composite laminates under low-velocity impac*t, Compos Struct, 132, 290-298.

[10] Chermoshentseva A S, Pokrovskiy A M, Bokhoeva L A 2016 *The behavior of delaminations in composite materials - experimental results*, IOP Conf Series: Materials Science and Engineering, 116, 1-5.

[11] Kovalovs A, Rucevskis S, Kulakov V, Aniskevich A 2016 *Delamination detection in carbon fibre reinforced composites using electrical resistance measurement*, IOP Conf Series: Materials Science and Engineering, 111, 1-6.

[12] Das S, Choudhury P, Halder S, Sriram P 2013 *Stress and free edge delamination analyses of delaminated composite structure using ANSYS*, Procedia Engineering, 64, 1364–1373.

[13] Silva M A G, Cismaşiu C, Chiorean C G 2005 *Numerical simulation of ballistic impact on composite laminates*, Int J Impact Eng, 31, 289-306.

[14] ANSYS® Documentation, ANSYS, Inc.

[15] Hudişteanu I, Ţăranu N, Isopescu D N, Enţuc I-S, Oprişan G, Ungureanu D 2018 *Numerical analysis of intralaminar damage evolution on various composite laminates*, International Conference ModTech 2018, IOP Conf Ser: Mat Sci Eng, 400, 042031.

[16] Barbero E J 2014 *Finite Element Analysis of Composite Materials using ANSYS*, Second edition, CRC Press, Taylor & Francis Group, Boca Raton.

IOP Conf. Series: Materials Science and Engineering **586** (2019) 012029 doi:10.1088/1757-899X/586/1/012029

Stability Analysis of a Scaled-Down Cold-Formed Steel Structure

G Țăranu, G Bunea, I Olteanu-Donțov, V M Venghiac, I Ovidiu Toma

Department of Structural Mechanics, The "Gheorghe Asachi" Technical University of Iasi, Iasi, 700050, Romania.

Corresponding e-mail: georgiana.bunea@tuiasi.ro

Abstract. The use of cold-formed steel CFS in building design has increased in the last period due to the numerous advantages it brings. However, there are relatively few research works on the behaviour of these types of structures to different loadings, especially compared to the hot-rolled steel structures. Unlike the hot-rolled steel elements, the CFS elements are prone to buckling failure. This paper investigates the stability of a scaled-down CFS structure considering the Eurocode EC and the American International Building Code IBC. First a comparison is done between the two codes in what concerns the stability analysis based on the theoretical aspects. Then, the results from the finite element program are presented, for both EC and IBC, and compared.

1. Introduction

Cold-formed steel CFS elements are used all around the world for building both residential and commercial structures. Having a high strength to weight ratio, they are easier to manufacture, to handle and to assemble on site, leading to a significant decrease of the structure's total cost. Taking into account their mechanical properties, the CFS structures have a different behaviour than the hot-rolled steel HRS ones, and they must be analysed according to their behaviour. Due to their slenderness ratio and, generally, thin walled open cross-sections, CFS elements fail by buckling before entering the plastic domain [1]. They can experience local, distortional and global buckling or even a combination of the mentioned phenomena which can lead to the failure of the entire structure. [2, 3, 4]

Taking into account the world-wide use of the CFS elements and the different national regulations, differences may appear in the structural design. Structural design using CFS is a laborious process especially due to the presence of local and / or global buckling modes, sometimes the two phenomena interacting with one another [5] This paper compares the stability analysis results of a CFS structure, for two of the most important building codes, namely the Eurocode (EC) and the American International Building Code (IBC). The comparison was carried out on a numerical model of a scaled down CFS structure.

2. Comparison between Eurocode EC and the International Building Code IBC

In this paper, only the differences which affect the stability analysis results will be presented. The regulations concerning the design of the CFS structures are found in *EUROCODE 3: Design of Steel Structures*, more specifically in *SR EN 1993-1-3-2007: Design of Cold-Formed Steel Structures*. The national annex considered in the analysis is the Romanian one. In the American norm, the code for designing the cold-formed steel structures is *AISI S100-2016: North American Specification for the Design of Cold-Formed Steel Structural Members*. The American code gives two design possibilities for the Ultimate Limit State ULS: Load and Resistance Factor Design (LRFD) and Allowable Strength Design (ASD).

The load combinations are an essential part of the stability analysis, the structure behaving in a different manner depending on the loading. Thus, the partial safety coefficients were taken into account in the analysis, and they are shown in Table 1, according to Table NA A1.2(B) from the Romanian national annex SR EN 1990-2004_NA-2006 [6] and to Chapter 2 from the American code ASCE/SEI 7-10 [7].

Table 1. Load combinations.

EUROCODE	ASCE/SEI 7-10
1.35·DL	1.40·DL
1.35·DL+1.50·LL+1.05·SL	1.20·DL+1.60·LL+0.50·SL
1.35·DL+1.50·SL+1.05·LL	1.20·DL+1.60·SL+0.5·LL

where DL is the dead load, LL – live load and SL – snow load.
One major difference between the two standards is the sets of units which are used. The EC adopted the metric units, which are largely used in Europe, in comparison with the IBC which uses the imperial format. This leads to a slight difference in material properties and sectional dimensions. In what concerns the permanent and the imposed loads, both norms evaluate them in a similar manner.
For the purpose of comparison, it is mandatory to first correlate the symbols used in each norm with the definition. Table 2 presents a list of these symbols based on the EC and the American National Standard ANSI/AISC 360-16 [8], along with their corresponding definitions [9].

Table 2. List of symbols found in EC and ANSI/AISC 360-16.

Definition	Symbol	
	EUROCODE	ANSI/AISC 360-16
Minimum yield stress	f_y	F_y
Minimum tensile stress	f_u	F_u
Design resistance – bending moment	M	M
Design resistance – axial force	N	P
Cross-sectional area	A	A
Elastic section modulus	W	S
Load factor	γ	γ

This paper analyses the stability of a CFS structure, buckling being one of the ways in which the structure could reach failure [10]. Therefore, the formulas regarding the elements' buckling resistance must be analysed for both EC and ANSI/AISC 360-16 [8]. Any difference could lead to distinct stability analysis results.
According to SR EN 1993-1-1-2006 [11], the buckling resistance of a compressed element is computed depending on the class of the cross-section, as presented in Equations 1 and 2. The assigning of an element to a certain class is established depending on the way the element behaves when it reaches local buckling.

$$N_{b,Rd}=\frac{\chi \cdot A \cdot f_y}{\gamma_{M1}} \text{ for Class 1, 2 and 3 cross-sections} \quad (1)$$

$$N_{b,Rd}=\frac{\chi \cdot A_{eff} \cdot f_y}{\gamma_{M1}}, \text{ for Class 4 cross-sections} \quad (2)$$

where χ is the buckling reduction factor, A – the cross-sectional area, A_{eff} – the effective area of the cross-section subjected to uniform compression, γ_{M1} – partial safety coefficient.
The buckling reduction factor χ for elements centrically compressed is computed with Equation 3.

$$\chi=\frac{1}{\phi+\sqrt{\phi^2-\bar{\lambda}^2}},\ \chi\leq 1 \tag{3}$$

in which

$$\phi=0.5\cdot[1+\alpha\cdot(\bar{\lambda}-0.2)+\bar{\lambda}^2] \tag{4}$$

$$\bar{\lambda}=\sqrt{\frac{A\cdot f_y}{N_{cr}}},\ \text{for Class 1, 2 and 3 cross-sections} \tag{5}$$

$$\bar{\lambda}=\sqrt{\frac{A_{eff}\cdot f_y}{N_{cr}}},\ \text{for Class 4 cross-sections} \tag{6}$$

α – the imperfection factor

N_{cr} – the critical buckling force, for the considered buckling mode.

The critical buckling force N_{cr} can be computed using Euler's formula, where L_{cr} is the critical buckling length, E is the modulus of elasticity and I – the minimum moment of inertia of the cross-section.

$$N_{cr}=\frac{\pi^2 EI}{L_{cr}^2} \tag{7}$$

Unlike the EC, the American norm ANSI/AISC 360-16 [8] considers in the analysis the nominal compressive strength P_n, computed, in this case, with the help of Equation 8.

$$P_n = F_{cr}\cdot A_g \tag{8}$$

The critical stress F_{cr} is computed with Equations 9 and 10 [8].

- $$\text{When } \frac{L_c}{r}\leq 4.71\cdot\sqrt{\frac{E}{F_y}},\ F_{cr}=\left(0.658^{\frac{F_y}{F_e}}\right)\cdot F_y \tag{9}$$
- $$\text{When } \frac{L_c}{r}>4.71\cdot\sqrt{\frac{E}{F_y}},\ F_{cr}=0.877\cdot F_e \tag{10}$$

where A_g is the gross cross-sectional area of the element, E – the modulus of elasticity, r – the radius of gyration and F_e – the elastic buckling stress determined with Formula (11).

$$F_e=\frac{\pi^2 E}{\left(\frac{L_c}{r}\right)^2} \tag{11}$$

Equation 11 given by the American norm is an equivalent of the Euler's critical buckling force, computed with Equation 7, but in terms of stresses. Therefore, although the two norms have different approaches in what concerns the buckling analysis, i.e. forces and stresses, the considered effects of the compressive force are similar [9].

3. Structural model

The structural elements were manufactured from a DX51D+Z steel sheet and, using a cold formed procedure, they were bent into a C89/41/12/1 profile. The connections of the elements were made using 2x Ø4x22 mm self-drilling self-tapping screws. Although the sectional dimensions are different in the United States and the materials used have slightly different values regarding the mechanical properties, the structure will not be changed and the analysis will be done on the initial structure having the above-mentioned geometrical dimensions.

The structural model dimensions are 2.30 x 2.70 x 2.60 m. The CFS scaled-down structure is presented in Figure 1. For the analysis, the structure was modelled in the SCIA Engineer v 18.1 software, where the connections were considered to be rigid. The structural model introduced in the finite element program is shown in Figure 2. The structure was subjected to a roof permanent load of 0.5 kN/m^2 and a snow load of 2 kN/m^2. Although the value of the snow load is different in the United States, for this analysis it will be considered the same as in the EC case, the purpose being to observe the possible difference in the obtained results based on the design philosophies of the two codes.

Figure 1. Scaled-down CFS structural model

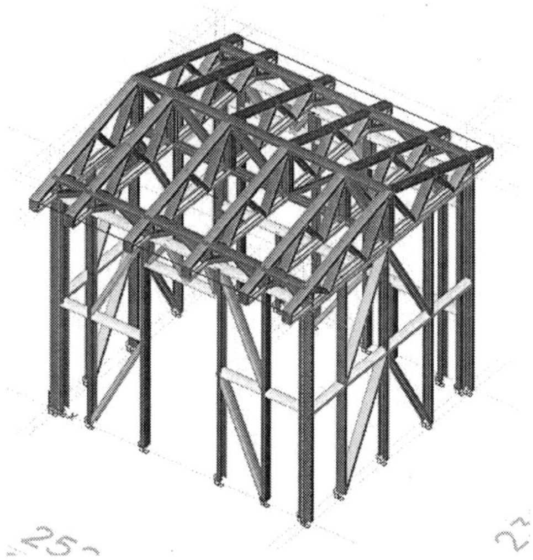

Figure 2. CFS model in SCIA Engineer

The analysis within SCIA Engineer is done according to the selected national code and national annexes. For the current analysis, two national codes are selected, namely the Eurocode EC and the International Building Code IBC. For the EC, the national annexes considered are the Romanian ones, which contain some additional information regarding EC parameters and establish national values. In what concerns the IBC, there are preset codes as follows:

- Code for loads: IBC 2012 (ASCE 7-10);
- Code for steel structures: AISC 360-16;
- Code for cold-formed steel structures: AISI S100-2016.

Table 3. Stability combinations.

Load case	Eurocode coefficients		International Building Code coefficients	
	S1EC	S2EC	S1IBC	S2IBC
Dead load	1.35	1.35	1.20	1.40
Roof dead load	1.35	1.35	1.20	1.40
Snow	1.50	-	1.60	-

In SCIA Engineer software, in order to run the stability analysis, it is required firstly to define the stability combinations. These are based on the linear combinations assembled automatically by the program based on SR EN 1990-2004_NA-2006 [6] for the European norm and ASCE/SEI 7-10 [7] for the American one. Two stability combinations, i.e. S1 and S2, were considered for both EC and IBC, as presented in Table 3.
Also, each element is assigned to a buckling group, which defines the buckling constraints and the span settings.

4. Stability analysis results

IOP Conf. Series: Materials Science and Engineering **586** (2019) 012029 doi:10.1088/1757-899X/586/1/012029

4.1 Critical load factor

The first step in a stability analysis is to compute the critical load factor α_{cr}, as presented in Equations 12 and 13. The principle of calculus is the same in both EC and IBC. Based on this value, it can be established if the deformations are significant enough to cause an increasing of the forces or a change in the global behavior of the structure. In case of large deformations, it is recommended to run a 2nd order calculus and, if necessary, to include global and/or local imperfections, in order to obtain more accurate results. The limit values of α_{cr} up to which a 2nd order analysis is required are presented in Equations 12 and 13, for elastic and plastic analysis, respectively [11].

$$\alpha_{cr}=\frac{F_{cr}}{F_{Ed}}<10, \text{ for elastic analysis} \tag{12}$$

$$\alpha_{cr}=\frac{F_{cr}}{F_{Ed}}<15, \text{ for plastic analysis} \tag{13}$$

where F_{Ed} is the design load which acts upon the structure, F_{cr} – elastic critical buckling load.

Table 4 presents comparatively the critical load factors α_{cr} of the considered structure obtained by using European and American codes, for the first three buckling modes.

Table 4. Critical load factors α_{cr}

Buckling mode	Eurocode		International Building Code		Comparison (IBC relative to EC)	
	S1EC	S2EC	S1IBC	S2IBC	S1	S2
1	218.58	914.94	213.41	882.27	-2.37%	-3.57%
2	232.52	947.57	227.28	913.73	-2.26%	-3.57%
3	245.56	1023.04	239.80	986.5	-2.35%	-3.57%

By comparing the critical load factors based on the EC with the ones based on the IBC, according to Table 4, it results slightly smaller values using IBC. The difference is given mainly by the partial safety coefficients used in the load combinations, the codes using different values, as shown in Tables 1 and 3. Therefore, for the first stability combination S1, which includes the snow load, the IBC values are smaller with approximately 2.30%, whereas for the second stability combination S2, made up only from permanent loads, the values are smaller with 3.57%.
Moreover, taking into account that the critical load factors α_{cr} are greater than 15, according to Equation 13 [11], it can be established that no 2nd order calculus is required, the structure itself failing due to the failure of its members when the strength condition is no longer fulfilled.

4.2 Nodal displacements

Figure 3 shows the deformed shape of the structure corresponding to the stability combinations S1 and S2. The roof has been filtered out in order to have a better view upon the structural elements which are prone to buckling. Both EC and IBC have similar results in what concerns the deformed shape, therefore, only one situation is presented.

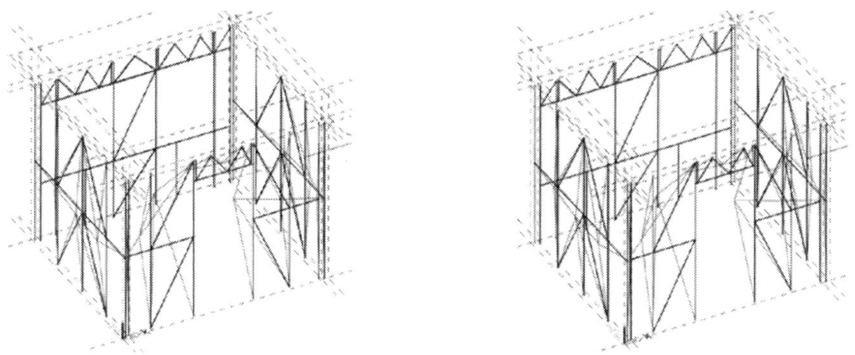

a) b)

Figure 3. Deformed CFS structure
a) Stability combination S1 b) Stability combination S2

Figures 4-7 present the displacement of nodes in Ox and Oy direction, obtained from stability combinations S1 and S2 and based on EC and IBC. The values are computed only for the first buckling mode. The buckling shapes and the values are normalized so that the maximum nodal displacement component is 1 m. The analysis results shown that this maximum value was obtained for the rotation about Oz axis of column bordering the door opening.

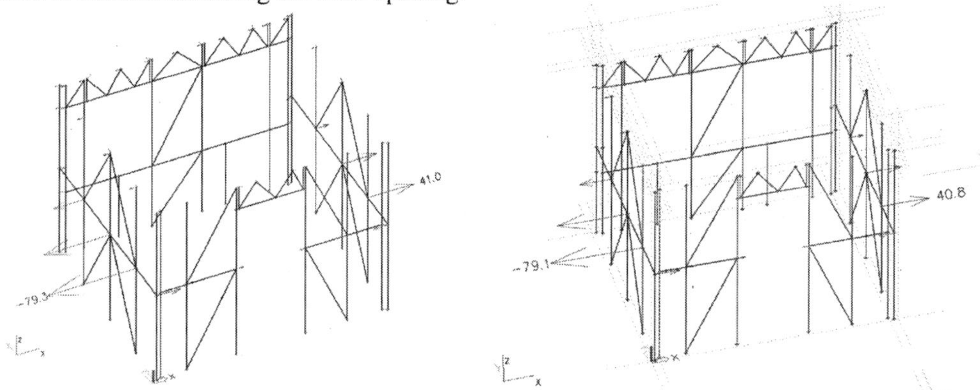

a) b)

Figure 4. Global maximum displacement of nodes on Ox direction - stability combination S1
a) Eurocode; b) International Building Code

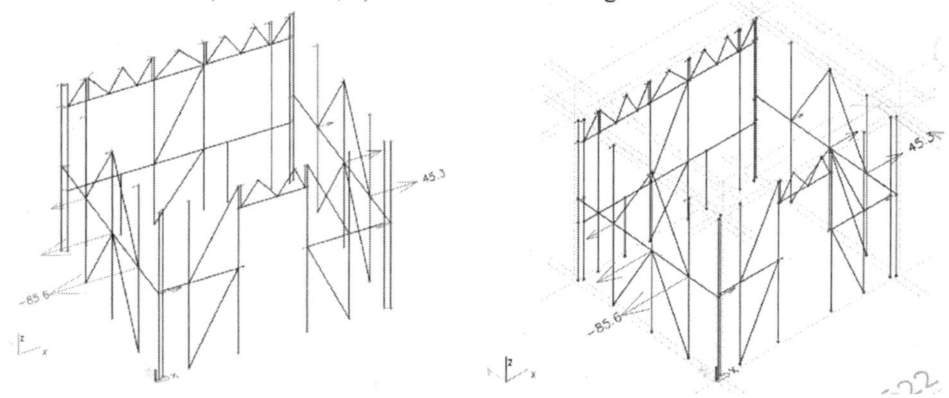

a) b)

Figure 5. Global maximum displacement of nodes on Ox direction - stability combination S2
a) Eurocode; b) International Building Code

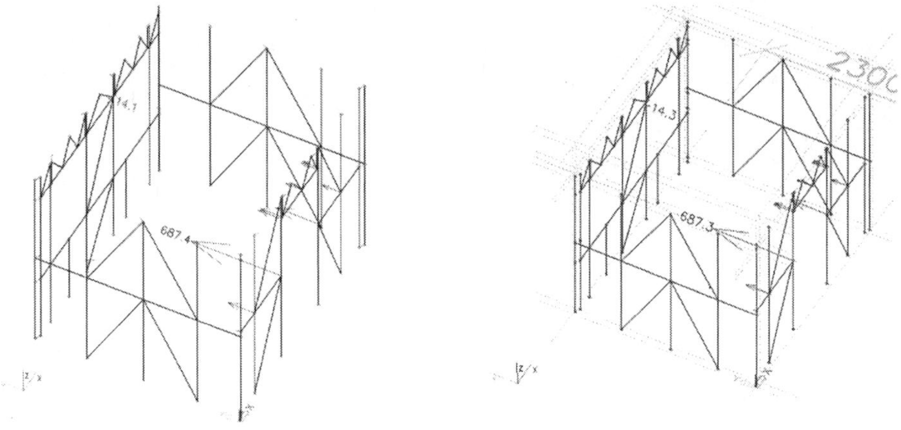

a) b)

Figure 6. Global maximum displacement of nodes on Oy direction - stability combination S1
a) Eurocode; b) International Building Code

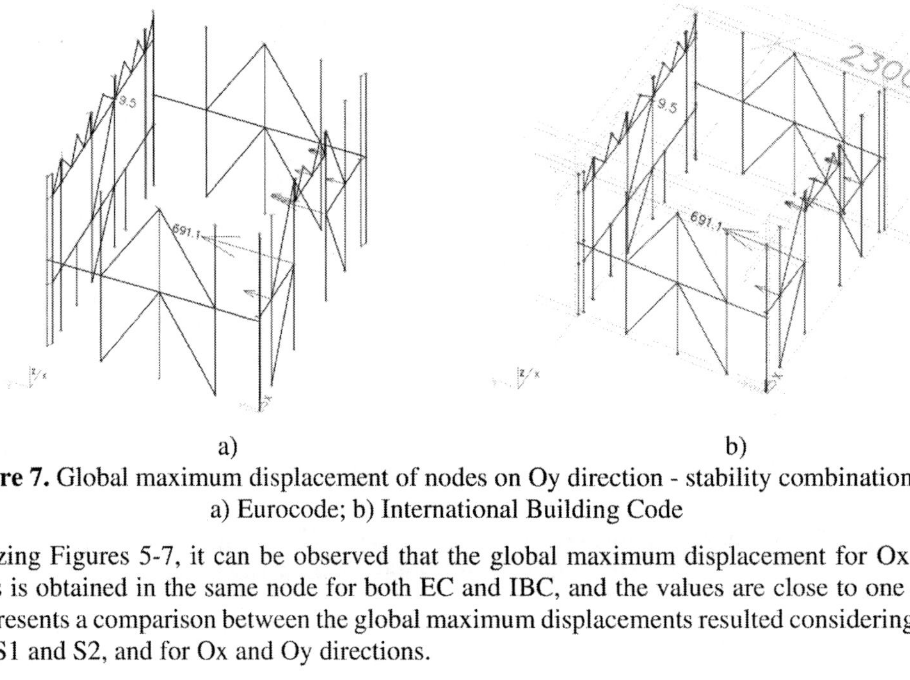

a) b)

Figure 7. Global maximum displacement of nodes on Oy direction - stability combination S2
a) Eurocode; b) International Building Code

By analyzing Figures 5-7, it can be observed that the global maximum displacement for Ox and Oy directions is obtained in the same node for both EC and IBC, and the values are close to one another. Table 5 presents a comparison between the global maximum displacements resulted considering EC and IBC, for S1 and S2, and for Ox and Oy directions.

Table 5. Global maximum node displacements

Direction	Eurocode [mm]		International Building Code [mm]		Comparison (IBC relative to EC) [%]	
	S1EC	S2EC	S1IBC	S2IBC	S1	S2
Ox	-79.3	-85.6	-79.1	-85.6	-0.25	0
	41.0	45.3	40.8	45.3	-0.49	0
Oy	-14.1	-9.5	-14.3	-9.5	1.42	0
	687.4	691.1	687.3	691.1	-0.02	0

The difference between the maximum node displacements obtained according to EC and to IBC is very small, so it can be stated that the codes lead to approximately the same result.
The large values for the maximum lateral displacements support the observations made in terms of the critical load factor that is the elements of the structure fail due to the values of stresses exceeding the strength of steel. The buckling, in the case of the investigated structure, occurs in the plastic range of the material behavior [4].

4.3 Steel slenderness
Taking into account that the two codes have similar approaches in what concerns the computation of the parameters related to buckling, as described in Chapter 2, the steel slenderness results are the same. Table 6 presents the global extreme values of the steel slenderness corresponding to the y-axis obtained in SCIA Engineer, where L_y is the system length for buckling around y-axis, k_y – the buckling ratio for buckling around y-axis, l_y – buckling length for buckling around y-axis and λ_y – the slenderness around y-axis.

Table 6. Steel slenderness parameters

Member	L_y [m]	k_y [-]	l_y [m]	λ_y [-]
B97	0.090	10.00	0.900	25.40
B34	2.495	1.91	4.762	134.41
B82	0.318	10.00	3.179	89.71
B174	1.125	5.98	6.732	**190.01**
B35	2.495	1.47	3.674	103.71

Taking into account the restraint conditions, most elements have a buckling ratio k_y between 1 and 2. However, there are some elements, shown in Table 6, for which k_y is greater than 2. In this case are members B97 and B82, for which the resulted buckling ratio k_y is 10, this being the maximum allowed value, have a small length, namely 9 cm and 31.8 cm. Thus, despite of this high buckling ratio value, the slenderness is relatively small. The maximum slenderness ratio was obtained for member B174, which represents a bottom chord of one of the trusses.

4.4 Stability check
The stability check is done according to SR EN 1993-1-1 and SR EN 1993-1-3 for the European regulations and to AISI S100-16 [12] LRFD for the American regulations. For comparison, two elements will be considered, associated to the global extremes of the unity check obtained by running the stability check considering both codes. The unity check is defined by Equation 14, which checks that the design value of a force does not rise above the corresponding design resistance and, if more forces act upon the element, their combined effects do not overpass this resistance [11]. In case of the American regulations, the unity check is defined similarly, but in terms of stresses in AISI S100-16 [12].

$$\frac{N_{Ed}}{N_{Rd}}+\frac{M_{y,Ed}}{M_{y,Rd}}+\frac{M_{z,Ed}}{M_{z,Rd}}\leq 1 \tag{14}$$

where N_{Ed} is the design axial compressive force, $M_{y,Ed}$ – the design bending moment with respect to y-y axis, $M_{z,Ed}$ – the design bending moment with respect to z-z axis, N_{Rd} – design resistance to axial force, $M_{y,Rd}$ – design resistance to bending moment with respect to y-y axis, $M_{z,Rd}$ – design resistance to bending moment with respect to z-z axis.

From the stability check according to EC, the global extreme value of the unity check corresponds to element B131, whereas according to IBC, it corresponds to element B149, as presented in Figure 8.

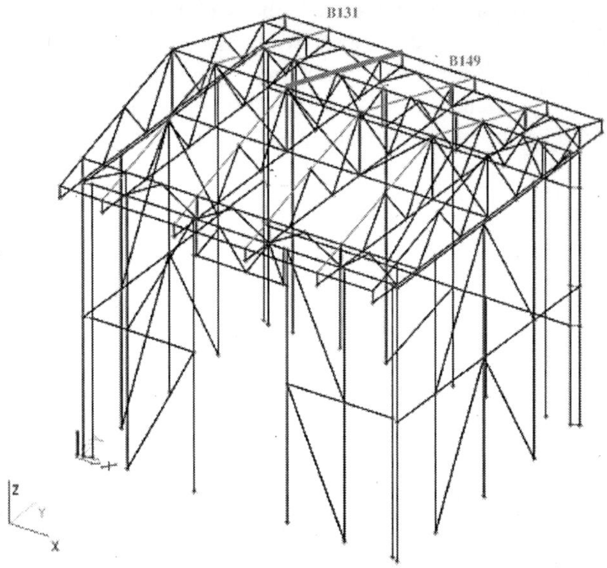

Figure 8. Global extreme of the unity check – elements

5. Conclusions

The paper presents the results in terms of critical buckling factor and lateral displacements obtained from the numerical analysis of a scaled-down clod-form steel structure. The results are obtained by applying the buckling analysis as defined by the two major design codes in use nowadays: The Eurocode and The American International Building Code.

Based on the obtained results, the following preliminary conclusions can be drawn:

- The two design codes, although based on two different approaches, lead to similar results in terms of the effects of the compressive forces that may lead to the occurrence of buckling.
- The scaled down model used in the present research exhibited large values of the critical buckling coefficient which may lead to the conclusion that the failure of the steel section due to large stresses occurs before the loss of stability happens. This observation is later supported by the large lateral displacements of the nodes corresponding to the buckling modes along the two in-plane directions.
- The numerical model should be further calibrated by means of laboratory tests. The authors are aware that further research in the field of loss of stability of CFS member should be conducted before reliable conclusions can be reached.

6. References

[1] Camotim D and Dinis P B 2011 Coupled instabilities with distortional buckling in cold-formed steel lipped channel columns *Thin-Walled Structures* **49** 562-575

[2] Laím L, Rodrigues J P C and Simões da Silva L 2014 Experimental analysis on cold-formed steel beams subjected to fire *Thin-Walled Structures* **74** 104-117

[3] Bae S-W, LaBoube R A, Belarbi A, Ayoub A 2008 Progressive collapse of cold-formed steel framed structures *Thin-Walled Structures* **46**:7-9 706-709

[4] Ye J, Mojtabaei S M and Hajirasouliha I 2018 Local-flexural interactive buckling of standard and optimized cold-formed steel columns *Journal of Constructional Steel Research* **144** 106-118

[5] Schafer B W 2011 Cold-formed steel structures around the world – A review of recent advances and applications, analysis and design *Steel Constructions* **4**:3 141-149

[6] SR EN 1990-2004_NA, 2006 *Eurocod: Bazele proiectarii structurilor. Anexa nationala*, Asociația de Standardizare din Romania ASRO, București, România

IOP Conf. Series: Materials Science and Engineering **586** (2019) 012029 doi:10.1088/1757-899X/586/1/012029

[7] ASCE Standard ASCE/SEI 7-10, *Minimum Design Loads for Buildings and Other Structures*, American Society of Civil Engineers
[8] ANSI/AISC 360-16, 2016 *Specification for Structural Steel Buildings*, American Institute Of Steel Construction, Chicago, Illinois
[9] Alber R, Rubik B, Vasonis A 2017 Acceptance of Steel and Aluminum Structures Designed per the Eurocodes at Fermilab. Eurocode Structural Review,
[10] Bernuzzi C, Gobetti A, Gabbianelli G and Simoncelli M 2015 Simplified Approaches to Design Medium-Rise Unbraced Steel Storage Pallet Racks. I: Elastic Buckling Analysis *Journal of Structural Engineering* **141**:11
[11] SR EN 1993-1-1, 2006 *Eurocod: Proiectarea structurilor din otel. Reguli generale si reguli pentru cladiri*, Asociația de Standardizare din Romania ASRO, București, România
[12] AISI S100, 2016 *North American Specification for the Design of Cold-Formed Steel Structural Members* American Iron and Steel Institute, Washington, D.C

Assessing the earthquake performance of existing buildings of various material types and configurations in New Zealand

L Tuleasca[1,2], W Loo[1], I Wardak[1]
[1] UNITEC Institute of Technology – Department of Civil Engineering, 139 Carrington Road, Mt Albert, Auckland, *New Zealand*,
[2] Beton Evolution Limited – Structural Engineers, Te Atatu South, Auckland, *New Zealand*

E-mail: ltuleasc@unitec.ac.nz

Abstract. The Initial Evaluation Procedure, conducted as part of Initial Seismic Assessment, is the main topic of this paper. An overview is provided of the eight key steps which define this procedure, and how they are used to determine ratio of New Building Standard (%NBS). The key steps are encapsulated in a programmed EXCEL worksheet (developed and current used in New Zealand) which takes into account various parameters and applies an algorithm to arrive at a preliminary assessed value. Actual examples of various buildings assessed using this procedure in recent years by authors are presented, and the results and some areas of interest are discussed.

1. Introduction

In the wake of the 2011 Canterbury earthquakes, assessing the structural performance of existing buildings has become a major focus for New Zealand structural engineers. To encourage consistency of evaluation, over the past few years the New Zealand Society for Earthquake Engineering, the Structural Engineering Society, the New Zealand Geotechnical Society, in conjunction with the Ministry of Business, Innovation and Employment, and the Earthquake Commission have developed a publication entitled "The Seismic Assessment of Existing Buildings, Technical Guidelines for Engineering Assessments". This document sets out the requirements for structural engineers to follow when evaluating the earthquake performance for existing buildings. A two-stage evaluation process is recommended: (1) Initial Seismic Assessment ISA, "which enable a broad indication of the seismic rating of a building", and (2) Detailed Seismic Assessment DSA, "which provide a more comprehensive assessment of the likely seismic rating of a building". Both stages quantify seismic performance as a seismic rating expressed as a percentage standard achieved from application of the new building standards requirements (%NBS).

It must be underlined that there are major differences between the design and assessment processes. For the design of a new building, structural engineers are generally required to consider a single design imperative, typically the Ultimate Limit State, and to ensure that this building meets this minimum required standard for new buildings and therefore, the minimum acceptable performance assumed attained at any level of earthquake actions. A minimum earthquake standard for new buildings, with a well-defined use, is achieved by determining design actions for the required limit state, and applying these to a chosen structural system to determine the critical stresses that could possibly occur within the members. Appropriate design and detailing of the individual elements is then carried out in accordance with the respective material standards.Applying these steps from the traditional design process are not always valid when assessing an existing building, unless the behaviour of the existing structure remains predominantly elastic during a design level seismic event. Because this cannot always be assumed, the approach for seismic assessment must be different from the traditional design approach with the latter possibly leading to a result that can be "either excessively conservative or non-conservative" [1].

2. Seismic assessment of existing buildings

The scope of the SAEB guidelines is to assist all parties involved in buildings "life" to respond to challenges involved in understanding, managing and reducing seismic risk for people using these existing buildings. Having unique guidelines, engineers can assess the seismic behaviour of existing buildings and elements of these buildings, in similar way, and report the assessment results to building owners and organisations responsible for their management. The SAEB guidelines are for all types of existing buildings covering all eras and all types of construction materials. However, structures such as bridges, towers, masts, and retaining walls are excluded. These guidelines might be applied for buildings that undergo seismic retrofit, alteration or when a change of use is required.

The guidelines underline the expectations regarding the expertise of engineers who undertake seismic assessment of existing buildings:

- They must have relevant experience in structural design and evaluation of buildings for earthquake effects in order to exercise the degree of judgement required for this process. Therefore, every assessment shall be carried out by or under the direction of a New Zealand Charter Professional Engineer (CPEng).
- They must have specific training in the objectives of and processes involved in the assessment procedures in order to provide consistent reporting; template covering letter and engineering assessment summary report are available.

2.1. Initial Seismic Assessment

Initial Seismic Assessment (ISA) is the first recommended step of seismic assessment of any existing building and is considered an important "first look" at the building performance. ISA is a "qualitative procedure that involves observing building attributes, and then using these to develop a holistic understanding of how the building will respond to an earthquake. It provides an initial assessment of its earthquake rating" [2]. Depending on the purpose of ISA and the time when the building was erected, this could involve:

- Basic ISA which will involve collecting basic building information, provided usually by local council, an exterior inspection, and completing an Initial Evaluation Procedure (IEP).
- Comprehensive ISA, which in addition to the above-mentioned information will require an interior inspection, available structural drawings (i.e. foundations, stairs, columns, walls, floors) review and if necessary supplementary calculations. Comprehensive ISA is recommended especially "if building's earthquake rating is around the threshold levels of 34%NBS and 67%NBS" [2].

The report with the results of the ISA would be appropriate for certain situations and can include information such as: building structural system description, the level of the information available at the time when the report was produced, limitations of the process and the results of the IEP.
IEP is a fundamental part of ISA and consists in "making an initial assessment of the standard achieved for an existing building against the minimum life safety standard required for a new building (the percentage of new building standard, or %NBS)" [2]. More details and examples of IEP produced by the authors, for buildings made of different types of materials and erected at different time, will be presented in this paper.

2.2. Detailed Seismic Assessment

Despite of the fact that there is no prerequisite for buildings to have an ISA before proceeding to DSA, it is strongly encouraged that engineers develop first a qualitative procedure (ISA) in order "to gain a holistic view of the building's potential structural weaknesses" [3].
Detailed Seismic Assessment (DSA) is "a quantitative procedure" used to confirm an earthquake rating for an existing building, "to identify retrofit needs and provide a benchmark for proposed upgrading strategy to be tasted against" [3]. Understanding the behaviour of structural systems and the likely behaviour of buildings in earthquakes must be done by quantifying the strength and deformation

capacities of existing building and by checking the building's structural integrity against the actual loads/deformations.

Depending on the circumstances and stated objective (client requirements) the approach taken to carry out a DSA may have considerable differences depending on the actual condition of building. Some buildings will require the use of lengthy and detailed analyses, some will not. For buildings with visible structural problems it will be better for the time to be spent in completing an appropriate solution for retrofit, rather than gaining a detailed understanding on how the existing structure might perform.

The objectives for the DSA recommended by NZ Technical Guidelines for Engineering Assessments are to [3]:

- Assess the level of earthquake shaking at which the ultimate capacity (seismic) is reached and identify the point at which the conditions are met for a significant life safety hazard.
- Determine an earthquake rating for the investigated building in terms of %NSB that is more reliable than the rating from ISA.
- Provide details on the expected mode of failure to assist territorial authorities to make decisions on life safety issues to determine earthquake prone building status.
- Determine whether or not a building is an earthquake risk building (i.e. < 67%NBS).
- Provide guidance on the likely needs for retrofit.

3. Initial Evaluation Procedure

As mentioned above IEP is an integral part of ISA, and it was designed to accommodate a varying level of information about structural characteristics regarding the building under investigation. It involves making an initial assessment of the standard achieved for an existing building against the minimum life safety standard required for a new building (the percentage of new building standard, or %NBS) [1].

With as few resources as possible, the IEP will identify, to an acceptable level of confidence, most of those buildings that fall under the earthquake-prone building threshold. As some of these aspects are unknown to the engineer, when the IEP is completed for the first time, it is a good practice for this process to be repeated when new information regarding the investigated building becomes available; information which results in any changes to the original IEP must be recorded so it can be referred to by anyone considering or reviewing the results. An outline of the IEP procedure is shown in figure 1 [2].

There is a standard form used for the IEP procedure and this has eight steps as follows:

- General Information: photos of the building exterior, building plans, list of relevant features, notes on any building strengthening work, recording the characteristics of any adjacent buildings, noting all information sources to complete the assessment.
- Determine Baseline Percentage of New Building Standard for each orthogonal direction as a product of: nominal %NBS, near fault scaling factor (factor E), hazard scaling factor (factor F), return period scaling factor (factor G), ductility scaling factor (factor H), structural performance scaling factor (factor I).
- Assessment of Performance Achievement Ratio (PAR) for longitudinal and transverse directions as a product of factors: plan irregularity (factor A), vertical irregularity (factor B).
- Short columns (factor C), pounding potential (factor D), site characteristics (factor E), other factors (factor F).
- Percentage of New Building Standard (%NBS) – Seismic Rating: the lower value obtained for longitudinal and transverse directions.
- Earthquake-prone: is %NBS < 34?
- Potential Earthquake Risk; is %NBS < 67?
- Provisional Grading for Seismic Risk based on IEP.

- Identification of Potential Sever Structural Weaknesses (SSWs) that could result in significant risk to a significant number of occupants: number of storeys above ground, presence of heavy concrete floors and/or roof.

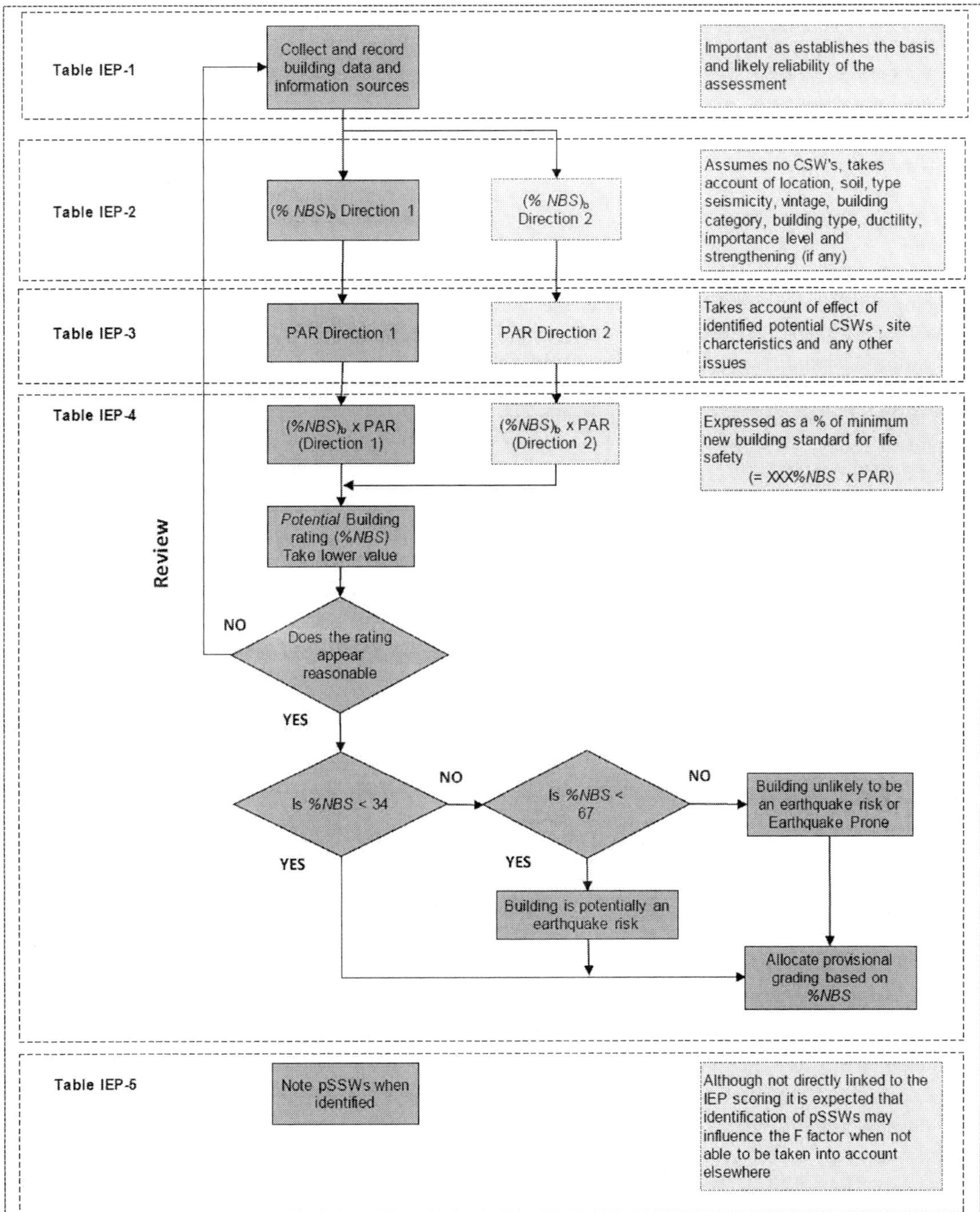

Figure 1. Initial Evaluation Procedure [2].

IOP Conf. Series: Materials Science and Engineering **586** (2019) 012030 doi:10.1088/1757-899X/586/1/012030

The IEP is a largely qualitative, score-based assessment based on the current standard for earthquake actions for new buildings in New Zealand, NZS 1170.5.2004. It is based on generic building characteristics and is dependent upon knowledge available at the time of the assessment. Due to its qualitative nature it should not come as a surprise that an IEP procedure of the same building carried out by two or more experienced engineers may have different results; this is acceptable, especially if the level of available information was different. If differences in opinion regarding the IEP results (from ISA) cannot be settle through professional discussion, the guidelines recommend being resolution by completion of a DSA [2].

4. Case study: Reinforced concrete building

The first publication on earthquake design was written by C. Reginald Ford, in 1926, a few years before the 7.8 magnitude earthquake (Napier 1931) that resulted in significant changes to New Zealand construction practice. After the devastating earthquake in Napier, a Building Committee was set up and a draft earthquake building by-law was presented to the NZ Parliament in 1931. This ended up being published by NZ Standards as the 1935 New Zealand Standard Model Building By-Law (NZSS 95:1935) and the 1939 NZS Code of Building By-Laws (NZSS 95:1939) [4].

The 1955 revision of NZS Standard Model Building By-Law (NZSS 95:1955) introduced changes but lacked improvement in seismic structural design. The NZS 1900:1964 code was a significant evolution from the previous one; it was based on knowledge from the American codes, ACI318-63, 1963 and the European codes, CEB-1964, 1964. NZS 3101:1982 provided improvement requirements in the detailing of plastic regions, including shear, confinement and anti-buckling reinforcement. This was reviewed and updated in 1995 and 2006 to incorporate further knowledge from research and the revision of earthquake actions standard NZS 1170.5:2004 [4].

Seismic assessments for existing buildings are done assuming that all buildings have been designed and built in accordance with buildings standards current at that time and with good practice. The changes which have occurred within the New Zealand standards have been taking into account within the IEP working spreadsheet; this includes information tables and required figures for assessment.

4.1. General information about building

A multi-storey reinforced concrete building situated in the Auckland Central Business District (CBD), was designed and built around 1999 on top of the previous building with two level reinforced concrete frame car park. The building has six levels between axis C09 and C22 and two levels between axis C02 and C09. The structural system is comprised of reinforced concrete frames (columns and beams) and prestressed precast floor elements (Hollow Core). The roof consists of lightweight metal cladding on purlins over structural steel.

Information sources were visual inspection of the exterior and interior of the building, and Auckland Council files for property and geotechnical report. The report was completed for a private company interested in buying some space in this property.

4.2. Results from IEP

Introducing all information regarding this building on the standard IEP working spreadsheet the following results were obtained:

- Nominal %NBS: for both longitudinal and transverse directions $(\%NBS)_{nom} = 28\%$: taking into account near fault, hazard, return period, ductility and structural performance scaling factors.
- Baseline %NBS for building: for both directions $(\%NBS)_b = 152\%$: taking into account near fault, hazard, return period, ductility and structural performance scaling factors.
- Performance Achievement Ratio: for both directions PAR = 0.59: taking into account plan and vertical irregularity, short columns and pounding effects, site characteristics and other factors for allowance of all other characteristics of the building.

- Percentage New Building standard or seismic rating based on the above factors: %NBS = 90%.
- Provisional Grading for Seismic Risk based on IEP: Seismic Grade A - seismic grading from IEP should be considered provisional and subject to confirmation by detailed assessment.

As the building was design and constructed around 1999, the results from IEP are closed with what shall be expected.

5. Case study: Unreinforced masonry building

Most of New Zealand's URM buildings were built between the late 1870s and 1940, and the construction methods are relatively uniform with only a few variations [5]. Foundations were typically shallow strip footings: often thin and unreinforced for small buildings and deeper and nominally reinforced with plain reinforcing bars, flats or train/tram rails. The commonly used nominal thickness of brick walls in New Zealand are 230 mm (two wythes), 350 mm (three wythes) and 450 mm (four wythes). This is in addition to any outer veneer of 110 mm (one wythe). Common types for walls were [6]:

- Solid walls - used generally for industrial buildings and buildings on the outskirts of town, and for partition walls and walls in lower storeys that are not visible.
- Cavity walls – used in buildings to control moisture ingress.

Unreinforced masonry (URM) building can be vulnerable to earthquake actions because of their high centre of mass, lack of integrity between elements and lack of deformation capacity. Assessing the performance of these buildings can be complex as the mechanism of failure is different from those occurring in other buildings types [6]. For URM buildings built prior to 1935, Steps 2 to Step 4 of IEP can be carried out using the attribute scoring method. The %NBS is determined directly from the total assessed score for each of the eight items: structural continuity, configuration, condition of structure, wall (URM) proportions, diaphragms, engineering connections between floor/roof diaphragms and walls, and walls and diaphragms capable of spanning between, foundations and separation from neighbouring buildings [2].

5.1. General information about building

The building is constructed of URM around 1930, generally up to 300 mm thick around the exterior, and comprising 200 mm thick walls with 100 mm thick leaf fixed as cladding with an air gap between. The building is placed on a significant slope from the road down to the rear, dropping by a building storey in height (i.e. two storeys at the street and three storeys at the rear). The building comprises five separate units with an URM full height internal wall approximately 200 mm thick separating them. The roof is of timber construction with lightweight metal cladding. The floors are typical tongue and groove floor boards on timber joists. From the original drawings the internal foundations appear to step down the slope from the road; these comprise URM piles on concrete pads. The exterior perimeter foundations are URM foundation walls on concrete strip footings.

From a visual perspective the property appears to be constructed as one building. However, from our inspection and the data provided in council files there are four units constructed around 1930 and a unit constructed later (unknown date).

5.2. Results from IEP

Introducing all information regarding this building on the standard IEP working spreadsheet the following results were obtained:

- Nominal %NBS: for both longitudinal and transverse directions $(\%NBS)_{nom}$ = 2.90%.
- Baseline %NBS for building: for both directions $(\%NBS)_b$ = 28%.
- Performance Achievement Ratio: for both directions PAR = 1.0.
- Percentage New Building Standard: %NBS = 28%.
- Provisional Grading for Seismic Risk based on IEP: Seismic Grade D.

Using attribute scoring method, the following results were obtained:

- Percentage New Building Standard: %NBS = 23%.
- Provisional Grading for Seismic Risk based on IEP: Seismic Grade D

The building was designed and constructed around 1930, with multiple cracks in the walls noted, and an approximate 800 mm high parapet: the results using the IEP procedure and scoring methods are close with what would be expected. While there is a difference between results for the %NBS, the results for seismic grade are the same.

6. Case study: Timber buildings

In New Zealand timber is a readily available material and it has been used widely for many building types since the earliest European settlement. The two main categories of timber buildings are [7]

- Timber framed structures such as those designed using non-specific design guides and standards.
- Engineered buildings as halls, commercial and industrial buildings.

Timber has been used as well in other buildings types that are primarily constructed of other materials – with timber used for roof framing, floor joists, flooring and sarking under roofs etc.

It has been known that timber framed, residential and small commercial, buildings generally perform well in earthquakes, and the 2010/2011 Christchurch earthquake confirmed this. Timber framed structures, known as stick framing, are made of small sections timber such as 45 x 90 mm that combined with timber studs create wall frames. These are equipped with top and bottom plates.

Before the introduction of NZS 3604:1978 bracing was commonly provided by addition of let-in diagonal braces or cut-in diagonal braces between studs. Some older buildings with timber framed walls rely on an internal lathe and plaster lining to provide the bracing instead of employing diagonal members; this construction system was used in the building for which IEP is presented in this paper. Modern timber framed walls are likely to be reliant on their lining materials for providing bracing resistance. Plasterboard, plywood, particle board and sometimes fibre cement are linings used now.

6.1. General information about building

The example we refer to is a two storey timber building in Wellington, with a steep heavy tiled roof and timber and concrete partial basement. From information provided by the Wellington City Council Archives, the original property/dwelling was constructed in 1893. Major alterations to the dwelling are recorded as being completed in 1927. The exterior walls cladding is a type of weatherboard and the interior walls are of particleboard. In 1953 several structural walls were demolished with an adverse impact on building bracing capacity. At the time of inspection some of the walls were found in poor condition and water infiltration and damage on the second story ceiling were observed.

In 1959/1960 parts of the building were demolished during the construction of an adjacent reinforced concrete building. The actual layout is different from the one presented on the initial plan.

6.2. Results from IEP

Introducing all information regarding this building on the standard IEP working spreadsheet the following results were obtained:

- Nominal %NBS: for both longitudinal and transverse directions $(\%NBS)_{nom} = 3.63\%$.
- Baseline %NBS for building: for both directions $(\%NBS)_b = 20\%$.
- Performance Achievement Ratio: for both directions PAR = 1.05.
- Percentage New Building Standard: %NBS = 21%.
- Provisional Grading for Seismic Risk based on IEP: Seismic Grade D

For this building the owner was proceeding with a request for DSA. The required calculations for DSA were done for the structural timber elements. More than half of the structural walls were found with inadequate bracing; this was mainly because of the removal of some of the structural walls during a building alteration carried out in 1953.

The building was found to have an unsatisfactory seismic protection level, achieving a lowest 25 NBS% for the ground floor on the transversal direction, and 36 NBS% for the ground floor on the longitudinal direction.
The recommendations for this property taking into account the DSA results, the age, the structural system and the usage of the building, was to carry out seismic strengthening remedial works in order to address deficiencies that were found. These would address:

- Inadequate bracing units required for earthquake.
- Inadequate connection of the subfloor framing with foundation.
- Damages to external building walls to avoid future deterioration of structural elements.

7. References

[1] 2017 *The Seismic Assessment of Existing Buildings. Technical Guidelines for Engineering Assessments. Assessment Objective and Principles Part A* (New Zealand: www.eq-assess.org.nz) p 71
[2] 2017 *The Seismic Assessment of Existing Buildings. Technical Guidelines for Engineering Assessments. Initial Seismic Assessment Part B* (New Zealand: www.eq-assess.org.nz) p 80
[3] 2017 *The Seismic Assessment of Existing Buildings. Technical Guidelines for Engineering Assessments. General Issues. Part C, Section C1* (New Zealand: www.eq-assess.org.nz) p 49
[4] 2017 *The Seismic Assessment of Existing Buildings. Technical Guidelines for Engineering Assessments. Concrete Buildings. Part C, Section C5* (New Zealand: www.eq-assess.org.nz) p 189
[5] Russell A P, Ingham J M 2010 *Bulletin of the New Zealand Society for Earthquake Engineering Prevalence of New Zealand's unreinforced masonry buildings* (New Zealand) Vol 43 p 183-202
[6] 2017 *The Seismic Assessment of Existing Buildings. Technical Guidelines for Engineering Assessments. Unreinforced Masonry Buildings. Part C, Section C8* (New Zealand: www.eq-assess.org.nz) p 195
[7] 2017 *The Seismic Assessment of Existing Buildings. Technical Guidelines for Engineering Assessments. Timber Buildings. Part C, Section C9* (New Zealand: www.eq-assess.org.nz) p 50
[8] 2004, *NZS 1170.5:2004 Structural Design Actions Part 5: Earthquake actions – New Zealand* (Standard New Zealand) p 80

IOP Conf. Series: Materials Science and Engineering **586** (2019) 012031 doi:10.1088/1757-899X/586/1/012031

Numerical modelling of RC floor slabs

J Mikula[1], J Szolomicki[2]
[1] Mestil Engineering Studio, Wroclaw, Kwidzynska 71, Poland,
[2] Wroclaw University of Science and Technology, Wroclaw, Wybrzeze St. Wyspianskiego 27, Poland

E-mail: Jerzy.Szolomicki@pwr.edu.pl

Abstract. The aim of the paper is to conduct a comparative analysis of four calculation models of reinforced concrete floor slabs, that differ in geometry and the way of supporting. The investigation focuses primarily on the analysis of deflections of cracked reinforced concrete slabs, while monitoring the degree of determined reinforcement and the width of crack opening. The used programs were analyzed in terms of their capabilities, necessary in the work of a modern construction engineer, such as modeling of concrete rheology, determining the real reinforcement, finite element mesh variation, taking into account changes in stiffness of the cracked section, etc. The computational methods used in the software were also taken into account, including the possibility of performing non-linear analysis (mainly in the case of material non-linearity), updating the stiffness matrix (inelastic method) or considering possible stiffening of the concrete under tension. Each model was calculated taking into account the influence of shrinkage and without analyzing it, that required additional analysis of the ways of modeling rheology of concrete with use of computer programs.

1. Introduction

The finite element method (FEM) is currently one of the most commonly used computational methods used for the computer analysis of various structural systems in building construction. Its effectiveness and universality results from the fact that it is possible to replace the analytical problem, which is formulated with differential equations, with an algebraic problem. Three computer FEA programs were used to conduct research on the calculation methods of reinforced concrete slab floors: Autodesk Robot Structural Analysis Professional, Dlubal Software RFEM and ABC Plate.

RC floor slab is one of the structural elements which is extensively used in civil engineering structure. Design of the floor slabs is based on limit states and requires an estimation of the ultimate limit state and knowledge of their deflections and cracking zones. The estimation of deflection under loads is complicated because of the influence of factors like cracking, creep and shrinkage effects.

The main aim of the paper was to conduct a comparative analysis of the calculation of slab floor deflections with reference to the most common software programs used in design offices in Poland. The paper focuses primarily on the analysis of deflections of reinforced concrete floor slabs in the cracking stage, while also monitoring the degree of the determined reinforcement and the crack opening width. The software programs were examined in terms of their capabilities, which are necessary in the work of a modern civil engineer, such as considering concrete rheology, real reinforcement modeling, finite element mesh variation, etc. Attention was also paid to the calculation methods used in the software, including the possibility of performing nonlinear analysis (mainly in the case of material non-linearity), changing the stiffness of the cracking section (elastic method), or taking into account possible hardening of the concrete in tension.

2. Characteristics of used computer programs

2.1. ABC Plate software

ABC Plate software is a product of the Polish company PRO-SOFT and is based on the finite element method that allows the calculation of reinforced concrete floor slabs according to the European standard

PN-EN 1992-1-1: 2008 [1]. ABC Plate software uses linear analysis to calculate given problems, and loads are therefore defined as consecutive simple cases. From the obtained results, further combinations of the superposition method are put together, optionally scaling individual components by increasing factors. ABC Plate software calculates the necessary reinforcement due to the ultimate limit state and determines deflections in the cracking stage. In the case of calculations compatible with the Eurocode, it is also possible to take into account the effects of shrinkage, but only in the analysis of deflections.

2.2. Robot Structural Analysis Professional software

Robot Structural Analysis Professional software of the American company Autodesk uses the finite element method for calculations [2,3]. The software has comprehensive tools for modelling and calculating reinforced concrete structures in the field of linear and nonlinear analysis. However, it should be clearly stated that the material non-linearity assumption does not take into account the change in material stiffness due to on external factors such as temperature. Rheological issues are also not taken into account. However, in the software, the stiffness is updated depending on the cracking of the floor slab, and is adequate to the conditions prevailing in the given node of the plate element (reinforcement ratio, cracking, support conditions).

2.3. Dlubal RFEM software

RFEM software of the German company Dlubal uses the finite element method in the modular structure for calculations [4]. The RF CONCRETE DEFLECT module was used for the analysis of reinforced concrete floor slabs, which allowed calculations in the linear-elastic range and RF CONCRETE NL, which in turn enables non-linear material analysis. The main assumptions of the RF CONCRETE DEFLECT program are calculations of the stiffness of the floor slab element by interpolating between the uncracked and cracked stage, and also consideration the effect of shrinkage and creep on the value of deformations, including deflections. However, the use of the RF CONCRETE NL module makes it possible to perform calculations in a non-linear range when determining deformations, crack opening width and internal forces.

3. Characteristics of the analyzed models of floor slabs

In order to compare the engineering possibilities and computational methods of the adopted computer software, it was decided to subject them to a determination of deflections of reinforced concrete slabs in the cracking mode, while also monitoring the degree of determined reinforcement and crack opening width.

Four different calculation models were used, and they are characterized in detail below. Each of the floor slab models was calculated while taking into account shrinkage without impact, which required analysis of the ways of modelling concrete rheology by the used computer software. Table 1 presents the material data and computational characteristics of the analyzed floor slab models. Table 2 shows load cases and load combinations.

Table 1. The material data and computational characteristics of the analysed floor slabs.

General	
Floor slab thickness	200 mm
Admissible crack width	0,3 mm
Mesh size of element	0,1 m x 0,1 m
Concrete	
Concrete class	C25/C30
Environment class	XC1
Cement class	N
Structure class	S4
Reinforcement cover	30 mm
Reinforcing steel	

Reinforcing steel	B500SP
Steel ductility class	C
Minimum diameter of reinforcing steel	12 mm
Rheology of concrete	
Concrete age (first loading moment)	90 days
Relative environment humidity	40%
Time calculation for the creep	18250 days
Time calculation for the shrinkage	18250 days
Concrete age at the moment of shrinkage start	14 days

Table 2. Load cases and combinations.

Load case	Load type	Characteristic value of load (kN/m^2)
P1	Self weight	Automatically included in the computer program
P2	Life load with category C3	5
Loads combination	Limit state	Combination formula
K1	Ultimate Limit State	1,35·P1+1,50·P2
K2	Serviceability Limit State	1,00·P1+1,00·P2

3.1. Numerical models of reinforced concrete floor slabs

Model M1

The floor slab modelled in rectangle plan as a 5m x 7m, pivotally supported on all four edges, figure 1.

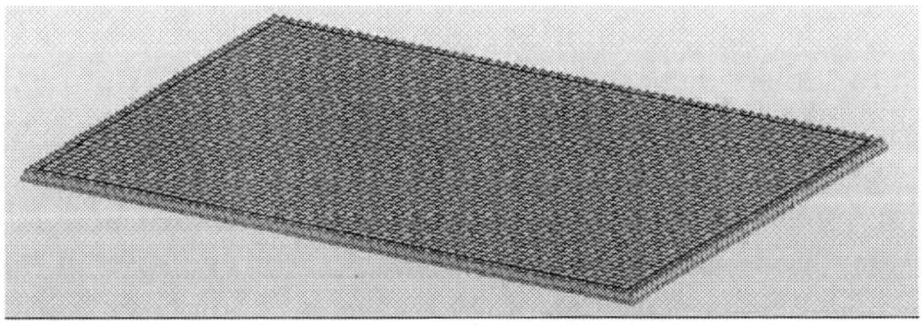

Figure 1. Geometry and boundary conditions of model M1.

Model M2

The floor slab modelled in rectangle plan as a 5m x 7m, fixed supported on all four edges, figure 2.

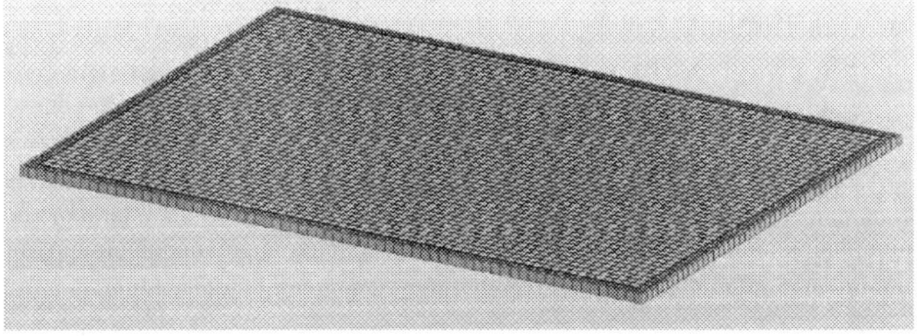

Figure 2. Geometry and boundary conditions of model M2.

Model M3
The floor slab modelled in rectangle plan as a 5m x 7m, fixed supported on one short edge and pivotally supported on two longer edges, figure 3.

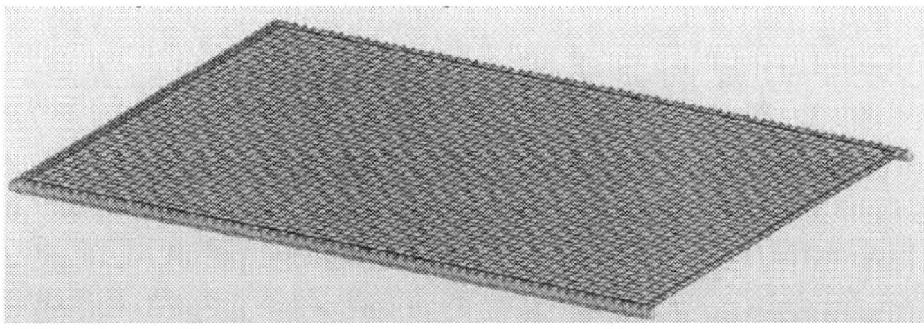

Figure 3. Geometry and boundary conditions of model M3.

Model M4
The floor slab modelled in rectangle plan as a 8m x 12m, pivotally supported on all four edges and additionally supported on a reinforced concrete binder of 0.3m x 0.4m made of C25 / 30 concrete and placed in 1/3 and 2/3 of its span, figure 4. The beams were offset, in relation to the center of gravity of the floor slab, in such a way that the slab lay on them and an eccentric connection between the elements was formed. The Binders were modelled as rods with reinforced concrete beam characteristics.

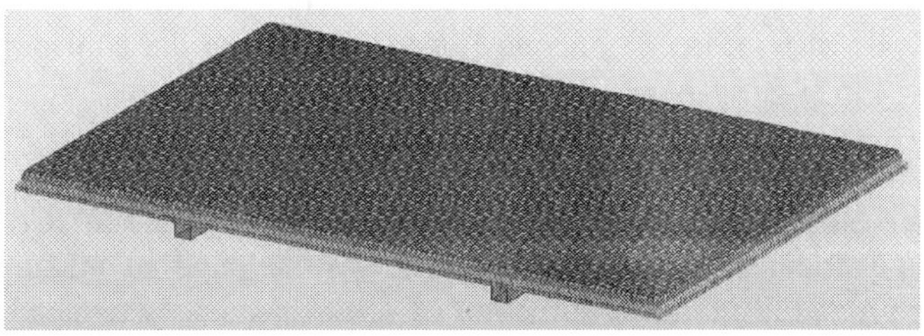

Figure 4. Geometry and boundary conditions of model M4.

4. Methods of rheology modeling in numerical analysis

4.1. Creep
Concrete creep is a phenomenon related to the deformation increase caused by constant long-term stress. It is mainly visible when the element has a guaranteed freedom of deformation. An important factor is the fact that the increase in deformation is caused by creep decreases with the passage of time. The effect of the phenomenon depends, among other things, on the class of concrete, its composition, age at the first loading moment and environmental conditions - mainly humidity and temperature [5]. All the applied software programs included the influence of the creep on the structure of the analysed floor slabs, Tab. 3.

Table 3. Creep coefficient value for the analysed floor slab models .

Software	RFEM	ROBOT	ABC PLATE
The creep coefficient	2,294	2,315	2,295

4.2. Shrinkage
Concrete shrinkage is a phenomenon consisting of intrinsic rheological deformations that are not related to mechanical loads. It is caused by structural changes occurring in the cement paste as a result of the physico-chemical processes of water loss as a result of the setting, hardening and drying of concrete. Total deformation due to shrinkage is the sum of two components: autogenous shrinkage and drying.

IOP Conf. Series: Materials Science and Engineering **586** (2019) 012031 doi:10.1088/1757-899X/586/1/012031

4.2.1. ABC Plate software

ABC Plate software only takes into account the effect of shrinkage when defining the deflection curve of elements. With this phenomenon, an additional deflection curvature is created, which should be taken into account when determining the total vertical deformation. In the case of cracked floor slabs, the slab deflection will occur towards the tension zone due to gravitational loads. ABC Plate software enables the adding of total deflections that are determined on the basis of assumed loads.

4.2.2. Robot Structural Analysis Professional software

Robot Structural Analysis Professional software does not take into account the influence of shrinkage in the designing of RC elements, even in the range of the serviceability limit state. Therefore, it is necessary to model the effects of shrinkage using the equivalent of another interaction, which would lead to similar stresses in the element. The effect of shrinkage in RC structures can be seen to be equivalent to lowering the temperature by 15ºC, and this method was used to model this phenomenon with the Robot Software. It should be taken into account that shrinkage modeled in this way is only an approximation, and results in a significant overdesigning of the structure. In fact the factors of deformation and strength of concrete that are variable over time are not taken into account. Moreover, an important issue in this type of modelling is that the temperature effect is taken into account as in an isotropic, linear-elastic material, which corresponds to the uncracked concrete structure [6]. The flexural stiffness of the analyzed cross-section is decreased at the moment crack appears, and this results in a lowering of the internal forces. This assumption leads to a significant over-designing of the structure and the determination of reinforcement which exceeds the required reinforcement. The issue is visible in the numerical analysis of reinforced concrete floor slabs, where the consideration of shrinkage in Robot Software resulted in significant additional reinforcement of the structure.

4.2.3. Dlubal RFEM software

Dlubal RFEM software takes into account the effect of shrinkage in the *RF CONCRETE NL* and *RF CONCRETE DEFLECT* modules by reducing the stiffness of the material and applying the distribution factor. Stiffness of the material in any ϕ direction of reinforcement is reduced by the coefficient of influence of shrinkage $k_{sh,\phi,c}$. The longitudinal axial force and bending moment are only determined for shrinkage deformation in both cracked and uncracked conditions.

$$n_{sh,\phi} = -\varepsilon_{sh} \cdot E_s \cdot (a_{s1} + a_{s2}), \tag{1}$$

$$m_{sh,\phi} = n_{sh} \cdot e_{sh}, \tag{2}$$

where:

$n_{sh,\phi}$ - aditional axial force from shrinkage in the ϕ –direction of reinforcement;

$m_{sh,\phi}$ - additional bending moment from shrinkage, defined for the center of gravity for the ideal cross-section, in the ϕ-direction of reinforcement;

a_{s1}- bottom surface of reinforcement;

a_{s2}- top surface of reinforcement;

E_s - Young's modulus for steel;

ε_{sh} - shrinkage deformation;

e_{sh} - eccentricity of shrinkage forces.

For internal forces determined in this way due to shrinkage, the additional curvature induced by these forces is calculated at the analysed point without any additional external forces.

Then the new coefficient $k_{sh,\phi,c}$ is determined as follows:

$$k_{sh,\phi,c} = \frac{\kappa_{sh,\phi,c} - \kappa_{\phi}}{\kappa_{\phi}}, \tag{3}$$

where:

$\kappa_{sh,\phi,c}$ - curvature caused by external loads without the influence of shrinkage on the ϕ-direction reinforcement;
κ_{ϕ} - curvature caused by shrinkage (and reinforcement arrangement) without the influence of creep on the ϕ-direction reinforcement.

4.2.4. Distribution coefficient
Distribution coefficient is used to account for concrete hardening in tension.Its calculation is initially carried out with the assumption of a linear-elastic behaviour. In this case the maximum stresses in the concrete, taking into account shrinkage can be written in the following form:

$$\sigma_{max,\phi} = \frac{n_{\phi}+n_{sh,\phi}}{A_{phi,l}} + \frac{m_{\phi}-n_{\phi}\cdot\left(x_{\phi,l}-\frac{h}{2}\right)+m_{sh,\phi,l}}{I_{\phi,l}}\cdot\left(h-x_{\phi,l}\right). \quad (4)$$

where:
n_{ϕ} - axial force from the external load in the ϕ-direction of reinforcement;
$n_{sh,\phi}$ – an additional axial force from shrinkage in the ϕ-direction of reinforcement;
m_{ϕ} - bending moment from the external load in the ϕ-direction of reinforcement;
$m_{sh,\phi,l}$ - an additional bending moment from shrinkage in the ϕ-direction of reinforcement;
$x_{\phi,l}$ - height to the neutral axis of the uncracked cross-section in the ϕ-direction of reinforcement;
h - height of cross-section in ϕ-direction of reinforcement;
$A_{ph,i,l}$ - cross sectional area in uncracked stage in the ϕ-direction of reinforcement;
$I_{\phi,l}$ - moment of inertia of the cross-section in uncracked stage in the ϕ-direction of reinforcement.
Distribution coefficient can be written as:

$$\xi = 1-\beta\cdot\left(\frac{f_{ctm}}{\sigma_{max,\phi}}\right)^{n} \quad (5)$$

where:
β - coefficient depending on the effect of the load term or the influence of repetitive loads on average strain;

$$\xi = 0\ for\ \sigma_{max,\phi} \leq f_{ctm} \quad (6)$$

$\xi = 1$ - without taking into account hardening in tension.

5. Numerical analysis of floor slabs
The procedure for determining the required reinforcement by using analyzed software was carried out after accepting the same material and computational assumption given in Table 1,2. The calculations were carried out for two cases, without the influence of shrinkage and by modelling its impact. On this basis, the minimum reinforcement of the floor slab and the required reinforcement that should be adopted to fulfil standard conditions were obtained. The results of these calculations were related to the results obtained in the Robot software (reference software). After determining the provided reinforcement in the Robot Structural Analysis software for each calculation model (M1, M2, M3, M4) which correspond to the demand due to ULS and SLS and at regular arrangement, the obtained reinforcement was introduced to the other two software models: ABC Plate and Dlubal RFEM. This procedure aimed to compare the deflection results in the next step. When analyzing the serviceability limit state, attention should be paid to cracking of the reinforced concrete floor slab. The formation of a crack is connected with reaching the ultimate tensile strength. According to the current standard (PN-EN 1992-1-1), the crack width should be limited to a level that does not impair the functioning or durability of the structure and nor make the appearance of the element unacceptable. In order to unify the method of selection of additional reinforcement against excessive cracks, the limit admissible crack width in all the analyzed software was set as $w_{k,max} = 0.3mm$.

The Dlubal RFEM and Robot Structural Analysis software give the obtained results depending on the analyzed plane of the floor slab (upper or lower) and the main directions of the determined reinforcement. The ABC Plate software is limited to only providing results for the upper and lower level of the floor slab. One of the main conditions of the serviceability limit state that all structural elements must fulfill, and above all, floor slabs, is that the permissible deformations, including deflections, are not exceeded. When designing, the analysis of deflections of cracked floor slab determines their thickness and its reinforcement. The basic parameter which should be defined is the value of permissible deflection.

Table 4. The maximum deflection of the analysed calculation models.

Model	Span L (mm)	Limit criterion	maximum deflection f (mm)
M1	7000		28
M2	7000		28
M3	7000	L/250	28
M4	8000		32

5.1. Algorithms for calculation of deflections in floor slabs

5.1.1. ABC Plate software

- Local stiffness method

This method consists of calculating the curvature in cross-sections that are densely distributed along the element, and then on defining the deflection by numerical integration. Static calculations are performed for a particular load scheme with the same assumptions made as for the elastic plate. As a result, the values of internal forces in the form of bending moments in each finite element of the structure are obtained. On the basis of the previously adopted reinforcement and internal acting forces, the flexural stiffness of the cross-section is determined independently in each direction of reinforcement. The algorithm recalculates the floor slab for the assumed load scheme, however, it takes into account the determined flexural stiffnesses in both directions in the finite elements. As a result, a new distribution of internal forces is obtained in the form of bending moments, and a re-determination of independent flexural stiffnesses is obtained with the assumed reinforcement, but with new values of moments. The algorithm is repeated until the difference between consecutive values of maximum deflections is minimal.

5.1.2. Robot Structural Analysis software

- Method of updating stiffness (FEA)

In this algorithm, after determining the required reinforcement for the ULS and SLS standard requirements, the software determines the appropriate stiffness for all finite elements of the structure, independently in both directions. Then, the weighted average of both stiffness values is determined, where the weight is the ratio of the acting bending moments.

$$B_{x,y} = c_f \cdot B_x + \left(1 - c_f\right) \cdot B_y \tag{7}$$

where:

B_x, B_y - real stiffnesses calculated for two reinforcement directions;

c_f – weight factor, calculated from the formula, depending on the value of the ratio of bending moments.

The effect of using the above formulas is the fact that in the case of a large difference between bending moments in both directions, the stiffness in the orientation of the action of the larger one is assumed in the calculations. The ratio of elastic stiffness D to the average weighted stiffness B is calculated according to the formula below and should be less than one.

$$\left(\frac{D}{B}\right)_{x,y} = \frac{D}{c_f \cdot B_x + (1-c_f) \cdot B_y} \tag{8}$$

where:

$$D = E \cdot I \tag{9}$$

The final stiffness ratio, which is taken to determine the total deflections of the analyzed floor slabs, is obtained from averaging the average stiffness ratio and stiffness ratio of the element, where extreme values of bending moment were determined in any direction. These values have additional weights with coefficients of 0.25 and 0.75, respectively.

$$\left(\frac{D}{B}\right) = 0{,}25 \cdot \left(\frac{D}{B}\right)_{x,y} + 0{,}75 \cdot \frac{D}{B(M_{max})} \tag{10}$$

5.1.3. Dlubal RFEM software

- Algorithm of the *RF CONCRETE DEFLCT* module

For elements subjected to bending, the deformation can be determined as follows:

$$\alpha = \xi \cdot \alpha_{II} + (1-\xi) \cdot \alpha_I \tag{11}$$

where:
α - is a deformation parameter;
α_I - the value of deformation in uncracked stage;
α_{II} - the value of the deformation in the cracked stage;
ξ - distribution coefficient, used to account for hardening in tension.

$$\xi = 1 - \beta \cdot \left(\frac{\sigma_{sr}}{\sigma_s}\right)^2 \tag{12}$$

where:
$\xi = 0$ for uncracked sections; $\xi = 1$ for fully cracked section;
β - coefficient depending on the influence of the term load or the impact of the repetitive load on the average strain;
σ_s - stresses in tensile reinforcement, calculated on the assumption that the cross-section is fully cracked;
σ_{sr} - stress in tensile reinforcement, calculated on the assumption that the cross-section is fully cracked, which is caused by the load initiating the cracking.
According to eq. (*11*), the deformation ε_{Φ}- and the curvature of the deflection $\kappa_{\varphi}$are determined by interpolation between a cracked and uncracked stage:

$$\varepsilon_{\phi} = \xi_{\phi} \cdot \varepsilon_{\phi,II} + \left(1-\xi_{\phi}\right) \cdot \varepsilon_{\phi,I} \tag{13}$$

$$\kappa_{\phi} = \xi_{\phi} \cdot \kappa_{\phi,II} + \left(1-\xi_{\phi}\right) \cdot \kappa_{\phi,I} \tag{14}$$

5.2. Comparative analysis of calculated deflections

After determining the provided reinforcement in the Robot Structural Analysis reference software, the obtained results, in the form of bars of a fixed diameter and at an assumed spacing, were modeled in the other two calculation software models: ABC Plate and Dlubal RFEM.
For each considered computational model, two cases were analyzed: including shrinkage and without taking shrinkage into account. Then, individually in all the programs, the maximum deflection value of the floor slab was determined.
Within one software model, the available calculation methods were used, which differed depending on the system selection.
Based on the results obtained in all the analyzed calculation models, can be noticed that deflections in the *Robot Structural Analysis* program are smaller when using the non-elastic method (with updating the stiffness matrix) than in the simplified elastic method when assuming an average elasticity of the

material of 20-40% . This is due to the fact of the taken into account partial hardening in tension. These differences are clearly visible in the M4 model, where the deflection values, taking into account the update stiffness, are half as much as in the case of the equivalent stiffness method. This results from the use of additional spring supports in the form of binders, as seen in Table 5.

Table 5. Comparison of elastic and non-elastic methods to determine deflection.

Model M4	Elastic Method	Non-elastic Method
maximum deflection (mm)	31,33	14,71

Particular attention should be paid to the comparison of deflection results in various computer programs, in particular in relation to those obtained with regard to shrinkage. In simple models M1-M3 (figure 5,6,7), where there are no additional cross-beams, the convergence of the results of Robot Structural Analysis and *ABC Plate* software can be seen.

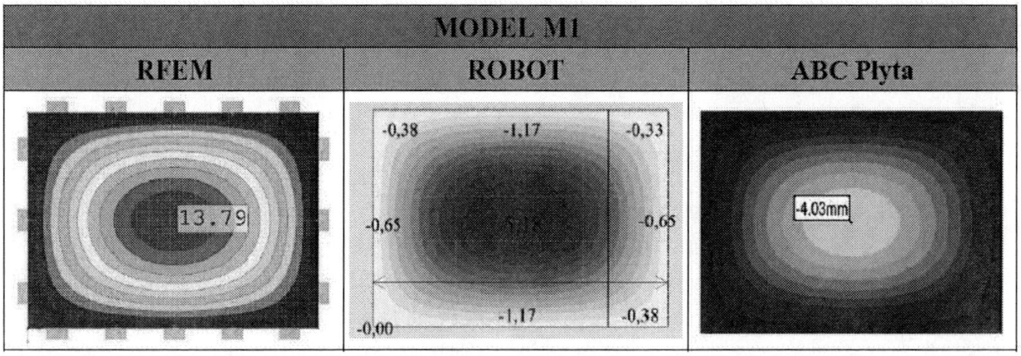

Figure 5. Maps of deflection considering the shrinkage effect for the M1 model obtained in the applied software.

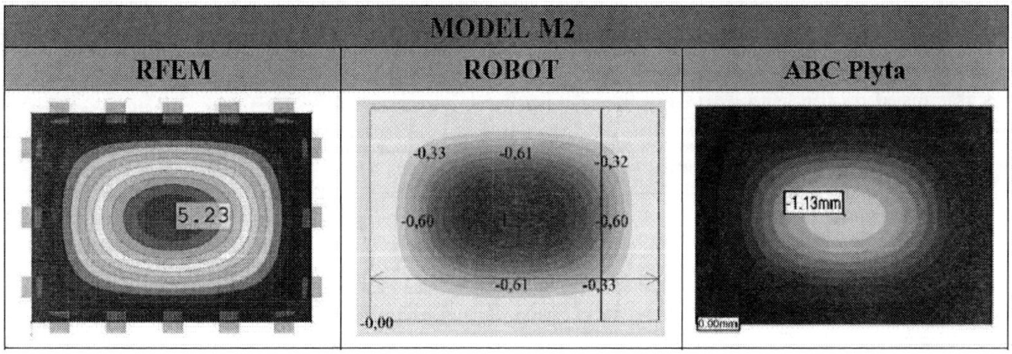

Figure 6. Maps of deflection considering the shrinkage effect for the M2 model obtained in the applied software.

In relation to the *Dlubal RFEM* software, these values differ significantly, which is caused by the lack of a clear way to define shrinkage deformations.
There are similar deflection values for Dlubal RFEM and ABC Plate software in the M4 model (figure 8), whereas the determined deflection in *Robot Structural Analysis* differs by 17-28%. This difference is caused by a different way of modelling the binder, and also by its impact on the analyzed floor slab.

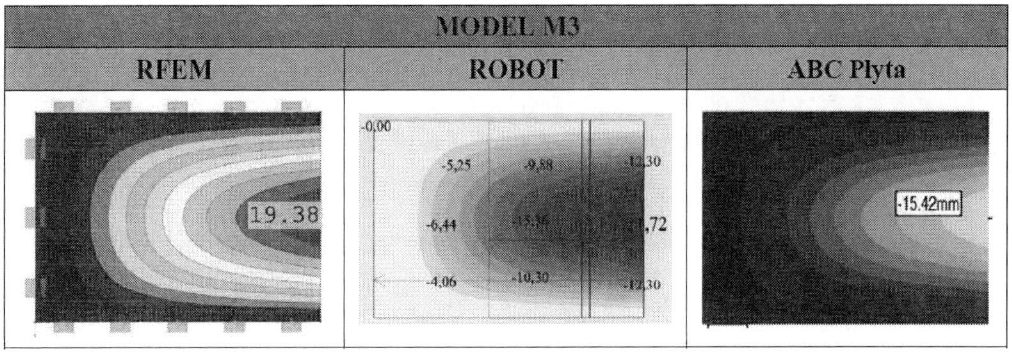

Figure 7. Maps of deflection considering the shrinkage effect for the M3 model obtained in the applied software.

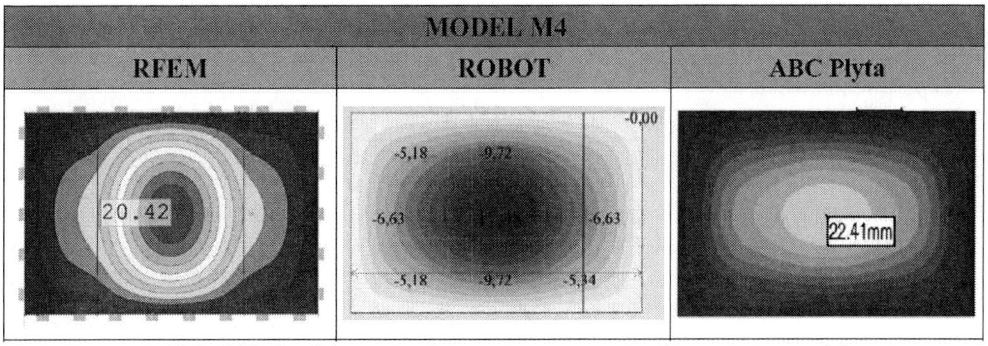

Figure 8. Maps of deflection considering the shrinkage effect for the M4 model obtained in the applied software.

6. Summary

By performing calculations with three leading computer programs (*Robot Structural Analysis, ABC Plate and Dlubal RFEM*), which are available in design offices in Poland, many problems resulting from their use were found. Despite the fact that these software programs are based on FEA and have the same iterative algorithm of the incremental Newton-Raphson method used in non-linear analysis, the obtained deflection results differ significantly. This is all the more interesting because the analyzes were carried out in accordance with the procedure described in the applicable standards. An important issue that appeared during the modeling of a reinforced concrete slab was the way of describing rheological phenomena and their inclusion by various computer programs. The effect of shrinkage and creep has a significant impact on the cracking and deflections of concrete elements. Large value of shrinkage and creep which is experienced by materials of slab structure will be one of the factor which cause enormous deflection. While the analysis of the impact of creeping proved to be consistent for all the software programs (it consisted of determining the creep coefficient, which then reduced the modulus of elasticity of the concrete), so the adoption of shrinkage effects was definitely different. The *ABC Plate* software takes into account the effect of shrinkage, but only when defining the deflection curve of elements. *Robot Structural Analysis* software does not take into account the influence of shrinkage in the designing of reinforced concrete elements, even in the range of the serviceability limit state, and its effect in reinforced concrete structures can be taken to be equivalent to a temperature reduction of 15°C. However, *Dlubal RFEM* software takes the impact of shrinkage on the designed concrete element into full account, both in the area of linear and non-linear analysis.

Differences in results are appeared when determining the minimum reinforcement in computational programs. This resulted from different interpretations of the standard recommendations, where it was

IOP Conf. Series: Materials Science and Engineering **586** (2019) 012031 doi:10.1088/1757-899X/586/1/012031

not specified when exactly the minimum reinforcement should be taken into account when analyzing the serviceability limit state. In the *ABC Plate* algorithm, each numerical calculation is defined as an exact analysis, whereas *Dlubal RFEM* considers it only those made using the non-linear method. Computer programs use linear and nonlinear analyzes to determine the effects of various factors. When analyzing reinforced concrete slabs, the material nonlinearity taking into account the behavior of specific materials during loading is decisive. The impact of the use of nonlinear analysis is most clearly visible in the *Dlubal RFEM* software and its *RF CONCRETE DEFLECT* and *RF CONCRETE NL* modules. This software was the only one of the analyzed programs which, has an algorithm that uses non-linear analysis that includes shrinkage. The use of this type of analysis resulted in significantly higher crack values and a different distribution and direction of the cracks (rotational cracking model). Differences in the applied analyzes can also be observed in the determination of deflections, where lower results were obtained in the case of nonlinearities, due to the consideration of hardening in tension with residual concrete strength. Summing up, it should be stated that the conducted numerical analyses, which are related to the determination of deflections in various models of floor slabs, require correction and confirmation by experimental methods on real structures. However, the numerical analysis carried out by the authors, which takes into account many material aspects, is necessary for the correct estimation of slab floor deflections. The next stage of the analyzes of deflections of reinforced concrete floor slabs will be experimental research.

7. References

[1] Grajek K 2016 ABC Plate description program – Deflection of floor slab (in Polish)

[2] Autodesk 2016 Robot Structural Analysis – user guide *Autodesk*

[3] Ambroziak A, Kłosowski P 2016 Design steel and reinforced concrete structure (in Polish) ed. *Gdansk University of Technology* pp 1-178

[4] Dlubal Software 2013 RF-CONCRETE Surfaces, reinforced concrete design – program description

[5] Neville A 2012 Properties of concrete (in Polish) ed. *Association of cement producers*

[6] Starosolski W 2013 Computer modelling of concrete engineering structures. Selected issues (in Polish) vol. 1 ed. *Silesian Technical University*

Evaluation of the thermal performance of composite insulated panels with metallic skin through steady-state numerical analysis – Part 1

A M Măgurean[1]
[1] Department of Civil Engineering and Management, Technical University of Cluj-Napoca, 25 Gh. Barițiu Street, Cluj-Napoca, 400027, Romania

E-mail: ancuta.magurean@cif.utcluj.ro

Abstract. The energy performance of the buildings is a wide spreaded area on which many research is done nowadays. However, because of the relative new large scale interest on it, there still are subjects that have limited research data available. One of this topics is the thermal performance of composite insulated panels with metallic skin, usually a closure solution for industrial halls, production buildings, commercial buildings or even office buildings. Because of the mandatory request of the Energy Performance Certificates (EPCs) in the European countries and in Romania as well, is highlighted that is a lack of informations regarding the thermal behaviour of this type of technical solution and also, synthetic working data for building energy auditors, as linear thermal transmission coefficients values (Ψ-values) for the usually constructive details that compose a regular industrial hall is missing. Therefore, this paper aims to provide an insight about the thermal behaviour of this type of solution, by offering results of Ψ -values for four different dimensions of sandwich panels, which represents appropriate data for a thermal bridges catalogue that may be used directly by the building energy auditors.

1. Introduction

Due to the high rise of CO_2 emissions at global level – figure 1, in the last years a high interest for the energy efficiency of the buildings exists, which is growing.

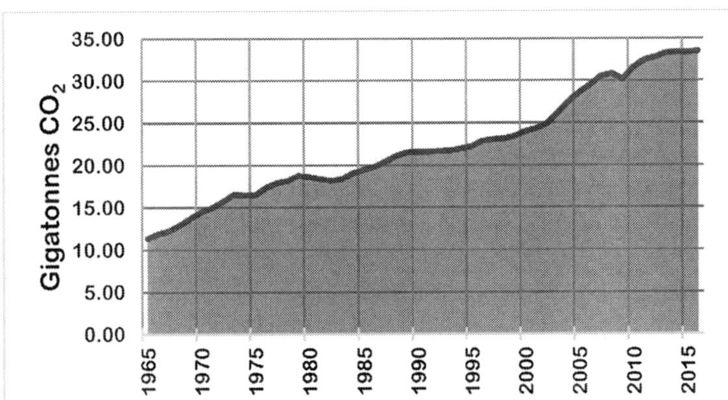

Figure 1. Global CO_2 emissions in the period 1965 - 2016 (based on statistical data from [1])

One of the measures taken at large scale in the European Union countries is the introduction of Energy Performance Certificates (EPCs), whose role is to prepare the market, to increase the awareness, respectively to inform the occupants of the building of how the energy is consumed in the building, and also to build the energy profile of the country's built-up fund by centralizing EPCs results in the national data bases (as is the case in Romania, for example). However, at least as regarding Romania, the number of specialists who are doing this type of analysis is relatively small, and also the methodology of calculation is limited. Although, especially for residential buildings and for usually constructive details

IOP Conf. Series: Materials Science and Engineering **586** (2019) 012032 doi:10.1088/1757-899X/586/1/012032

there are many detailed studies [2], [3], [4] regarding the heat flux transfer under certain conditions and the impact of the thermal bridges that occurs on the thermal performance of the constructive detail, there are relative new technical envelope solutions for which thermotechnical scientifical studies are limited [5], aspect which may have as effect that the energy performance certificate results to be inaccurate.
As regarding state of the art for the investigated subject in this paper, in order to reduce energy consumption, worldwide research is carried out in an integrated manner, starting from the physical and chemical properties of materials. Thus, in Greece, Mastrapostoli et al. [6] investigated sandwich panels on which fluorocarbon-based "cold" coatings were applied, reflecting solar radiation and rejecting solar gains at the opaque exterior surfaces of the building. After application by spraying, these materials reduce the temperature at the surface of the panel, also reducing the indoor temperature inside the hall or the energy requirement to maintain a constant temperature inside. For a temperate climate, these materials can be used in industrial halls that have high internal gains caused by the technological processes, with the intended effect during the summer to reduce the cooling demand of the building.
Another direction of research in this area is represented by the use of phase change materials (PCMs), used in wall panels to increase the energy storage capacity. An experiment conducted in Spain by Castellon et al. [7] studied the behaviour of microencapsulated phase change materials (Micronal BASF) in conventional sandwich panels made of metallic outer skin and injected polyurethane foam, in order to increase thermal inertia of panels, therefore to reduce energy demand in operation of buildings. The study has shown the possibility of increasing the thermal inertia of the panels by adding PCMs during the manufacture process of sandwich panels, with the possibility of improving the heat transfer behaviour by increasing thermal inertia of the panels.
Carbonari et al.[8] studied sandwich panels with an air layer between the PCM (encapsulated in rigid containers) and the metallic layer, and achieved energy behaviour similar to high thermal mass walls.
Regarding composite insulated panels with metallic skin, under the assumption that is not excluded that studies have been carried out, the author did not find research undertaken from a thermal performance point of view or thermotechnical studies related to this constructive solution, either at national or international level, highlighting the necessity of studies in this direction.

Nomenclature

T	temperature (K)	Φ_{2D}	bidimensional heat flux (W)
λ	thermal conductivity (W/m·K)	ΔT	temperature difference (K)
q	heat flux density (W/m^2)	δ	thickness (m)
Ψ	linear thermal transmission coefficient (W/m·K)	h	height (m)
χ	point thermal transmission coefficient (W/K)	L_i, L_e	dimension (interior / exterior) (m)
U	unidirectional thermal transmittance (W/m^2·K)	S_i	area (interior overall dimensions) (m^2)
h_i, h_e	superficial heat transfer coefficient (interior / exterior surface) (W/m^2·K)	R	unidirectional thermal resistance (m^2·K/W)
		R'	corrected thermal resistance (m^2·K/W)
L_{2D}	bidimensional thermal coupling coefficient (W/K)	r	the ratio of corrected thermal resistance to unidirectional thermal resistance (-)
$L_{2D.a}$	bidimensional thermal coupling coefficient calculated with EN ISO 10211: 2017 (W/K)		

2. Theoretical considerations

2.1. Theoretical considerations regarding steady-state bidimensional heat flux calculation through FEM analysis

In order to study the influence on the heat flux transfer of geometrical and material thermal bridges that occurs in several constructive details of composite thermal insulated panels with metallic skin, specific for walls and roofs of non-residential buildings, FEM steady-state numerical analysis is undertaken.

For the steady-state FEM analyzes, software THERM 7.2 was used, which is a freeware elaborated within University of California, Berkeley, funded by U.S. Department of Energy (DOE). In the undertaken study, the heat flux perturbation caused by the existence of thermal bridges, either caused by the geometry or by the material inhomogeneity, is numerically calculated for several common constructive details, which are currently representing the closures of an industrial hall.
The partial differential equation solved in this case is the Laplace type equation or the heat equation without internal heat sources in stationary regime:

$$\lambda \cdot \left(\frac{\partial^2 T}{\partial x^2} + \frac{\partial^2 T}{\partial y^2}\right) = 0 \tag{1}$$

In the mathematical model of the software is also used Fourier's law:

$$q = - \lambda \cdot \text{grad } T \tag{2}$$

2.2. Numerical analysis conditions used for the constructive details

2.2.1. Geometry. All lengths were measured by reference to the total interior dimensional system. Geometrical 2D modeling of the analyzed details was performed in accordance with the requirements of EN ISO 10211: 2017 [9] The analyzed details were sectioned with adiabatic limits at d_{min} = 1 m towards the position of the thermal bridge.
Regarding the details of the wall coupling with slab on the ground, the geometrical modeling was made considering the inferior limit of the model being located at 20 m below the ground level.

2.2.2. Material properties. Design values for the thermal conductivities of the materials used in the constructive details studied in this paper are presented in table 1.

Table 1. Properties of the used materials for different constructive elements

Building element	Used construction materials	λ (W/mK)	Reference source
Wall / Roof sandwich panels	Galvanized steel	58	[10]
	Rigid cellular polyurethane	0.038	[11]
	Rubber band	0.25	[12]
Slab on ground	Reinforced concrete	1.74	[10]
	Gravel	0.70	[10]
	Soil layer	2.00	[10]
	Extruded polystyrene	0.042	[10]
4-16-4 window, without applied films	PVC frame	0.17	[12]
	Float glass	1.00	[12]
	EPDM seal	0.25	[12]
	Silicagel (desiccant)	0.13	[12]
	Aluminum	160	[12]
	Rigid polyurethane (seal)	0.25	[12]

2.2.3. Boundary conditions. For the most of the analyzed details, Fourier type boundary condition is applied, which involves knowledge of the temperature of the fluid (air) that is surrounding the solid (the building element). Therefore, the analyzes are made by using T_i = 20 °C and T_e = -18 °C for the indoor air temperature, respectively for the outdoor air temperature. Fourier condition is obtained by using the law of conserving the heat flux density, therefore for the surface limit of the element superficial heat transfer coefficients are used, with the values presented in table 2.

Table 2. Superficial heat transfer coefficient values

Superficial heat transfer coefficient type (W/m^2K)	Direction of thermal flow		
	Upward	Horizontally	Downward
h_i	10	8	6
h_e	24	24	24

2.3. Analytical relations for the calculation of heat transfer coefficients

The results obtained through numerical analysis are integrated in analytical calculation, in order to determine the linear thermal transmission coefficients values (Ψ-values), which regularly are used by building energy auditors in heat losses calculation, and which often are available in thermal bridges catalogues for many type of constructive details.

The analytical relations used in order to obtain linear thermal transmission coefficients values are those indicated in EN ISO 10211:2017 [9].

$$\Psi = L_{2D} - \Sigma U_j l_{j\cdots} \quad (W/m{\cdot}K) \tag{3}$$

As regarding the analysed details of the slab on ground, the used analytical relation is (4).

$$\Psi_g = L_{2D} - h_{wall} \cdot U_{wall} - L_{2D,a} \ldots (W/m{\cdot}K) \tag{4}$$

3. Thermal performance of common constructive details of composite insulated panels with metallic skin

The studied details include regular constructive details of the building envelope for walls and roof, the interaction between them, the interaction between window and wall, with a comprehensive model for the air cavities of the window frame and glass, the interaction between wall and ground, also with a comprehensive model for the slightly ventilated grooves with small cross section, which are constructively specific for the wall panel's joint with the slab on ground.

For each type of constructive detail, four different dimensions were analyzed, which cover the main regular dimensions usually found on the market. The analyzed details followed by the obtained results, both numerically and by using the above presented analytical relations are presented below.

3.1.Detail of wall panels joint to the corner

The geometry of the detail is shown in figure 2 and the results, as isothermal surfaces and synthetic values of linear thermal transfer coefficients, are shown in figure 3, respectively table 3.

Table 3. Linear thermal transmission coefficients for the wall panels joint detail to the corner

Detail identifier	$\delta_{insulation}$ (mm)	Φ_{2D} (W)	**R** (m^2K/W)	**R'** (m^2K/W)	L_i (m)	**Ψ** (W/mK)
Wall panel (left)	60	23.312	1.746	1.636	1.00	0.039
Wall panel (right)	60	24.881	1.746	1.533	1.00	0.080
Wall panel (left)	80	18.332	2.272	2.080	1.00	0.041
Wall panel (right)	80	19.868	2.272	1.920	1.00	0.081
Wall panel (left)	100	15.137	2.799	2.516	1.00	0.040
Wall panel (right)	100	16.611	2.799	2.292	1.00	0.079
Wall panel (left)	120	23.312	3.325	2.932	1.00	0.039
Wall panel (right)	120	24.881	3.325	2.640	1.00	0.080

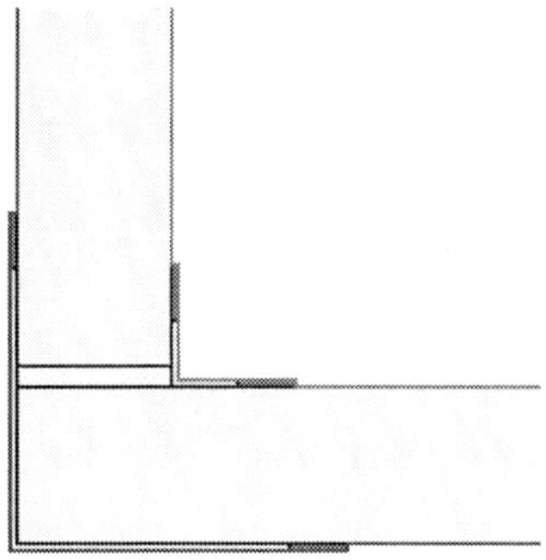

Figure 2. Detail of geometric modelling wall panels joint to the corner

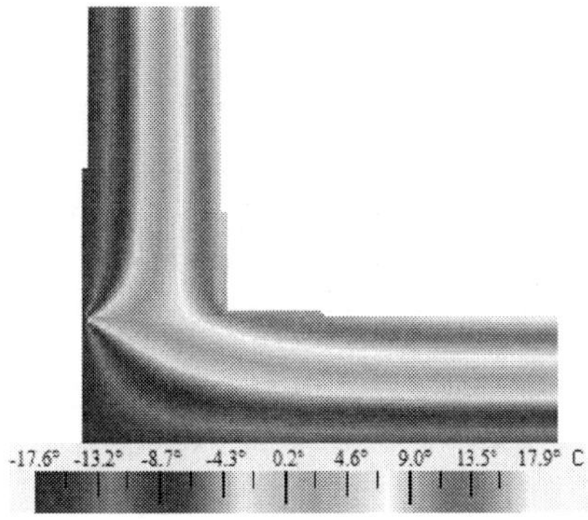

Figure 3. Isothermal surfaces for detail 3.1

3.2. Details of wall panels horizontal joints

3.2.1.Detail of wall panels horizontal joint. Specifically to the composite panels with metallic skin are joints between wall panels, usually at a distance of 1.00 m, joints that can be horizontal or vertical. In figure 4 the geometric detail of a horizontal joint is shown and the simulation results are presented graphically in figure 5, respectively in numerical form in table 4.

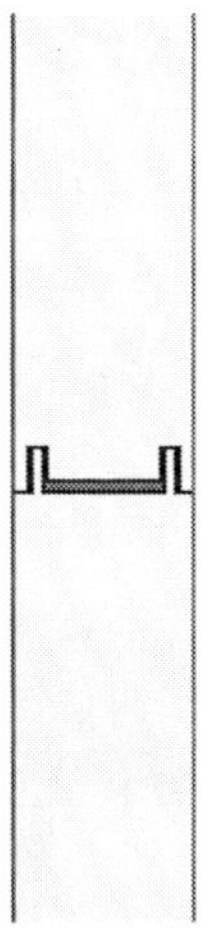

Figure 4. Detail of geometric modelling joint between wall panels

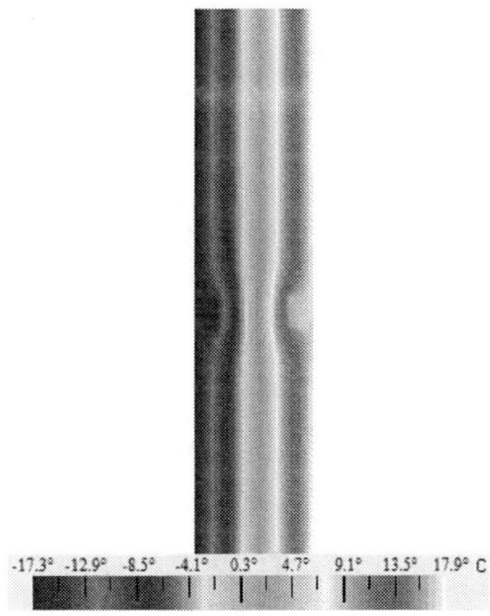

Figure 5. Isothermal surfaces for detail 3.2.1

Table 4. Linear thermal transmission coefficients for the horizontal joint detail between wall panels

Detail identifier	$\delta_{insulation}$ (mm)	Φ_{2D} (W)	**R** (m^2K/W)	**R'** (m^2K/W)	L_i (m)	Ψ (W/mK)
Wall panel (down)	60	22.45	1.746	1.693	1.00	0.018
Wall panel (up)	60	22.38	1.746	1.698	1.00	0.016
Wall panel (down)	80	17.17	2.272	2.213	1.00	0.012
Wall panel (up)	80	17.11	2.272	2.221	1.00	0.010
Wall panel (down)	100	14.11	2.799	2.693	1.00	0.014

Wall panel (up)	100	14.05	2.799	2.705	1.00	0.012
Wall panel (down)	120	12.03	3.325	3.160	1.00	0.016
Wall panel (up)	120	11.98	3.325	3.173	1.00	0.014

3.2.2.Detail of wall panels horizontal joint – hidden connection. For the "hidden" area an equivalent thermal conductivity of the non-ventilated air cavity was calculated.

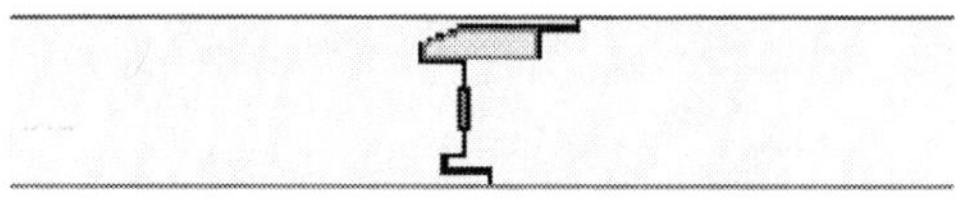

Figure 6. Detail of geometric modelling joint between wall panels

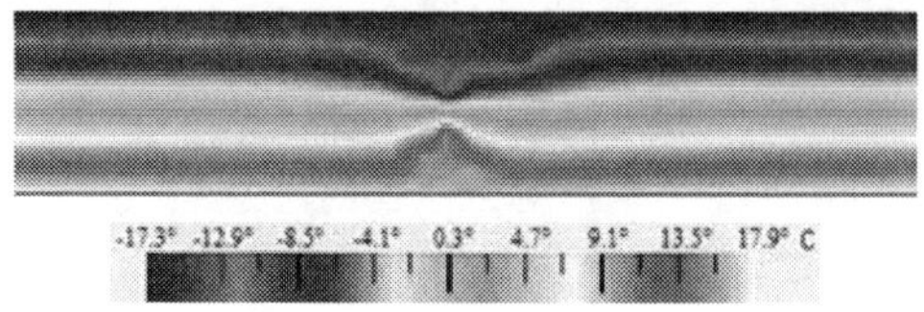

Figure 7. Isothermal surfaces for detail 3.2.2

Table 5. Linear thermal transmission coefficients for the hidden horizontal joint between wall panels

Detail identifier	$\delta_{insulation}$	Φ_{2D}	R	R'	L_i	Ψ
	(mm)	(W)	(m^2K/W)	(m^2K/W)	(m)	(W/mK)
Wall panel (left)	80	17.52	2.272	2.169	1.00	0.021
Wall panel (right)	80	17.59	2.272	2.170	1.00	0.023

It is noticed that the results obtained for panels with a thickness of 80 mm have higher Ψ-values in relation to the first type of joint, but these values are also low. For other thicknesses of this type of joint, no other numerical simulations were made -due of the lack of technical details.

3.3. Detail of wall panel wall joint with roof panel

This detail represents the connection between a wall panel and a roof panel with a roof slope of min. 7°. The thermal section was simulated by using thermal insulation material with $\lambda = 0.05$ (W/mK).

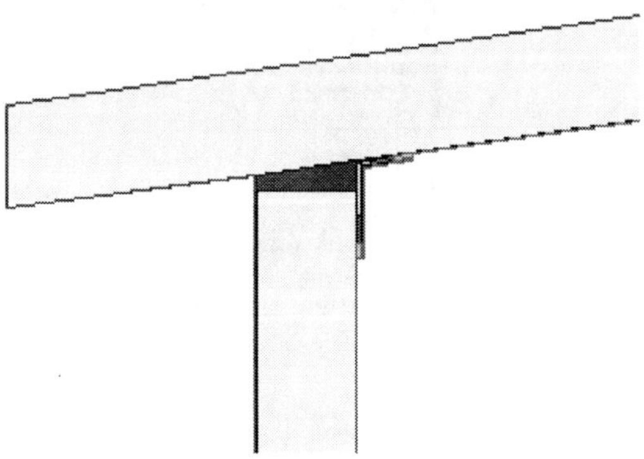

Figure 8. Detail of geometric modelling joint between wall panel and roof panel

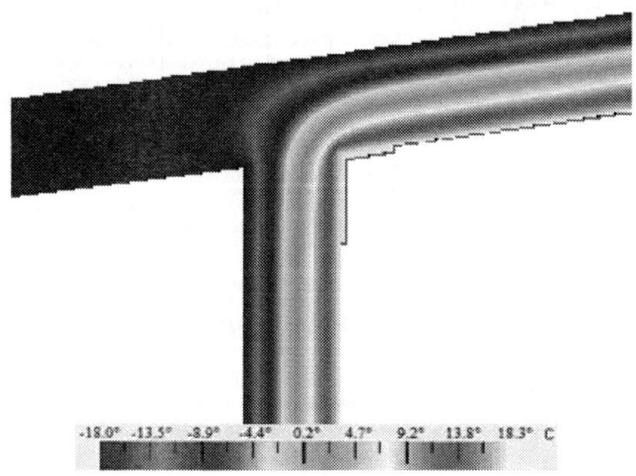

Figure 9. Isothermal surfaces for detail 3.3

In table 6 it can be noticed that the values obtained for the linear thermal transfer coefficients are very low. For this type of constructive detail, the effect of geometric thermal bridge and material is almost inexistent.

Table 6. Linear thermal transmission coefficients for the joint between wall panel and roof panel

Detail identifier	$\delta_{insulation}$ (mm)	Φ_{2D} (W)	R (m^2K/W)	R' (m^2K/W)	L_i (m)	Ψ (W/mK)
Roof panel	60	22.74	1.721	1.697	1.00	0.009
Wall panel	60	22.34	1.746	1.737	1.00	0.003
Roof panel	80	17.74	2.247	2.175	1.00	0.011
Wall panel	80	17.24	2.272	2.250	1 .00	0.004
Roof panel	100	14.37	2.774	2.685	1.00	0.013
Wall panel	100	14.06	2.799	2.759	1.00	0.005
Roof panel	120	12.17	3.300	3.171	1.00	0.015
Wall panel	120	11.89	3.325	3.263	1.00	0.006

3.4. Detail of wall panel joint with PVC window frame

The air cavities inside the PVC profile are considered as non-ventilated air cavities (figure 10). The simulation of this constructive detail reveals that Ψ-value related to the wall is relatively high and the influence of the thermal bridges on the wall panel increases with the increase of the thermal insulation thickness, as can be seen in table 7.

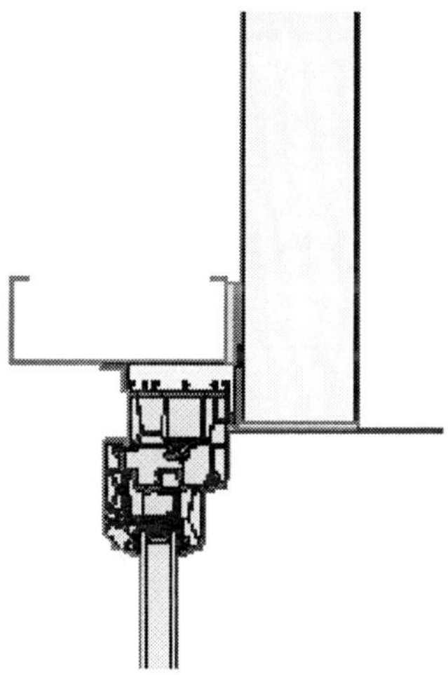

Figure 10. Detail of geometric modelling joint between wall panel and window frame

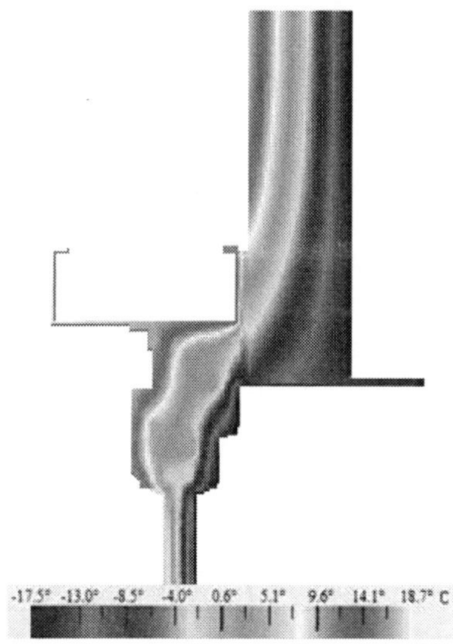

Figure 11. Isothermal surfaces for detail 3.4

Table 7. Linear thermal transmission coefficients for the joint between wall panel and window

Detail identifier	$\delta_{insulation}$ (mm)	Φ_{2D} (W)	ΔT (K)	R (m^2K/W)	R' (m^2K/W)	L_i (m)	Ψ (W/mK)
Wall panel	60	25.66	38	1.746	1.340	1.00	0.157
Wall panel	80	21.30	38	2.272	1.614	1.00	0.162
Wall panel	100	18.60	38	2.799	1.849	1.00	0.166
Wall panel	120	16.77	38	3.325	2.051	1.00	0.169

3.5. Detail of wall panel joints with slab on ground

3.5.1. Detail of wall panel joint with slab on ground – type 1. It should be noted that all interstitial areas in this constructive detail with an inward or outward slot less than 10 mm were considered to be poorly ventilated air cavities (figure 12).
According to the obtained numerical and analytical results, it can be concluded that the joint of the wall panel with the slab on ground and the foundation beam, with a 10 cm overlap with the concrete slab and without the insulation of the base, is insufficient given the high Ψ-values obtained in table 8.

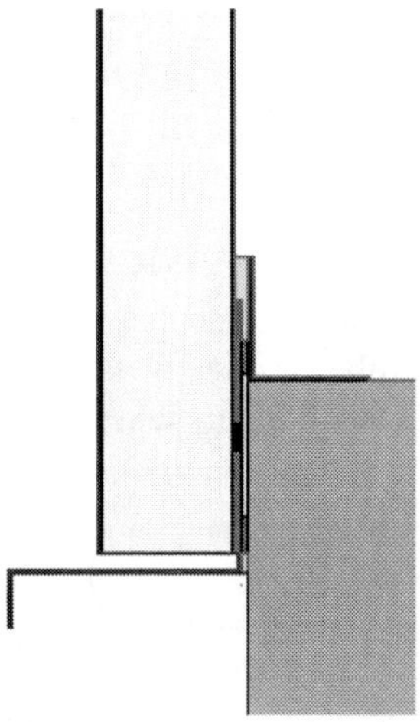

Figure 12. Detail of geometric modelling joint between wall panel and slab on ground

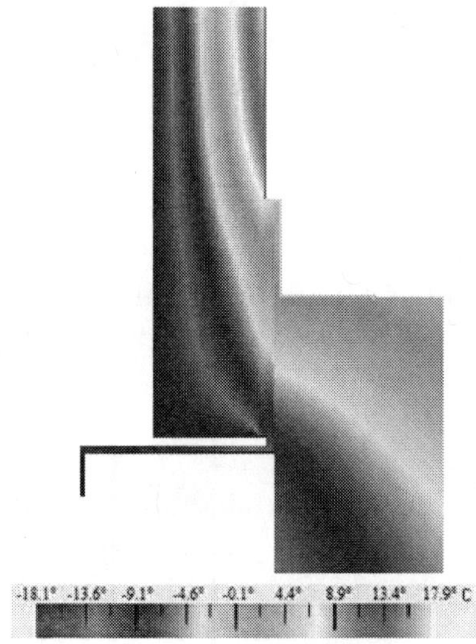

Figure 13. Isothermal surfaces for detail 3.5.1

Table 8. Linear thermal transmission coefficients for wall panel joint with slab on ground – type 1

Detail identifier	$\delta_{insulation}$	Φ_{2D}	ΔT	L_{2D}	h_w	U_w	$L_{2D.a}$	Ψ_{detail}	Ψ
(-)	(mm)	(W)	(K)	(W/K)	(m)	(W/m²K)	(W/K)	(W/mK)	(W/mK)
Wall panel	60	36.137	38	0.951	1.00	0.573	2.011	1.124	0.343
Slab on ground	--	106.104	38	2.792	-	-			0.781
Wall panel	80	31.600	38	0.832	1.00	0.440	2.011	1.142	0.364
Slab on ground	--	105.995	38	2.789	-	-			0.778
Wall panel	100	28.781	38	0.757	1.00	0.357	2.011	1.154	0.378
Slab on ground	--	105.931	38	2.788	-	-			0.776
Wall panel	120	26.861	38	0.707	1.00	0.301	2.011	1.163	0.388
Slab on ground	--	105.890	38	2.787	-	-			0.775

3.5.2. Detail of wall panel joint with slab on ground – type 2. For another type of detail, the thermal performance was evaluated with an overlap of the wall panel on the reinforced concrete element of 6 cm with a simpler type of fastening also – as can be seen in figure 14.

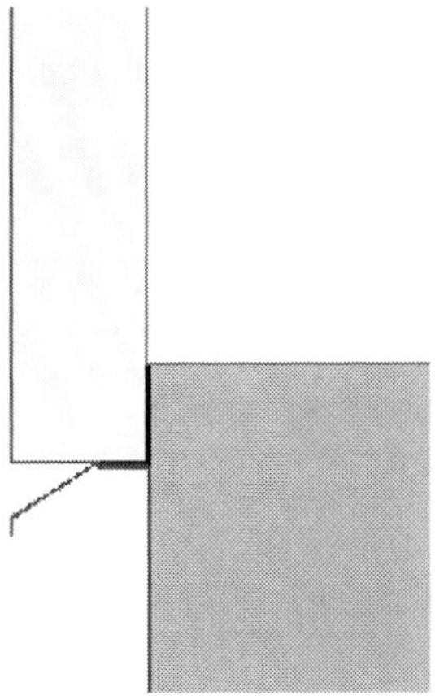

Figure 14. Detail of geometric modelling joint between wall panel and slab on ground

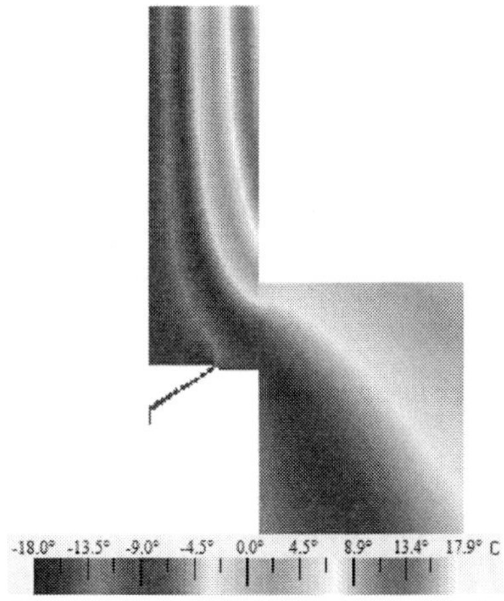

Figure 15. Isothermal surfaces for detail 3.5.2

Table 9. Linear thermal transmission coefficients for wall panel joint with slab on ground – type 2

Detail identifier	$\delta_{insulation}$	Φ_{2D}	ΔT	L_{2D}	h_w	U_w	$L_{2D.a}$	Ψ_{detail}	ψ
(-)	(mm)	(W)	(K)	(W/K)	(m)	(W/m²K)	(W/K)	(W/mK)	(W/mK)
Wall panel	60	30.624	38	0.806	1.00	0.573	2.002	1.134	0.233
Slab on ground	60	110.310	38	2.903	-	-			0.900
Wall panel	80	25.929	38	0.682	1.00	0.440	2.002	1.141	0.242
Slab on ground	80	110.248	38	2.901	-	-			0.899
Wall panel	100	23.022	38	0.606	1.00	0.357	2.002	1.146	0.249
Slab on ground	100	110.216	38	2.900	-	-			0.898
Wall panel	120	21.049	38	0.554	1.00	0.301	2.002	1.151	0.253
Slab on ground	120	110.195	38	2.900	-	-			0.897

Compared to the first studied model, by analysing results from table 9, an increase in the thermal flux for this type of detail is observed (with an increase of the linear thermal transfer coefficients related to the slab on ground, respectively a reduction of those related to the wall panels), suggesting the influence of the length of the overlapping of the panel on the reinforced concrete element. The same tendency, to increase the effect of the thermal bridge in the wall as the thickness of the thermal insulation in the panel increase is noticeable.

3.6. Detail of roof panel joint at ridge

In the assumption of ensuring the continuity of thermal insulation, applied in situ after the positioning of the panels at the projected slope – figure 16, it can be noticed in table 10 that the values of linear thermal transmission coefficients are negligible.

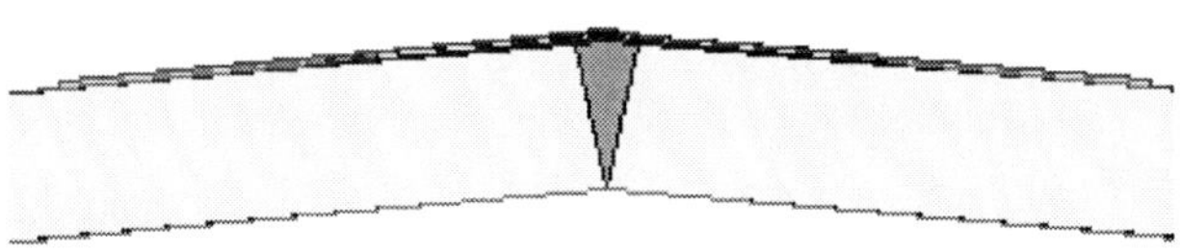

Figure 16. Detail of geometric modelling joint between roof panels at ridge

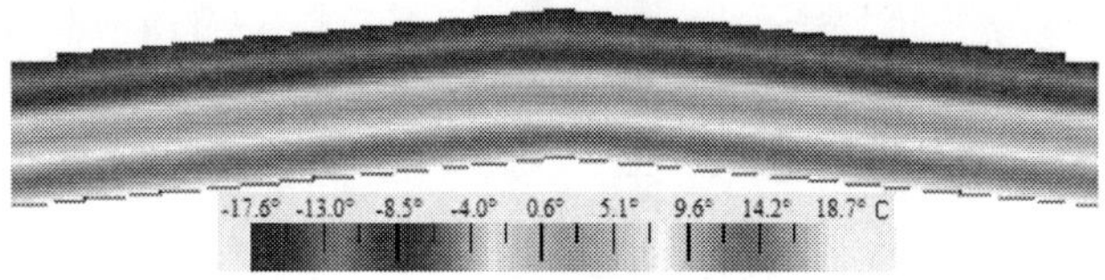

Figure 17. Isothermal surfaces for detail 3.6

Table 10. Linear thermal transmission coefficients for roof panel joint at ridge

Detail identifier	$\delta_{insulation}$ (mm)	Φ_{2D} (W)	R (m^2K/W)	R' (m^2K/W)	L_i (m)	Ψ (W/mK)
Roof panel (left)	60	23.52	1.721	1.728	1.00	- 0.002
Roof panel (right)	60	23.52	1.721	1.728	1.00	- 0.002
Roof panel (left)	80	18.06	2.248	2.251	1.00	- 0.001
Roof panel (right)	80	18.06	2.248	2.251	1 .00	- 0.001
Roof panel (left)	100	14.65	2.774	2.774	1.00	0.000
Roof panel (right)	100	14.65	2.774	2.774	1 .00	0.000
Roof panel (left)	120	12.33	3.300	3.296	1.00	0.000
Roof panel (right)	120	12.33	3.300	3.296	1.00	0.000

3.7. Details of roof panels joints

3.7.1. Details of roof panels joint – type 1. Also specific to the composite insulated panels with metallic skin are joints between roof panels, usually at the distance of 1.00 m. In figure 18 is a type of detail of the geometrical modelling of the joint; the simulation results are presented graphically in figure 19, respectively in numerical form in table 11.

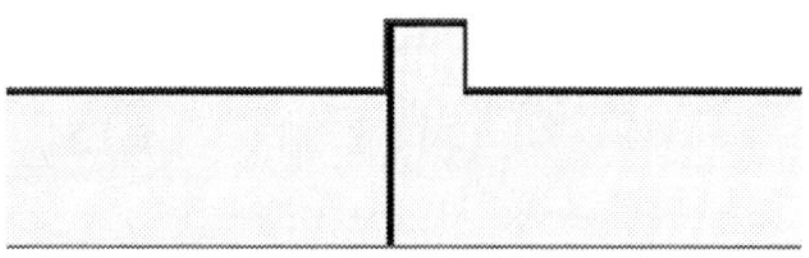

Figure 18. Detail of geometric modelling joint between roof panels

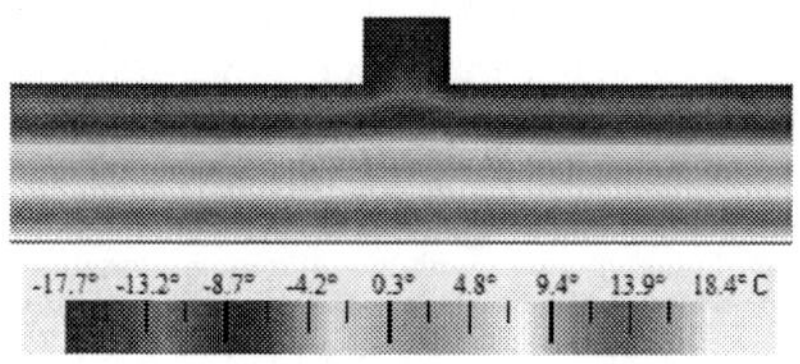

Figure 19. Isothermal surfaces for detail 3.7.1

Table 11. Linear thermal transmission coefficients for joint detail between roof panels – type 1

Detail identifier	$\delta_{insulation}$ (mm)	Φ_{2D} (W)	R (m^2K/W)	R' (m^2K/W)	L_i (m)	Ψ (W/mK)
Roof panel (left)	60	21.88	1.738	1.733	1.00	0.002
Roof panel (right)	60	21.93	1.738	1.723	1.00	0.005
Roof panel (left)	80	16.81	2.264	2.245	1.00	0.002
Roof panel (right)	80	16.90	2.264	2.255	1.00	0.004
Roof panel (left)	100	13.65	2.791	2.779	1.00	0.002
Roof panel (right)	100	13.72	2.791	2.766	1.00	0.003

Roof panel (left)	120	11.50	3.317	3.297	1.00	0.002
Roof panel (right)	120	11.54	3.317	3.287	1.00	0.003

3.7.2. Details of roof panels joint – type 2. In the current practice, have been identified roof panels where the metal sheet is continuous in the joint area (see figure 20); thus such a detail of roof panels joint has been simulated. The results obtained are presented in figure 21, respectively in table 12.

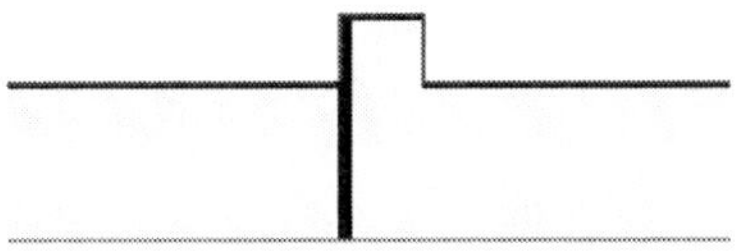

Figure 20. Detail of geometric modelling joint between roof panels

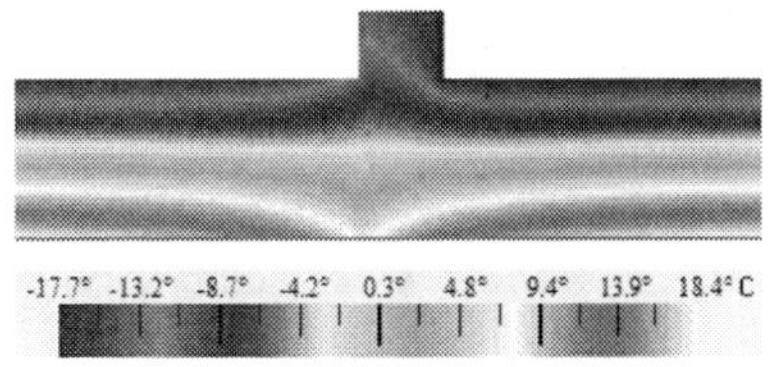

Figure 21. Isothermal surfaces for detail 3.7.2

Table 12. Linear thermal transmission coefficients for joint detail between roof panels – type 2

Detail identifier	$\delta_{insulation}$ (mm)	Φ_{2D} (W)	**R** (m^2K/W)	**R'** (m^2K/W)	L_i (m)	Ψ (W/mK)
Roof panel (left)	60	29.45	1.738	1.287	1.00	0.201
Roof panel (right)	60	29.63	1.738	1.281	1.00	0.205
Roof panel (left)	80	23.68	2.264	1.602	1.00	0.182
Roof panel (right)	80	23.79	2.264	1.595	1.00	0.185
Roof panel (left)	100	19.99	2.791	1.897	1.00	0.168
Roof panel (right)	100	20.06	2.791	1.892	1.00	0.170
Roof panel (left)	120	17.39	3.317	2.181	1.00	0.157
Roof panel (right)	120	17.43	3.317	2.177	1.00	0.158

As regarding the analysed constructive details of wall panels, it is noticed that a low thermal performance is observed for the detail of the wall panel joint with the window and for the wall panel joint with the slab on ground.

As regarding roof panels regular constructive details, it is noticed that linear thermal transmittance coefficients (Ψ –values) are lowest than $\Psi \leq 0.015$ (W/mK), excepting the case when in the joint between roof panels the metal layer is continuous.

The effect of the analysed thermal bridges in Part 1 of the paper is further studied from the point of view of its occurrence; an additional analysis being made in Part 2 of the paper, for building elements (facades and roofs).

4. Conclusions

The results of this paper offers a comprehensive knowledge about the thermal performance of several constructive details using composite insulated panels with metallic skin.

Although that the constructive details for this type of technical envelope solution have a great diversity, already currently used in existent buildings, which is impossible to be covered, the results of numerical simulations presented in this paper provide an insight into the effect of linear thermal bridges that occurs, and also provides a working paper for building energy auditors, which are using linear thermal transmittance coefficients in the EPCs elaboration or in building energy audits.

The results may be applied directly in the evaluation of the energy performance for a range of non-residential buildings such as: conditioned industrial halls, commercial buildings, office buildings, etc. for which currently is used this type of building envelope solution. The results can be used as working data by building energy auditors and by civil engineering specialists, in order to obtain correct information regarding the energy consumption of the building during exploitation.

5. References

[1] BP Statistical Review of World Energy [Online]. Available: www.bp.com
[2] Infomind Sarl 2003 *Catalogue des ponts thermiques*, ed. Suisse Energie (Zurich: Office federal de l'energie OFEN) p 128
[3] Anexa la Ordinul nr. 1590/2012 pentru modificarea și completarea Reglementării C107/3-2005: *Catalog cu punți termice specifice clădirilor* p 837
[4] Iacob A 2012 *Atlas de punți termice. Teză de doctorat.* (Iași: Universitatea Tehnică "Gheorghe Asachi")
[5] Măgurean A M 2018 An analysis of the influence of emissivity in the heat transfer for constructive details specific to insulated sandwich panels under summer and winter conditions *Proc. C65 Int. Conf. "Tradition and Innovation - 65 Years of Constructions in Transilvania"* (Cluj-Napoca: U.T. Press)
[6] Mastrapostoli E, Karlessi T, Pantazaras A, Kolokotsa D, Gobakis K, and Santamouris M 2014 On the cooling potential of cool roofs in cold climates: Use of cool fluorocarbon coatings to enhance the optical properties and the energy performance of industrial buildings 2014 *Energy Build.* **69** pp 417–425
[7] Castellon C, Martorell I, Roca J, Lazaro A, Zalba B, Fernandez I, Navarro M E, Cabeza L F 2009 Comparison of manufacturing techniques to use microencapsulated phase change materials in sandwich panels - experimental study *Abstract Book and Proc. 11th Int. Conf. on Thermal Energy Storage for Energy Efficiency and Sustainability (Effstock 2009)*
[8] Carbonari A, De Grassi M, Di Perna C, and Principi P, Numerical and experimental analyses of PCM containing sandwich panels for prefabricated walls 2006 *Energy Build.* **38** pp 472–483
[9] EN ISO 10211:2017 2017 *Thermal bridges in building construction -- Heat flows and surface temperatures -- Detailed calculations*
[10] C107/5-2005 2005 *Normativ privind calculul termotehnic al elementelor de construcție ale clădirilor, Partea a 5-a Normativ privind calculul termotehnic al elementelor de construcție în contact cu solul*
[11] EN ISO 10456:2007 2007 *Building materials and products - Hygrothermal properties - Tabulated design values and procedures for determining declared and design thermal values*
[12] EN ISO 10077-2:2017 2017 *Thermal performance of windows, doors and shutters - Calculation of thermal transmittance - Part 2: Numerical methods for frames*

Acknowledgments
This paper represents the dissemination of research undertaken in the Ph.D. thesis with the title: "Analiza performanței energetice a clădirilor nerezidențiale prin tehnici de modelare numerică și inteligență artificială aplicată" (in Romanian), author: Ancuța Maria Măgurean.

Evaluation of the thermal performance of composite insulated panels with metallic skin through steady-state numerical analysis – Part 2

A M Măgurean
Department of Civil Engineering and Management, Technical University of Cluj-Napoca, 25 Gh. Barițiu Street, Cluj-Napoca, 400027, Romania

E-mail: ancuta.magurean@cif.utcluj.ro

Abstract. Although the constructive detail type approach analysis already offers an extensive image about the influence of thermal bridges that occurs in the composite insulated panels with metallic skin, studies of the subject go further, by integrating the detail obtained results into parts of the building analyses, with the purpose to identify how factors such as occurrence frequency of the thermal bridge, or the percent of the glazed surfaces existent in regular facades influence the thermal behaviour of the building part, along with the numerical value of the linear thermal transfer coefficients obtained in Part 1 of the paper. Also, for the identified constructive detail with the lowest thermal performance, framework solutions were identified and analysed through steady-state FEM analysis in order to offer a working tool for building energy auditors.

1. Introduction

In a companion paper, the thermal performance of several common constructive details of composite insulated panels with metallic skin was evaluated, for four different thicknesses of this type of building enclosure solution. The analysed details represent the regular details which constitute the assembly of an industrial hall.

In order to obtain a global perspective on the thermal performance of the composite insulated panels with metallic skin, the study continues with a transition from the constructive detail analysis to the building envelope element analysis.

Otherwise formulated, starting from the local analysis of the effect of thermal bridges, which may cause the intensification of the thermal flux values, analysis which is specific to the constructive detail, with the determination of the linear thermal transmission coefficients that was made in Part 1, study goes further in Part 2 of the paper with the assessment of the impact of thermal bridges in the thermal performance of the facade, respectively of the roof, as building part analysis. This approach allows the identification of the regular constructive details that have a high impact on the energy performance of the studied building element.

The study also aims to optimize the constructive details identified in Part 1 of the paper as being with the lowest thermal performance, by using steady-state FEM analysis.

2. Parametric study of the impact of thermal bridges in the building envelope

2.1. Considered case studies

One of the main steps in the energy performance analysis of a building is the calculation of the heating and cooling demand, in order to determine the heating and the cooling energy consumptions. To calculate that, the building envelope thermal performance analysis is necessary to be done. Therefore, by using the results obtained on constructive detail level in previous paper (Part 1) and also, by using results of paper [1], building envelope analysis level is further pursued, in different scenarios (table 1). Thus, the corrected thermal resistances for a few facade dimensions were determined, assuming the use

IOP Conf. Series: Materials Science and Engineering **586** (2019) 012033 doi:10.1088/1757-899X/586/1/012033

of 10 cm thick wall sandwich panels, a thickness considered to be common for this type of solution, for the outer wall length of 30 m and the height 4 m and 6 m, respectively.

Table 1. Parametric studies depending on the building element type

Building element type	Case
Facade	A1, A2, B1, B2
Roof	C1, C2

The metallic sheet being really thin (0.0006 m for the exterior side of the panel, respectively 0.0004 m for the interior side of the panel), is considered negligible regarding the thickness of the panel, being considered that the thickness of the panel equals the thickness of the insulation layer, as notation. However, the metallic sheet was taken into consideration in the FEM analysis. Several different scenarios were used with regard to the existence or not of glazed surfaces, respectively with regard to the percentage of glazed surfaces relative to the analyzed facade surface. Also, for the type of constructive detail on which in Part 1 of the paper were analyzed two distinct variants, as is the case of the joint between the wall panels, the wall panel joint with the slab on ground and the joint between roof panels, for the further studies one was chosen. Input data used for the considered facades are centralized in table 2, respectively in table 3 for the roofs.

Table 2. Input data for the analyzed facades

Case	Used peculiar constructive details	$\delta_{insulation}$ (cm)	L_e (m)	L_i (m)	h_i (m)	S_i (m^2)
A1	3.5.1 | 3.2.1	10	30	29.80	4.00	119.20
A2	3.5.2 | 3.2.1	10	30	29.80	4.00	119.20
B1	3.5.1 | 3.2.1.	10	30	29.80	6.00	178.80
B2	3.5.2 | 3.2.1	10	30	29.80	6.00	178.80

Corrected thermal resistances were determined for roof slab, also for 10 cm panel thickness and outer roof size arbitrary chosen (horizontal projection) of 30x30 m.

Table 3. Input data for the analyzed roofs

Case	Used peculiar constructive details	$\delta_{insulation}$ (cm)	L_e projection (m)	L_i projection (m)	L_i slope 7° (m)	S_i (m^2)
C1	3.7.1	10	30	29.80	30.22	900.56
C2	3.7.2	10	30	29.80	30.22	900.56

The influence of each thermal bridge on the building elements was analyzed for several undertaken scenarios, depending on the value of the linear thermal transmission coefficient obtained by the type of detail and the length to which it applies ($\psi \cdot l$), as component of the relation (1), and which represents the impact of each thermal bridge that occurs, simultaneously with the frequency of its occurrence in regular constructive facades or roofs. It is also specified that punctual thermal bridges (represented by χ – the punctual thermal transmission coefficient), such as screws used for mechanical fastening, for which quantitative assessment requires three-dimensional analysis, are not the subject of current investigation.

$$R' = \frac{1}{\frac{1}{R} + \Sigma\frac{\Psi \cdot l}{S_i} + \Sigma\frac{\chi}{S_i}} \ (m^2 \cdot K/W) \quad (1)$$

$$r = \frac{R'}{R} \ (-) \quad (2)$$

2.2. Parametric study of the impact of thermal bridges in the building envelope

The obtained results for the facades are presented in figure 1, figure 2, figure 3 and figure 4.

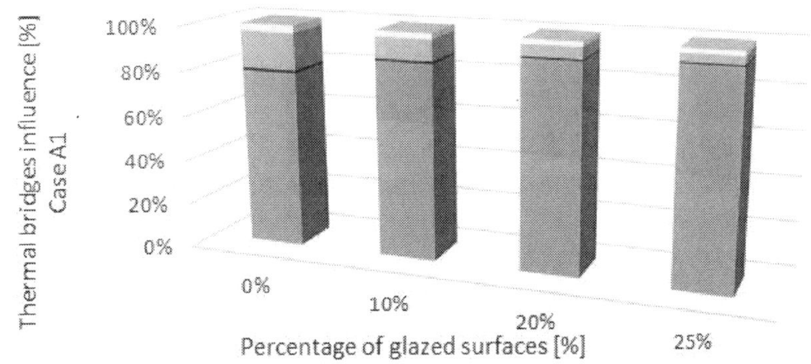

Thermal bridge type	0%	10%	20%	25%
Detail 3.1: Wall corner	3,36%	2,62%	2,21%	2,09%
Detail 3.2: Wall - Wall	16,58%	9,46%	5,06%	3,44%
Detail 3.3: Wall - Roof	1,10%	0,86%	0,73%	0,69%
Detail 3.4: Wall - Window	0,00%	25,43%	39,92%	44,63%
Detail 3.5: Wall - Slab	78,96%	61,62%	52,08%	49,15%

Figure 1. Analysis of influence of thermal bridges on analyzed façade – Case A1

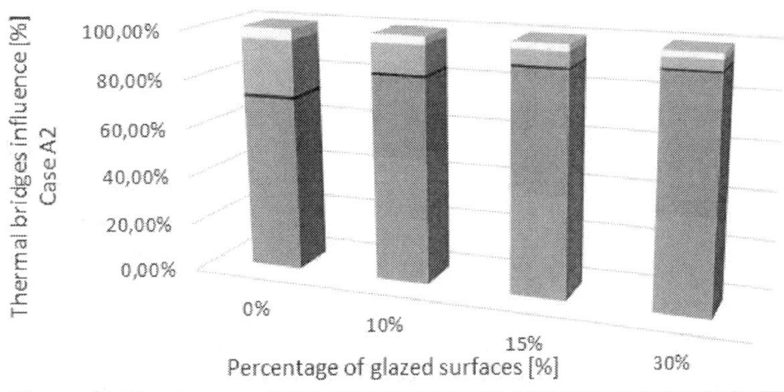

Thermal bridge type	0%	10%	15%	30%
Detail 3.1: Wall corner	4,60%	3,32%	2,69%	2,51%
Detail 3.2: Wall - Wall	22,73%	12,00%	6,16%	4,14%
Detail 3.3: Wall - Roof	1,51%	1,09%	0,88%	0,82%
Detail 3.4: Wall - Window	0,00%	32,24%	48,59%	53,67%
Detail 3.5: Wall - Slab	71,16%	51,35%	41,67%	38,85%

Figure 2. Analysis of influence of thermal bridges on analyzed façade – Case A2

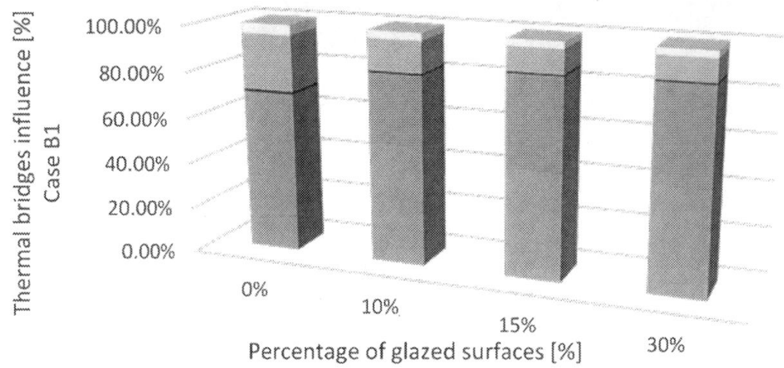

Thermal bridge type	0%	10%	15%	30%
Detail 3.1: Wall corner	4.47%	3.30%	2.96%	2.90%
Detail 3.2: Wall - Wall	24.51%	13.73%	10.35%	9.56%
Detail 3.3: Wall - Roof	0.98%	0.72%	0.65%	0.64%
Detail 3.4: Wall - Window	0.00%	30.51%	39.66%	41.36%
Detail 3.5: Wall - Slab	70.05%	51.74%	46.39%	45.54%

Figure 3. Analysis of influence of thermal bridges on analyzed façade – Case B1

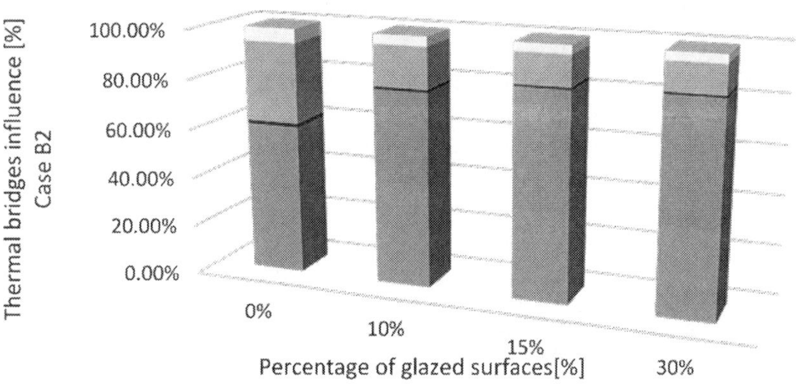

Thermal bridge type	0%	10%	15%	30%
Detail 3.1: Wall corner	5.88%	4.01%	3.52%	3.44%
Detail 3.2: Wall - Wall	32.25%	16.69%	12.30%	11.33%
Detail 3.3: Wall - Roof	1.29%	0.88%	0.77%	0.75%
Detail 3.4: Wall - Window	0.00%	37.08%	47.15%	49.00%
Detail 3.5: Wall - Slab	60.59%	41.34%	36.26%	35.47%

Figure 4. Analysis of influence of thermal bridges on analyzed façade – Case B2

It is noted that for the building element analysis, for determining the horizontal joint length between the wall panels, the width of each panel was considered to be 1 m. Further, the results of corrected thermal resistances, R', and of r-values, which is calculated with relation (2), are centralized in table 4.
Analyzing the results presented in table 4 and, subsequently, in figure 1, figure 2, figure 3 and figure 4, it is observed that the influence of the thermal bridges existing at the joints between the panels (the joints between wall panels, wall panels with roof panels, corners of the building facade) have a relatively low impact in the thermal performance of the facade, but there are two types of common constructive details (joint of the wall panel with window and with the slab on ground), whose geometric and material thermal

bridges have a major impact on the thermal performance of the building element, significantly reducing it.

Table 4. Thermal performance of the analyzed facades

Case	Results	Percentage of glazed surfaces			
		0%	**10%**	**20%**	**25%**
A1, A2, B1, B2	**R**	2.799	2.799	2.799	2.799
A1	**R'**	2.096	1.958	1.856	1.820
	r	0.75	0.70	0.66	0.65
A2	**R'**	2.249	2.091	1.975	1.933
	r	0.80	0.75	0.71	0.69
Case	**Results**	**0%**	**10%**	**15%**	**30%**
B1	**R'**	2.031	1.852	1.782	1.770
	r	0.73	0.66	0.64	0.63
B2	**R'**	2.175	1.970	1.891	1.878
	r	0.78	0.70	0.68	0.67

Similar analysis was undertaken for roofs, with 10 cm panel thickness and outer roof size arbitrary chosen (horizontal projection) of 30x30 m. It was analyzed the influence of each thermal bridge on the roof as building element, for two considered cases, as can be seen in figure 5 and figure 6.
In order to determine the joining length between the roof panels, it is stated that the width of each panel equal to 1 m, and the joining of the panels is considered parallel to the span of the hall (parallel to the width of the considered floor).

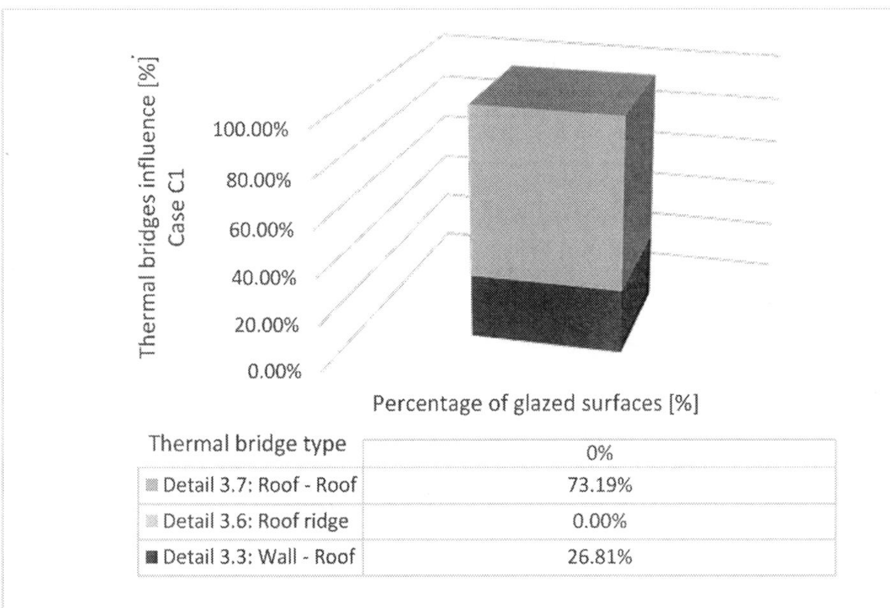

Figure 5. Analysis of the influence of thermal bridges on the analyzed roof floor - Case C1

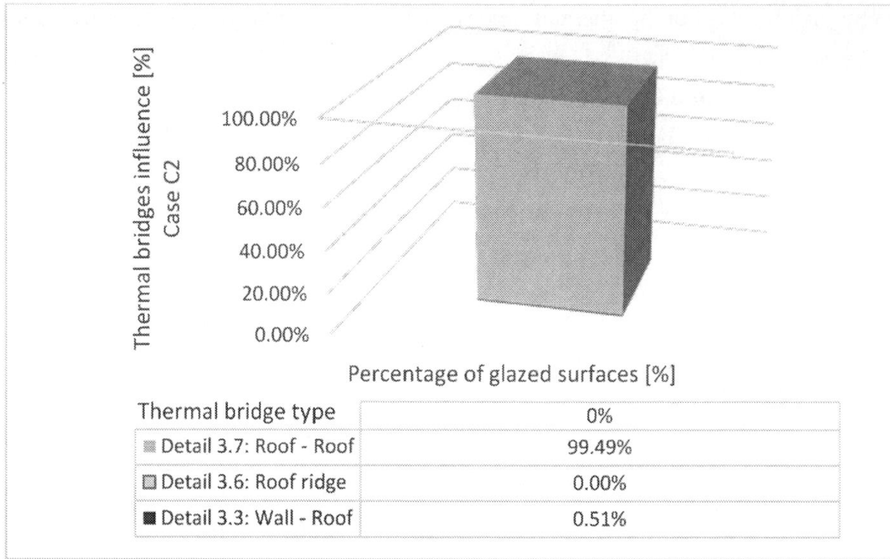

Figure 6. Analysis of the influence of thermal bridges on the analyzed roof floor - Case C2

Corrected thermal resistances and r-values were also determined for roof. According to the results obtained in table 5, figure 5 and figure 6, in the case C1 there is a negligible influence of the thermal bridges on the studied building element, but the corrected thermal resistance decrease significantly in case C2, leading to a very low thermal performance for the building element because of the high occurrence of the roof between joints, concurrently with relatively high Ψ-values.

Table 5. Thermal performance of the analyzed roofs

Case	**R** (m^2K/W)	**R'** (m^2K/W)	**r** (-)
C1	2.791	2.743	0.98
C2	2.791	1.460	0.52

3. Energetically optimized constructive details

After analyzing several constructive details of sandwich panels and the interpretation of the results from the first part of the paper, respectively in the above presented study, it is concluded that the constructive details that have the potential to significantly improve heat transfer behavior by reducing the effect of thermal bridges are the joining details of the wall panel with the slab on ground. For the identified constructive details with low thermal performance, FEM analysis in the conditions already detailed in Part 1 of the paper is undertaken.

Thus, variants of optimization of this type of detail with the purpose to reduce the effect of the thermal bridge were analyzed by: increasing the length of the overlap of the wall panel on the concrete element with the optimization of the fastening (measure types 1), respectively by insulating the base, including the introduction of a horizontal insulation on the perimeter of the building (measure types 2).

For this study, the numerical and analytical calculations for the determination of linear thermal transfer coefficients were performed for the wall panel of 80 mm thickness.

Main solutions and results are presented in the following tables. As can be seen in table 6, following the application of measures type 1, there is a significant reduction of linear thermal transfer coefficients: around 50% in the wall panel, over 40% for the slab on ground. However, there are situations when this type of solutions is not applicable, because of the low difference between the upper part of the slab on ground and the ground that surrounds the building.

Table 6. Linear thermal transfer coefficients for detail 3.5.1, optimized by using measures type 1

No.	Detail 3.5.1 Measures type 1	Element (-)	L_{2D} (W/K)	Ψ_{detail} (W/mK)	Ψ (W/mK)	$\Psi_{reduction}$ (%)
-	Reference model - overlap 10 cm	Wall	0.832	1.142	0.364	-
		Slab	2.789		0.778	-
1.a	Overlap 30 cm	Wall	0.628	0.622	0.183	49.85
		Slab	2.436		0.440	43.49
1.b	Overlap 30 cm + injected PUR	Wall	0.617	0.593	0.173	52.53
		Slab	2.416		0.420	46.03
1.c	Overlap 30 cm + EPDM sealing – figure 7	Wall	0.620	0.618	0.176	51.60
		Slab	2.437		0.441	43.30

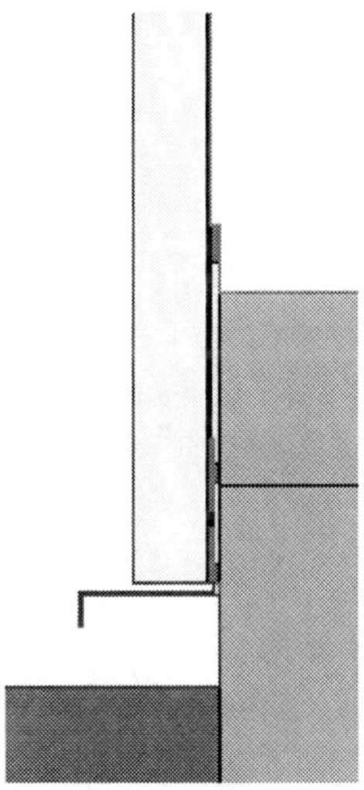

Figure 7. Geometrical modeling detail 3.5.1 – solution 1.c

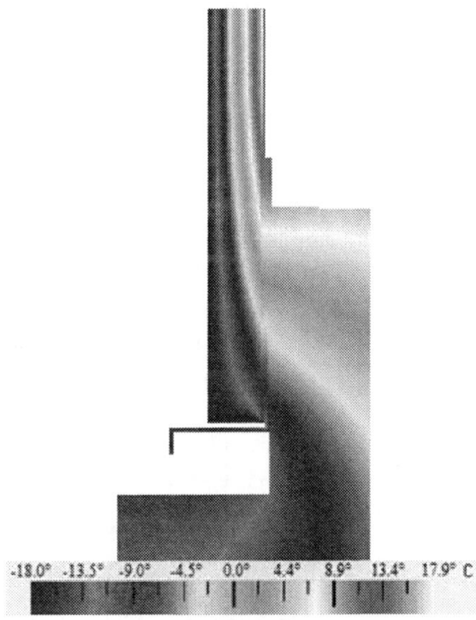

Figure 8. Isothermal surfaces detail 3.5.1 – solution 1.c

The results of implemented measures type 2 are centralized in table 7. For optimized details 2.a, 2.b and 2.e (figure 10) there was found a reduction of approx. 25-30% of the linear thermal transfer coefficients relative to the values calculated in the initial detail, for both the wall panel and slab on ground. Regarding the technical solutions 2.c, 2.d (figure 9), respectively 2.f and 2.g, the horizontal thermal insulation layer was proposed with the intention of ensuring a deviation of the temperatures in the perimeter area of the wall-to-ground intersection, in the meaning of executing a "barrier" in passing the heat flow beneath the slab on the ground to the outside, aiming to reduce the flow that dissipates to the surface of the ground. However, the values of linear thermal transfer coefficients obtained by numerical simulations are not much lower than those of solutions 2.b and 2.e, so it is observed that this type of technical solution is not necessarily justified given the quantities of the works and additional materials necessary for their implementation.

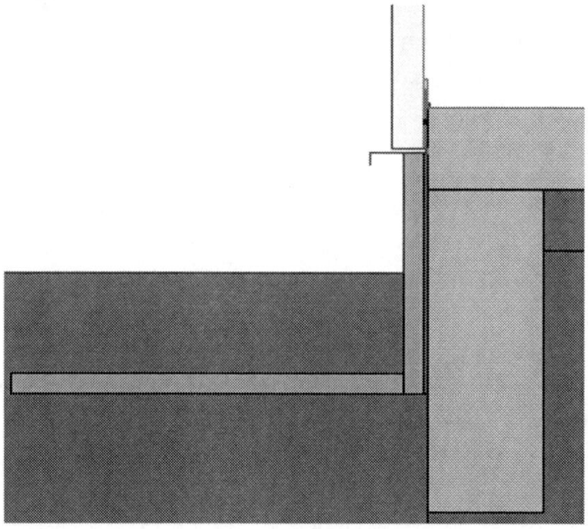

Figure 9. Geometrical modeling detail 3.5.1 – solution 2.d

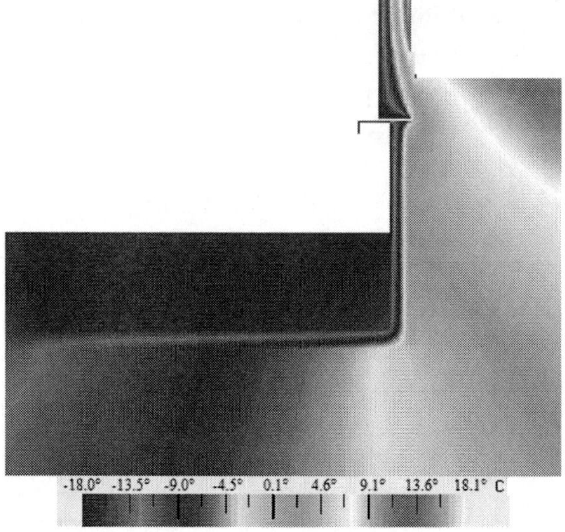

Figure 10. Isothermal surfaces – detail 3.5.1 – solution 2.d

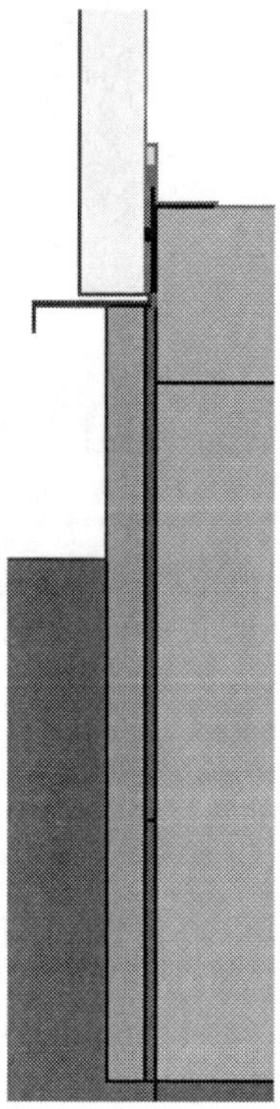

Figure 11. Geometrical modeling detail 3.5.1 – solution 2.e

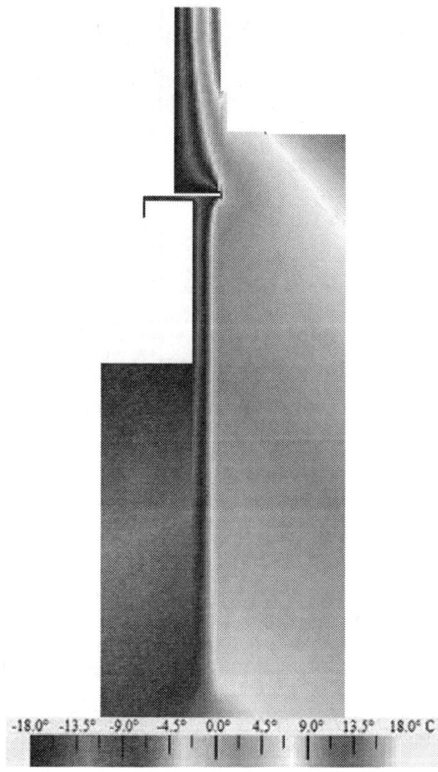

Figure 12. Isothermal surfaces – detail 3.5.1 – solution 2.e

Table 7. Linear thermal transfer coefficients for detail 3.5.1, optimized by using measures type 2

No.	Detail 3.5.1 Measures type 2	Element (-)	L_{2D} (W/K)	Ψ_{detail} (W/mK)	Ψ (W/mK)	$\Psi_{reduction}$ (%)
	Reference model - Overlap 10 cm	Wall	0.832	1.142	0.364	-
		Slab	2.789		0.778	-
2.a	Base insulation with 5 cm XPS down to ground	Wall	0.806	0.845	0.273	25.04
		Slab	2.421		0.572	26.52
2.b	Base insulation with 5 cm XPS + 30 cm vertical insulation of foundation beam	Wall	0.793	0.813	0.260	28.61
		Slab	2.259		0.553	28.97
2.c	Base insulation with 5 cm XPS + 30 cm vertical insulation of foundation beam + 50 cm length horizontal insulation below the ground	Wall	0.790	0.809	0.257	29.46
		Slab	2.172		0.552	29.02
2.d	Base insulation with 5 cm XPS + 30 cm vertical insulation of foundation beam + 100 cm horizontal insulation below the ground – figure 9	Wall	0.789	0.809	0.256	29.87
		Slab	2.120		0.554	28.86
2.e	Base insulation with 5 cm XPS + 60 cm vertical insulation of foundation beam – figure 11	Wall	0.789	0.806	0.256	29.85
		Slab	2.173		0.550	29.26
2.f	Base insulation with 5 cm XPS + 60 cm vertical insulation of foundation beam + 50 cm length horizontal insulation below the ground	Wall	0.788	0.808	0.254	30.18
		Slab	2.128		0.553	28.90
2.g	Base insulation with 5 cm XPS + 60 cm vertical insulation of foundation beam + 100 cm length horizontal insulation below the ground	Wall	0.786	0.806	0.253	30.48
		Slab	2.085		0.552	29.03

Finally, the proposals through which was obtained the highest reduction of the thermal bridges effect from each type of proposed solution, were analyzed and the measures have been cumulated. The identifier of combinations of proposed measures and the obtained results are presented in table 8.
By combining the proposed solutions, a significant reduction of linear thermal transfer coefficient values is observed, for both the wall panel (over 60%) and the slab on ground, where the Ψ-value reduction is between 58% and 63%, compared with the initial detail of which thermal performance was studied in Part 1 of the paper.
It is noticed the significant influence of the length of the overlap of the wall panel with the reinforced concrete perimetral base, so as the length of the overlap increases, the effect of the geometric and material thermal bridge is reduced into the constructive detail.
Considering the specificity of this type of constructive detail, characterized by the fixing of the wall panel with the L-shaped metal plate, metallic fastening which is necessary by reasons of strength and stability, the results obtained are considered to be satisfactory.

Table 8. Linear thermal transfer coefficients for detail 3.5.1 using measures type 1+2

Detail 3.5.1 Measures type 1+2	**Element** (-)	$\mathbf{L_{2D}}$ (W/K)	Ψ_{detail} (W/mK)	ψ (W/mK)	$\Psi_{reduction}$ (%)
Reference model	Wall	0.832	1.142	0.364	-
	Slab	2.789		0.778	-
1.c+2.b	Wall	0.587	0.466	0.142	61.10
	Slab	2.174		0.324	58.35
1.b+2.b	Wall	0.585	0.445	0.140	61.49
	Slab	2.154		0.304	60.88
1.b+2.e	Wall	0.580	0.423	0.135	62.87
	Slab	2.076		0.287	63.06

4. Conclusions

The results presented in Part 2 of the paper offers a comprehensive knowledge about the thermal performance of composite insulated panels with metallic skin from a building element integrated perspective. The results of the undertaken parametric studies highlights that the impact of the thermal bridges on the thermal performance of the building elements, especially for the walls, is significant, similar to the walls of brick masonry with traditionally arranged external thermo-insulation.

The paper highlights that an important effect of the thermal bridge can appear also from the frequent occurrence of a thermal bridge, even though the Ψ-values are low or relatively low.

Constructive details optimization for the initial solution with the lowest thermal performance is also made, offering ready to use data for building energy auditors, in case that the energy audit of the building is performed.

The results of the full paper (Part 1, Part 2) offers a great view of the thermal performance of composite insulated panels with metallic skin, which is relative a new technical solution, starting from steady state FEM analyses of common constructive details, which are integrated in studies of real scale parts of building (facades and roofs).

By using advanced calculation methods, the results presented in this paper may be used as working data by building energy auditors and by civil engineering specialists, in order to predict correct information regarding the energy consumption of the building, considering as well the case that energy improvements are being made for an existing building.

5. References

[1] Măgurean A M, Lupan L M, and Moga I 2014 Insulated Sandwich Panels - Thermal performance *Proc. of the Second Int. Conf. for PhD Students in Civil Engineering and Architecture (CE-PhD 2014)* pp 477–482

Acknowledgments

This paper represents the dissemination of research undertaken in the Ph.D. thesis with the title: "Analiza performanței energetice a clădirilor nerezidențiale prin tehnici de modelare numerică și inteligență artificială aplicată" (in Romanian), author: Ancuța Maria Măgurean.

IOP Conf. Series: Materials Science and Engineering **586** (2019) 012034 doi:10.1088/1757-899X/586/1/012034

Numerical analysis of heat exchanger with uniform heat flux used in heat pumps systems

C G Popovici [1], S V Hudişteanu [1]
[1] "Gheorghe Asachi" Technical University of Iaşi, 67 Dimitrie Mangeron Bvd, *Romania*

E-mail: sebastian.hudisteanu@tuiasi.ro

Abstract. The paper presents a numerical analysis of an innovative heat exchanger with uniform heat flux, integrated as a cold source, for heating and air-conditioning systems which are equipped with heat pumps. The ground-water or water-water surface heat exchanger type is modelled, having variable geometry, uniform efficiency and heat flux along the system. The purpose of the work is to analyse a heat exchanger with uniform thermal flux, in order to offer an alternative to actual technical solutions using polyethylene pipes mounted in parallel on horizontal or vertical drilling. For typical geothermal heat exchangers, both the surface and the depth are characterized by non-uniform exploitation of the soil. The uniformity of the thermal load of the earth mass can be an energy efficiency solution in terms of optimizing the storage capacity and consequently the reduction of the land used, respectively the length of the drilling needed for mounting the source. The analysed heat exchanger with uniform thermal flux, designed to achieve low or medium sources of thermal energy, offers an efficient energy consumption, being environmentally friendly and has a reduced relative cost. It can be an attractive alternative for the residential buildings or Nearly Zero-Energy Buildings (NZBE).

1. Introduction

The heat pump domain has been of interest in Europe since the 1970s in Switzerland and the 1980s in Sweden, when an intense program of tracking behaviour over time and the life span of these systems has begun [1]. Shortly, it was found that the best quantification of heat pump quality is given by their coefficient of performance (COP), and the following studies, research and tests led to its progressive increase. Research from the University of Ontario, 2011, on the thermodynamic behaviour of hybrid geothermal heat pump systems, has led to the conclusion that they have superior performance over conventional systems [1, 2].

When the intrinsic efficiency of the system has reached a considerable level of performance, the studies from literature have focused on improving the conditions of the heat extraction / injection into the storage material [3]. Thus, in 2013, at the University of Technology in Nicosia, Cyprus, studies are started on the geothermal properties of the soil used for heat pump systems. Also, studies have been carried out at the Perpignan Via Domitia University in France on the improvement of geothermal drills with vertical boreholes, controlling the thermal conductivity of the bentonite used [4].

Available data on the European geothermal heat pump (GSHP) market reveals an interesting picture showing the existence of stable markets and the development of some new ones - as shown in Figure 1 and Figure 2 [5].

EU legislation on heat pump systems is not fully resolved yet, with large differences between countries, due to a number of specific factors such as: geological conditions, level of development, financial incentives offered by state institutions etc. [13-16]. This is the main reason why legislation cannot be fully harmonized [17-19].

Most studies from literature have focused on system analysis using geothermal heat pumps equipped with different types of heat exchangers with soil [8-10]. The main conclusion consists of that the most important element that ensures the operation of the heat pumps at optimal parameters is the geothermal

IOP Conf. Series: Materials Science and Engineering **586** (2019) 012034 doi:10.1088/1757-899X/586/1/012034

heat exchanger [6-10]. Most often, geothermal heat exchangers are characterized by the global heat transfer coefficient (K) ($W/m^2 \cdot K$) [11, 12].

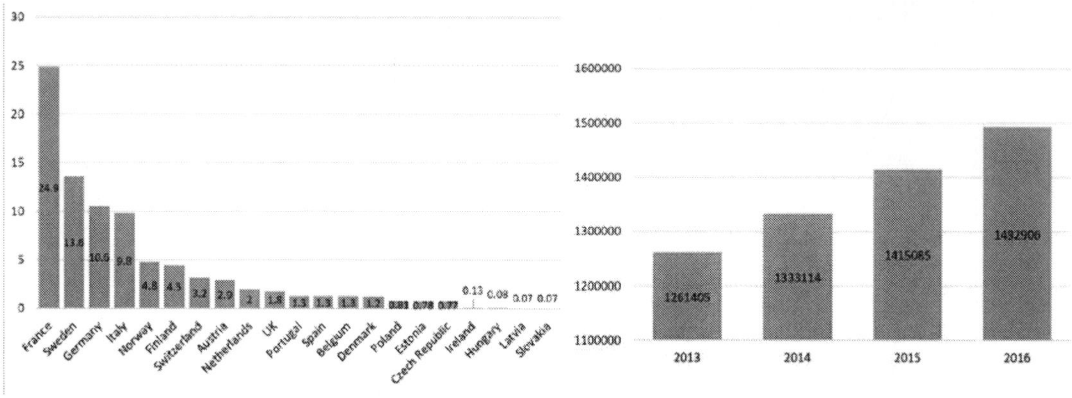

Figure 1. a) Energy produced (TWh) with heat pumps in Europe in 2014 [20]; b) Number of geothermal heat pump units in operation in Europe 2013-2016 [21].

During the present study, a heat exchanger with uniform thermal flux is analysed, in order to offer an alternative to actual technical solutions using polyethylene pipes mounted in parallel on horizontal or vertical drilling.

2. Problem description

The purpose of this research is to realize the numerical analysis of an innovative heat exchanger in order to obtain conclusive data regarding the velocities (magnitude, spectra, vectors and path lines) and temperatures (spectra and values) recorded during operation. Using these results, the operating parameters and efficiency of the heat exchanger are evaluated. The heat exchanger subjected to analysis is part of the schematic diagram of the experimental stand presented in Figure 2.

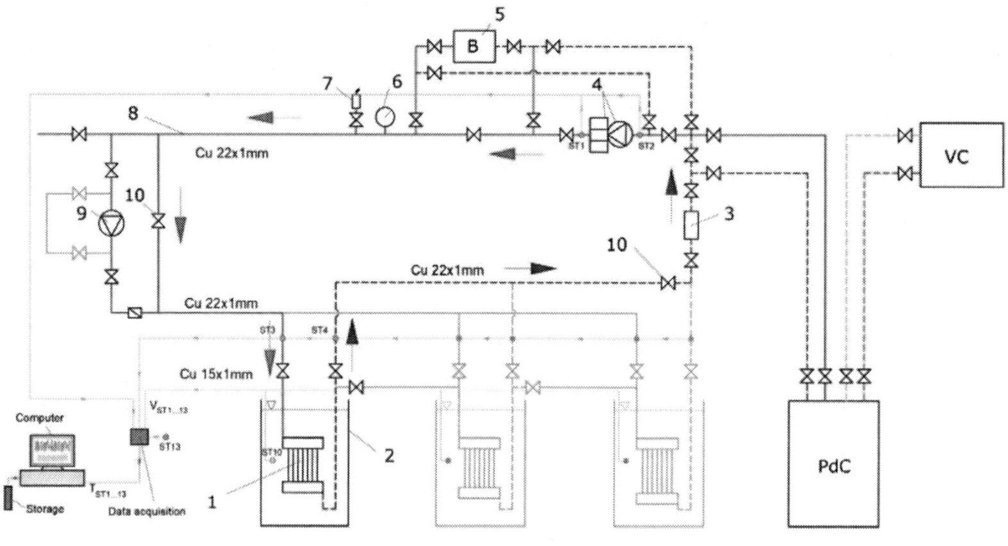

Figure 2. Functional scheme of the experimental stand.

The main elements of the system are: 1 - heat exchanger, 2 - storage tank, 3 - rotameter, 4 - heat source - electric heater with circulation pump, 5 – hot water accumulator, 6 - manometer, 8 - distribution

pipelines, 9 – backup circulation pump, 10 - valves, ST/1 ... 13 - temperature sensors, PdC - heat pump, VC - fan coil, data acquisition station.
The analysed heat exchanger has a special geometry, consisting of a distributor (D) and a collector (C), connected to each other by means of 20 copper pipes, Cu15x1 mm. Figure 3 shows an overview of the model analysed and some images of the real heat exchanger.

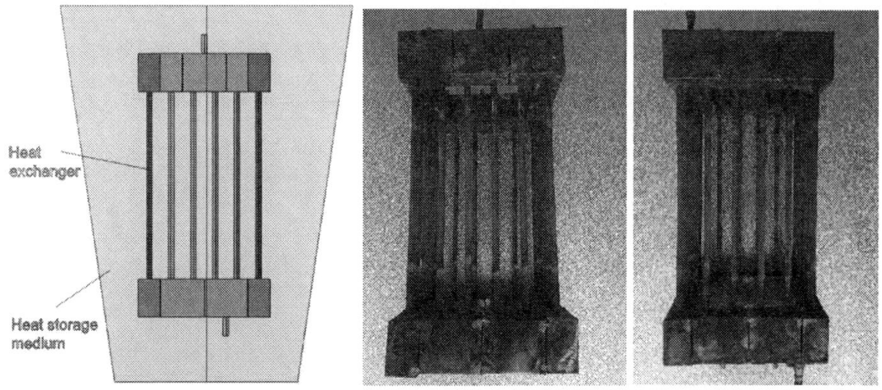

Figure 3. Heat exchanger: a) heat exchanger-heat storage ensemble; b) images of the real model.

The particularity of this exchanger consists in the circulation of the thermal agent, determined by the configuration of the distributor and the collector which are separated by alternating sectors. Thermal agent circulation is forced to take place in an imposed direction between the input and output sectors, respectively from top to bottom and from bottom to top, alternatively. Thus, the average temperature of the heat transfer is evenly distributed throughout the surface of the exchanger.
Consequently, the thermal flux dependent on the product of the two parameters $Qi = \varphi \cdot (Si \cdot Ti)$, is constant.
The heat exchanger analysed is made of high thermal conductivity material (D/C and pipes made of copper), with a transfer capacity superior to the exchangers made of polyethylene pipes. It is designed to be used as a cold low and medium depth source, required for systems equipped with geothermal heat pumps with mechanical vapor compression. It works with hot water or glycol between the heat pump and the source used. The heat flux transmitted by small scale heat exchangers is calculated according to the following relationship (1).

$$Q = G \cdot \rho \cdot c\,(T_{in} - T_{out}) \tag{1}$$

The global heat transfer coefficient (*K*) is determined using the following mathematical expression:

$$K = \frac{1}{\frac{1}{\pi\alpha_i D_i} + \frac{1}{2\pi\lambda} ln\frac{D_i}{De} + \frac{1}{\alpha^*}}\ [\mathrm{W/m^2 \cdot {}^\circ C}] \tag{2}$$

where: ρ [kg/m^3] - density of thermal agent;
c [k$_{cal}$ /kg$\cdot^\circ$C] – specific heat of thermal agent;
$T_{in} - T_{out} = \Delta T$ [$^\circ$C] – temperature difference at heat pump;
λ [W/m$\cdot^\circ$C] – thermal conductivity of the pipes;
D_i, D_e – inner, outer diameter of the pipes;
α_i – convective heat transfer coefficient from thermal agent to the pipe wall;
α^* – convective heat transfer coefficient from pipe wall to the soil.

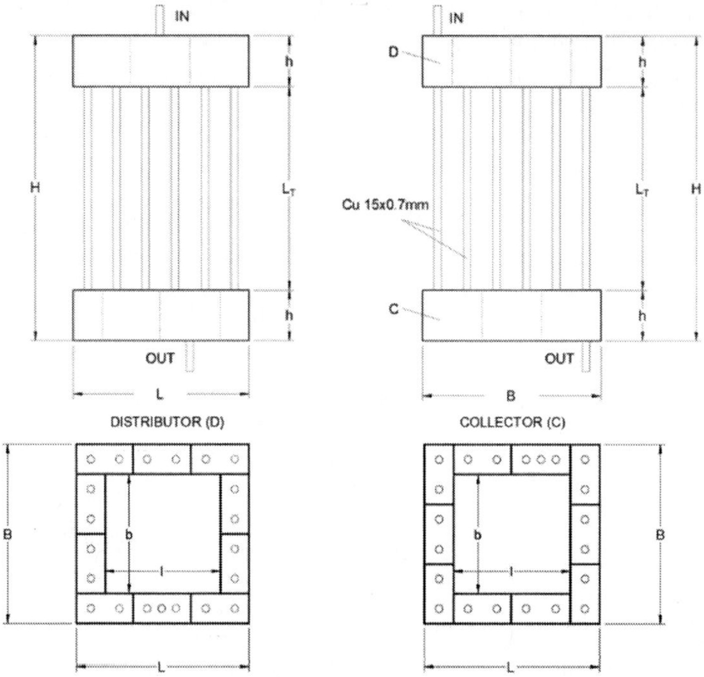

Figure 4. Details of heat exchanger: a) front view; b) side view; c) top view; d) bottom view.

Figure 4 (c, d) shows the offset between the distributor and collector sectors, causing alternative circulation of the agent through the fascicle of pipes.
The main constructive characteristics of the analysed heat exchanger are presented in Table 1.

Table 1. Description of the model of the heat exchanger

Dimensions [mm] – height (H) – width (L) – depth (B)	$H = 600$ $L = 360$ $B = 360$
Material	Copper
Structure	Distributor (D) Fascicle of pipes Collector (C)
Structure of D/C	10 separate sectors for thermal agent circulation
Fascicle of pipes	Cu 15x0.7mm
	20 pieces
Length of pipes L_T [mm]	400
Heat exchange surface [m^2]	0,918
Volume of heat exchanger [litre]	15,81
Mass of empty heat exchanger [kg]	6,17
Length of thermal agent path inside the exchanger [m]	8,24

3. Numerical approach

The geometry of the model, Figure 5, was realized using AutoCAD 3D and imported into ANSYS-Fluent software. The mesh was realized in ANSYS Meshing with different refinement for distributor,

collector and fascicle of pipes, with a particular interest given to the compartmentation of the D/C. Therefore, a total of 511874 cells were used for meshing.
The flow inside the heat exchanger was considered turbulent, taking into account the velocities imposed, dimensions of the pipes and the resulting Reynolds number, with a turbulent intensity of 5%. The numerical simulation was realized using k-ε Re-Normalization Group (RNG) model of turbulence, with enhanced wall treatment, which is more accurate and reliable for a wider class of flows than the standard k-ε model [22].
During simulations, the inlet mass flow rate (0.2 m^3/h) and temperature (50 °C) were imposed, while the initial temperature of the storage medium (water) was 15 °C. Also, a 8 $W/m^2 \cdot K$ exterior convective heat transfer coefficient was used.

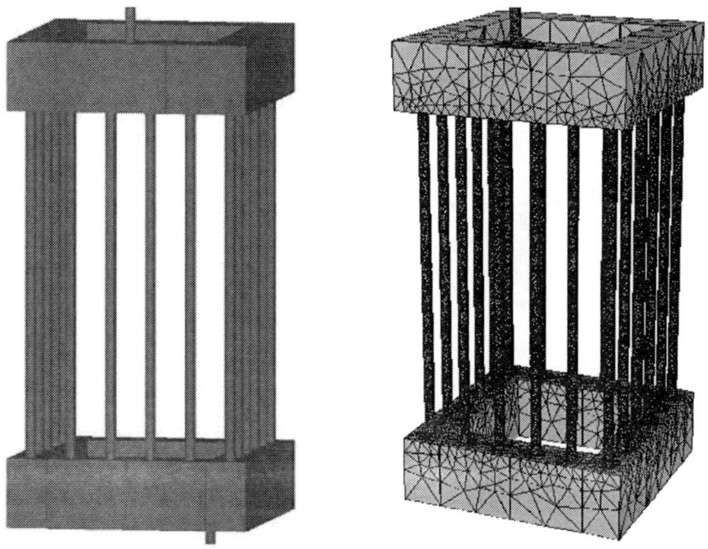

Figure 5. Geometry (a) and mesh (b) of the studied model.

The calculation is iterative, with hybrid initialization and convergence criteria of 10^{-6} for the energy equation and 10^{-3} for the pressure, velocities and continuity equations.

4. Results
Results of simulations consist of the heat exchanger inside temperature variation and on its outer surface and velocity magnitude and spectra for the front, lateral and back sections.
Figures 6-14 exported from the ANSYS-Fluent software show the results of numerical modelling on the proposed reduced scale heat exchanger. Qualitative and quantitative aspects regarding the flow inside the heat exchanger and the distribution of operating temperatures can be observed. There are also obtained the quantitative data of the thermodynamic process which will be used in further calculations, Table 2.

4.1. Temperatures
The temperatures spectra, Figures 6-9, revealed a quasi-uniform distribution of values both inside the heat exchanger and on its surface. Taking into account the uniformity of the storage medium, it can be assumed that the heat flux transferred from the heat exchanger is also uniform on its entire surface.
It can be observed that the temperature of the thermal agent in the front section has an average value of 50 °C, while the same value for the back section reaches values of 48 °C. Thus, despite the heat transfer along the thermal agent path inside the heat exchanger, the operating temperature near the outlet section is comparable to the inlet one, determining a smooth loading of the storage medium.

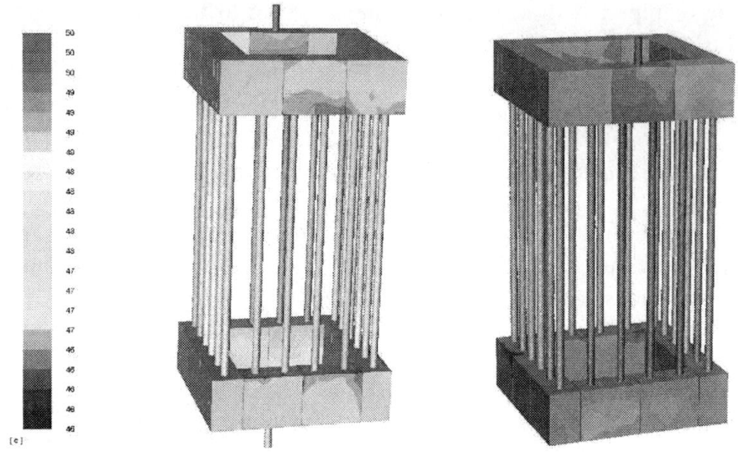

Figure 6. Distribution of temperatures on the surface of heat exchanger.

Figure 7. Contours of temperature – inside overview.

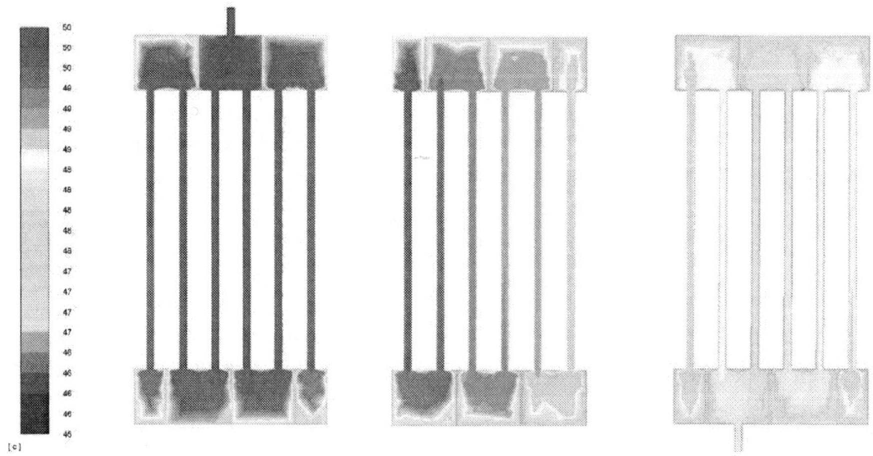

Figure 8. Contours of temperature: a) front section; b) lateral section; c) back section.

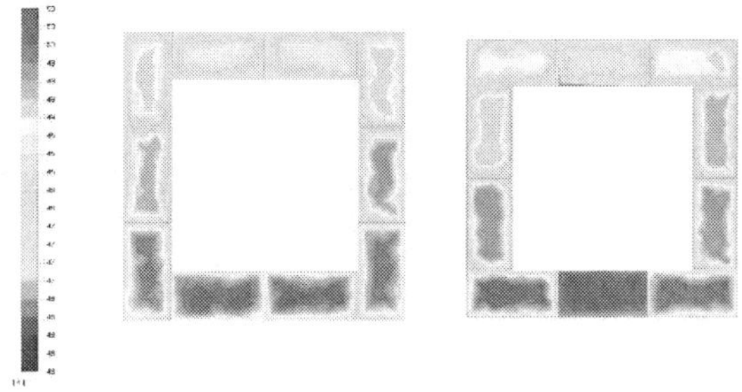

Figure 9. Contours of temperature: a) distributor section; b) collector section.

4.2. Velocities

In terms of velocities, Figures 10-14 emphasize the efficiency of the heat transfer because of the amplified turbulence phenomena determined by the compartmentation of the distributor and collector, where there is usually a high tendency of stagnation for the thermal agent.

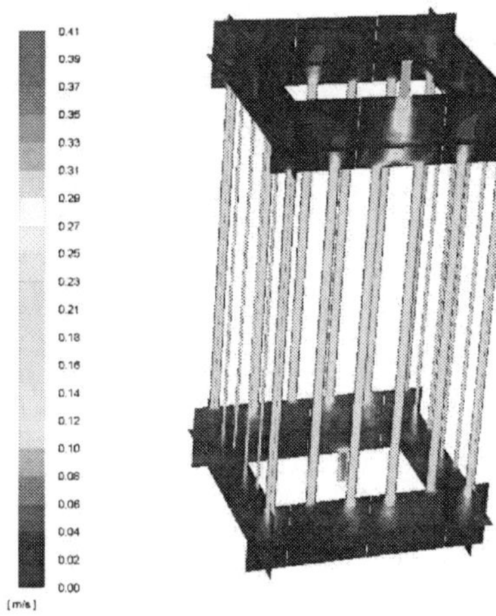

Figure 10. Contours of velocity – inside overview.

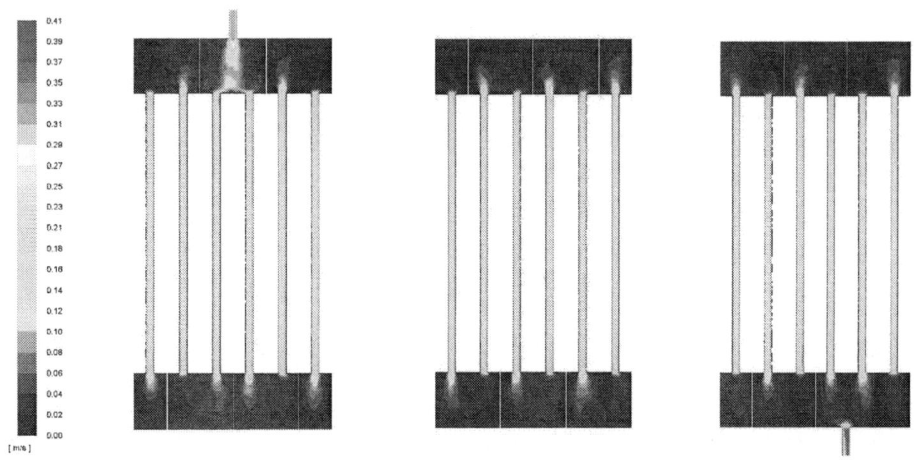

Figure 11. Contours of velocity: a) front section; b) lateral section; c) back section.

Figure 12. Velocity vectors: a) front section; b) lateral section; c) back section.

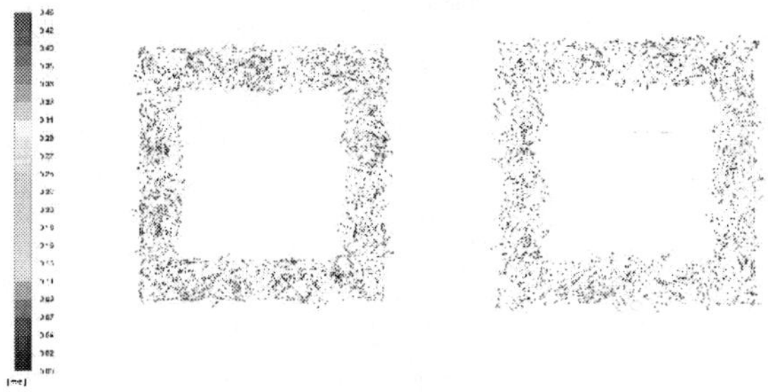

Figure 13. Velocity vectors: a) distributor section; b) collector section.

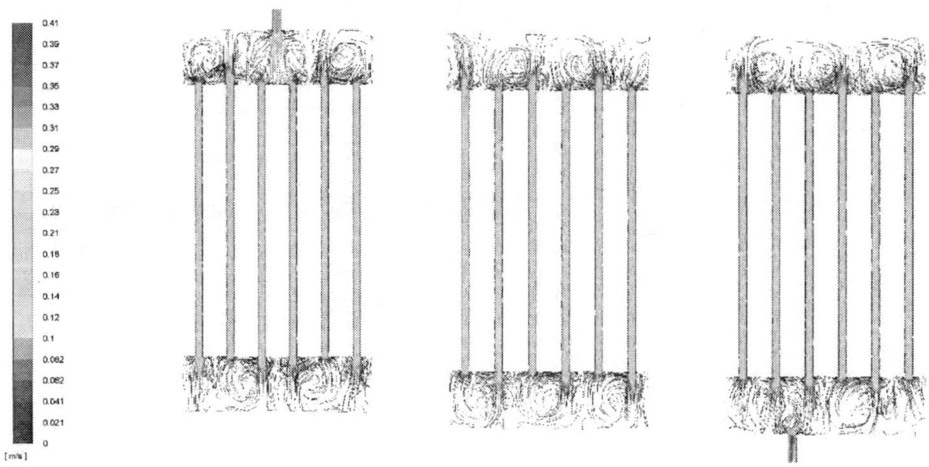

Figure 14. Velocity path lines: a) front section; b) lateral section; c) back section.

Numerical modelling provides a good view of the qualitative aspects of flow and distribution of heat agent temperatures through the proposed heat exchangers. Also, a good correlation can be observed between literature values and those obtained by numerical modelling.

The performance of the heat exchanger is analysed from the point of view of the heat transfer efficiency as global heat transfer coefficient. During the operation of the system, a uniformity of the temperatures on the surface of the heat exchanger is noted.
The results of the numerical modelling were extrapolated for an example close to the daily practice (T_{ag_term} = 32 °C, T_{st_init} = 18 °C) and the values of the heat exchanger parameters are determined. Table 2 shows the values of the thermal flux (Q) and the global heat transfer coefficient (K) for the heat exchanger analysed for different operating time intervals (τ).

Table 2. Calculation of global heat transfer coefficient

q [m^3/h]	τ [h]	T_{ag_term} [°C]	T_{st_init} [°C]	T_{st_fin} [°C]	Q [W]	K [W/m^2·K]
0.200	1	32	18	21.2	883,1	88,0
	2			22.6	744,3	78,2
	3			23.7	726,9	76,9

The proposed heat exchanger is a viable solution for use in systems equipped with heat pump and geothermal sources. The resulting global heat transfer coefficient has values between 76.9 ... 88.0 W/m^2·K. It is noted that this coefficient varies during the heat injection into the storage medium, decreasing once the temperature of the storage medium begins to increase.
According to the results, it has been found that by using the proposed heat exchanger a quasi-uniform loading of the storage medium is achieved, which is an important advantage in pre-dimensioning and correct evaluation of a system to be installed.

5. Conclusions

The main result of the study is that the heat exchanger achieves a constant heat flux ensuring uniform heating of the earth mass.
The proposed heat exchanger model is very well suited to new buildings with low availability for low-depth groundwater capture, equipped with ground-water or water-water heat pumps. Also, the proposed solution can successfully replace ground energy capture systems with deep drillings (100 to 200 m) at a reduced price and superior efficiency.
With regard to existing buildings equipped with cold-water heat pumps in the classical variant, the advantage of integrating the proposed model lies in the flexibility, the reduced amount of necessary works and the limitation of the intervention area.

6. References

[1] Cazacu I 2016 Utilizarea formelor de energie regenerabile în sisteme echipate cu pompe de căldură, Iași
[2] Prică G 2015 Teza Doctorat: Cercetări privind optimizarea pompelor de căldură care au solul drept sursă termică-București, pag 09-30
[3] Popovici CG, Hudişteanu S, Ciocan V, Burlacu A, Verdeș M, Bărbuță M 2018 Principiu constructiv și model de calcul pentru schimbătoare geotermice cu geometrie variabilă cilindrică sau tronconică, Conferinta Nationala de Instalatii, Sinaia, Ed. Matrix Rom, pp 179-184
[4] Popovici C G, Hudişteanu S, Ciocan V, Burlacu A, Verdeș M, Bărbuță M 2018 Algoritm de calcul pentru geoschimbătoare cu geometrie specială, Conferinta Nationala de Instalatii, Sinaia, 2018, Ed. Matrix Rom, pp 167-170
[5] Gavriliuc R 2015 Regulatory Aspects Referring To Gshps Systems, Conferința „Instalații Pentru Construcții Și Economie De Energie" Iași, pp 07-14
[6] Gavriliuc R 1999 Pompe de căldură - de la teorie la practică, Editura MatrixRom

[7] Gavriliuc R, Cucueteanu D, 2016 First data base for ground source heat pumps systems in Romania – Romanian Conference on the Energy Performance of Buildings RCEPB VIII, Bucharest, June 2-3, 2016
[8] 2013 ASHRAE Handbook, Fundamentals, American Society of Heating, Refrigerating and Air-Conditioning, Atlanta, USA
[9] 1995 ASHRAE Commercial/institutional ground-source heat pump engineering manual, American Society of Heating, Refrigerating and Air-Conditioning Engineers, Atlanta
[10] 2012 ASHRAE Handbook- HVAC System and Equipment
[11] Sebarchevici C 2013 Optimizarea instalațiilor termice din clădiri în scopul reducerii consumului energetic și a emisiilor de CO2, utilizând pompa de căldură cuplată la sol, teza doctorat
[12] Societatea Română GeoExchange, 2016, ‘Baza de date sisteme GSHP Technocon Engineering Baza date sisteme GSHP - SC Termoline Oradea
[13] Bonin J 2015 Heat Pump Planning Handbook, Routledge. Available at: https://books.google.com/books?id=LEOsCQAAQBAJ&pgis=1
[14] Ekman B et al 2008 Handbook on indirect refrigeration and heat pump systems, 46(0), pp. 1–23
[15] EurObserv’ER 2011 The State of Renewable Energies in Europe, 33(December 2011), pp. 4–7 Available at: http://www.eurobserv-er.org/pdf/bilan11.asp
[16] Forsén M 2008 Market status for ground source heat pumps in Europe Deliverable 2 Ground-Reach Market status for ground source heat pumps in Europe, July
[17] McQuay International 2002 Application Guide AG 31-008 Geothermal Heat Pump Design Manual, pp. 1–70, Available at: http://www.15000inc.com/wp/wp-content/uploads/Geothermal-Heat-Pump-Design-Manual.pdf
[18] NRC 2004 Natural Resource Canada’s Office of Energy Efficiency: Heating and cooling with a heat pump, EnerGuide
[19] Rehman, U 2011 Heat Transfer Optimization of Shell-and-Tube Heat Exchanger through CFD Studies Heat Transfer Optimization of Shell-and-Tube Heat Exchanger
[20] https://www.ehpa.org
[21] https://www.statista.com
[22] ANSYS-Fluent Documentation

Acknowledgement: This work was supported by a grant of the Romanian Ministry of Research and Innovation, CCCDI - UEFISCDI, project number PN-III-P2-2.1-CI-2018-1491, within PNCDI III.

IOP Conf. Series: Materials Science and Engineering **586** (2019) 012035 doi:10.1088/1757-899X/586/1/012035

Comparative analysis of heating systems used in Orthodox churches

S V Hudişteanu[1], V Ciocan[1,*], M Verdeş[1], N C Chereches[1], M C Bălan[1], E F Ţurcanu[1]

[1]Gheorghe Asachi" Technical University of Iaşi, Faculty of Civil Engineering and Building Services, 700050 Iaşi, Romania

E-mail: vasilica.ciocan@tuiasi.ro

Abstract. This work displays a numerical examination of the improvement of indoor atmosphere conditions inside an Orthodox church of Iasi. The objective consists in proposing a solution for reducing the risk of condensation for the case of the real heating system with hot air, used in Three Holy Hierarchs Monastery of Iasi. The solution was to use the local ventilation at towers, that generates air flow in these zones and evacuate humidity. This configuration was tested numerically and the results highlighted its effects. For comparison, the local ventilation was studied for two other heating solutions that are largely used in this type of buildings, underfloor heating and static heaters. The simulations demonstrated that the local ventilation represents a suitable solution when it is combined with the hot air heating.

1. Introduction

The importance of churches consists in their wealth in sacred and liturgical items, as well as the patrimony preserved in museums, historical and heritage buildings. It is specific for the churches to preserve many kinds of valuable artworks, many of them with high vulnerability, like paintings on canvas or wooden panels, which are subject to cracking, swelling, blistering, and soiling. The frescoes are mostly exposed to efflorescence and blackening, while wooden artefacts to cracking, metals to corrosion and textiles to fading and soiling [1]. Therefore, the ventilation and air conditioning system have a decisive role in preserving this worth.

For heritage buildings, the thermal indoor climate is influenced by following parameters: air temperature, temperature of surfaces, relative humidity and air velocity.

For controlling the indoor climate, there is important to have physical and quantitative understanding of the complex interaction in the building between air and the building elements, objects and people.

The comfort represents a subjective and debatable factor that describes if the occupant precept the indoor climate acceptable. Usually, the occupants are very sensitive to temperatures, but not so sensitive to relative humidity. The comfort temperature range depends mainly on clothing, activity and duration of stay in the building; a typical range is 18–22°C. Relative humidity is important for humans when it is very high, over 80%, or lower than 30% [2].

The indoor climate assures the conservation of materials in the building and minimizes aging and degradation of the materials that are to be preserved. These effects are influenced by on the materials and the type of degradation processes that are prevalent in the building. In the case of the materials inside heritage buildings, the relative humidity is often the most important climate parameter [2].

Another important aspect is the cost of a solution, which is a limiting factor and must be taken into account from the beginning, because a solution that is too expensive is most often useless [2].

In reference [3], Ulf Christensen presented several theories and tests referring to the traditional heating systems and products/components, such as electrical heating, low temperature electrical sheets, panel-heaters at walls or more recent ones, like bench-heaters or mobile radiant heaters. Another solution for newer or older churches is the gas radiant heating or water-based heating, with some different sorts of heating sources or distribution networks [3].

The literature also presents the particularities of HVAC systems installed in four Lithuanian churches and their impacts [4].
An example of the principle of blowing out warm air with gas heating system has been applied for heating-ventilation process at the Cathedral of Sts. Apostles Peter and Paul. An automatic control system allows the warming up the Cathedral up to the comfort temperature during ceremonies, and to maintain the standard relative air humidity as well when necessary. Thus, an economic efficiency was assured and the degradations inside this church are minimally [4].
An electrical underfloor heating system has been introduced at the St. Virgin Mary church and the church of St. Trinity. The churches were continuously heated during the cold season and the electrical underfloor heating system has not damaged the interior of the churches [4].
The radiant heating system, which has been installed at the church of St. Virgin Mary, has damaged the interior of the church and did not produced the desired thermal effect in local places of people's presence. Therefore, this heating system has not been utilized in the church any longer [4].
The present work presents the analysis of the heating system of the Romanian place of worship "Three Holy Hierarchs" Monastery from Iasi. Since 1880, when a capital restoration was achieved, the building was equipped with hot air central heating system, which is partially functional in present.
The analysed solution has been engineered by F.R. Richnowski of Lemberg, between 1885 and 1886. Indoor heating was provided by air heated in a central station powered by wood, located in a specially designated place in the church basement. The fresh air intake and the flue were located outside the building. The natural air circulation was assured, through channels of stone and block brick work laid under the floor. In this area, there are connected the air intake and discharge gates, with a perimeter arrangement at the decking and distributed evenly in the middle, in front of the church and inside the altar [5].
The importance of the HVAC system in the conservation of unique ceremonial objects and works of art housed inside the cathedral is determinant, because the heat and humidity managed by these systems are the most aggressive factors of destroying paintings and other valuable works [6-9].
The main objective of the paper consists in proposing a technique to decrease the risk of condensation for the case of the actual heating system (by hot air) used in the Three Holy Hierarchs Monastery of Iasi. The solution consists in using the local ventilation at towers that generates an air circulation in this area and evacuate humidity. This configuration is numerically tested and the results are highlighted as temperature and velocity spectra and profiles.

2. Case description

The current HVAC system implemented in the Three Hierarchs Monastery consists the base case of the study. In figure 1 is presented the system and its main components.

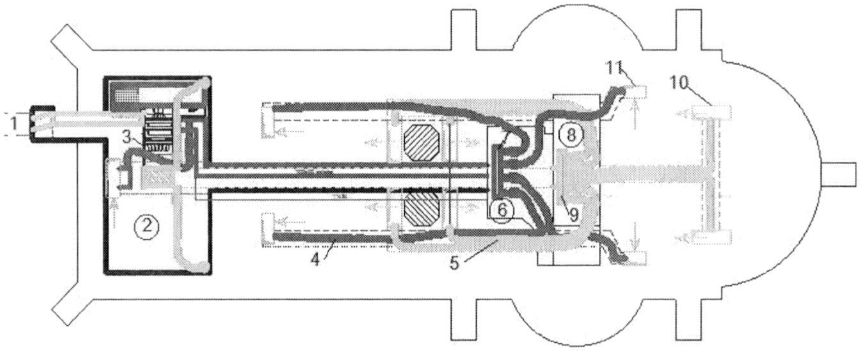

Figure 1. Current heating and ventilation system of the Three Hierarchs Monastery.
1 - fresh air inlet duct; 2 - subsoil of porch; 3 – air handling unit; 4 - flexible tube connected to the absorption grid; 5 - flexible tube connected to the exhaust grid; 6 - depression chamber; 7 - exhaust air manifold; 8 – under pressure chamber; 9 - treated air manifold; 10 – exhaust grid; 11 - inlet grid

The following cases are also studied in the paper. The second configuration, figure 2b, is obtained by improving the base case. The third and fourth cases, figure 3a and figure 3b, were analysed in order to compare the proposed solution with another two largely used heating systems suitable for this type of buildings. In these figures, T_i and v_i represent the inlet temperature and velocity for each admission zone.

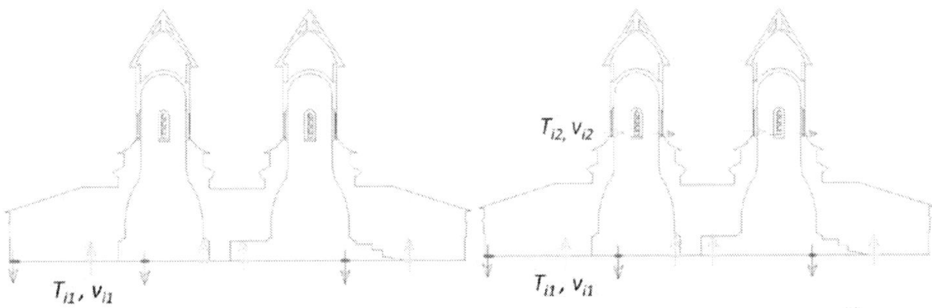

Figure 2. Grid positions for (a) base case; (b) second case: - with tower ventilation.

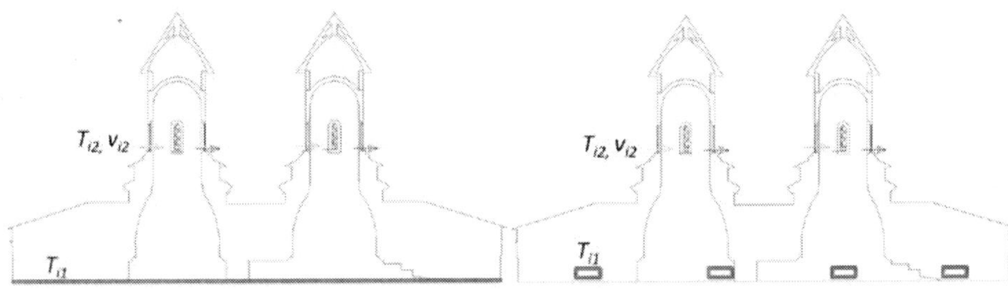

Figure 3. (a) Underfloor heating system; (b) Static heaters system.

3. Numerical analysis

The numerical analysis is achieved by using ANSYS-Fluent software. The fluid flow in steady state regime is used, while the type of flowing is the turbulent one: k- ε RNG model.
The numerical analysis was realized for a 2D geometry of the entire longitudinal section of the building. The mesh of the model and the operating conditions are presented in figure 4.

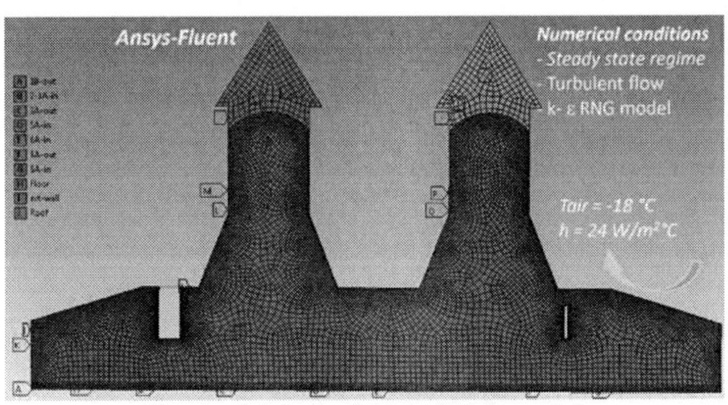

Figure 4. Mesh of the building model.

The external conditions imposed to the walls and windows in simulations were the temperature of air of -18 °C and the convective heat transfer coefficient of 24 W/m^{2}K.

4. Results

The results obtained from numerical simulations are represented as temperature and velocity spectra and profiles. The values for qualitative analysis of the flowing are presented in the following images - figures 5-15.

The currently implemented solution that consists of heating and ventilation of the church by air represents the base case (Case1 – figure 5a), with the inlet and outlet air grilles placed at the floor level. The air recirculation at the lower zone of the church leads to a homogenous temperature, determined by the three air inlets and two outflows. The inlet velocities were of 0.5 m/s, while the outflow velocities reach a maximum of 1.4 m/s. Simultaneously, the poor air circulation inside the two towers is influencing this zone, determining a high danger of condensation. Thus, the local ventilation is recommended for moisture evacuation and condensate removal, figure 5b. This configuration represents the Case2 of the presented analysis. Under the effect of ventilation, the velocities raise inside the towers and evacuate the excess of humidity.

The other two cases studied are taking into account two heating solutions that are largely used in churches: Case3 – Underfloor Heating (figure 7a) and Case4 – Static Heaters (figure 7b), combined with the solution of ventilation the towers. In terms of air circulation, the two solutions are almost the same. The quantitative data are merged in figure 11 for the velocity at 1 m height from the floor in all studied cases.

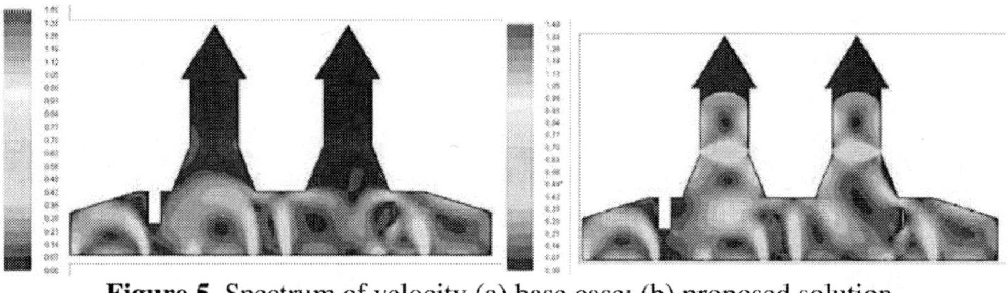

Figure 5. Spectrum of velocity (a) base case; (b) proposed solution.

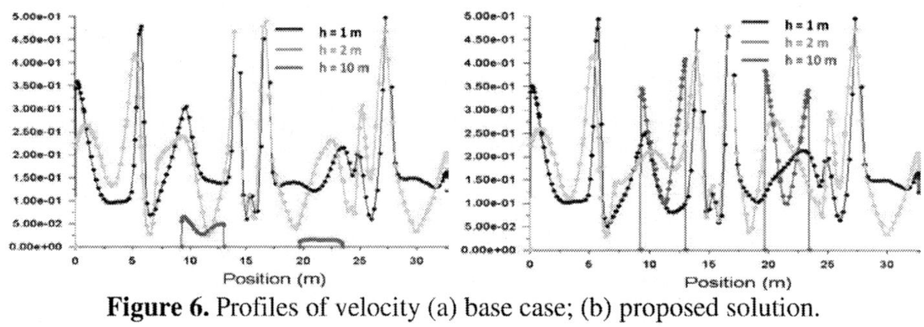

Figure 6. Profiles of velocity (a) base case; (b) proposed solution.

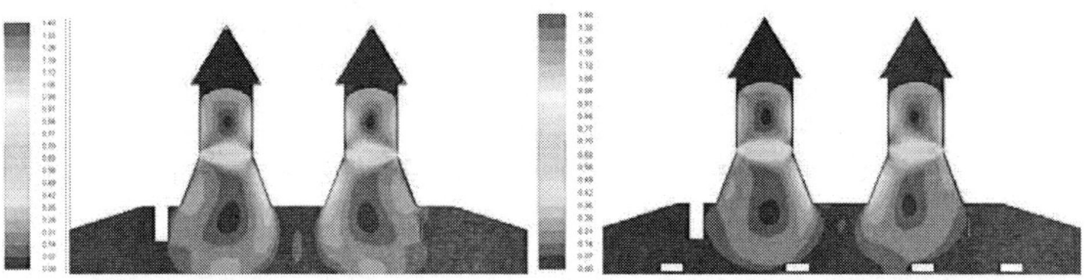

Figure 7. Spectrum of velocity (a) underfloor heating; (b) static heaters.

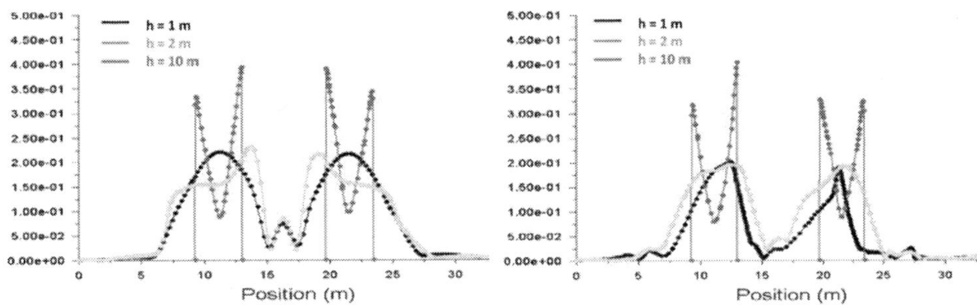

Figure 8. Profiles of velocity (a) underfloor heating; (b) static heaters.

The same behaviour of the air flow can be observed in figure 9 and figure 10, which represent the velocity vectors.

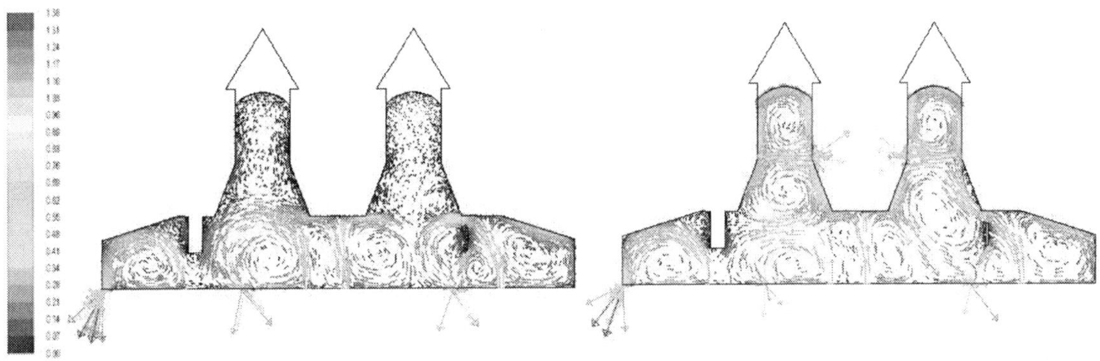

Figure 9. Vectors of velocities (m/s) (a) base case; (b) proposed solution.

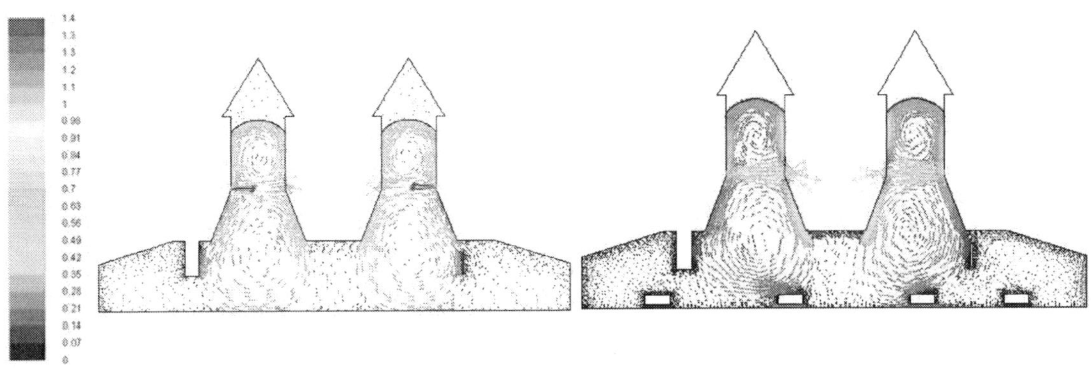

Figure 10. Vectors of velocities (m/s) (a) underfloor heating; (b) static heaters.

The analysis of the velocities for the studied cases revealed some important particularities. In the base case, when the heating and ventilation of the church consists in air-based solution, the air recirculation at the lower part of the building leads to a more uniform temperature. In the three inlets grids the velocities reach values of 0.45 m/s, while for the three outlets the maximum velocities are of 1.45 m/s. The low air circulation within the two towers is influencing this area, creating a high danger of condensation, figure 12a and figure 13a.

Thus, the local ventilation of the towers is justified, for removing the excess of humidity and eliminate the risk of condensation. The results for this configuration are presented in figure 12a and figure 12b. Under the effect of ventilation, the velocities rise inside the towers and evacuate the excess of humidity.

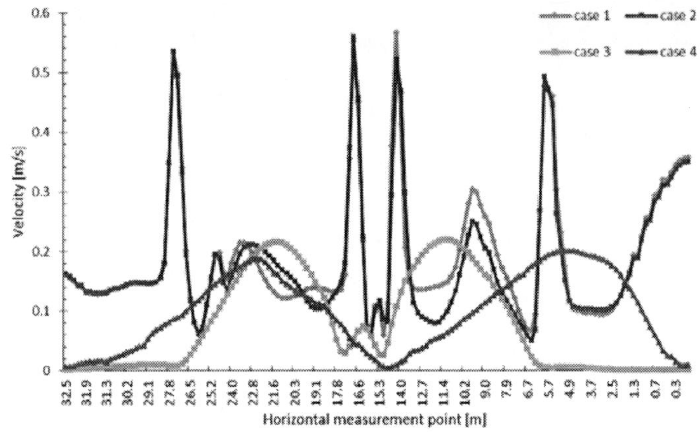

Figure 11. Velocity profiles at 1 m height from the floor.

Figure 14 presents the other two heating solutions that are largely used in churches: figure 14a – underfloor heating and figure 14b – static heaters. These configurations are combined with the ventilation solution of the towers. The results showed that the global air circulation for the last two solutions is almost similar, while the temperature profiles are slightly different, figure 15.

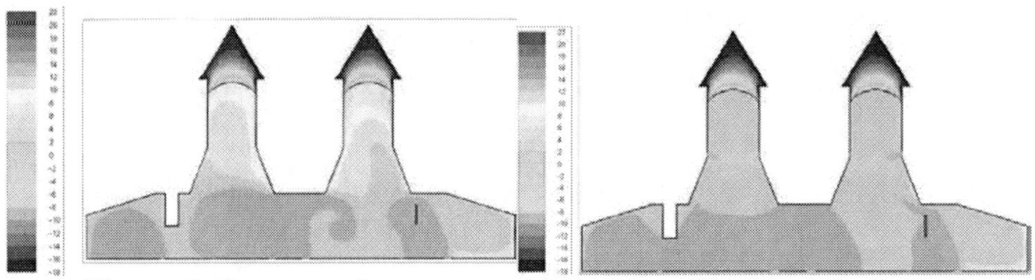

Figure 12. Spectrum of temperature (a) base case; (b) proposed solution.

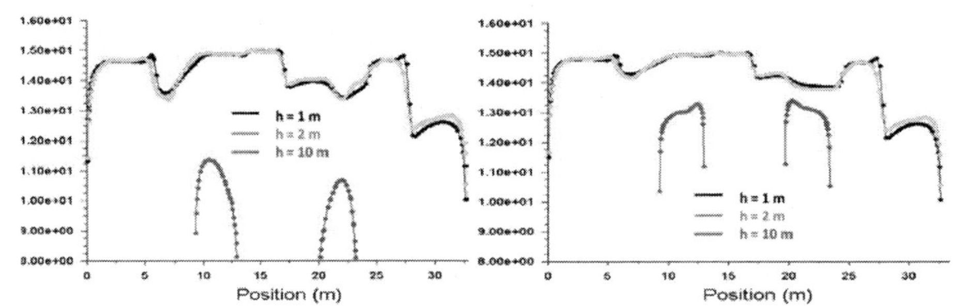

Figure 13. Profiles of temperature (a) base case; (b) proposed solution.

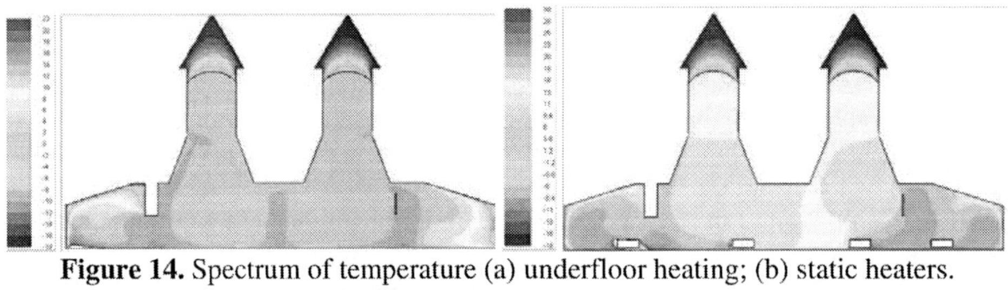

Figure 14. Spectrum of temperature (a) underfloor heating; (b) static heaters.

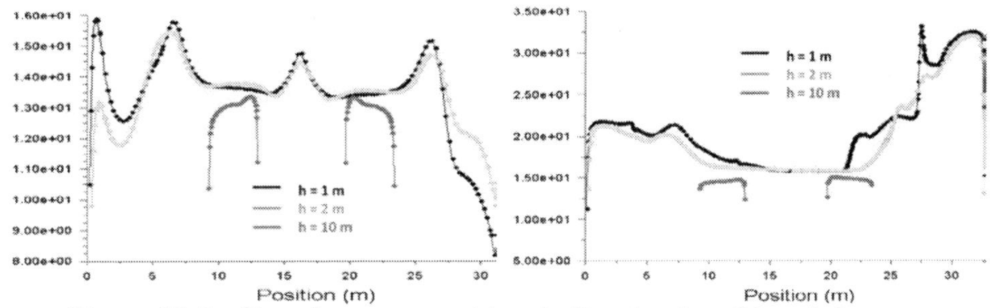

Figure 15. Profiles of temperature (a) underfloor heating; (b) static heaters.

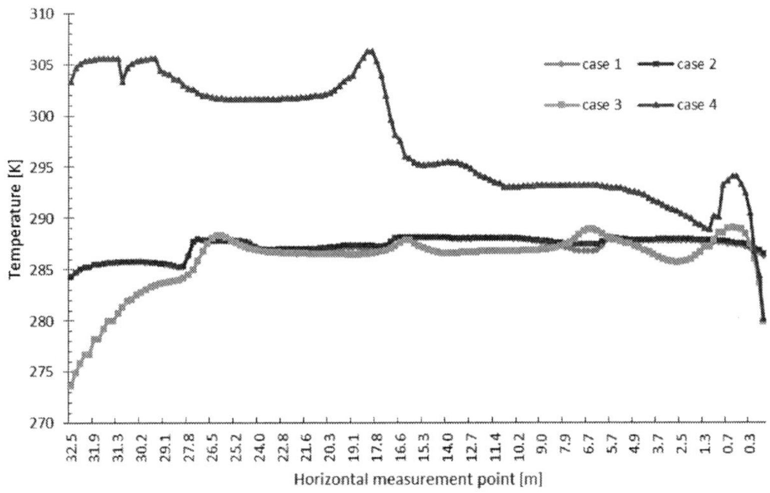

Figure 16. Profiles of temperatures at 1 m height from the floor.

The effect of the air flowing for each case is emphasized by the temperature distribution. The major deficiency is registered in the base case, when a lower temperature of the air in the towers and their walls is recorded. Instead, for the second case, the advantage of using the local ventilation can be noticed in the raise of temperatures adjacent to these zones.

The underfloor heating system is characterized by lower velocities at the church's external walls, that determines lower temperatures on their surface. In the case of static heaters, it can be observed a non-uniform distribution of temperatures, especially in the occupied zone.

For the first three cases studied, the temperature at 1 m height has approximately the same value of 15 °C, figure 16. The graph of temperatures for the static heater case shows the influence of their presence. However, the mean value of the temperatures in the occupation zone is also about 15 °C.

5. Conclusions

The new configuration of the HVAC system, with local ventilation in towers, enhanced the exhaust of humidity and also reduces the risk of condensation on the walls and paintings.

An important aspect consists in the small effect of the tower ventilation over the distribution of temperatures and velocities on the occupied zone.

When underfloor heating system and static heaters are used together, the tower ventilation creates some recirculation zones below them which determines a gradient of temperatures that rises towards the sides of the church.

The main conclusion is that the second case studied, the hot air HVAC system, is the most appropriate for assuring the comfort parameters in the occupational zone of this church.

6. References

[1] Camuffo D and della Valle 2007 A Church heating: A balance between Conservation and Thermal Comfort, *Experts Roundtable on Sustainable Climate Management Strategies held in Tenerife*, Spain

[2] Bronström T 2004 Fundamentals of indoor climate *Seminar in Riga regarding Indoor Climate in Churches – Problems and solutions*

[3] Chriestensen U 2004 Heating strategies in Norway *Seminar in Riga regarding Indoor Climate in Churches – Problems and solutions*

[4] Piksriene D 2004 Heating devices and their influence on the interiors of cold churches *Seminar in Riga regarding Indoor Climate in Churches – Problems and solutions*

[5] Mateescu T 2014 *Ensuring the microclimate in religious buildings – Historical testimonies*, XXIVth National Conference Building Services and Energy Economy, Iaşi, 3-4 July 2014, vol I

[6] Hudisteanu V S, Baran A I, Balan M C, Chereches N C, Mateescu T, Verdes M and Ciocan V 2014 Improvement of the indoor climate conditions inside orthodox churches, *International Scientific Conference CIBV 2014*, Braşov, pp. 61-66

[7] Ţurcanu F E, Verdeş M, Şerbănoiu I, Ciocan V and Bălan M C 2017 Assessment of Indoor Climate Environment via Hygrothermal Simulation in Historical Churches, *INTER-ENG 2017*; 10(10): 412-417, 5-6 Oct 2017 Târgu Mureş, Romania

[8] Şerbănoiu I, Verdeş M and F E Ţurcanu 2015 Churches Heating: A Choice Between Conservation and Thermal, *The Bulletin of the Polytechnic Institute of Jassy*, 2015; LXI (LXV) (3): 115-122

[9] Ţurcanu F E, Verdeş M, Bălan M C and Şerbănoiu I 2015 The Impact of Underfloor Heating in a Historical Church: A Case Study, *Proceedings of International Scientific Conference CIBV*, 2015; 457-462, Braşov, România

IOP Conf. Series: Materials Science and Engineering **586** (2019) 012036 doi:10.1088/1757-899X/586/1/012036

Microclimate analysis in Gheorghe Asachi auditorium: showcase for the indoor climate performance

M Verdeş[1,*], V Ciocan[1], E F Ţurcanu[1], R S Luciu[1], M C Bălan[1], S V Hudişteanu[1], A Burlacu[1]
[1]Gheorghe Asachi" Technical University of Iaşi, Faculty of Civil Engineering and Building Services, 700050 Iaşi, Romania
E-mail: verdesmarina2003@yahoo.com

Abstract. The subject of this paper is to the study the microclimate inside a heritage building. The Gheorghe Asachi Auditorium is one of most important building in the university and in the history of Jassy. Build in 1897, it was inaugurated by King Carol I. The auditorium serves as meeting place and manifestations for different occasions. Auditorium is built in baroque style with multiple paintings, artworks and wooden elements sensitive to moisture and temperature variations. CFD analysis is the proper tool for understanding the phenomena's regarding the movement of large volume of warm and moisture air. The analysis allowed us to make some recommendations to the conservation of such historical place but with regard to main human comfort.

1. Introduction

The Auditorium of the Technical University Gheorghe Asachi from Jassy is found inside the Palace University and thought its classic baroque style is distinguished by its beauty and elegant architecture. On its inaugural stone - 22 October 1897, King Carol I make the inauguration. Since than it known a cruel history, it has suffered massive structural degradation during the war and then rebuilt. The sketches of the University Palace were drawn by Petru Poni, professor at the university and the final project were finished by the renowned architect Louis Blanc.

The purpose of all factors concerning heritage and historic building is to preserve them and their contents. If the external factors as ground humidity, roof seepage or earthquakes, can be excluded we can conclude that damage depends mainly on the internal factors strictly that are connected with the indoor microclimate or that are creating the interior climate. The main subject of the indoor climatic studies concerns only the relations between occupants and indoor microclimate, such as the physiological response, hygiene and thermal comfort of the humans to the microclimate. [1], [5]

2. Heritage building and microclimate: point of view

In the case of heritage building such as, auditorium of Technical University, figure 1, figure 2 and figure 3, our opinion is that the main important aspect is regarding to the historical heritage of the build so that the indoor microclimate evaluation and monitoring should be to preserve the building and artefacts and not the thermal comfort in the events that are hold in the auditorium.

Figure 1. View from Auditorium of Technical University.

In historic buildings, we need to change the view that has occurred until now so: the "object" such as – artefact, pictures, statues, artwork, building, are more essential than people are. Therefore, the microclimate for the indoor of the auditorium has to have the comfort range calibrated and adapted in order to privilege the "historical heritage artwork" instead of wellbeing. Historic heritage buildings are used for a very various purposes that put in the front the pure conservation tin instead of reuse for specific functions: this can be seen in religious building such as churches, museums or libraries. Indoor microclimate in historically heritage buildings is the centre of the paper research, so the comfort studies, that have high temperature and relative humidity (RH) could damage plaster, wall that have frescoes, wood objects, brick and other building materials and objects contained in the building: papers, canvas, old but even new manuscripts and other types of artworks. [2], [3], [4]

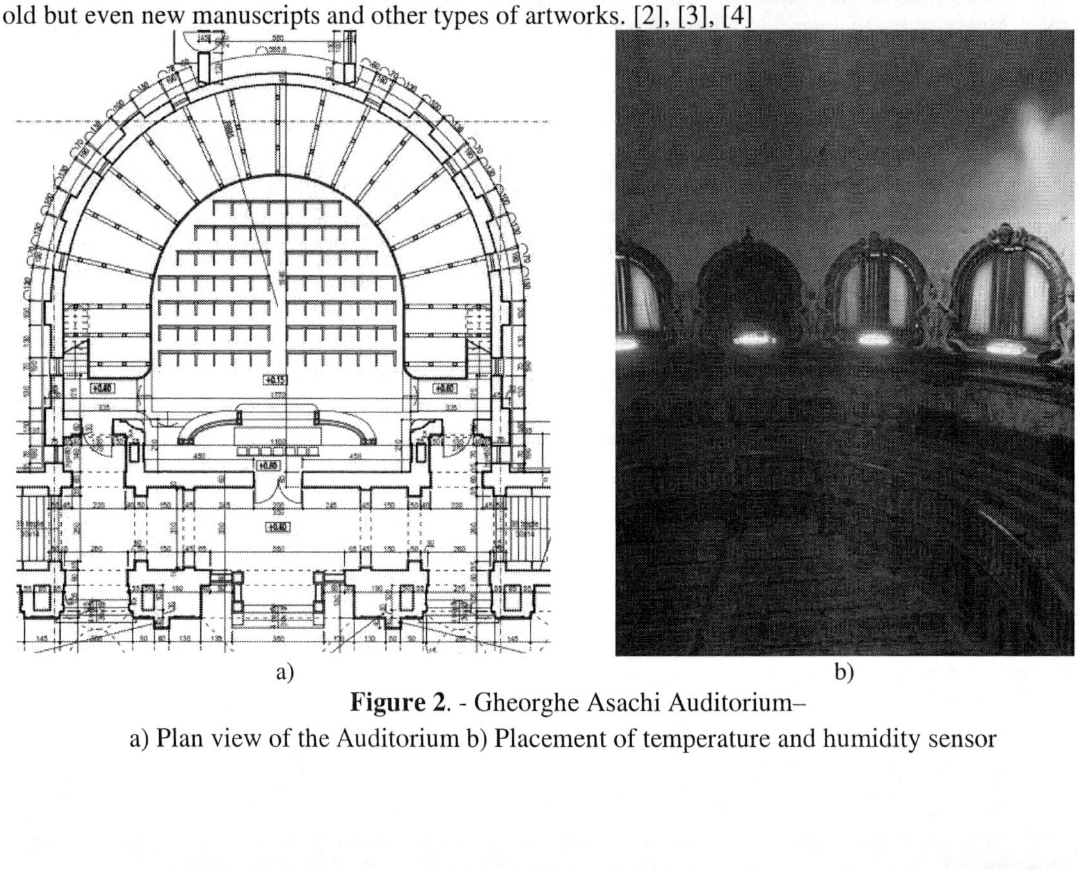

a) b)

Figure 2. - Gheorghe Asachi Auditorium–
a) Plan view of the Auditorium b) Placement of temperature and humidity sensor

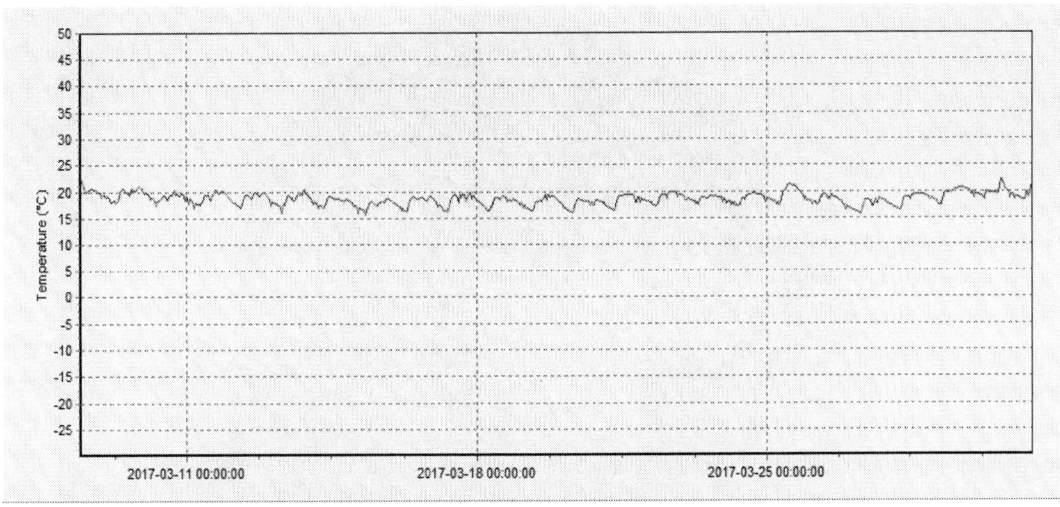

Figure 3. Variation of temperature in the plan view of the Auditorium.

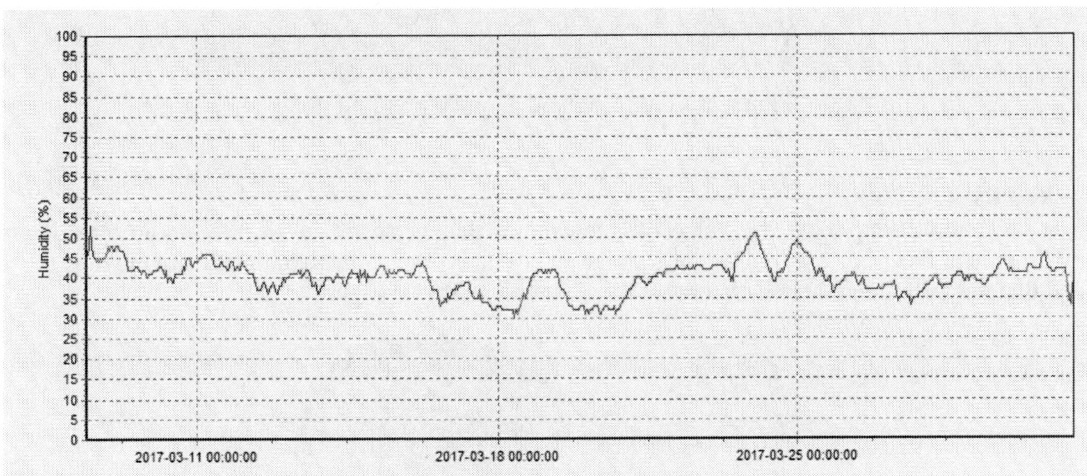

Figure 4. Variation of humidity in the plan view of the Auditorium.

3. Aim of this research

In this paper we describe results of the indoor microclimate modelling, simulation CFD is used since there is a large volume, parameter studies must be performed in order to have a better computation domain. The software used is Autodesk CFD 2017 – capable of solving heat, air and water vapour simulation, the model has been made in Autodesk Inventor 2017 in 3d space in order to have a model closed to the investigation site. Modelling had been carried out in steady state regime, in a climatic zone is III-rd, with -18°C air temperature. Now, the heating system is a classical one with static heaters. The boundary conditions imposed for walls are:

- the film coefficient for exterior walls is U=0.8 W/m^{2}K
- the film coefficient for interior walls is U=0.6 W/m^{2}K
- the film coefficient for windows is U=0.5 W/m^{2}K
- the film coefficient for floor is U=1 W/m^{2}K

The boundary condition used for static heater are taken from the site, cast-iron material with an approximated temperature of 80oC. Model discretization has a step of 0,5 meter. (figure 5) in a volume of air of 2736 m3.

IOP Conf. Series: Materials Science and Engineering **586** (2019) 012036 doi:10.1088/1757-899X/586/1/012036

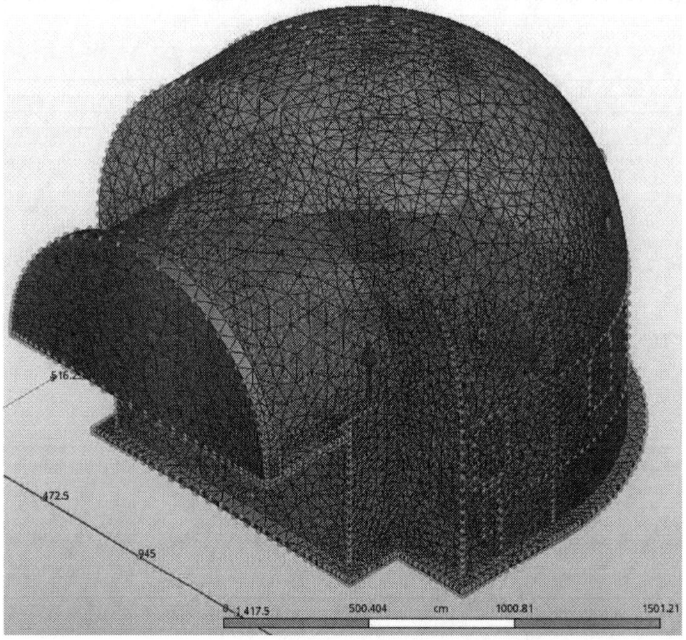

Figure 5. Auditorium 3d geometry and meshing.

4. Results

The CFD modelling show the value of temperature that is create inside the Auditorium interior environment. It can be observed that near to the static heaters the air has a high temperature than the rest and that the walls behind the radiators are washed with warm air (figure 6) [6].

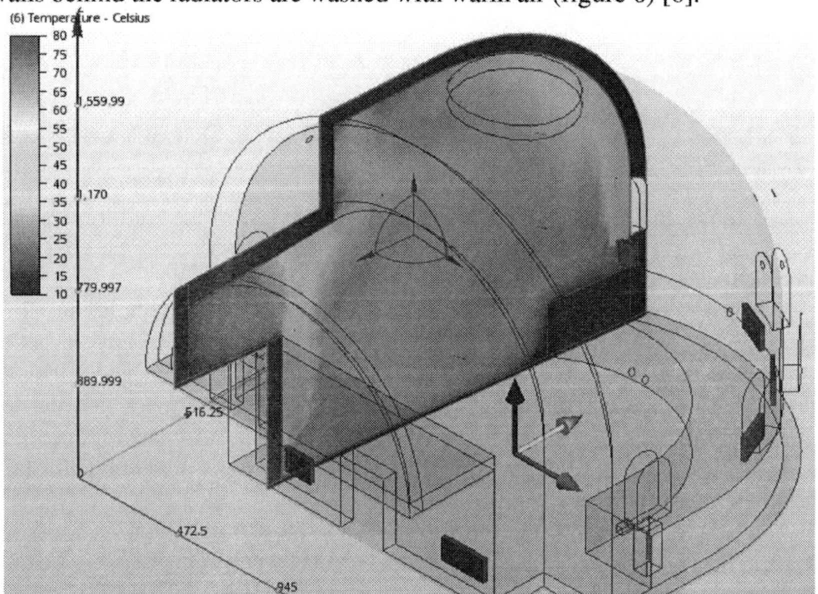

Figure 6. Range of temperature and heat dissipation viewed in cross section of Auditorium.

Standard thermal comfort developed by P.O. Fanger or predicted mean vote (PMV) is in good range despite the interior is heated with static heaters, meanwhile the temperature is uniform (figure 7).[5], [7]

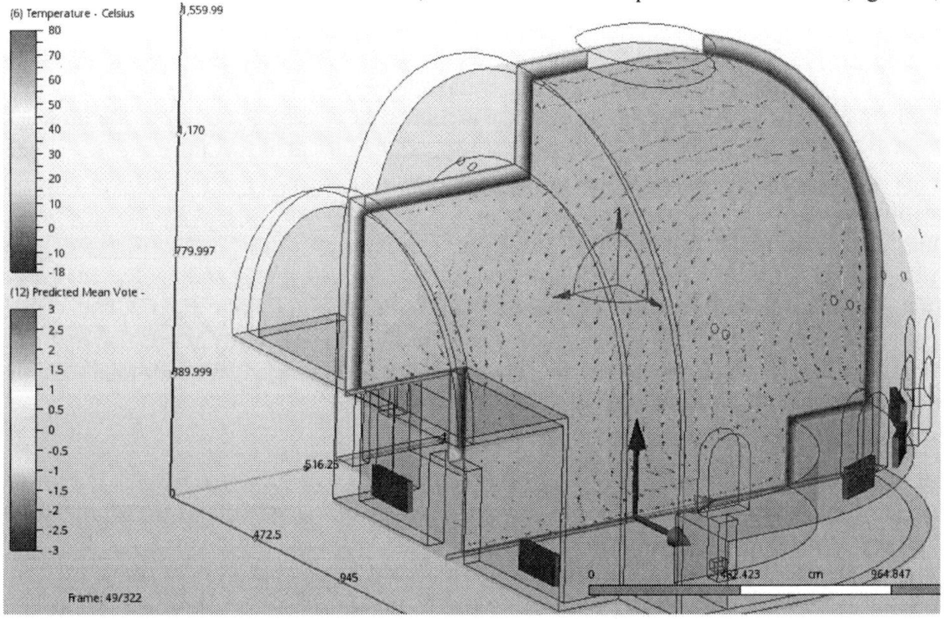

Figure 7. Predicted mean vote inside de interior climate created in Auditorium.

The air distribution it can be observed in figure 8, where the pattern of air movement is the best way to see that elements are affected by the speed of air near the walls thus affecting the paint and artworks.

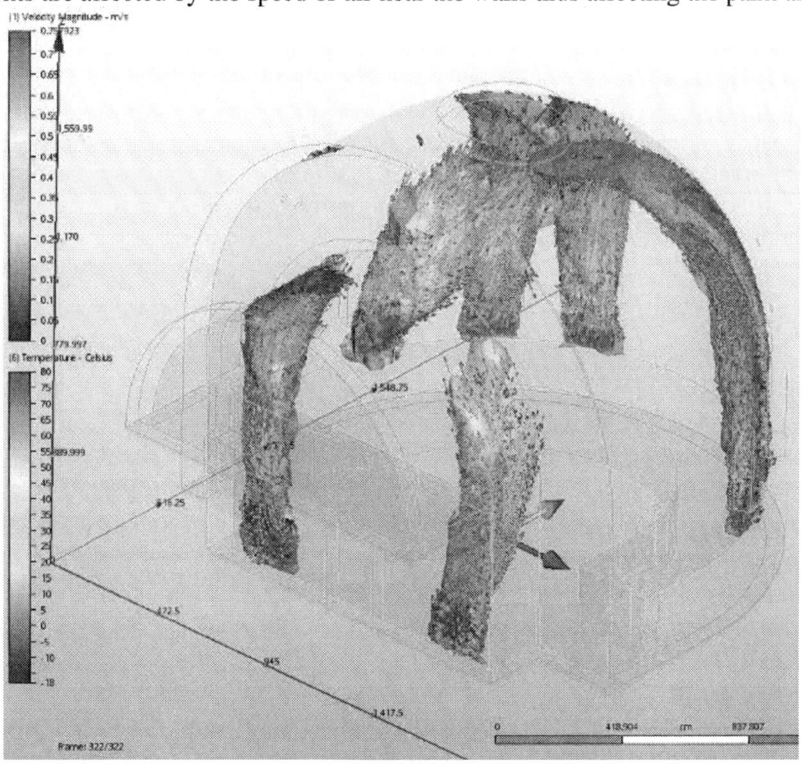

Figure 8. The model of trace for temperature and air movement in Auditorium.

It can be observed the difference in temperature distribution in the occupation zone. In case of classical solution with central heating and static heaters, the whole volume of air is heat figure 6. The warm air tends to rise to the upper part of church, which contain the art works. [8]
The air velocity of air is another important aspect. The system using static heats figure 6, figure 7 produce a high speed of air, that rises and in its move carry the dust, water vapor that has a strong destructive effect on paints and art works because is their condensing. In addition, the high speed of air causes a superficial drying of objects made from wood, painting of canvas, old book, etc. The movement of large mass of air has as result the movement of smoke from the candles that affect the paintings and fresco paintings.[9]
A building, and in particularly a historical heritage, can be viewed as a thermo-physical system that affects different scientific are of knowledge in the thermal comfort and archaeological artwork. The aim is in the knowledge to sustain the key factors conditions in the indoor microclimate: an important role is in this understanding the relationship between building and heating systems, also in the ventilation and air conditioning, this interaction beeing responsible for achieving ideal conditions for conservation of historically heritage buildings. [10]

5. Conclusion

The solution of heating with local heating elements can be effective for the occupants but for the interior artworks elements and paints it has a destroying effect. Also this system heats the whole volume of air after all, and because of that it isn't too efficient. In order to maintain the artworks heritage from the Auditorium, it must be carry out a research the see, which is the proper heating system with the minimum effects on the art work.
The simulation software Autodesk CFD is a proper tool in thermal comfort modelling and indoor climate simulations with good results in understanding of fluid dynamics and heat transfer.
Creating an indoor microclimate, different from the outdoor climate, is the inherent condition of most buildings. The indoor microclimate is meant to provide thermal comfort and wellbeing for people who live and work in climate created by the envelope of the building volume. Also, indoor microclimate induced degradation and this is one of the major hazards to our cultural heritage. The best conservation strategy is to act in order to prevent damages and degradation rather than reacting afterwards. Indoor microclimate control, when is properly used and design can act as an efficient and cost-effective method for heritage conservation. Often the discussion on indoor climate control has is focus on the technical solutions whereas the real difficulty lies in establishing proper climate criteria.
The thermal indoor microclimate is defined by:

- Air temperature
- Surface temperatures
- Relative humidity
- Air movements

The process of degradation, is basically described as a progressive and cumulative process of materials, that has different response to the values given by indoor microclimate and their changes, both in a temporal and spatial way. Rapid changes in a short period of time (or strong spatial gradients) in temperature and/or relative humidity, or exchanges of heat and mass in different direction in the area of the artwork (pictures, frescoes, wood, brick, stone), cause internal stress in materials with irreversible and cumulative effects that accelerate the deterioration process.

References

[1] Burlacu A et al 2014. Experimental study for data validation regarding the flow movement in natural convection in an asymmetrical heated vertical channel. Applied Mechanics and Materials, Volume 659, pp. 313-318
[2] Camuffo D 1998 Microclimate for Cultural Heritage. Amsterdam: Elsevier Science BV
[3] Camuffo D et al 2011 Humidity and environmental diagnostics in Palazzo Grimani, Venice. In: D. D. Curto, ed. Indoor environment and preservation – Climate control in museums and historic buildings. Florence: Kermes Quaderni, Nardini Florence, pp. 45-50

[4] Corgnati P S, Fabi V, Filippi M 2009 A methodology for microclimatic quality evaluation in museums: Application to a temporary exhibit. Building and Environment, pp. 1253-1260
[5] Fanger P O 1970 Thermal comfort. Analysis and applications in environmental engineering. Copenhagen: Copenhagen: Danish Technical Press
[6] Zuo W et al 2016 Coupling Indoor Airflow, HVAC, Control and Building Envelope Heat Transfer in the Modelica Buildings Library. Journal of Building Performance Simulation, 9(4), pp. 366-381
[7] Tian W, Xu H, Wangda Z, Sohn M D 2018 Building energy simulation coupled with CFD for indoor environment: a critical review and recent applications. Energy & Buildings, Volume 165, pp. 184-199
[8] Schellen H, Lambertus H 2002 Heating monumental churches - indoor climate and preservation of cultural heritage. Eindhoven(NL): Technische Universiteit Eindhoven
[9] Rohdin P, Moshfegh B 2011 Numerical modelling of industrial indoor environments: a comparison between different turbulence models and supply systems supported by field measurementss. Building and Environment, Volume 46, pp. 2365-2374
[10] Marco P, Kristian F 2018 Indirect investigation: building simulation in Historic indoor microclimate of the heritage buildings. Bologna: Springer, pp. 100-102

IOP Conf. Series: Materials Science and Engineering **586** (2019) 012037 doi:10.1088/1757-899X/586/1/012037

Contributions to simulation of the non-permanent movement in sewerage collectors

M Luca[1], A L Luca[2], C F Scripcariu[3], S Chirica[4]
[1]*Hydroamelioration and Environmental Protection Department, "Gheorghe Asachi" Technical University of Iasi, Iasi City, 700050, Romania*
[2]*Polias-Company, Iasi City, 700585, Romania*
[3]*Doctoral School, "Gheorghe Asachi" Technical University of Iasi, Iasi City, 700050, Romania*
[4]*Doctoral School, "Gheorghe Asachi" Technical University of Iasi, Iasi City, 700050, Romania*

E-mail: mluca2015@yahoo.com

Abstract: The water movement of the sewer collectors is of a non-permanent type because of the variation in flow over time. Non-permanent movement in collectors is mathematically modelled by the Saint Venant equations. To solve the equation system, the characteristic method was used, provided that both equations are quasi-linear in the unknown h and Q. Non-dynamic contouring conditions are given by the hydraulic feature of the collectors / tanks located on the collector. Dynamic contour conditions are rendered by the collector-specific equations (river level effluent hydrograph, manoeuvring parameters of the boiler, manoeuvring characteristics at pumping stations, etc.). For the mathematical model implementation, a computational program was developed in the Matlab programming environment (Variation Gradual Collector Program). The program was used to simulate the non-permanent movement in the main sewerage collector "Cacaina" from the sewerage network of the city of Iasi. The simulation of the non-permanent movement was done in two collector evacuation scenarios (gravity and pumping). The analysis of the results obtained for scenario 2 highlights the differential operation of the eight collector sections separated by the side feeders. Simulation of the movement has highlighted the exceeding of the minimum and maximum values of the parameters imposed by the operating rules.

1. Introduction

A sewerage network consists of a succession of collectors (closed section channels) on which hydro-urban public work structures are placed with the purpose of taking over and transporting waste water. The water movement in sewer networks in Romania is generally free flow (gravitational type) [1]. The main form of water movement in sewer collectors is non-stationary (non-permanent) but it can be considered as stationary (permanent) for longer time frames. This aspect is used in the sizing of sewer collectors [1], [2], [3], [4].

Most waste water sewage systems in Romania were designed and built for combined operation [1]. Various technical concepts, specific to Romania's economic and political development phases, were used during their design. The running of combined sewerage network systems has a series of problems determined by the rainwater regime. Climate change in the last time frame has aggravated the collectors' operation during torrential rains. A particular problem is presented by the main sewer collectors, with visitable section, which take over the waste water (domestic + pluvial) and transport it to the discharge structure or the wastewater treatment plant.

The operational problems of the visitable sewage collectors that have occurred in the last time frame, especially sewage spills on the streets surface, have required research on the hydraulic regime and simulation of permanent and non-permanent movement in the longitudinal profile of the collectors [5]. The simulation of water movement in the collector requires precise knowledge of the initial and contour conditions existing in the longitudinal profile [6], [7], [9]. A large part of the main sewer collectors have sections which were modified from the original ones, and differ geometrically and hydraulically from the standardised sections. This situation requires special computational programs to design and verify flow parameters. The purpose of this paper is to present the research conducted in order to make a computational program necessary to verify the non-permanent water movement in the main collectors.

2. The hydraulic – mathematical model

The water movement in a sewer collector is of non-permanent type. This type of movement is present especially in the takeover and transport of torrential rains. The drainage of the rainwater through the connection chambers located along the sewer collector amplifies the parameters of the non-permanent movement.

For the analysis and simulation of the non-permanent movement type in a visitable sewage collector, equipped with connection chambers, a hydraulic-mathematical model was developed, to which a numerical computational program was attached. The hydraulic model simulates the "waste water + pluvial water" movement in a collector in accordance to the boundary conditions and contour conditions required by the operational process [10].

Rapid and gradually varied non-permanent movements in sewer collectors can be analysed and / or simulated through complex mathematical models generally composed of: 1 - motion equations; 2 - the continuity equation; 3 - initial conditions; 4 - limit conditions [2], [3].

Non-permanent movement in collectors is mathematically modelled through the Saint - Venant equations, consisting of the mass conservation (the continuity equation) and the moment conservation equations. Saint - Venant equations can be mathematically rendered in both integral and differential forms, which are equivalent only if the dependent variables are differentiable [3], [8].

The differential forms of the equations were used in order to carry out the analysis model, respectively [2], [3],

- the continuity equation

$$B\frac{\partial h}{\partial t}+\frac{\partial Q}{\partial x}-q=0 \quad , \tag{1}$$

where Q is the flow; q - the specific inflow / outflow; B – the surface width of the bed; h – water depth; x – length; t – time;

- the convergent form of the moment conservation equation (the dynamic equation of the wave)

$$\frac{\partial Q}{\partial t}+\frac{\partial}{\partial x}\left(\beta\frac{Q^2}{A}\right)+g\,A\left(\frac{\partial h}{\partial x}+S_f-S_0\right)=0\,, \tag{2}$$

where S_0 is the slope of the bed; S_f – the hydraulic slope; A – the area of the flow section; β Boussinesq coefficient. The S_f hydraulic slope, assuming that in the non-permanent movement the flow direction can be reversed, has the expression:

$$S_f=\frac{|Q|\cdot Q}{K^2}, \tag{3}$$

where K is the flow modulus.

From a mathematical point of view, (1) and (2) equations make a system of partial derivative equations, with t, time and x, spatial coordinate independent variables. The main dependent variables are h, water depth and Q, flow. A method of solving the Saint-Venant equation system, provided that both equations are quasi-linear for the h and Q unknown terms, is the characteristics method [2], [3]. The continuity equation (1) is quasi-linear; but to also prove the quasi-linearity of the dynamic wave equation (2), it was transcribed under the following form [3]:

$$\left(gA-\frac{\beta}{A^2}Q^2\cdot B\right)\cdot\frac{\partial h}{\partial x}+\frac{\partial Q}{\partial t}+\frac{2\cdot\beta}{A}\cdot Q\cdot\frac{\partial Q}{\partial x}+gA(S_f-S_0)=0 \quad . \tag{4}$$

In general, a system of two quasi-linear equations has the form [11]:

$$\begin{cases} a_{11}\dfrac{\partial h}{\partial t}+a_{12}\dfrac{\partial h}{\partial x}+a_{13}\dfrac{\partial Q}{\partial t}+a_{14}\dfrac{\partial Q}{\partial x}+a_{15}=0, \\ a_{21}\dfrac{\partial h}{\partial t}+a_{22}\dfrac{\partial h}{\partial x}+a_{23}\dfrac{\partial Q}{\partial t}+a_{24}\dfrac{\partial Q}{\partial x}+a_{25}=0, \end{cases} \tag{5}$$

and is characterised by A matrix, defined as follows:

$$\mathbf{A}=\begin{pmatrix} a_{11} & a_{12} & a_{13} & a_{14} & a_{15} \\ a_{21} & a_{22} & a_{23} & a_{24} & a_{25} \end{pmatrix}. \tag{6}$$

The elements of the A matrix can be determined through the coefficients in (6.1) and (6.4) equations, as follows:

$$\left.\begin{aligned} &a_{11}\equiv B,\ a_{12}=a_{13}\equiv 0,\ a_{14}\equiv 1,\ a_{15}=-q \\ &a_{21}\equiv 0, a_{22}=g\,A-\beta Q^2\frac{B}{A^2}, a_{23}\equiv 1, a_{24}=\frac{2\cdot\beta}{A}\cdot Q, a_{25}=g\,A\left(S_f-S_0\right) \end{aligned}\right\} \tag{7}$$

It is found that, in general, A = A (*t, x, H, Q*), so the equations (1) and (4) first order equations as well as the equation (5) system are quasi-linear. For concrete analysis, the system of Saint - Venant partial derivatives equations must be solved considering a series of appropriate *initial and limit conditions.*

The assumption of permanent and non-permanent gradually varied movement is accepted at the t_I initial moment. The initial conditions follow the (9) Equation form and can be numerically rendered by the following triplets:

$$(x_{0i}, h_{0i}, Q_{0i}),\ \text{cu } i = 1, 2, K, \dots, N_{tI} \tag{8}$$

The x_{0i} values for the x spatial coordinate are designed to include all significant variations for both the constructive and functional parameters of the collector. The (8) conditions can be expressed by the following three N_{tI} - dimensional vectors:

$$\mathbf{X}_0=\{x_{0i}\},\mathbf{h}_0=\{h_{0i}\},\mathbf{Q}_0=\{Q_{0i}\}. \tag{9}$$

There are two types of contour conditions: non-dynamic and dynamic. Non-dynamic contour conditions are based on the hydraulic characteristic of the hydro-urban public work structures present on the network. The dynamic contour conditions are given by specific equations, and the most common are: a - the hydrograph of levels in the effluent; b - the parameters for manoeuvring a weir valve; c - the pumping stations operational manoeuvring characteristics [10].
The limit conditions, defined in $x = x_k$, (k = 1, 2,..., N_{SC}) sections, can be put into the following general form [11]:

$$\mathrm{F}_k\left(t, H_k^{\mathrm{am}}(t), H_k^{\mathrm{av}}(t), Q_k^{\mathrm{am}}(t), Q_k^{\mathrm{av}}(t), \mathrm{u}_k(t)\right)=0,\ \text{for } t\in[t_0, t_f], \tag{10}$$

where:

$$H_k^{am}(t)=H(t,x_k-\varepsilon), H_k^{av}(t)=H(t,x_k+\varepsilon),\ \text{for } t\in[t_0, t_f] \text{ and } \varepsilon\to 0, \tag{11}$$

$$Q_k^{am}(t)=Q(t,x_k-\varepsilon), Q_k^{av}(t)=Q(t,x_k+\varepsilon),\ \text{for } t\in[t_0, t_f] \text{ and } \varepsilon\to 0, \tag{12}$$

and $\mathbf{u}_k(t)$ is *the control* vector from $x = x_k$ section, having $m_k \geq 0$ components (control units); $\mathbf{F}_k$ – operator of differentiated nature (system of algebraic or functional equations, tables, graphs etc.), having the following number of components [11]:

$$m_k + n + 1,\ \text{for } k \in \{1, N_{SC}\};\ m_k + n,\ \text{for } k \in \{2, N_{SC} - 1\}. \tag{13}$$

When the variation laws of the m_k control units are predetermined, that is, the following functions are known:

$$u_k^{\nu} = f_{uk}^{\nu}(t),\ (\nu = 1, \ldots, m_k),\ \text{for } t \in [t_0, t_f], \tag{14}$$

then this type of contour conditions are known as *non-dynamic*, otherwise the contour conditions are *dynamic*.

Due to the complexity of the mathematical model, flow problems in rapid varied non-permanent movements, gradually varied in the collectors of the sewage systems, can be solved only by numerical methods, of which the most used due to their efficiency are [2], [3], [7]: 1 - the characteristics method; 2 - the finite difference method. The characteristic method was used in the making of the hydraulic-mathematical model. Part of the elaboration details of the hydraulic - mathematical model are presented in the [5], [10] and [12] papers.

The algorithm designed was used to develop the **TGCProg** (Transient Gradually Varied Collector Program) computer program, necessary to numerically simulate non-permanent water movement in sewer collectors. The **TGCProg** computer program was validated through a representative test application in the field of prismatic beds free flow [10]. This computer program has the following features:

A. Solves verifying problems for complex sewer collectors in the Transient (Non-Permanent) **G**radually **V**aried **M**ovement.

B. The collector, considered an enclosed channel with sections with constant flow and slope, is made of standardised or non-standardised tubes with a complex cross-section consisting of line segments and / or circular arcs.

C. The main program enables the following operations:

- entering input data directly from the keyboard and saving them to specialised data files;
- entering input data from previously created and stored data files;
- processing the output data and saving them into data files on the collector operating specialties;
- appropriate graphical representations based on output data and in accordance to the requirements for the collector validation.

D. The main program calls, both directly and / or indirectly, the following six user functions for which the main features are specified [10]:

1. **F1** - evaluates the coefficients of the base, processed and resulting equation systems, which define the transient water movement in the collector;
2. **F2** - evaluates the value of a discontinuous function, with variations in stages;
3. **F3** - generates a differences grid divided with a variable step along the x-x axis;
4. **F4** - evaluates contour conditions at the upstream end of the hydraulic system (type II nodes);
5. **F5** - evaluates contour conditions at the downstream end of the hydraulic system (type III nodes);
6. **F6** - evaluates contour conditions in intermediate sections of the hydraulic system (type V nodes).

TGCProg computer program has the following working structure:

A. Menu with the codes used to define initial and contour conditions.

B. Input data:

B.1. Program constants

B.2. General variables of the problem

B.3. General variables on finite difference digitisation

B.4. Sizes and functions that define the equation system's coefficients

B.5. General notations which appear in the contour conditions

B.6. General notations which appear in the initial conditions

C. Output data:

C.1. Values for independent and dependent variables

D. Specific notations for the computation scenario

D.1. Constants of the application

D.2. Specific data from system coefficient expressions

D.3. Specific notations from contour conditions expressions
D.4. Specific notations from initial conditions expressions

3. Results and discussion

The **TGCProg** computation program was applied to the analysis of non-permanent flow in a collector located in the sewerage network of Iaşi city in Romania (figure 1). The "C Main Sewer Collector", referred in the paper as "C Collector", is an enclosed channel with a length of 2.416 km. It transports the Cacaina stream waters, the waste water and the rainwater from the served area. The shape of the flow section is differentiated on the length of the collector (figure 2). The ST weir is placed at the upstream end of the collector, which allows controlled access of the water from the Cacaina stream. "C Collector" discharges downstream through the CD8 discharge chamber into the "D Main Sewer Collector". On the "C Collector" route six chambers (CA1 - CA6) are located, in which lower order collectors with unilateral inflow discharge.

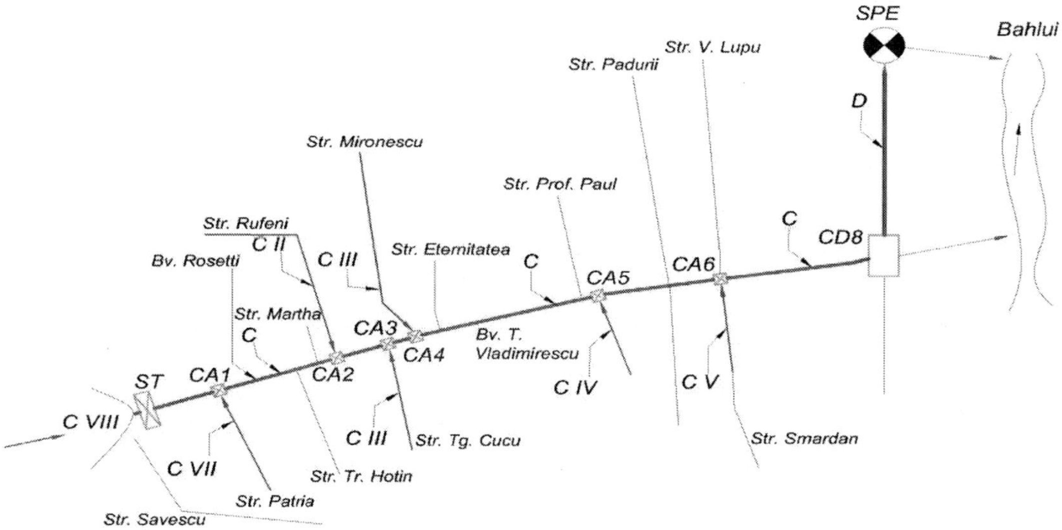

Figure 1. The structural network of "C Collector": "C" and "D" – main sewer collectors; ST – weir; CA1 – CA6 and CD8 discharge chamber; SPE – evacuation pumping station; C II – C VIII sewer collectors.

The "C main sewer collector" is located on C.A. Rosetti Boulevard - first section and continues on Tudor Vladimirescu Boulevard - the second section (figure 3). The first section, with a limited length of about 300 m, has a circular shape (Dn 700 reinforced concrete tubes). The second section is located on C.A. Rosetti Boulevard and T. Vladimirescu Boulevard, with a length of 2.116 km and a mixed flow section (reinforced concrete box made of a semicircle and a rectangle with the dimensions of 3450/2250 mm) (figure 2).

The "C" sewer collector functions in an "unitary system", which implies a non-permanent movement, especially in the case of torrential rains [14], [15]. The collector was rehabilitated on a series of sections, which changed the geometry of the flow section on a series of sectors [13]. The modified section no longer complies with the original (standardised) shape, due to the transition to a mixed type non-standard shape (curves + lines) [5], [10], [17], (figure 2).

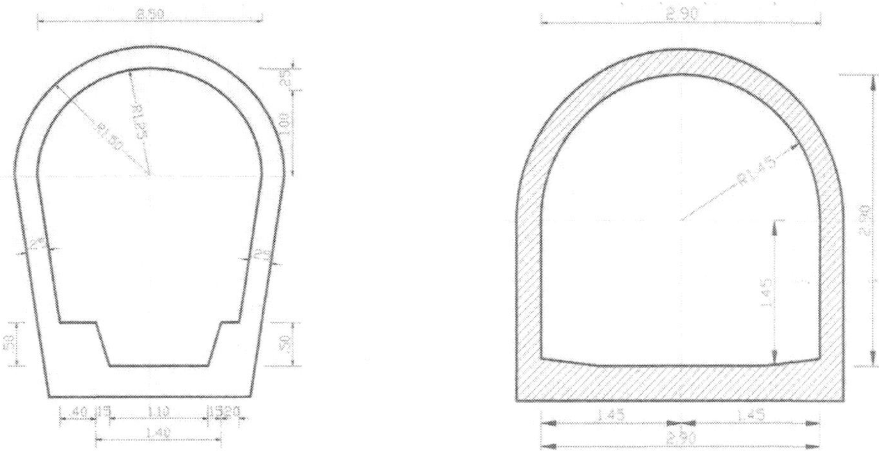

a b

Figure 2. "C collector" cross section: a – standardised section; b – non standard section.

Two operating scenarios were defined for the "C" Main Collector, with 1 - 2 analysis alternatives:
A - Gravitational discharging through the CD8 discharge chamber scenarios:
Scenario A1 - Bahlui River level is under the elevation of CD8 discharge chamber foundation.
B – "Tudor Vladimirescu" pumping station discharge scenarios:
Scenario B1 - The Bahlui River level is above the CD8 discharge chamber foundation;
Scenario B2 – Shut down of the pumping station caused by "power failure".

a b

Figure 3. The location of "C collector": a – C. A. Rossetti Boulevard; b – Tudor Vladimirescu Boulevard.

The simulation of collector running in the non-permanent movement was accomplished in two work stages:

Stage I: Pre-processing of constructive and hydraulic data. The geometric and hydraulic data of the collector, considered as input data into the program, were processed in uniform and non-uniform permanent movement for the adopted analysis scenario (details in [5], [10], [17]). Data processing was done by categories, using specialised computing programs on each type of water movement [5], [10]. The processed data was entered into the files as "Input data" in the **TGCProg** computation program. The technical data was stored in the **M** matrix, **M = [hC; BC; AC; RHC; zGC; WC; KC]** in data file *Hydraulic features Semicircle-rectangle.mat*. The ST weir is located at the upstream end of "C" collector, which allows the controlled access of the water from the Cacaina stream. Downstream, the "C" collector discharges into the CD8 discharge chamber. The constructive and functional data of the "C" collector were centralised in the calculation tables (T1 – "C" collector parameters in longitudinal profile, T2 – Constructive parameters of unilateral supply chambers, T3 - Hydraulic - functional parameters of the collector). Figure 4 and figure 5 shows the water depth /share variation at initial moment ($t = 0$) on the "C Main Collector" (h_N - normal depth, h_{CR} - critical depth, h_{GV} - non-uniform movement depth).

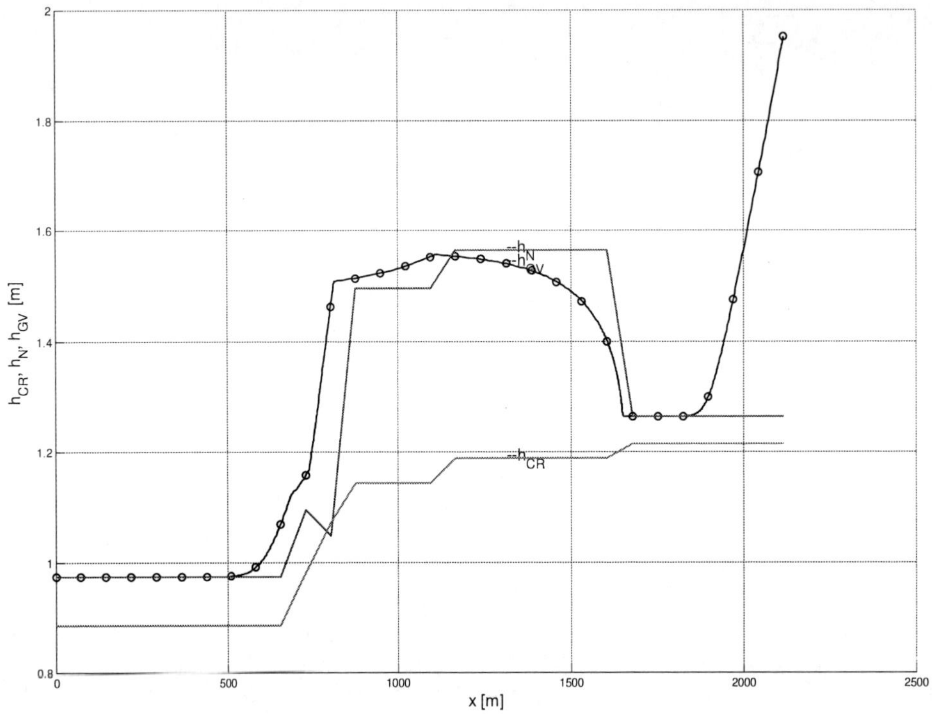

Figure 4. Depths of water at the initial moment on the „C" Collector": h_N – normal depth; h_{CR} - critical depth; h_{GV} – non-uniform movement depth.

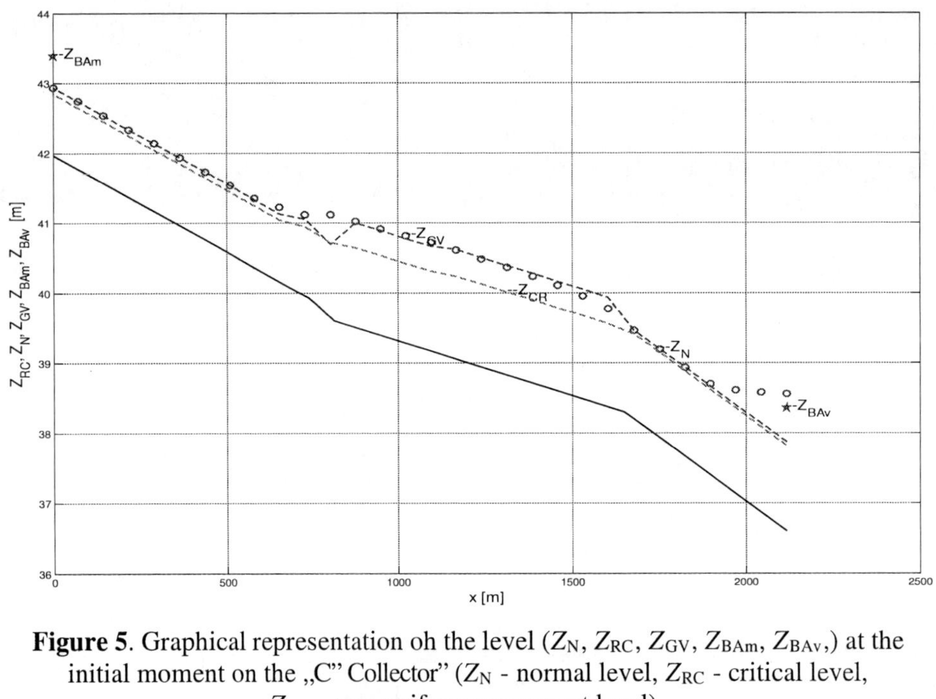

Figure 5. Graphical representation oh the level (Z_N, Z_{RC}, Z_{GV}, Z_{BAm}, Z_{BAv},) at the initial moment on the „C" Collector" (Z_N - normal level, Z_{RC} - critical level, Z_{GV} - non-uniform movement level).

The analysis of the data in Figure 4 shows the variation of the water height at the initial moment in the collector at the length of 2116 m. Comparative analysis of normal depths and critical depths indicates the presence of a subcritical movement state ($h_N > h_{CR}$). The analysis of the water depth variation mode in Figure 5 shows a continuous non-uniform variation movement in the collector in the length of 2116 m. The varied gradual non-uniform movement is represented by a set of curves of the free surface of type a_1 (ascending) and b_1 (descending).

Stage II: Simulation of the computation scenario (scenario B2 is presented in the paper). Stage II has two main work phases:
II.1. Contouring conditions processing phase.
II.2. Basic data processing phase.
Since all six chambers on the "C Collector" path are coaxial, displaying the same values for the width and foundation elevation as it, the assumption of gradually and rapid varied non-permanent movement can be accepted in the collector. On some sections of the collector there may be a gradually varied permanent movement. Non-dynamic contour conditions are determined by the hydraulic characteristic of the hydro-urban public work structures present on the network (the six connection / discharge chambers, CA1-CA6). For the $Q_{af}\ (t)$ time variation of inflows in the CD1 - CD6 chambers, the following equation has been considered:

$$Q_{af}(t)= Q_{af\,0}\cdot Q_r(t), \qquad (15)$$

where $Q_r(t)$ is the relative inflow and has the same variation law for all six inflows.
Constructive, functional and general base data for any "C Collector" computational scenario are stored in the "Input Data" files. The files are FIS2_Collector and FIS3_Collector, respectively FIS4_Collector for the initial conditions. The files have been developed by computational programs for the permanent movement. Thus, all the base data can be entered into the **TGCProg** program by simply reading the three data files.
After running the program, the following output data resulted, which define the collector's functioning in the operational scenario adopted:

1° - the value of the depth $h = \text{h}(t,\text{x})$, for $0 \le t \le 4{,}496"$ and $0 \le x \le 2{,}115.75$ m.

To obtain a greater precision, the interval for x, [0, 2115.75] m was subdivided by $N_{ph} = 100$ equidistant division points. For the time interval t, [0, 4,496"], $N_{pk} = 1{,}200$ quasi-equidistant division points resulted. Thus, the solution $h = \text{h}(t,\text{x})$ is represented by the following discrete values for t, x and h:

$$(\text{t}_i,\ \text{x}_j,\ \text{h}_{i,j}),\ \text{with } i = 1, 2, \ldots, \text{N}_{pk} = 100 \text{ and } j = 1, 2, \ldots, \text{N}_{ph} = 1200 \qquad (16)$$

The graphic representations were done with the values from equation (16). For tabular results, a lower number of equidistant division points was considered both along the x axis, $N'_{ph} = 30$ ($\Delta x = 72.96$ m step), and the t axis, $N'_{ph} = 30$ ($\Delta t = 150"$ step),

$$(\text{ti, xj, hi,j}),\ \text{with i} = 1,2,..\text{NIpk} = 30 \text{ and j} = 1, 2, \ldots, \text{NIph} = 30 \qquad (17)$$

For three representative values of the x (m) spatial coordinate,

$$\text{x} \in \{687{,}00;\ 1108{,}25;\ 1649{,}00\} \qquad (18)$$

the following explicit solutions have been selected from equations (16) and (17) solutions:

$$h_{0.7km} = h(t_j, 687), h_{1.1km} = h(t_j, 1108.25), h_{1.6km} = h(t_j, 1649) \qquad (19)$$

$$\text{with j} = 1, 2, \text{K}\ldots, \text{Nph} = 1200,$$

respectively,

$$h_{0.7km} = h(t_j, 687), h'_{1.1km} = h(t_j, 1108.25), h'_{1.6km} = h(t_j, 1649) \qquad (20)$$

$$\text{with } j = 1, 2, K\ldots, N_{\text{ph}} = 30.$$

The equation (19) solutions are shown in the figure 6 graphs and the equation (20) solutions are centralised in table 1. The overall figure 6 curve rate corresponds from a quality point of view to the physical aspect of the phenomenon formed in the collector.

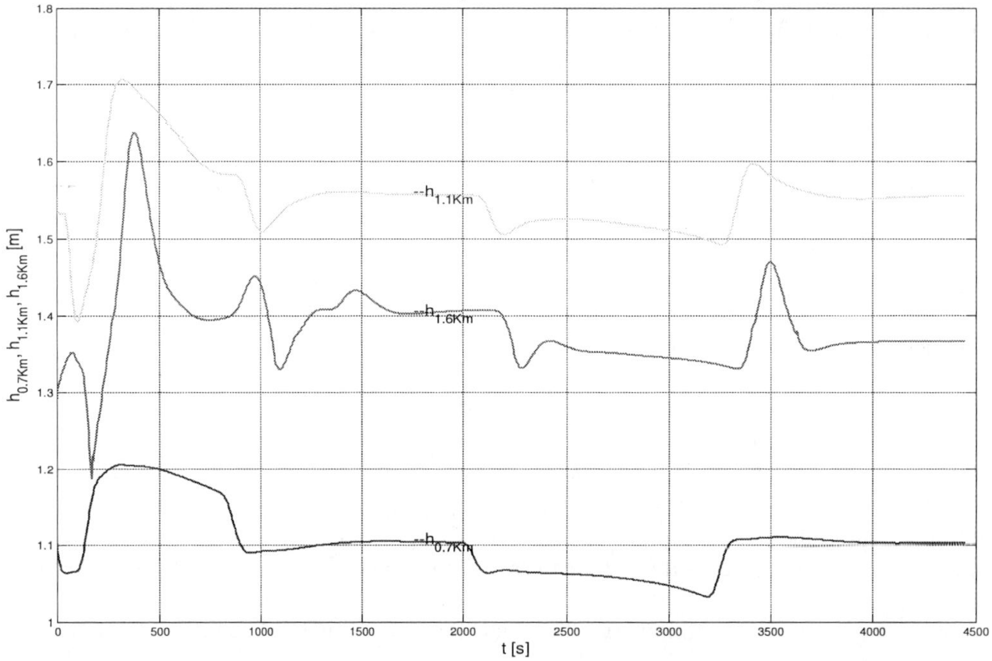

Figure 6. The maximum depth variation chart in three sections located on „C" Collector": $h_{0.7\,km} = 1.21$ m, blue line; $h_{1.1\,km} = 1.71$ m, red line; $h_{1.6\,km} = 1.64$ m, green line.

2° - depth minimum values

$$h_{\min}(x) = \min_{0 \le t \le 4496''} \{h(t,x)\}, \text{ for } 0 \le x \le 2115.75 \text{ m} \tag{21}$$

for $0 \le x \le 2115.75$ m.

3° - depth maximum values,

$$h_{mAX}(x) = \underset{0 \le t \le 4496''}{mAX} \{h(t,x)\}, \tag{22}$$

for $0 \le x \le 2115.75$ m.

The variations of the minimum and maximum depths were presented tabular (table 1) and graphical on characteristic analysis lengths (figures 6 and 7).

Table 1. Maximum and minimum values for depths and levels in sections with contour conditions

Section		x	h_{min}	h_{GV0}	h_{max}	Z_{MIN}	Z_{MAX}
No.	Symbol	(m)	(m)	(m)	(m)	(m)	(m)
1	2	3	4	5	6	7	8
1	Upstream	0.00	***0.94***	0.97	1.02	42.90	42.98
2	0.7 km	687.00	1.04	1.11	1.21	41.11	41.28
3	-	739.55	1.08	1.19	1.29	41.01	41.22
4	-	760.25	1.12	1.27	1.36	40.96	41.20
5	-	812.50	1.31	1.49	1.57	40.91	41.17

6	1.1 km	1,108.25	1.39	1.55	1.71	40.53	40.85
7	1.6 km	1,649.00	1.18	1.29	1.63	39.48	39.93
8	Downstream	2,115.75	1.87	1.95	***2.60***	38.47	39.20

The extreme depths (marked with italic symbols in the table), which correspond to extreme filling degrees appear in the same sections as in the initial gradually varied movement: the maximum depth $h_{max} = 2.60$ m with $U_{max} = 1.00$ in the downstream section (complete filling of the section); the minimum depth $h_{min} = 0.94$ m, with $U_{min} = 0.36$ in the upstream section. According to the technical literature, the filling capacity of the main collector must be maximum 0.80 [14], [15].

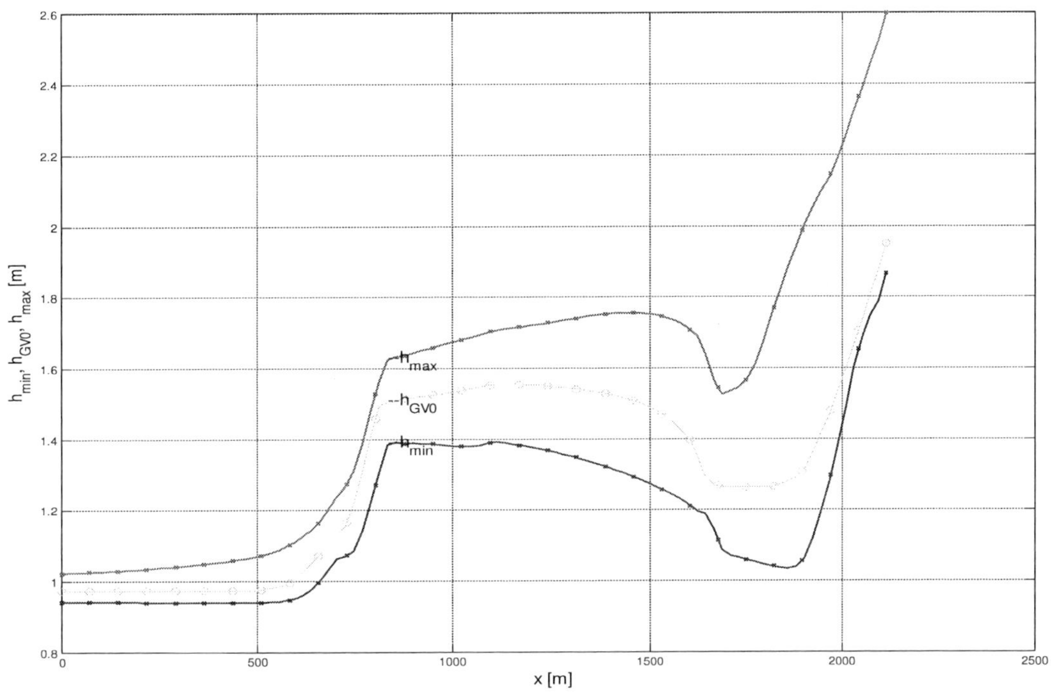

Figure 7. Maximum depths envelope curves (h_{max}, h_{GV0}, h_{min}) on „C" Collector"

From the analysis of the obtained data, it is possible to analyse and interpret the forms of movement which appear in the longitudinal profile of the collector. The shutdown of the pumping station which discharges the torrential rain transported by the collector determines the exceeding of the admissible filling capacity, and also the pressurising of the flow section on the final sector (figure 7). The exceeding of the filling level indicates unsatisfactory functioning of the collector, as confirmed by field research.
The simulation conducted showed an important variation in the maximum depth values on some sections of the "C Collector". The numerical results obtained with the simulation program were highlighted as follows: the maximum depth h_{max} varies over a range of 1.02 - 260 m (progressive increase); the filling capacity of the collector varies from 0.36 to 1.00 (progressive increase) at the length of 2116 m; the maximum filling degree, $U_{max} = 1.00$, appears in the downstream section of the collector (so full section filling is achieved, situation which is unfavorable to the normal operation of the collector); the minimum depth varies over an interval of 0.94 - 1.87 m at the length of 2116 m.

Thus, on the final section it resulted the exceeding of the flow section height ($h_{max} = 2.60$ m), in which the filling degree became maximum ($U_{max} = 1.00$). It results that the movement of the free-flowing water is transformed into pressurised movement on that collector section.

By simulating the operation of the collector in the chosen scenario, elements can be taken into account to detect, in real time, the causes that cause the collector's faulty operation, and then, by eliminating or at least mitigating these causes, the collector safety can be increased.

4. Conclusions

The main form of water movement in the main sewer collectors is non-stationary (non-permanent), but over longer time frames it can be considered as stationary (permanent), which influences the hydraulic computation.

The **TGCProg** computer program solves problems consisting of verifying complex sewage collectors with flow sections in gradually in varied non-permanent movement, depending on various contour conditions existing in the longitudinal profile.

The **TGCProg** computer program is applicable to sewer collectors which transport waste waters + pluvial waters, and especially to the intake, transport and discharge of torrential rainwater.

The case study solved for a main sewer collector with a mixed section (rectangle + semicircle, with the dimensions of 3450/2250 mm), by using the **TGCProg** computer program, presents the water height maximum and minimum values , determined by the direct and the indirect wave, characteristic to the non-permanent movement.

The results of the research show the applicability of the **TGCProg** computer program in the study of non-permanent movements in sewer collectors, with the condition of knowing the initial geometric and hydraulic parameters in the cross section and the longitudinal profile for the permanent movement (**TGCProg** program works together with other 2 - 3 adjacent computation programs).

References

[1] Blitz E 1970 *Proiectarea canalizărilor*, Editura Tehnică, Bucureşti

[2] Luca M 1998 Hidraulică tehnică. Mişcarea permanentă în canale, vol. I, Edit. Tehnopress, Iaşi

[3] Hâncu S, Rus E, Dan P, Teodoreanu Gh, et. all. 1985 *Hidraulică aplicată, Simularea numerică a mişcării nepermanente a fluidelor*, Edit. Tehnică, Bucureşti

[4] *NP 133-2013 2013 Normativ privind proiectarea, execuţia şi exploatarea sistemelor de alimentare cu apă şi canalizare a localităţilor. Vol. 2 Canalizări.* M.D.R.A.P., Edit. Matrix Rom, Bucureşti

[5] Luca M, Scripcariu C F, Luca Al L 2016 Consideraţii privind calculul hidraulic al colectoarelor de canalizare vizitabile. *Conferinţa Tehnico-Ştiinţifică A.I.I.R., Ediţia a XXVI-a, Universitatea Tehnică „Gheorghe Asachi" Iaşi, Edit. Matrix Rom, pp. 226-233*

[6] Cazanescu S, Mărăcineanu F, Tadjer J Sewer network design using urbane Canalis version 7 releases 2010. *Bulet. Ştiinţ. Univ. "Politehnica" Timişoara, tom 55 (69), Fasc. 1, 2, seria Hidrotehnica, p. 81-86*

[7] Hapurne T 2002 Fuzzy control of urban drainage systems, *"Ovidius" University Annals, Constanza, year III-IV, Special Proceedings, Series-Constructions, p. 632-638*

[8] Carlier M, 1973 *Hydraulique générale et appliqueé,* Editure Eyrolles, Paris

[9] Fuchs L, Beeneken T, Sponemann P, Scheffer C, 1997 Model based real-time control of sewer system using fuzzy logic, *Water Science and Technology, vol. 36, no 8-9, p. 343-347*

[10] Scripcariu C F 2018 *Contribuţii la optimizarea reţelelor de canalizare în scopul creşterii siguranţei în exploatare*. Teza de doctorat, Scoala Doctorală, Universitatea Tehnică „Gheorghe Asachi" din Iaşi

[11] Popescu Şt 1999 *Aplicaţii informatice în hidraulica sistemelor hidrotehnice,* Edit. Cermi, Iaşi

[12] Luca M, Scripcariu C F, Luca Al L 2018 Consideraţii asupra caracteristicii hidraulice a căminelor de pe colectoarele de canalizare vizitabile. *Conferinţa Ştiinţifică A.I.I.R., „Instalaţii pentru construcţii şi economia de energie", Ediţia a XXVI-a, A.I.I.R., Universitatea Tehnică „Gheorghe Asachi" Iaşi, Edit. Matrix.Rom, Bucureşti, p. 200-210*

[13] S.C. APAVITAL S.A., 2016 *Raport tehnic 2016*

[14] Giurconiu M, Mirel I 1989 *Hidraulica construcţiilor şi instalaţiilor hidroedilitare*, Editura Facla, Timişoara, 1989

[15] Negulescu M, *Canalizări*, Ed. Did. şi Pedag. Bucureşti, 1988

[16] Safari M J S, Aksoy H, Unal N E, Mohhamadi M, 2016 Self-cleansing Design of Circular Urban Drainage Channels, *Conference proceedings "The 3rd International Conference Water Across Time in Engineering Research" Constanța, June 23-25, 2016 pp. 245-252*

[17] Luca M, Tamaşanu F, Luca Al L, 2012 Flow modeling with free level in the polycentric channels, *Ovidius University Annals, year XII, Issue 14, series Civil Engine., Constanta, p. 155-162*

IOP Conf. Series: Materials Science and Engineering **586** (2019) 012038 doi:10.1088/1757-899X/586/1/012038

Study of natural gas pipeline behaviour

L Radermacher[1], T Mateescu[2]
[1]Petroşani University, Petroşani, str. Universităţii nr. 20, 332006, Romania
[2] Gheorghe Asachi Technical University, Bulevardul Profesor Dimitrie Mangeron 67, Iaşi 700050, Romania

E-mail: santierpetrosani@yahoo.com

Abstract:
In order to prevent incidents due to uncontrolled gas leakage and explosion/fires were devastating look at the factors that determine the safety in operation of gas networks made of polyethylene. Sustainability and safety in operation, lines of distribution networks and installations for use in natural gas represents a complex issue with implications for economic, social and environmental. Underground natural gas pipes are subjected to a series of prompts both static and dynamic which could lead to. damage. The complexity of the problems stemming from the diversity of stress factors acting on the related pipelines supply natural gas.
In this context, the authors analyse the behaviour of pipelines under the effect of the stresses to which they are subject by using finite element modelling.
Key words: gas, gas networks, urban grids, polyethylene.

1. Introduction

Polymeric materials today are used widely in industry pipelines. Polyethylene pipes are widely adopted in many countries for distribution networks for natural gas.
PEHD pipes offer many benefits compared to metal pipes. The advantages of these pipes include lightweight, flexibility, high ductility, weldability and low-cost installation.
Use pipes buried natural gas represents a common solution. Mechanical behaviour of underground pipes is influenced directly by the different properties of material filling and the surrounding soil.
The higher the fill material and rigidity of the structure around the pipe is higher so its resistance to external requests.
Underground pipelines requests include the weight of the soil above the pipe, additional tasks due to internal and external actions-the pressure in the pipeline, hydrostatic pressure-and request traffic. For an underground pipe support material is sand where it buried pipeline excavated rocks, respectively.[4],[5]
The state of tension induced in a pipe buried by the applied loads was first determined by Marston [3]. Later, Spangler used linear elasticity and experimental studies to propose the Iowa formula [8].
Watkins and Anderson (2000) have developed more formulas to calculate the modulus of soil elasticity around a buried pipe that gives better results than Sprangler's formula [11].
Using the finite element method (FEA) allows the investigation of soil-pipe interaction of buried pipes. In this respect under article considering the cumulative effect of factors that influence the behaviour of underground pipes from HDPE with actual conditions and track technology.

2. Design of piping

Design of piping buried natural gas is based on the principle of soil-pipe interaction. Thus, the pipe and the surrounding field influence each other's physical and mechanical performance. Tensions induced in polyethylene gas pipelines are due to a combination of permanent and dynamic internal or external loads due to traffic. Internal strength comes from gas pressure.
The permanent static load acting on the underground pipeline is determined by the weight of the overhead filling, which can be composed of the weight of the sand, the soil, and the road clothing (if the pipe is located in the road)

Pressure on the traffic pipe depends on vehicle weight, tire pressure and size, drag speed, and other factors.
It is assumed that the maximum load applied to the outside of the pipe is equal to the load above it. In fact, the load applied to an underground pipe may be significantly lower than the one mentioned above because the shear resistance transfers part of the external load to the side walls of the trench.
Typically, pipeline design uses the calculated NP116-2005 load according to Figure 1

3. Buried pipeline Modelling

Modelling the tensions of pipe buried is used algorithm [7],[9]:

1.Declaration of the initial conditions:

Type and diameter of pipeline; depth of laying; the type and size of the trench;
Nature of the land; traffic type; hydrostatic level

2. Establishment of model of soil-pipe interaction

3. Determination of loads:

Soil, water, traffic, total

4. To specify the characteristics of the material and the criteria limit

5. Calculation of stresses and deformations of pipe walls

6. Sizing-checking the thickness of walls

By way of example, a case study for determining the stress and deformation state in a PE 100-SDR11 high density polyethylene pipe having a D = 63x3.6 mm and the physico-mechanical characteristics shown in Table 1 is presented below.

The pipeline is laid according to ANRE-2018 Technical Regulations, in clay soil, in a vertical wall excavation, on a compacted sand bed with physical-mechanical characteristics determined in the Rocky Mechanics Laboratory at U.P., presented in Table 2.

Table 1 Elastic properties of PE100 pipes

Static force reference N/m^2x10^6	**The modulus of elasticity to traction at 23°C** N/m^2x10^6	**Minimum required resistance** N/m^2x10^6	**Long-term hydrostatic strength at 20°C** N/m^2x10^6	**Poisson's coefficient**
10÷11.19	900÷1100	10	8	0,35

The additional traffic load was 544,780 N / m2. It was considered to be an average traffic on a street road.[3],[4],[5],[6]

Table 2 Properties of compacted sand

Specific weight Kg/cm^3	**Angle of inner friction grad**	**The modulus of elasticity** N/m^2x10^6	**Poisson's coefficient**
1900-2000	25-35	60	0,2-0,3

The ground in which the pipe was buried is the dense, yellow-earthed clay with the elastic parameters shown in Table 3.

IOP Conf. Series: Materials Science and Engineering **586** (2019) 012038 doi:10.1088/1757-899X/586/1/012038

Table 3 Elastic Properties of Initial Roof Roofs (Unperturbed Rock)

Specific weight Kg/m^3	**The modulus of elasticity, N/m^2 x10^6**	**Poisson's coefficient**
2000-2250	240	0,35

To model the pipeline behaviour, use the QuickFieldTMStudent Edition 6.3.2.2098 SP2 program. The stress distribution analysis is based on the Maximum Shaping Energy Rating Criterion (von Mises criterion). Von Mises criterion (stored energy of deformation):

$$\sigma_{Mi} = \sqrt{\frac{1}{2}[(\sigma_1 - \sigma_2)^2 + (\sigma_2 - \sigma_3)^2 + (\sigma_3 - \sigma_1)^2}$$

where σ_1, σ_2 and σ_3 denote the principal stresses in descending order.
The discretization of the model for applying the finite element method was done according to the scheme in figure 2.

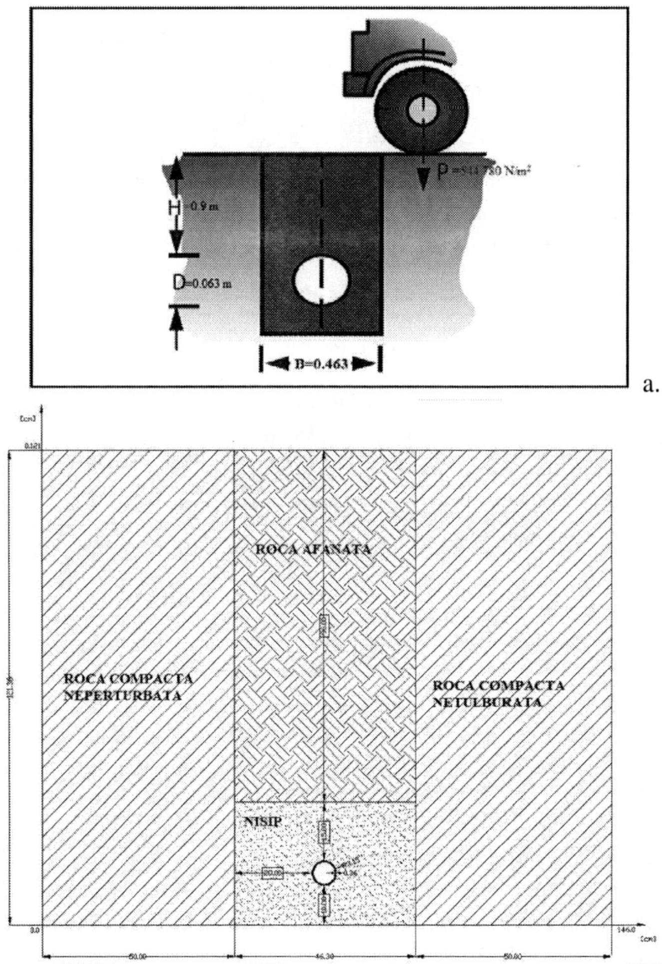

a.

b.

Figure 1. Schematic presentation of the model a) and dimensions model for finite element analysis b)

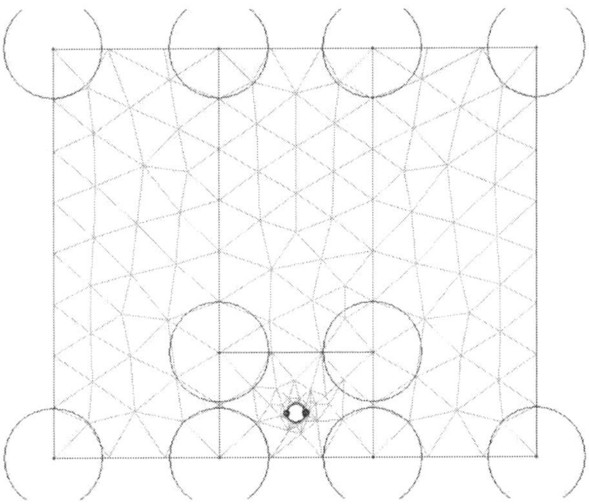

Figure 2. Discretization of the model for finite element analysis

The model was analysed comparatively in three application assumptions, corresponding to successive phases of execution:

a) Inserting the pipe into a sand layer, compacted to a height of 15 cm above the upper generator, the load is given by the weight of the sand, $P=\gamma_{ssand}\cdot h_{sand}$ N / m2;

b) Completing the soil filler filling, compacted to the initial state $P=\gamma_{ssand}\cdot h_{sand}+\gamma_{srock}\cdot h_{rock}$ N/m^2;

c) Loading of the filling with additional traffic load.

The state of tension and deformations are presented in tabular and graphical form in the figures 3,4,5 and the tables 4,5,6.

Case 1. Applying the model presented for the compacted sand situation, we obtain the outcomes of the von Mises stress state and the deformation according to Figure 3 and Table 4.

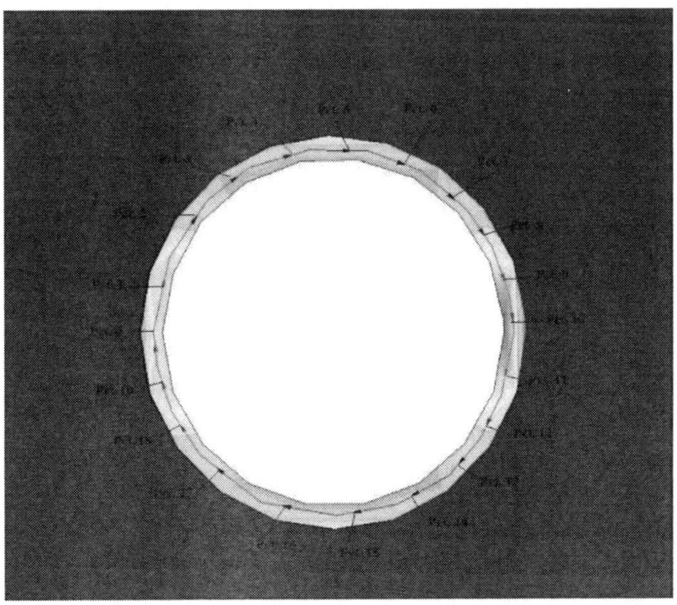

Figure 3. Distribution of Mises stress state in the pipe wall, compacted sand

Table 4 Mises distribution of the stress and deformation of the pipeline, compacted sand

Position	L (cm)	x (cm)	y (cm)	d (cm)	σ_{Mi} (N/m²)
0	0	70.536	12.8689	0.132252	2145230
1	0.922703	70.731	13.7706	0.13467	2472910
2	1.84541	71.143	14.5791	0.138632	1692440
3	2.7681	71.761	15.2481	0.144104	939449
4	3.69081	72.575	15.6495	0.149479	478791
5	4.61352	73.478	15.7839	0.151687	491391
6	5.53622	74.377	15.648	0.149477	591255
7	6.45892	75.178	15.2179	0.144048	878318
8	7.38162	75.833	14.5771	0.138019	1624550
9	8.30432	76.245	13.7607	0.133941	2508810
10	9.22703	76.431	12.8625	0.131675	2069350
11	10.1497	76.249	11.9587	0.128963	2294360
12	11.0724	75.851	11.1314	0.124487	1591040
13	11.9951	75.204	10.4872	0.118621	899783
14	12.9178	74.378	10.0885	0.113583	585162
15	13.8405	73.469	9.9449	0.111704	579640
16	14.7632	72.559	10.098	0.113917	599382
17	15.6859	71.776	10.5416	0.118886	1039430
18	16.6086	71.119	11.1734	0.124789	1690310
19	17.5313	70.706	11.9892	0.129488	2385600

Case 2. Taking into account that the sand is compacted and the rock is compacted to the initial state, obtain the results of the von Mises tension and the deformation state in Figure 4 and Table 5.

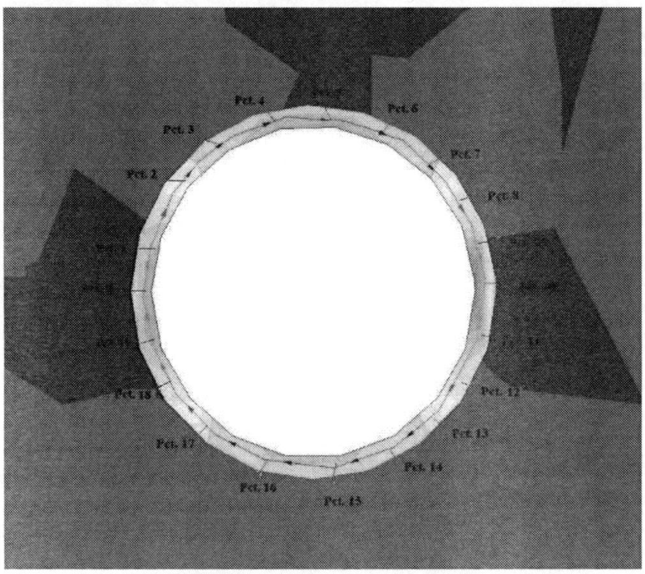

Figure 4. Distribution of Mises stress state in the pipe wall, compacted sand and rock is compacted to the initial state

Table 5 Mises distribution of the stress and deformation of the pipeline, compacted sand and rock is compacted to the initial state

Position	L (cm)	x (cm)	y (cm)	d (cm)	σ_{Mi} (N/m^2)
0	0	70.536	12.8689	0.043506	2067090
1	0.922703	70.731	13.7706	0.04554	2082700
2	1.84541	71.143	14.5791	0.047912	1604080
3	2.7681	71.761	15.2481	0.050597	1108400
4	3.69081	72.575	15.6495	0.052983	626235
5	4.61352	73.478	15.7839	0.054069	476723
6	5.53622	74.377	15.648	0.053407	584594
8	7.38162	75.833	14.5771	0.048551	1479970
9	8.30432	76.245	13.7607	0.046117	1979060
10	9.22703	76.431	12.8625	0.044044	2113500
11	10.1497	76.249	11.9587	0.042061	2144120
12	11.0724	75.851	11.1314	0.039317	1673850
13	11.9951	75.204	10.4872	0.036512	1182160
14	12.9178	74.378	10.0885	0.034059	797620
15	13.8405	73.469	9.9449	0.03278	575054
16	14.7632	72.559	10.098	0.033325	622721
17	15.6859	71.776	10.5416	0.035554	1131770
18	16.6086	71.119	11.1734	0.038389	1677500
19	17.5313	70.706	11.9892	0.041134	2129900

Case 3. Since the sand is compacted, the rock is compacted to the initial state and taking into account the additional load, obtain the results of the von Mises tension and the deformation state of Figure 5, 6 and Table 6.

Table 6 The distribution of the Mises stress state and deformation to the pipe wall when the excavated land is brought to the initial state.

Position	L (cm)	x (cm)	y (cm)	d (cm)	σ_{Mi} (N/m^2)
0	0	70.536	12.8689	0.019843	1032660
1	0.922703	70.731	13.7706	0.020737	1063340
2	1.84541	71.143	14.5791	0.022002	928282
3	2.7681	71.761	15.2481	0.023594	642271
4	3.69081	72.575	15.6495	0.025135	319510
5	4.61352	73.478	15.7839	0.026152	224242
6	5.53622	74.377	15.648	0.026066	137695
7	6.45892	75.178	15.2179	0.024943	338752
8	7.38162	75.833	14.5771	0.023307	675131
9	8.30432	76.245	13.7607	0.021819	965484
10	9.22703	76.431	12.8625	0.020662	1141200
11	10.1497	76.249	11.9587	0.019747	1176960
12	11.0724	75.851	11.1314	0.018378	838585

13	11.9951	75.204	10.4872	0.016854	562376
14	12.9178	74.378	10.0885	0.015517	330230
15	13.8405	73.469	9.9449	0.014635	185317
16	14.7632	72.559	10.098	0.014636	104601
17	15.6859	71.776	10.5416	0.015605	347752
18	16.6086	71.119	11.1734	0.017088	670786
19	17.5313	70.706	11.9892	0.018541	911689

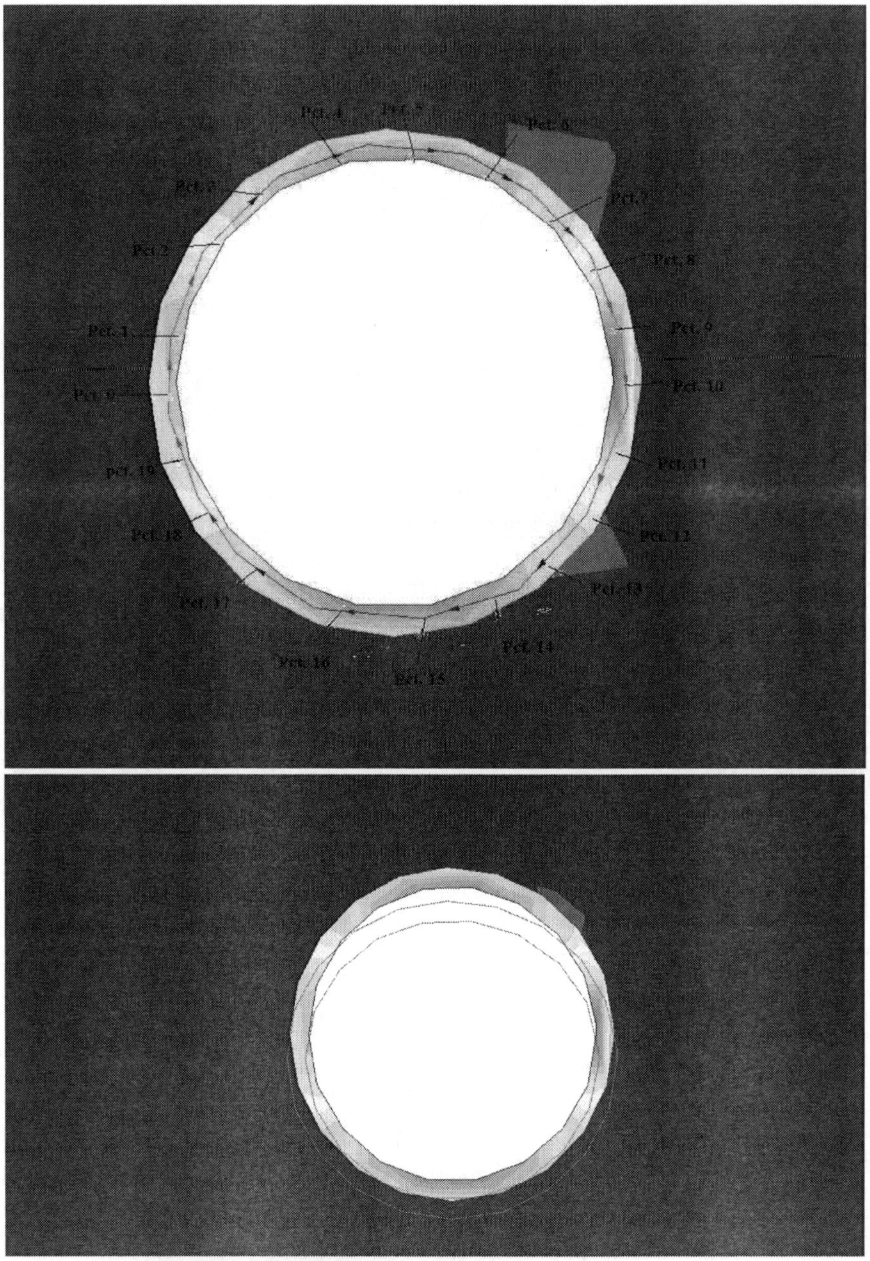

a.

b.

Figure 5. The distribution of the Mises stress state a) and deformation b) to the pipe wall when the excavated land is brought to the initial state.

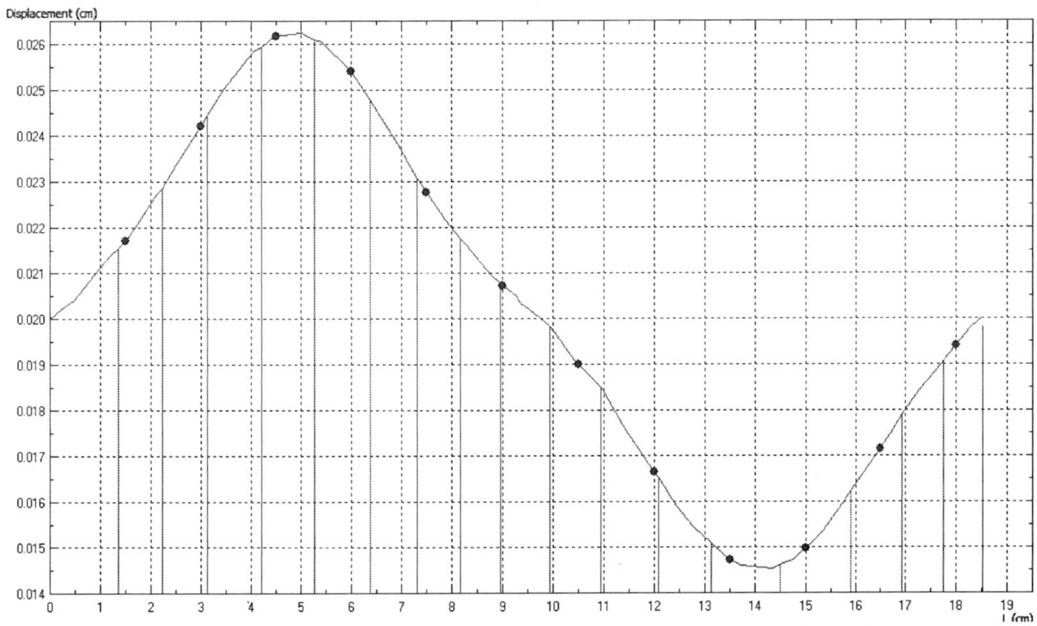

Figure 6. Graph of material deformation along the contour of the pipe

Conclusions

Polyethylene pipe systems have a significant and growing market share for low, medium and medium pressure piping systems for the transport of combustible gases.

The results of the finite element analysis indicate that the maximum values of the Von Mises tension appear on the inner surface of the tube.

The condition of stress and deformation decreases due to the increase in the crest of the sand and decreases with the increase of the compaction of the soil above the sand.

The results show that the mechanical properties of PE100 can be strongly influenced by the actual load conditions.

If the maximum value of Von Mises in the buried pipeline exceeds the allowable value for the material used in high density polyethylene pipes, shown in Table 1, the tube is damaged and must be replaced. [13],[1].

To avoid these problems, polyethylene pipes can be protected in rigid protective tubes that can take over the loads to which the pipe is subjected.

For all underground pipelines, account must be taken first of all of the need to properly compact the filler materials.

The results show that the mechanical properties of PE100 can be greatly affected by the load conditions used in the model.[2]

To complete the study, we propose to extend the analysis for other categories of stresses, types of pipe materials and diameters.

The pressure exerted on the pipeline due to a vehicle on the active surface (mainly the loads due to the wheels of trucks or trains) depends on vehicle weight, tire pressure and size, vehicle speed, and other factors.

Bibliography

[1] Avrigean E, Grecu V 214 *Market research regarding problems in using polyethylene pipe and fittings* (2014) Academic Journal of Manufacturing Engineering, VOL. 12, ISSUE 4/2014, pp. 6-11

[2] Băbuţ G B, Moraru R I 2018 *Occupational risk assessment: imperatives for process improvment,* Quality - Access to Success, Volume 19, Issue 166, pp. 133-144

[3] Chevron Philips Chemical Company LP 2003 *Buried pipe design. Bull* PP900, Book 2, pp. 81–115.

[4] Khademi-Zahedi R 2018 *Application of the finite element method for evaluating the stress distribution in buried damaged polyethylene gas pipes, Underground Space,* https://doi.org/10.1016/j.undsp.2018.05.002

[5] Khademi-Zahedi R, Shishesaz M, 2018 *Application of a finite element method to stress distribution in buried patch repaired polyethylene gas pipes, Underground Space,* https://doi.org/10.1016/j.undsp.2018.05.001

[6] Vrabie M, Băetu S A, Toma A 2017 *Studies Regarding The Vertical Soil Pressure Assessment On Rigid Pipes Buried In Straight Ditch,* BULETINUL INSTITUTULUI POLITEHNIC DIN IAŞI, Vol 63 (67), No 4, 2017, Secţia, CONSTRUCŢII. ARHITECTURĂ)

[7] Mateescu Th, Profire M, Pop A, Ciascai I 1996 *Conducte din mase plastice pentru sisteme de utilităţi urbane,* Bistriţa

[8] Moser A P, Folkman S 2008 *Buried pipe design* (3rd ed.. USA: Mc-Graw Hill Book Company pp. 387

[9] Radermacher L, Borş A-M, Lingvay D, Nicula Butoi N O, Voina A, Marin D. 2018 *Sustainable and safe in exploitation of gas networks. Part 1. Stress factors of plastic pipelines* EEA - Electrotehnica, Electronica, Automatica, 66 (4), pp. 66-72

[10] Serebrennikov A, Serebrennikov D 2018 *Mathematical model of polyethylene pipe bending stress state* IOP Conference Series: Materials Science and Engineering, 327 (4), art. no. 042099, DOI: 10.1088/1757-899X/327/4/042099

[11] Watkins R K, Anderson L R 2000 *Structural mechanics of buried pipes.* New York, CRC Press pp. 22–57

[12] Voina A, Niţă P, Luchian A-M, Lingvay D, Butoi N, Borş A-M, Lingvay I 2017 *Behaviour in exploitation of some buried pipelines insulated with heat applied polyethylene foil - case analysis* EEA - Electrotehnica, Electronica, Automatica, 65 (2), pp. 60-65

[13] Zhang Y, Ben Jar P-Y 2015 *Quantitative assessment of deformation-induced damage in polyethylene pressure pipe, Polymer Testing,*Volume 47, Pages 42-50, ISSN 0142-9418, https://doi.org/10.1016/j.polymertesting.2015.08.005.

IOP Conf. Series: Materials Science and Engineering **586** (2019) 012039 doi:10.1088/1757-899X/586/1/012039

The effect of wave crest speed in breaking waves

C C Craciunescu[1], M Christou[1]
[1] Department of Civil and Environmental Engineering, Imperial College London, SW7 2AZ, London, UK

E-mail: constantin.craciunescu12@imperial.ac.uk

Abstract. The present work focuses on the evolution of the crest speed around large focal type water wave events. The phenomenon of crest slowdown around the point of maximum elevation is studied numerically using a boundary element model. The model is used to generate large design waves based on realistic ocean spectra. The evolution in terms of coupling between crest speed reduction and increase in particle velocities is studied up until initiation of wave breaking. It is shown that this coupling is a viable mechanism that leads to the formation of large deep water unidirectional breaking waves regardless of spectral shape and bandwidth.

1. Introduction

The global oceans and seas have always been an indispensable resource for the world's population. The global economy is dependent on the ever-increasing human activity in the oceans. Structures that range from offshore wind farms, high-voltage direct current converter platforms, wave energy converters and oil and gas platforms ensure and satisfy our energy demands. Furthermore, bulk carriers, tankers, vessels and container ships all roam the oceans to maintain global trade and transport of goods.

The design of reliable, fit for purpose and cost-effective offshore infrastructure requires a grasp of the harsh environmental conditions and safety factors. An overconservative design represents a considerable increase in the cost of marine structures in the order of millions of dollars. This in turn can render a project commercially infeasible. To balance the costs of the project with acceptable levels of failure probabilities, modern codes of design and practice were formalised and are used today. However, the importance of wave breaking is unfortunately not emphasised enough in any of the current standards.

As breaking occurs at the crest of steep waves, it causes maximum loads from both the highest fluid particle kinematics and largest moment-arm about the sea bed. Slamming loads due to wave breaking pose a serious threat to sub-structures with sizeable vertical members as well as super-structures. For existing structures, especially those affected by subsidence, the main concerns related to wave breaking originate from loss of air gap and the resulting wave-in-deck loads as well as large dynamical excitations.

All the above emphasise the engineering importance associated with the understanding of breaking waves. However, the present understanding around the process of wave breaking is limited due to its complexity. The turbulent two-phase flow characterised by different length scales of interacting eddies is beyond the modelling capabilities of many numerical approaches and difficult to completely measure and describe from laboratory experiments. The prediction and detection of wave breaking remains one of the greatest challenges for metocean engineers.

The scientific community has adopted several approaches in an attempt to study and predict wave breaking. The study of [1] summarised that advancements have been made in describing wave breaking geometry, energy dissipation and wave breaking onset criteria. While a universal geometric criterion seems improbable, dynamic and kinematic criteria more in keeping with the physics of the process are showing promising results. Moreover, recent studies into the reduction of crest speed in large wave events [2] have prompted the reassessment of the classical kinematic wave breaking criterion which states that breaking occurs when the horizontal velocity of the free surface particle (u) exceeds the propagation velocity of the wave crest (C).

IOP Conf. Series: Materials Science and Engineering **586** (2019) 012039 doi:10.1088/1757-899X/586/1/012039

After this introductory section, the present paper continues with a brief outline of the background work in Section 2 around the effects of crest slowdown on the formation of wave breaking. Following this, the wave cases generated using the fully-nonlinear boundary element model (BEM) of [4] will be presented in Section 3 together with the methodologies applied in computing the crest speed. Finally, Section 4 presents the current findings while Section 5 draws the relevant conclusions.

2. Background

A kinematic breaking criterion is based on accurately determining the fluid particle velocity and comparing it to some threshold value above which wave breaking would occur. According to [5] the kinematic criteria are based on wave motions such as fluid particle velocity or wave phase speed. The basic kinematic criteria frequently used is that breaking occurs when the maximum horizontal water particle velocity, u_{max}, exceeds the phase velocity, c; essentially meaning that the water travels faster than the wave, leading to a jet. As the process of wave breaking is highly nonlinear, [1] states that in the context of random waves, the application of these criteria is difficult due to the need to accurately calculate particle velocities and define the crest speed. While for monochromatic waves the phase speed c is clearly defined, this is not the case for random waves. For random waves, the definition of a wave crest speed C is preferred, but accurate measurements of this quantity (especially in field conditions) is difficult.

There have been several studies that suggested that breaking initiation occurs well below the unity threshold. Numerical work of [6] found that the maximum horizontal velocity at the wave crest u_{max} computed from a third-order numerical formulation was 38% smaller than the crest velocity for waves which in the laboratory were seen to break. The study concluded that the horizontal components of the water particle velocity remained smaller than the crest propagation velocity for the studied cases. This would seem to suggest that the basic kinematic criteria defined at the beginning of this section does not always hold. On the other hand, this result could also be a limitation of the third-order numerical scheme implemented in [6], that uses a semi-Lagrangian method which cannot model breaking waves. The experimental studies of [7] backup the findings of [6]. Using both PIV measurements of laboratory generated breaking waves and field data from the GOTEX project, [7] conclude that the maximum particle velocity remains well below the local wave speed ($u_{max} < c$).

A recent set of studies show that the wave crest slows down near a maximum point such as focusing. The study of [8] includes the measurements after breaking initiation of the surface profile. Having the measured time series at two closely spaced measuring stations allowed them to estimate the speed of the breaking crests (C_{br}) and compare that to a prediction by the linear phase speed (c). Based on a sample size of 30 waves and measuring passing time of the crest between two stations, the relationship of $C_{br}/c = 0.9$ was established. Effectively, a linear prediction of the phase speed seems to overestimate the actual speed of the crest. This is because as the wave group approaches focusing, there is a well-known transfer of energy towards the high frequency range [9]. This range is characterised by short waves that travel at low speeds. The increased energy levels in this region coupled with the low propagation speed of these wave components results in the overall slowdown of the crest relative to linear theory.

Within the context of these new findings, it has been discussed that the classical kinematic criterion of unity might indeed be sufficient for breaking, but breaking initiation may occur at lower threshold values. This phenomenon of crest slowdown has been the focus of several numerical and experimental based publications [2,10–12]. The most promising result that stems from this recent finding related to the evolution of large wave crests is the dynamic breaking threshold proposed by [13]. This newly derived breaking criterion simplifies to a kinematic criterion at the free surface of water waves and is further detailed in Section 4.

On the topic of kinematic wave breaking criteria there is much research ongoing. A case in point is that there is still no clear understanding relating the effects of wave nonlinearity on the evolution of crest speed. Furthermore, the topic of the reduction in crest speed near large wave events has not been extensively studied in the context of realistic ocean wave spectra. This phenomenon has a clear implication in the formation of breaking waves, but the implications in the context of engineering design are less understood. This can be put in the context of engineering challenges such as wave-in-deck and slamming loads, where focused or NewWave [14] wave events are generally used in design. As such, this study will present a numerical investigation based on boundary element modelling into the evolution of steep waves to breaking from the perspective of a kinematic criterion.

3. Methodology

For the present study unidirectional deep water focused wind wave groups based on the NewWave theory developed by [14–16] were simulated. Describing the free wave components that are directly proportional to the autocorrelation function of the surface elevation, NewWave generates the most probable shape of the largest waves in the field. They typically represent design wave cases.
The BEM was employed to generate these wave groups within a numerical wave tank at a given location. Full details regarding the set-up of the numerical model can be found in [17]. The equivalent linear surface elevation was generated using linear random wave theory (LRWT), which employed the identical space-time domain as the BEM simulations.

3.1. Wave cases

Wind waves that are locally influenced by weather conditions (fetch limited) are typically described using the JONSWAP spectra of [18] while swell dominated seas are described using Gaussian spectra. Within the present study, focused waves were generated such that they replicate laboratory experiments corresponding to a length scale of 1:100 and a time scale of 1:10. The peak spectral component for both spectra had a period of T_p = 1.2 s. For the JONSWAP spectra, three typical values for the peak enhancement factor γ of 1, 2.5 and 5 were used. These effectively control the peakedness of the spectra and slightly modify bandwidth. For the Gaussian spectra the bandwidth was controlled using three values for standard deviation of σ = 0.15 Hz, 0.25 Hz and 0.45 Hz.
The depth of the numerical domain was fixed at 1.5 m, setting the depth regime of the cases at $k_pd=4.2$ ($\tanh(k_pd) = 0.999$ - deep water), where k_p is the wavenumber of the spectral peak. All the wave cases studied are summarised in Table 1. The linear amplitude sum (A - which is the largest elevation wave event within a linear wave group) was systematically increased until wave breaking was initiated. A typical example of the spatial surface elevation of focused events normalised by A obtained for the Gaussian spectra with σ = 0.15 Hz for increasing steepness is shown in Figure 1. The figure illustrates the typical nonlinear amplifications above the linear input sum A as waves evolve towards breaking.

Table 1. Non-breaking JONSWAP and Gaussian spectra wave groups investigated in present study.

Spectrum	Bandwidth	k_c [rad/m]	Ak_c range
	γ =1	4.8	
JONSWAP	γ =2.5	4.3	$Ak_c \leq 0.31$
	γ =5	3.9	
	σ = 0.15 Hz	2.9	$Ak_c \leq 0.26$
Gaussian	σ = 0.25 Hz	3.1	$Ak_c \leq 0.29$
	σ = 0.45 Hz	3.8	$Ak_c \leq 0.31$

An estimate of the nonlinearity of a given wave group that incorporates the bandwidth of the spectrum can be computed using a weighted average quantity termed the central wavenumber $k_c = \sum a_n k_n / \sum a_n$ (where a_n and k_n are the amplitude and linear wave number of individual components respectively). Hence, the global steepness of all wave cases is described using Ak_c.
To test convergence of the generated BEM focused wave events, the input location of the numerical wave paddle x_l in the computational domain was iteratively changed and the focused wave events compared. An example of the focused events obtained from four different domain sizes is shown in Figure 2a for the JONSWAP case with γ = 1 and Ak_c = 0.24 and in Figure 2b for the JONSWAP γ =

IOP Conf. Series: Materials Science and Engineering **586** (2019) 012039 doi:10.1088/1757-899X/586/1/012039

2.5 case with Ak_c = 0.215. The two figures show that for $x_l \leq$ -9 m there are no discernible differences in the focused event. The location of x_l = -9 m was chosen such that the computational effort was minimised, and more spatial information of the wave group was available, especially for the narrow-banded cases.

Additionally, the input was far away enough from the focus location so that interactions above second-order (which confer the asymmetry to the focused crest) were allowed enough time to develop. The horizontal and vertical asymmetry of the largest crest and downshift of the location of η_{max} is evident in Figure 2.

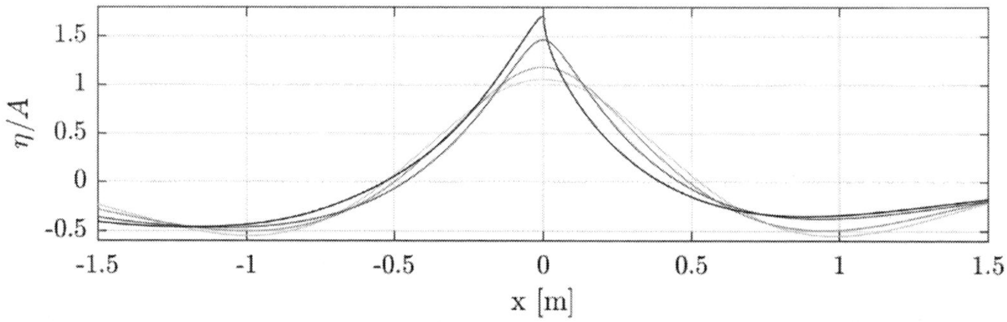

Figure 1. Spatial surface elevation of focused wave event normalised by the linear amplitude sum A for Gaussian spectra with σ = 0.15 Hz and increasing steepness Ak_c (shown as increasingly darker shades of grey). For comparison all focused events are shifted to x = 0 m.

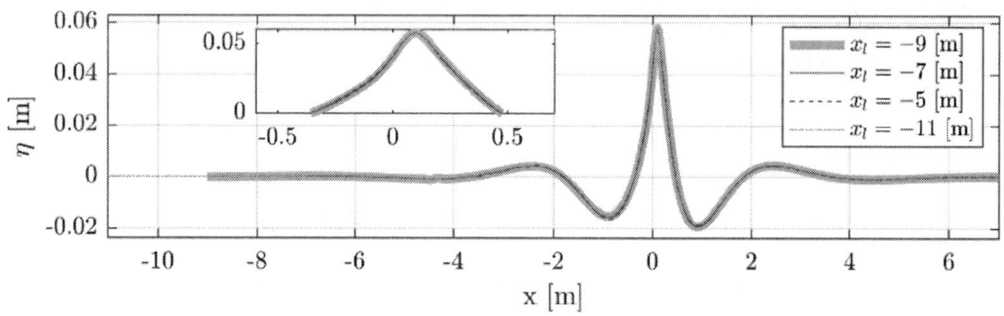

(a) JONSWAP γ = 1 Ak_c=0.240

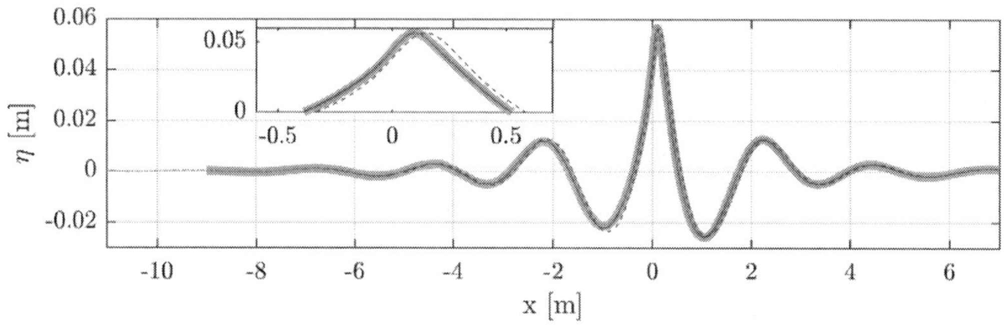

(b) JONSWAP γ = 2.5 Ak_c=0.215

Figure 2. Comparison between focused wave events for four different domain sizes.

As the wave group becomes steeper and approaches breaking, both [19] and [20] showed the importance of using a correct reference frame within the BEM. As such, the BEM employs a switching scheme between a semi-Lagrangian and a fully-Lagrangian reference frame [17]. This switch is done at a user specified time, and in the case of breaking waves, [20] notes that the switching time becomes essential in obtaining an appropriate nodal resolution. The fully-Lagrangian frame allows the BEM model to accurately describe steep waves and clear plunging breaking waves that form a jet up until the point of surface reconnection.

For a linear wave group [19] showed that the numerical predictions between a semi-Lagrangian and a fully-Lagrangian reference frame are identical. Furthermore, the fully-Lagrangian results are independent of the switching times. For the cases studied here, Figure 3a shows that for the JONSWAP case with $\gamma = 1$ and $Ak_c = 0.24$ the differences between the two reference frames are small. However, as the nonlinearity of the wave group is increased in Figure 3b, the changes in predictions become important and the fully-Lagrangian frame is needed to capture the appropriate shape of the wave event.

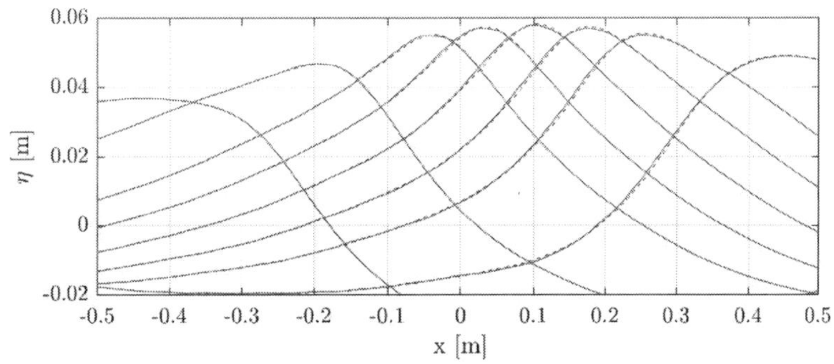

(a) Ak_c=0.240

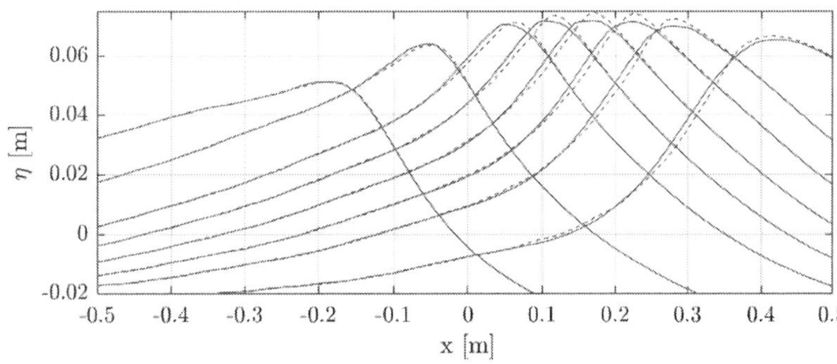

(b) Ak_c=0.288

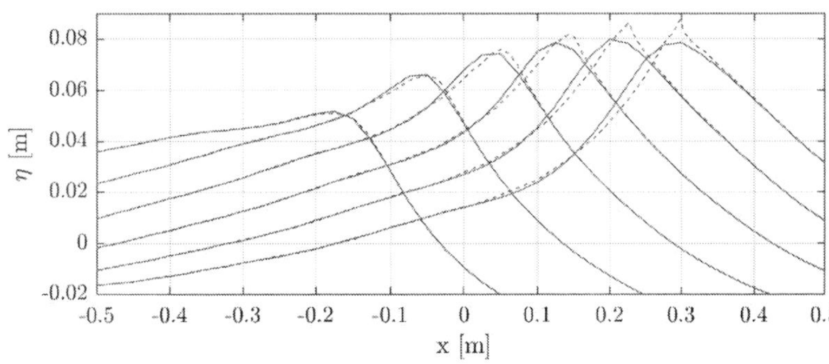

(c) Ak_c=0.312

Figure 3. Comparison of the BEM predicted surface elevation in the vicinity of the focused event in a fully-Lagrangian frame (dashed blue line) and a semi-Lagrangian frame (continuous red line). The surface profiles from both frames of reference are shown at the same simulation time.

It is clear that to obtain both accurate estimates of the crest propagation speed and particle kinematics, the fully-Lagrangian frame should be used for very steep waves. Figure 3c shows the evolution of the focused crest for the steep JONSWAP $\gamma = 1$ case with $Ak_c = 0.312$. The increased steepness and amplitude of the fully-Lagrangian simulation is mostly affecting the crest region, where significantly higher particle velocities will develop.

3.2. Calculation of crest speed

As the BEM simulations are characterised by a high temporal ($\Delta t \approx 0.002$s) and spatial discretisation ($\Delta x = 0.025$m), this allowed accurate calculations of the crest speed. A first step in the crest speed estimation is to identify the space-time trajectory of the wave crest. In this study, the location of the crest was determined from time-signals of the free surface $\eta(t)$ at each individual computational node in the domain. The instantaneous characteristic crest speed is thus given by the slope $\Delta x_c/\Delta t_c$ of the crest space-time trajectory. Using this approach, Δt_c measures the time taken for the crest to travel the distance between two measuring points Δx_c. It must be noted that with the use of a Lagrangian frame, the computational nodes are free to move both horizontally and vertically. To keep a constant Δx_c, interpolation of the spatial free surface profiles onto a fixed grid is done in the post-processing stage. An example of typical crest trajectories obtained is shown in Figure 4. The crest speed calculated using the above approach will be termed C_{BEM} when it is obtained from the BEM simulations and C_{LRWT} when it is obtained from the equivalent LRWT simulation.

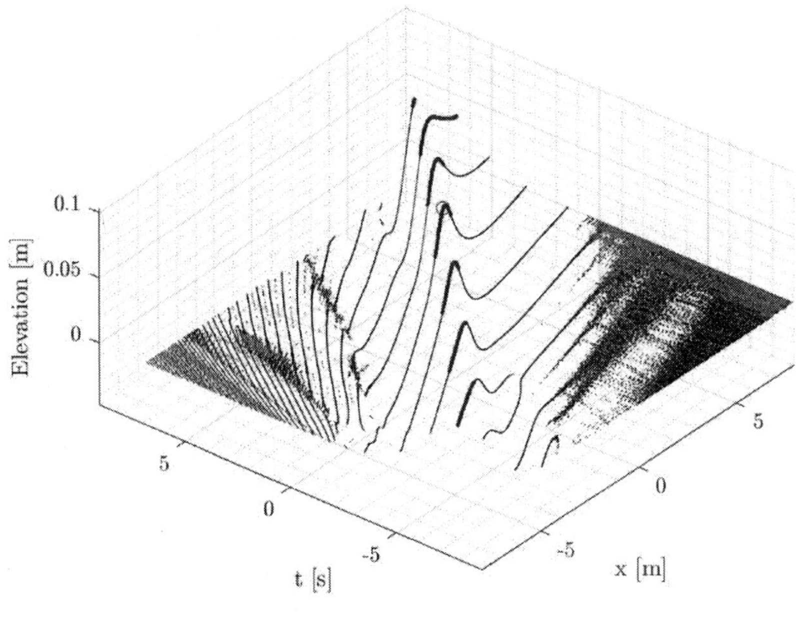

Figure 4. Example of identified crest trajectories with highlights for maximum elevation at a given location xi (thick continuous blue line) and elevation of focal event (red circle).

In addition to the crest speed, the location of the centroid of the wave above still water level (SWL) was also computed from the simulations. The x-coordinate of the centroid (X_G) shown in Figure 5 was calculated by geometric decomposition of the wave above SWL into small trapeziums based on the spatial discretisation of the BEM simulation. The coordinate was then used to compute an estimate of the propagation speed of the centroid of the wave to compare with the results obtained from the tracking of the actual crest. This was only done for the non-breaking waves.

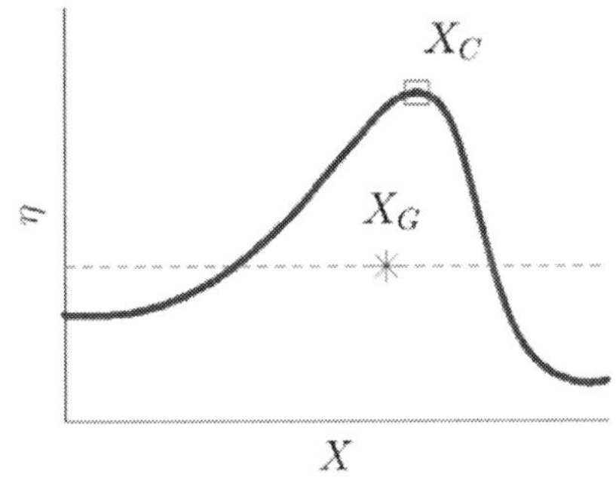

Figure 5. Schematic indicating crest (red square) and wave centroid (red star) locations. Still water level indicated as thin dashed line.

4. Results and discussion

Within the present study, the BEM simulations were used to observe the trajectory of the crest within the space-time domain that would lead to the focused event. The reasoning behind this was discussed in Section 2 in the context of the wave crest speed evolution and the application of the NewWave theory for the design of offshore structures.

Within a wave group, the variation in crest speed will be present within all wave crests as they evolve and decay [2,11]. Panel (a) of Figure 6 concentrates on the crest speed of the largest crest within the wave group as well as the two adjacent crests around the focal event for the JONSWAP case with $\gamma = 1$ and $Ak_c = 0.096$. The results of the wave crests obtained from both LRWT and BEM are illustrated. Immediately visible is a considerable reduction in the crest speed of the waves as they reach their maximum elevation.

Together with the crest speed, the speed calculated by tracking the centroid of the wave is presented, (as shown in Figure 5). It is noticeable that while the entire body of the wave slows down as a crest reaches its maximum elevation, there is a considerable additional slowdown in the crest region. This figure confirms the finding of [11] that the crest slowdown is a highly localised process, and this occurs in all crests within the wave group. As breaking is also a highly localised process that occurs high-up in the wave crest, the significance of this local crest slowdown becomes apparent. In this panel, as the steepness is low, the LRWT results are in good agreement with the BEM simulations.

To complement Figure 6a, Figures 6b and 6c illustrate the same results but for steeper cases of Ak_c= 0.192 and 0.29. In both figures it can be observed that in the focal region, the centroid of the wave crest has a significantly larger speed compared to the crest. Focusing only on the focal event at $t/T_p = 0$ and comparing all three panels of Figure 6 it can be seen that the minimum crest speed attained reduces from C = 1.2 m/s to 1 m/s. Therefore, there is evidence to demonstrate that steeper focused waves are characterised by smaller minimum crest speeds. Additionally, there is an asymmetry in the evolution of the crest speed profile with increasing nonlinearity. The reduction of the crest speed with increasing Ak_c is also noticeable by comparison with the LRWT results. This reduction of crest speed with increasing nonlinearity has also been observed experimentally by [21].

In addition to all previous observations, Figure 6c shows the BEM predicted maximum horizontal particle velocity at the free surface $U_{\eta_{max}}$ along the trajectory of the largest crest. A systematic slowdown of the crest as the elevation increases is associated with an increase in the particle velocity. The simultaneous occurrence of reduced crest speed and increased particle velocity has obvious implications in the formation of deep water breaking waves.

IOP Conf. Series: Materials Science and Engineering **586** (2019) 012039 doi:10.1088/1757-899X/586/1/012039

The well-known kinematic breaking criterion of $u = C$ at breaking inception can be employed here to study the effects observed in Figure 6 within the NewWave formulation of generating steep design waves. However, very recent studies such as [13] have promoted the use of dynamic breaking criteria based on the energy fluxes of the unsteady evolving wave.

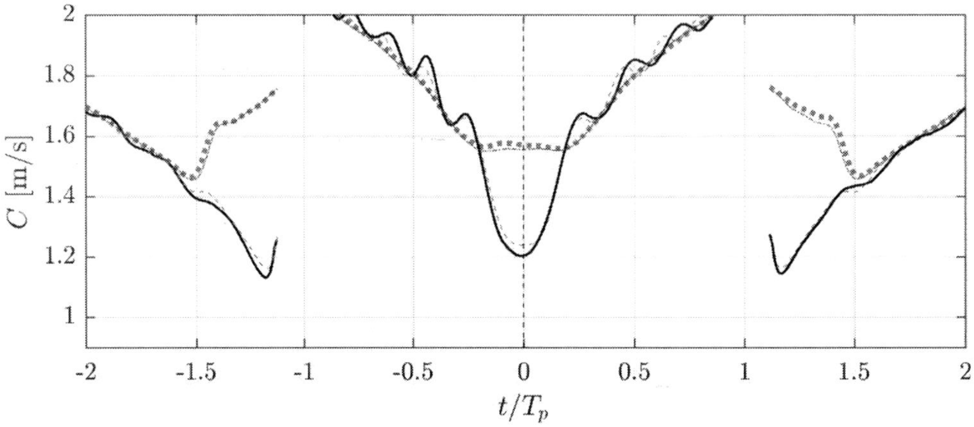

(a) Ak_c=0.096

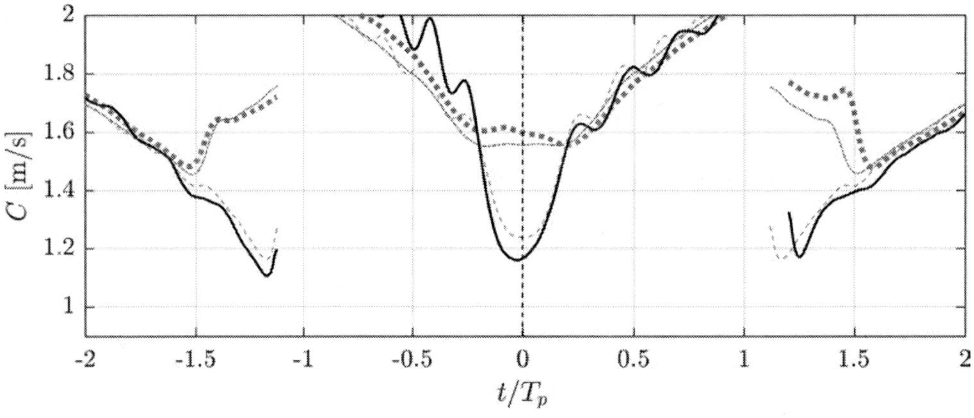

(b) Ak_c=0.192

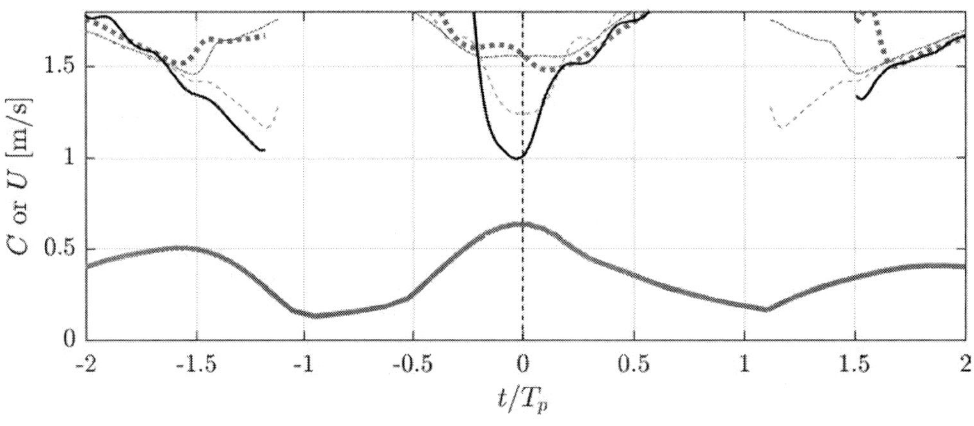

(c) Ak_c=0.288

Figure 6. Evolution of crest speed and centroid of wave speed for JONSWAP cases with γ=1. In all panels: thin continuous black line C_{BEM}, thin dashed green line C_{LRWT}, thin dash-dotted blue line LRWT centroid of wave speed and thick dotted red line BEM centroid of wave speed. In panel (c) thick continous red line shows $U_{\eta_{max}}$. In all panels BEM focused event was shifted at $t/T_p=0$.

Knowing that breaking occurs in the crests of very steep waves and is a highly localised process, [13] have studied the excess wave energy flux within the crest of steep waves in order to generate a breaking criterion. The starting point in deriving the breaking threshold proposed by [13] is the mechanical energy balance equation that relates the local rate of change of the energy density E shown in Equation (1) with the divergence of the local energy flux **F** in Equation (2).

$$E = \rho g(z - z_0) + \frac{1}{2}\rho\|\boldsymbol{u}\|^2. \tag{1}$$

$$\boldsymbol{F} = \boldsymbol{u}((p - p_0) + \rho g(z - z_0) + \frac{1}{2}\rho\|\boldsymbol{u}\|^2. \tag{2}$$

In the above equations, ρ is the fluid density, p and p_0 are the pressure and atmospheric pressure respectively, g is the gravitational acceleration, z and z_0 are the vertical coordinate and the origin of the z coordinate respectively and $\|\boldsymbol{u}\|$ is the modulus of the fluid velocity vector. Taking the atmospheric pressure as reference, the ratio of **F**/E can be computed as:

$$\frac{\boldsymbol{F}}{E} = \frac{\boldsymbol{u}(E + p)}{E}, \tag{3}$$

and [13] chose the instantaneous crest velocity as the normalising velocity $\boldsymbol{c}$ for this flux ratio. The reasoning behind this choice stems from the relevance of this quantity with respect to wave breaking as well as the fact that within an unsteady moving frame of reference with velocity $\boldsymbol{c}$, at the free surface, ($\boldsymbol{u}$-$\boldsymbol{c}$) is the flux velocity that transports the energy within the evolving crest. Using this result, the parameter proposed is:

$$\boldsymbol{B} = \frac{\boldsymbol{F}}{E\|\boldsymbol{c}\|}. \tag{4}$$

Additionally, on the free surface, the pressure is equal to the atmospheric pressure and hence from Equation (3), $\boldsymbol{F}/E = \boldsymbol{u}$. In applying this parameter, one can calculate the norm of the two vector quantities or project along the dominant propagation direction. In unidirectional waves at the free surface $F_x/E = u_x$, and hence the threshold simplifies to a kinematic criterion $B_x = u_x/c_x$. The threshold value given by [12] is $0.85 < B_x < 0.86$.

This parameter has been validated experimentally in the work of [22,23]. Additionally, [24] applied the parameter as a wave breaking onset criteria within their high-order spectral model. Within the present study, this threshold was applied to the BEM simulations with two purposes: to identify the limiting steepness waves and separate them from the incipient breaking cases and to validate the threshold of [13] for deep water wave groups based on realistic ocean spectra. The definition of limiting steepness is: a wave is said to be at its limiting steepness if any additional increase in the linear amplitude sum (A) used for its generation would cause it to break, by spilling.

The results are shown in Figure 7 for the γ = 1 JONSWAP spectra, for which the domain of the BEM simulations was kept the same as throughout the present study and the corresponding linear amplitude sum was incrementally increased. The horizontal velocity of the water particle at the free surface $u_x = U_{\eta_{max}}$ at the largest crest along the tracked trajectory was used within Figure 7. Note that, for very steep plunging waves, as shown in [18] and [19], the largest horizontal velocity U_{max} does not correspond to the horizontal velocity at the largest elevation $U_{\eta_{max}}$ due to the development of large Lagrangian drift

IOP Conf. Series: Materials Science and Engineering **586** (2019) 012039 doi:10.1088/1757-899X/586/1/012039

velocities. The choice of $u_x = U_{\eta_{max}}$ and not $u_x = U_{max}$ in this study is due to the definition of the B_x parameter. Furthermore, c_x was taken as C_{BEM}.
It can be seen in panels (a) and (b) of Figure 7 that the value of the B_x parameter is just under the breaking initiation threshold. In panel (c), for the case with Ak_c=0.312, the parameter exceeds the threshold, signifying wave breaking. Indeed, the BEM simulation crashes shortly afterwards due to node clustering, as the BEM model of [17] does not perform filtering, re-gridding or redistribution of the nodes. The present results for limiting wave steepness are in good agreement with the numerical work of [25] based on OpenFOAM as well as [19] using BEM and the laboratory investigations of [20]. The evolution of the B_x parameter with nonlinearity for the JONSWAP cases studied in the present work is plotted in Figure 8a.

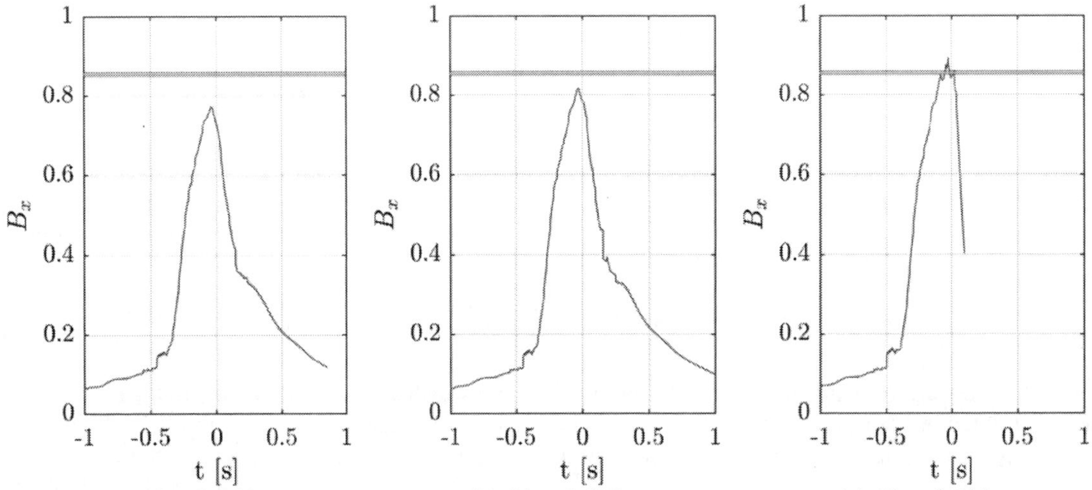

(a) Ak_c=0.303 (b) Ak_c=0.308 (c) Ak_c=0.312

Figure 7. Evolution of $B_x = U_{\eta_{max}}/C_{BEM}$ parameter of [12] for increasingly steeper JONSWAP waves with γ =1. In all panels thick grey line indicates threshold value 0.85 < B_x < 0.86.

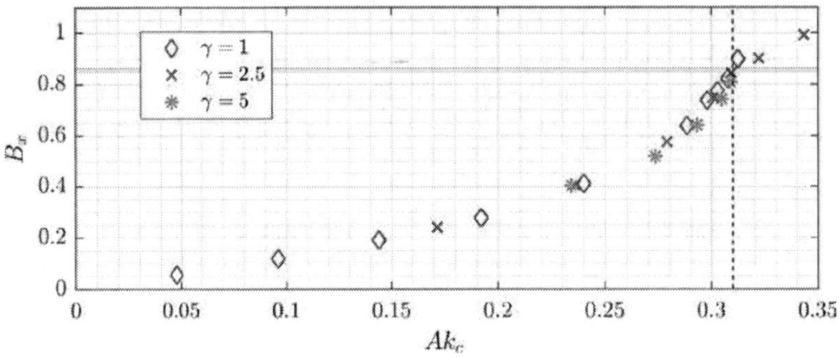

(a) JONSWAP spectra

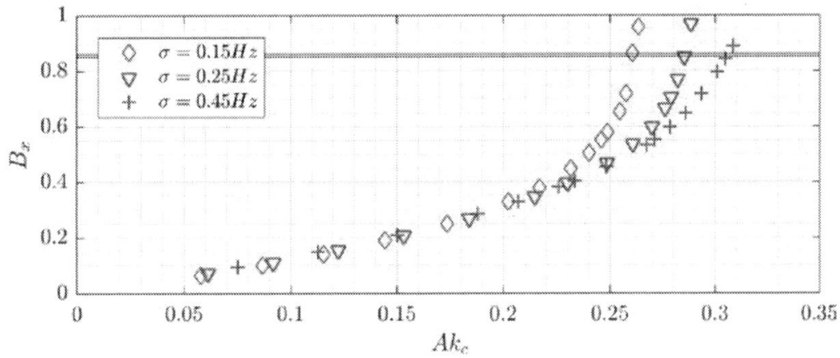

(b) Gaussian spectra

Figure 8. Evolution of B_x parameter with increasing nonlinearity for JONSWAP and Gaussian cases. In all panels thick grey line indicates threshold value $0.85 < B_x < 0.86$. In panel (a) dashed line indicates limiting Ak_c for JONSWAP spectra.

It can be seen that the evolution is consistent regardless of the parameter used to describe the spectral bandwidth of the JONSWAP spectra. At this point it would seem that the characteristic Ak_c value for a given wave group would also be a good indicator for breaking initiation for JONSWAP waves. As noted by studies such as [26], the global steepness Ak_c at breaking initiation is dependent on the spectral bandwidth. This is illustrated when studying the results for the Gaussian spectra. These results are plotted in Figure 8b and show a clear dependence on spectral bandwidth for the limiting global steepness values. The difference in results between the two spectra might be explained by the fact that the γ parameter does not alter the bandwidth of the JONSWAP spectra as much as the standard deviation does in a Gaussian spectrum. Hence, the Gaussian spectra is preferred when studying the effects of bandwidth.

Coupling the free surface wave profile obtained from the BEM with the estimation of B_x for a given wave event allows accurate separation between breaking and non-breaking events. Examples of wave events that have just surpassed the breaking threshold for Gaussian spectra are illustrated in Figure 9.

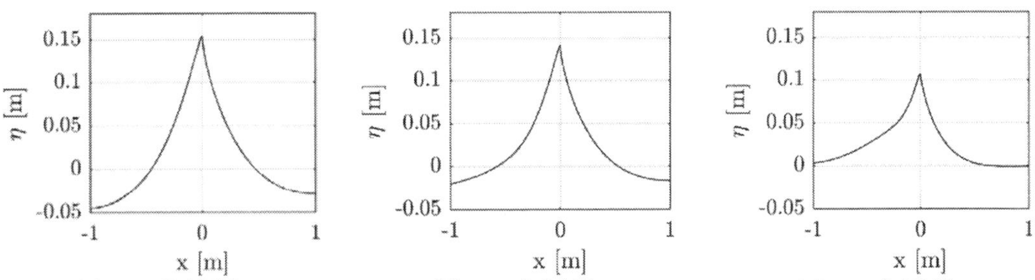

(a) σ = 0.15 Hz (b) σ = 0.25 Hz (c) σ = 0.45 Hz

Figure 9. Spatial surface profiles *η(x)* for Gaussian spectra waves past their breaking initiation point.

Having demonstrated in Figure 8 that the breaking threshold value of the B_x parameter is not sensitive to the bandwidth of the spectrum, it is interesting to highlight the evolution of the crest speed and horizontal particle velocity for waves characterised by different spectral bandwidths but the same value of the B_x parameter. Essentially, this would highlight the difference in spectral bandwidth for two wave groups that are at the same level of nonlinearity compared to their breaking limit. The comparison is made in Figure 10 for a narrow-banded Gaussian spectrum σ = 0.15 Hz and a broad-banded spectrum with σ =0.45 Hz characterised by a maximum B_x value of 0.717.

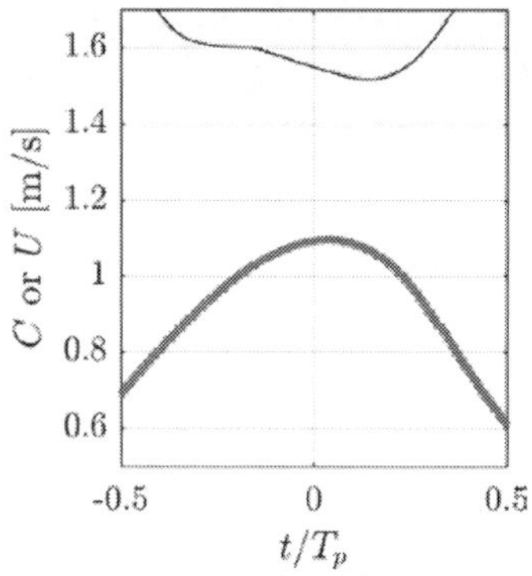

(a) $\sigma = 0.15$ Hz

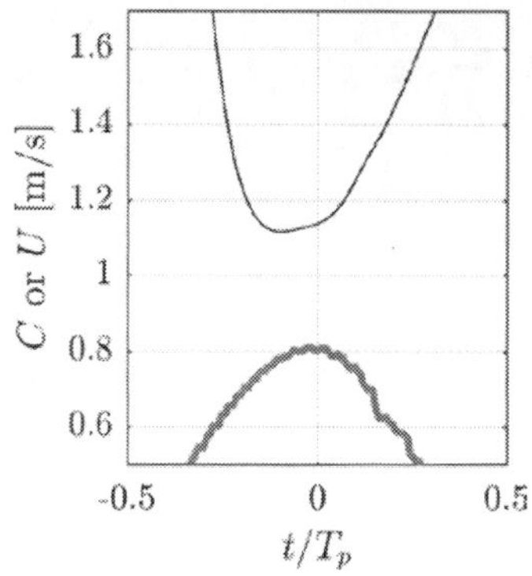

(b) $\sigma = 0.45$ Hz

Figure 10. Evolution of crest speed C_{BEM} (thin continuous black line) and $U_{\eta_{max}}$ (thick continuous red line) for narrow-banded and broad-banded Gaussian spectra wave events characterised by B_x=0.717.

While the minimum crest speed for the narrow-banded spectra is considerably larger than the broad-banded spectra, the horizontal velocity of the water particles is also significantly higher. Hence, a broad-banded spectrum generates slower crests but also generates smaller particle velocities. It is thus clear that taken individually, neither the increased particle velocities nor the reduction in crest speed are sufficient for the initiation of wave breaking. It is the combination of the two nonlinear behaviours that eventually results in wave breaking.

If the classical kinematic breaking criteria is used (essentially B_x = 1), a further increase in the linear amplitude sum compared to Figure 7 would be needed. For example, using this approach for the JONSWAP spectrum with $\gamma = 1$, an increase of the global steepness up to Ak_c=0.34 was needed before the criteria $u = C$ was met. As observed in both the studies of [25] and [19], further increases in the linear amplitude shift the location of the maximum elevation and breaking crest to the wave just before the focal event.

The combination of reduced crest speed and increased particle kinematics, present in all crests within the wave group, as shown in Figure 6c, can be used to explain this effect. The combination of these two mechanisms that leads to wave breaking was first suggested in [27] and is re-emphasized in the present study.

To generate a vigorous spilling or even plunging breakers, the linear amplitude sum needs to be increased. Two new cases for the JONSWAP spectrum with $\gamma = 1$ were generated. The cases presented are a spilling (Ak_c=0.35) and a plunging (Ak_c=0.44) breaker that occur in the same crest, which precedes the focused event. The only change needed for the new simulations to be accurate is an appropriate switching time between reference frames. The iterations needed in the selection of this switch time is only done to achieve satisfactory nodal resolution so that the breaking event is well described. For the present cases the selected switching time was t_{sw} = -2.3 s.

The comparison between the two wave cases in terms of both crest speed and particle kinematics are given in Figure 11. In both panels of the figure, the maximum horizontal particle velocity at any elevation at the location of the crest is plotted. Comparing the panels of Figure 11 with the results presented and 8c highlight the fact that as the linear amplitude sum is increased, the slowdown of the crest and the corresponding increase in the horizontal particle velocity results in the breaking events occurring in the crest preceding the focused event. For these cases, the theoretical focal location is t =0s. From Figure 11 it can also be observed that as the nonlinearity is increased, the breaking event occurs earlier in the space time domain and the drop in the crest speed immediately before the breaking event occurs much more abruptly. It was shown in studies such as [28] and [20], that for very steep waves the maximum horizontal velocity does not occur on the particle at the free surface, but rather at an elevation of around $0.9\eta_{max}$. This is a characteristic feature in the formation of the ejecting jet of fluid that gives the characteristic shape of an overturning wave.

The present results highlight an important limitation of the NewWave theory, especially if used to generate very steep waves. As such, if the linear amplitude sum used to generate the waves is large enough, the expected focused wave event will never occur, and energy will be dispersed via breaking in the wave crest preceding the main event. Additionally, it was observed that the focused event based on JONSWAP spectra had a reversed asymmetry compared to the preceding event. This is due to the different relationship between the phasing of the components in the focused event and the preceding spilling breaker.
Figure 12 compares the spilling event obtained in the case of $\gamma = 1$ and Ak_c=0.35 at the previous crest (shown in Figure 11a) and the spilling event obtained for a case with Ak_c=0.34, in the focused crest. While the comparison between the two cannot be made directly due to the different phasing of the components, it indicates the type of asymmetry needed to obtain clear spilling and even plunging breaking waves.
Compared to the focused wave event, the preceding crest has inverted asymmetry and the characteristic wave breaking profile shown in studies such as [26] and [1]. For comparison, the events have been shifted to $x = 0$ m. The difference in shape between the two waves is obvious, while difference in the maximum elevation is less than 2%.

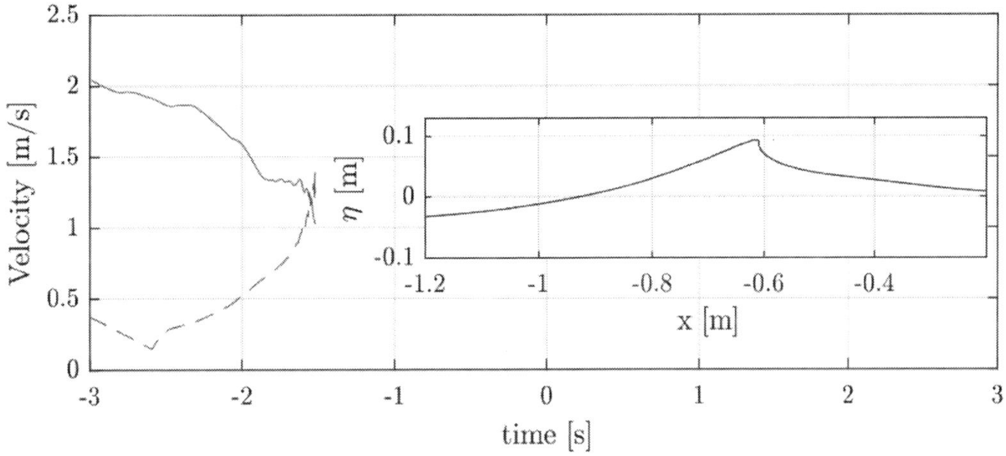

(a) Ak_c=0.35

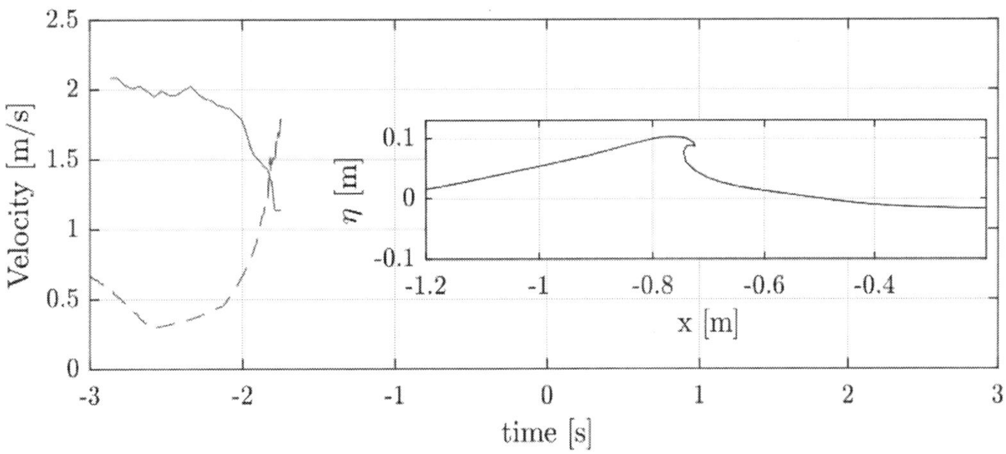

(b) Ak_c=0.44

Figure 11. Comparison between C_{BEM} (thin continuous red line) and U_{max} (thin dashed blue line) until breakdown of the BEM simulations for JONSWAP cases with γ=1. Shape of breaking crest is illustrated in the inset figures.

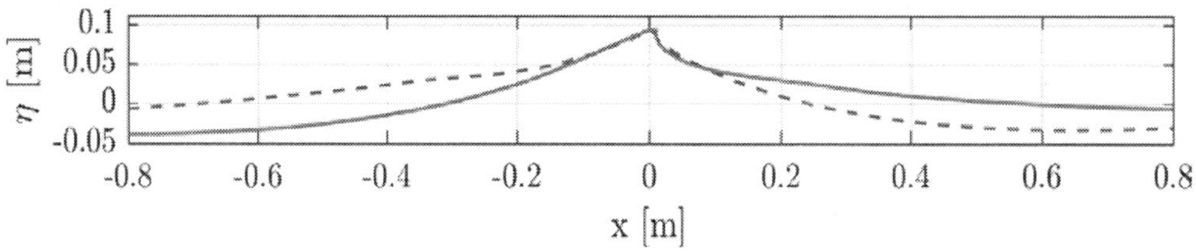

Figure 12. Spatial surface elevation of spilling events for cases with Ak_c=0.34 (dashed line) that correspond to the focused wave and Ak_c = 0.35 (continuous line) that corresponds to the wave preceeding focus for JONSWAP spectrum with γ =1. As the two surface profiles correspond to different waves they have been shifted to x = 0 m for comparison.

5. Conclusions

The work presented in the present study has addressed the effects of crest speed reduction in the formation of deep water breaking waves. Deep water unidirectional focused wave groups were generated using the BEM formulation of [4]. The numerical wave tank was used to generate waves based on realistic JONSWAP ocean spectra and Gaussian spectra.

When comparing the propagation speed of the wave centroid with the propagation speed of the point of maximum elevation in the crest, it was found that there is an additional slowdown of the top part of the crest compared to the main body of the wave. This is expected, as the highly unstable nature of breaking waves is localised high up in the water column. This small note confirms the findings of [11] who concluded that the noticeable slowdown observed is a localised crest effect.

Looking over the entire space-time domain of the wave group propagation, it was found that the generic slowdown of the crest combined with the increase in particle kinematics as a large crest develops is a potential mechanism that leads to the process of wave breaking. In this context, the newly proposed dynamic breaking parameter B_x of [13] was validated for deep water waves based on realistic ocean spectra. When comparing two waves characterised by a similar B_x value but with different underlying spectral bandwidth, the broad-banded spectrum gives a smaller crest speed but also lower horizontal particle velocities.

In terms of engineering relevance, the study offers further insight into the use of focused wave groups to generate very steep and even breaking waves that can be used as design waves in the context of deep-water offshore structures. Further work is needed to assess the implication of the coupling between crest speed reduction and particle kinematics increase in steep waves in the context of directionally spread sea-states.

6. References

[1] Perlin M, Choi W and Tian Z 2013 *Annu Rev Fluid Mech.* **45**(1):115–45
[2] Banner ML, Barthelemy X, Fedele F, Allis M, Benetazzo A, Dias F, et al. 2014 *Phys Rev Lett.* **112**(11):114502
[3] Vernimmen P. Breaking wave on L09FB1 platform. Pinterest
[4] Hague CH and Swan C 2006 *25th Int.Conf. on Offshore Mechanics and Arctic Engineering* American Society of Mechanical Engineers pp 167–73
[5] Wu CH and Nepf HM 2002 *J Geophys Res* **107**(C10):3177
[6] Shemer L and Noskowitz SH 2013 *Procedia IUTAM* **8**:205–12
[7] Grue J and Jensen A 2012 *J Geophys Res Ocean* **117**(7):1–16

[8] Banner ML and Peirson WL 2007 *J Fluid Mech.* **585**(2007):93
[9] Gibson RS and Swan C 2007 *Proc R Soc A Math Phys Eng Sci* ;**463**(2077):21–48.
[10] Fedele F 2014 *EPL (Europhysics Lett.)* **107**(6):69001
[11] Barthelemy X, Banner ML, Peirson WL, Fedele F, Allis M and Dias F 2015 On the localproperties of highly nonlinear unsteady gravity water waves. Part 1. Slowdown, kinematics and energetics arXiv:1508.06001 [physics.ao-ph]
[12] Khait A and Shemer L 2018 *Phys Fluids* **30**(5)
[13] Barthelemy X, Banner ML, Peirson WL, Fedele F, Allis M, and Dias F 2018 *J Fluid Mech* **841**:463–88.
[14] Lindgren G 1970 *Ann Math Stat.* **41**(6):1870–83
[15] Boccotti P. 1983 *Applied Ocean Research* **5** (3) 134–140
[16] Tromans PS, Anatruk AR and Hagemeijer P 1991 *Proc First Int Offshore Polar Eng Conf* pp 64–71
[17] Hague CH and Swan C 2009 *J Comput Phys* **228**(14):5111–28
[18] Hasselmann K, Barnett TP, Bouws E, Carlson H, Cartwright DE, Enke K, et al. 1973 *Erganzungsheft zur Deutschen Hydrographischen Zeitschrift Reihe* 12(8)
[19] Christou M 2019 *Fully nonlinear computations of waves and wave-structure interaction.* Imperial College London
[20] Haley JF 2016 *Fluid forcing in the crest of large ocean waves.* Imperial College London
[21] Johannessen TB and Swan C 2001 *Proc R Soc A Math Phys Eng Sci.* **457**(2008):971–1006
[22] Saket A, Peirson WL, Banner ML, Barthelemy X and Allis MJ 2017 *J Fluid Mech.* **811**:642–58
[23] Saket A, Peirson WL, Banner ML, and Allis MJ 2018 *Coast Eng.* ***133***(1994):159–65
[24] Seiffert BR and Ducrozet G 2016 *31st Int Work Water Waves Float Bodies* pp 6–9
[25] Hafizi Kasiman E 2017 *Numerical Modelling of Wave Interactions with a Vertical Surface piercing Column.* Imperial College London
[26] Rapp RJ and Melville WK 1990 *R Soc London A* **331**:735–800
[27] Seiffert BR, Ducrozet G and Bonnefoy F 2017 *Ocean Model* **119**(September):94–104
[28] Christou M, Swan C and Gudmestad OT 2007 *Volume 5: Ocean Space Utilization; Polar and Arctic Sciences and Technology; The Robert Dean Symposium on Coastal and Ocean Engineering; Special Symposium on Offshore Renewable Energy.* ASME pp 291–9

Acknowledgments
The present work has been supported by an Imperial College President's PhD Scholarship through the Engineering and Physical Sciences Research Council (EPSRC) grant no. 1855727.

IOP Conf. Series: Materials Science and Engineering **586** (2019) 012040 doi:10.1088/1757-899X/586/1/012040

Numerical simulation of a heat exchanger using Al_2O_3 nanofluid and STES

A I Baran[1], T Mateescu[2] , R S Luciu[2]
[1] PhD. Student, Department pf Building Services Engineering, Technical University of Iasi, Romania
[2] Professor, Department pf Building Services Engineering, Technical University of Iasi, Romania

E-mail: i.aandreea@yahoo.com

Abstract. The major energy crisis, predictable for the global economy, requires attentive research & development in order to grow energy performance, heat transfer and energy storage. It is a known fact that in the energy balance, the biggest share is in thermal energy and that justifies the continuous effort in the optimization of fossil fuel saving solutions and intense promotion of renewable energies. Use of the latter in heating and cooling systems as well as "waste heat" from different processes is consistent with the concept of sustainable development and results in reduction of conventional fuel consumption and emissions. Optimizing such a system can be made by replacing the usual storage material with phase changing materials and by using nanofluids as a secondary agent with the beneficial effect on heat transfer. In this paper we have studied a simplified multitubular heat exchanger, buried in the ground, having as a primary working fluid plain water and nanofluid, using Ansys. The nanofluid is composed of particles of Al_2O_3 dispersed in water at a concentration of 1%. The study indicates an increase in convective heat transfer coefficient of nanofluid compared to water.

1. Introduction

The national energy strategy, correlated with the EU directives, provides for important measures to improve the energy efficiency of buildings and related facilities, as well as for the use of advanced technologies by promoting appropriate solutions and equipment in full knowledge of the peculiarities and applicability limits to solve different categories of requirements with maximum functional and energy efficiency. [1]

The use of heating and cooling systems for the use of renewable forms of heat and heat from different processes is consistent with the concept of 'sustainable development' and directly contributes to reducing conventional fuel consumption and emissions.

The summer heat surplus can be accumulated in the seasonal storage, which replaces natural sources, ensuring the functional conditions and autonomy of systems.

Granular materials, phase or water change materials, well-sized and customized design conditions can be used as storage media. The solution is reasonable for systems with small and medium capacities applied to objectives located in isolated locations. [2]

Research programs have been supported within the Department of Building Services Engineering at the Technical University of Iasi and have the main objective to increase the overall efficiency of plant systems. The themes of two doctoral theses were preserved in the complex study of thermal storage (cold sources), with seasonal self-compensation and also one regarding the usage of nanofluids in thermal processes.

In the paper we have analysed the possibilities of functional and energetic optimization of the local systems of systems using nanofluid as a primary working fluid in heat exchangers embedded in storages with controlled thermophysical characteristics and with cyclic operation.

2. System description

2.1. Seasonal thermal energy storage

Thermal energy storage allows thermal energy to be collected for later use (hours, days or even months). It is a valuable asset and can be applied for individual buildings that are isolated and also for larger buildings such as schools, hospitals and even for an entire town. So, energy demand can be balanced between day and night or winter and summer. Geothermal heat pumps and nanofluids are the influencing factors to improve the system performance.

In the case of surface or deep geothermal sources, the quantities of heat absorbed under natural conditions are conditioned by the thermophysical characteristics of the soil and by the evolution of the ambient temperature influenced by the seasonal variation of the climatic factors.

Since the cooling load is higher than the heating load, for systems equipped with reversible heat pumps, the storage size must be adjusted according to the heating mode.

From a constructive point of view, the solution can be embodied in small or medium depths, preferably in the form of underground enclosures, wells or large diameter wells and depths up to 10 m, individual or parallel coupled, depending on the storage volume necessary.

2.2. Nanofluid

There are fluids that contain particles of nanometric size (1 × 10-9 m), called nanoparticles usually made of metals, oxides, carbon or carbon nanotubes. Common base fluids include water, ethylene glycol and oil.

The first studies on nanofluid behaviour and their performance were made in the Argon National Laboratory by Choi between 1995 and 1999.

The research that has taken place over the years has been aimed at determining the thermal conductivity of nanofluids with different volumetric concentrations (0.5-4%), but also different basic fluids. Between these volumetric concentrations, increases in thermal conductivity of about 25% compared to those of the base fluids used were observed. [3]

Results have shown that nanofluids have distinctive features such as: an improvement in convective transfer coefficient in the case of low particle concentration, an increase thermal conductivity and also in flow heat.

Experiments made by Wen&Ding, in laminar regime, using nanofluid composed of water and Al_2O_3 and a copper tube have revealed that an increased volume concentration of particles also increases thermal transfer, Reynolds number by 40% and thermal conductivity by 15%.

In the case of turbulent regime, Reynolds number is less influenced compared to laminar regime but there has been an increased thermal transfer as well.

So it is safe to say that by using nanofluids, we are able to obtain low-temperature heat transfer systems with low production costs but with increased energy efficiency, thanks to the physic-mechanical characteristics that they present: low pulse, increased mobility and high surface/volume ratio.

3. Numerical modelling

3.1. Geometry

Using Autodesk Inventor we were able to define a cube of quart sand with the following dimensions: Lxlxh = 1x1x1 m and a copper tube of 1 m in length with a diameter of 1".

The next step was to insert the copper tubes in the quartz sand cube, at equal distances.

The objective here is to compute heat transfer per tube row to the storage. So it is a physical system in which we have a complicated setup of several cascade of tubes. To start with and due to the fact that the length for a paper is limited, we proceeded to solve this case by simplifying it - taking a section for analysis and set-up the working conditions. More results have been obtained after, which are available upon request.

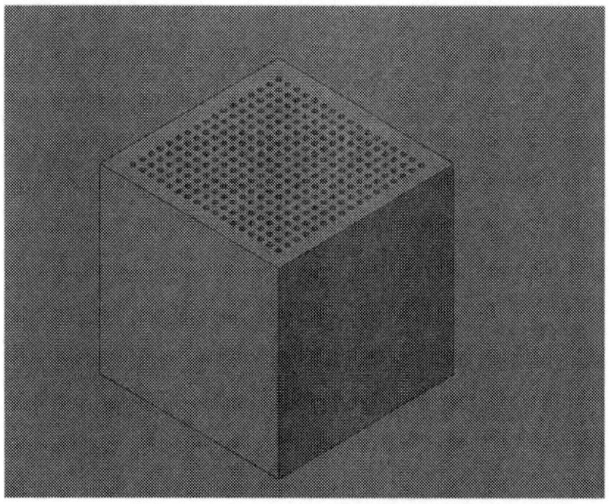

Figure 1. Quartz sand cube

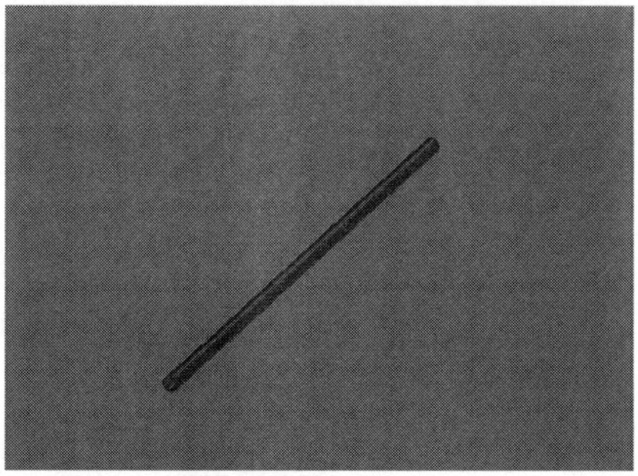

Figure 2. Copper tube

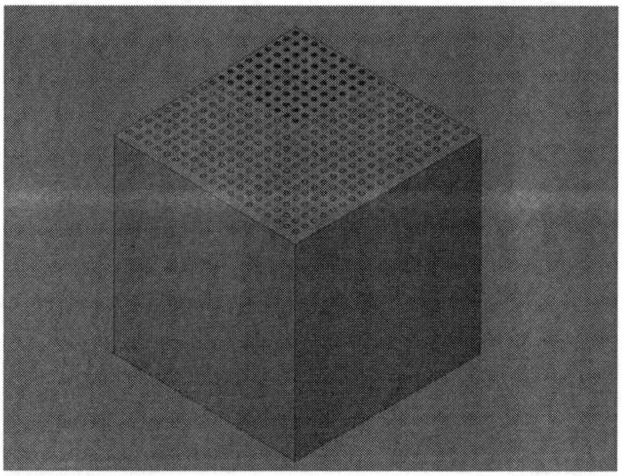

Figure 3. Copper tubes inserted in the quartz sand cube.

3.2. Ansys Fluent

Simulations of working conditions were done using the Ansys Fluent program. It uses CFD methodology, a science that predicts fluid flow, heat and mass transfer, chemical reactions and the resulting phenomena.
Thus CFD process in overall is a 3 step procedure:
Pre-processing: This step consisted in defining a geometry to our domain of interest. The domain of interest is then divided into segments, called as mesh generation step and the problem is set-up defining the boundary conditions.
Solver: Once the problem is set-up defining the boundary conditions we have solved it with the software on the computer. The numerical methods are also defined at this stage and we solve the whole problem.
Post-processing: Once we get the results as values at our probe points we have analysed them by means of colour plots, contour plots and appropriate graphical representations.

3.3. Working hypotheses

• Nanofluid working temperatures: 50, 60, 70, 80 90°C;
• Flows: 0.5, 0.6, 0.7, 0.8, 0.9 m^3 / h;
• Nanoparticle concentrations: 0% (water), 1%.

Water properties:
• ρ = 1000 kg/mc
• Cp = 4500 J/kg*K
• Λ = 0.9 W/m*K
• μ = 0.0015 kg/m*s

Quartz sand properties:
• ρ = 1600 kg/mc
• Cp = 795 J/kg*K
• Λ = 0.2 W/m*K

Also, the assumptions are:

- flow is laminar and translationally periodic (i.e. geometry repeats itself)
- flow approaching tube bank is steady with a known velocity
- body forces due to gravity are negligible

Table 1. Water properties with 1% nanofluid concentration.

Temp [°C]	ρ [kg/mc]	Cp [J/kg*K]	μ [kg/m*s]	Λ [W/m*K]
50	1036.5	4147.0	0.00059	0.66037
60	1032.0	4159.5	0.00050	0.67103
70	1027.0	4172.8	0.00043	0.68025
80	1021.2	4187.1	0.00038	0.68804
90	1014.8	4202.4	0.00034	0.69445

4. Results

The temperatures along the copper tube are shown in the images bellow, for different inherent temperatures and flows, for water (fig. 4,6, 8, 10, 12) and water with 1% nanofluid (fig. 5, 7, 9, 11, 13)

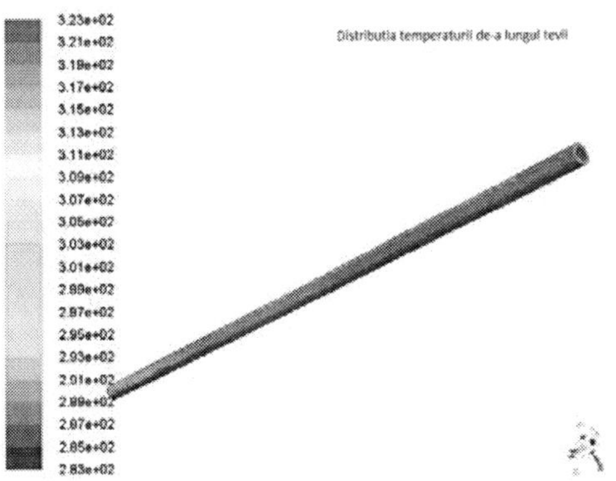

Figure 4. 50°C, 0.5 m^3/h (water)

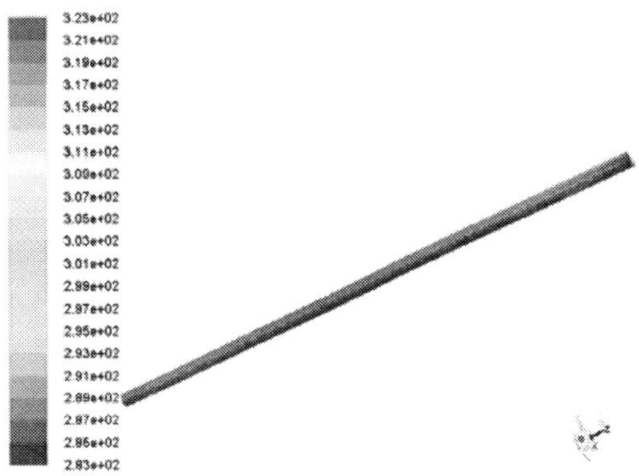

Figure 5. 50°C, 0.5 m^3/h (1% nano)

IOP Conf. Series: Materials Science and Engineering **586** (2019) 012040 doi:10.1088/1757-899X/586/1/012040

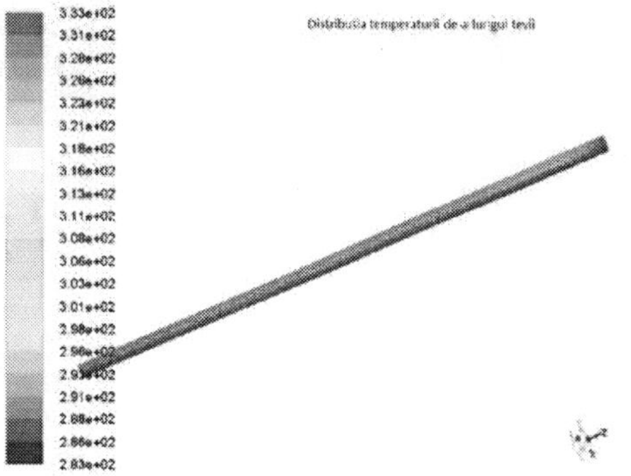

Figure 6. 60°C, 0.6 m^3/h (water)

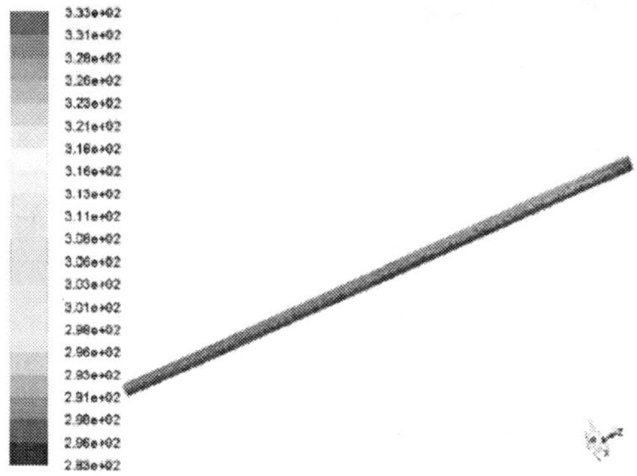

Figure 7. 60°C, 0.6 m^3/h (1% nano)

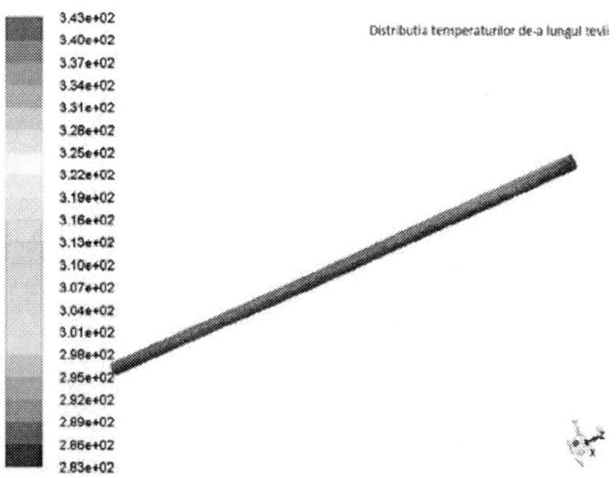

Figure 8. 70°C, 0.7 m^3/h (water)

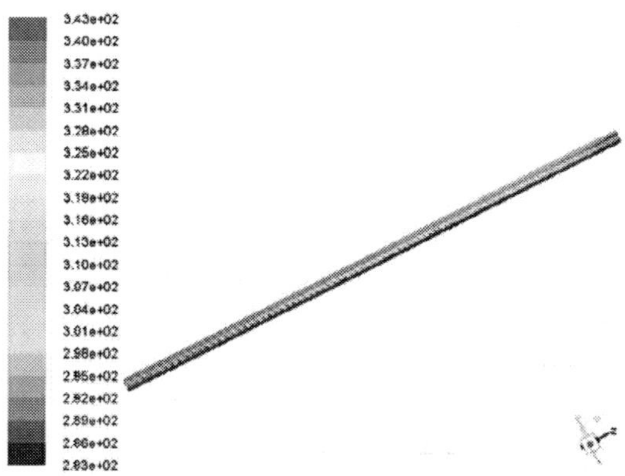

Figure 9. 70°C, 0.7 m^3/h (1% nano)

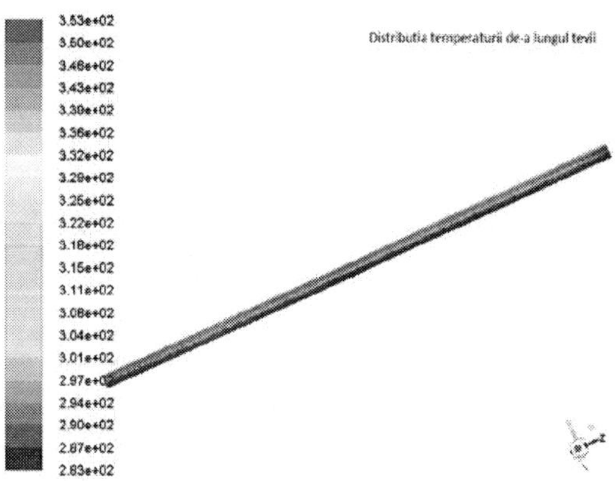

Figure 10. 80°C, 0.8 m^3/h (water)

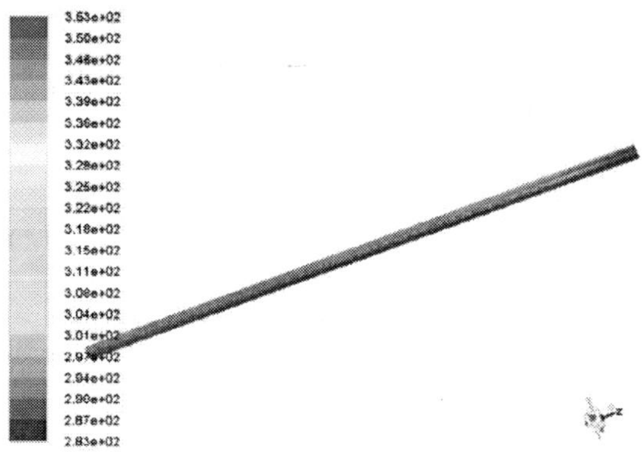

Figure 11. 80°C, 0.8 m^3/h (1% nano)

IOP Conf. Series: Materials Science and Engineering **586** (2019) 012040 doi:10.1088/1757-899X/586/1/012040

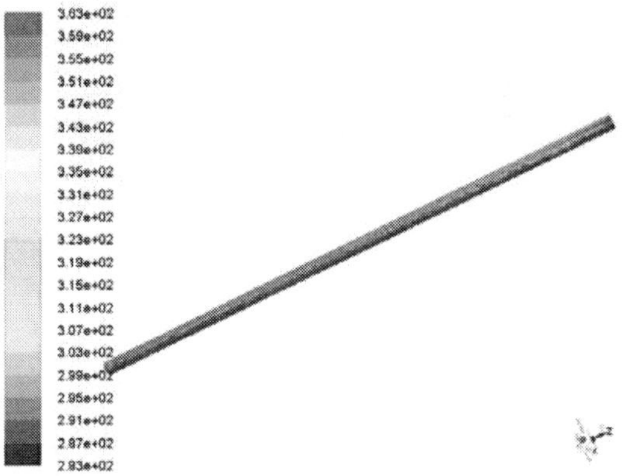

Figure 12. 90°C, 0.9 m^3/h (water)

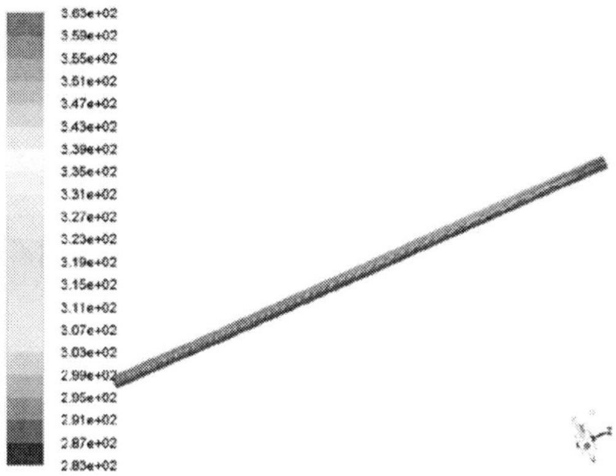

Figure 13. 90°C, 0.9 m^3/h (1% nano)

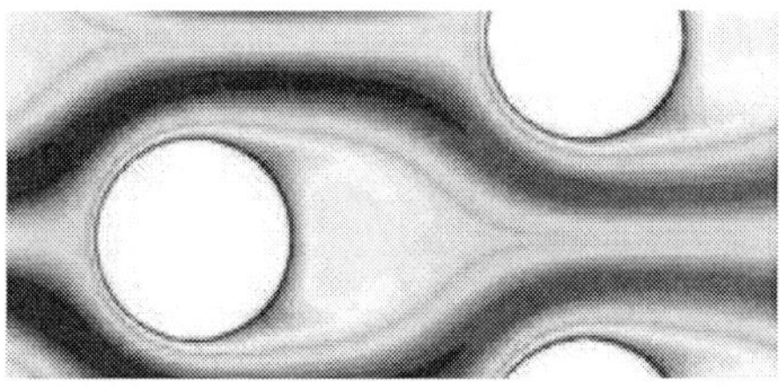

Figure 14. Temperature contours around tubes

5. Conclusions

As seen, the regions near the tube wall have re-circulation zones as a result of which there is a heat built-up (red colour) also shown by colour variations. So we see that we can have a very nice depiction of real life situation through a simple 2-D analysis.

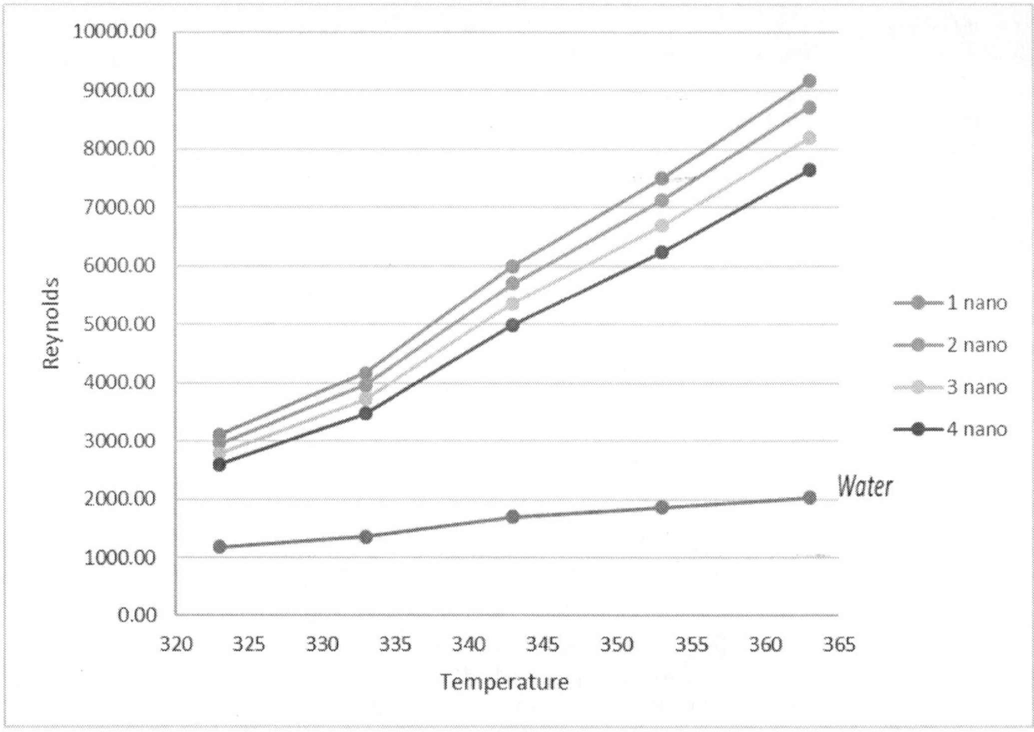

Figure 15. Comparison between different concentrations depending on the inherent temperature and Reynolds number

From the experimental system, the values that have been measured are, the temperatures of the inlet and outlet of the hot water as well as the inlet of the distilled water and the different concentrations of nanofluids at different mass flow rates.

The conclusions of the numerical study have highlighted the intensification of heat transfer, when using nanofluids, compared to water. [4][5]

6. References

[1] Baran A I, Cazacu I, Mateescu Th, Luciu R S *Surse reci autocompensate sezonier in sisteme de instalatii echipate cu pompe de caldura reversibile* (Timisoara: Instalatii pentru constructii si confortul ambiental)

[2] Cazacu I *Doctoral Thesis* (Iasi, 2016)

[3] Luciu R S *Doctoral Thesis (Iasi, 2011)*

[4] Baran A I, Mateescu Th, Luciu R S *Thermal convection analysis of heat pump systems, preprinting Clima 2019* (Bucharest)

[5] Baran A I, Luciu R S, Mateescu Th, *Studiul unui schimbator de caldura pentru valorificarea formelor regenerabile de energie, ICEE Conference 2016* (Iasi)

IR thermography applied in church heritage conservation

E F Țurcanu[1], V Ciocan[1], M Verdeș[1,*], R S Luciu[1], M C Bălan[1], S V Hudișteanu[1], A Burlacu[1]
[1]Gheorghe Asachi" Technical University of Iași, Faculty of Civil Engineering and Building Services, 700050 Iași, Romania

E-mail: verdesmarina2003@yahoo.com

Abstract. IRT – infrared thermography has met an extensive popularity among the non-destructive technologies. This paper is presenting some of the monumental church that had been examine with thermovision camera in order to assess the behavior of structure to ageing, earthquake damages, detect hidden characteristics or order anomalies. IRT investigation is based on principle that heat flowing in a material is altered by presence of all kind of anomalies, our job been to discover and interpreting. Thus heat bridges can be detected, heat loses and air leakage can be identify or the presence of mold, moisture is growing in a particular area.

1. Introduction

Discovered by William Herschel in 1800 and after the thermoelectric effect been discovered in 1821 infrared radiation or IR, have preoccupied the scientific world to find technique of interpretation and test methods for different building structures that were examined. [4]

The first infrared camera was invented in 1929 by Tihanyi been employed by the military in Britain for anti-aircraft defence. Also, significant progress has been made during the period between World War I and II. Today infrared thermography (IRT) is widely employed for buildings diagnostics. [3]

The first generation of commercial infrared cameras appeared in the end of 60's. Only in '90s a new generation of equipment with array detector appeared, the new equipment been capable both of temperature readings at different points and did not required special cooling systems. [2]

Finally, the thermography is based on the emission of objects and has three essential radiation law:

1. Kirchhoff's law of thermal radiation that define de relation between emission and absorption of energy. Accord to this principle, the emission coefficient, ε, is introduced in equation as the ratio of emissivity, E, of a real body to the emissivity, E_z, of the black body under the same temperature $\varepsilon = E/E_z$. ε can take value between 0 and 1, based to the temperature, surface texture and wavelength.

2. Planck's law of radiation describes the specific spectral radiation *I'*:

$$I'(\lambda,T) = \frac{2\pi hc^2}{\lambda^5} \times \frac{1}{e^{hc/\lambda kT} - 1} \quad (1)$$

Where: λ– is the wavelength, T – the absolute temperature, h – the Planck constant and c – the speed of light.

3. Stefan – Boltzman law, applied to the emission of a surface over all wavelengths, integrated the Planck law and discovered that the radiant power, I, [W/m^2], grows with the fourth power of the temperature, I = σ T 4, where $\sigma = 5.67 \times 10^{-8}$ W/m^{2}K^4. [4]

2. Infrared thermography applied to building physics

It is essential that studies be performed to evaluate the performance of building materials, especially when we talk about buildings that have century ahead. These studies are a step towards improving technical solutions and regulations to ensure the building's durability and to guarantee user comfort and satisfaction.

IOP Conf. Series: Materials Science and Engineering **586** (2019) 012041 doi:10.1088/1757-899X/586/1/012041

Most material pathologies are related with temperature action. Therefore, measuring a material's temperature is crucial for understanding the causes of those defects. The use of non-destructive techniques to test a building material may be very useful by making it possible to evaluate a material behavior without destroying it and without interfering with the users' life. [2]

3. Case study

The main objective of this work was to evaluate the applicability of thermography in order to study the behavior of building materials, to evaluate the building performance, age, heat loss through windows, dampness and "hidden details" (subsurface pipes, flues, ducts, wall ties, etc.). It can also be used for examination of heating systems and preventive maintenance. To do so, different heritage church was subjected to study. IR camera used for this study was a FLIR B335 (9Hz, 2010 model). Measures were made during winter.

4. Results

The main areas for using IR thermography in building diagnostics are presented in detail by [1]. In this study, it is suggested that building IR thermography on external building elements should be performed either at night or during a cloudy day. This was found to be important in order to avoid the problem of temperature increase which occurs as a result of the incident solar radiation, and the impact from the absorbed solar energy, which presents a time lag of a few hours. Additionally, in this work it is pointed out that measurements should be carried at low wind speeds in order to minimise as much as possible the influence of convective heat losses.

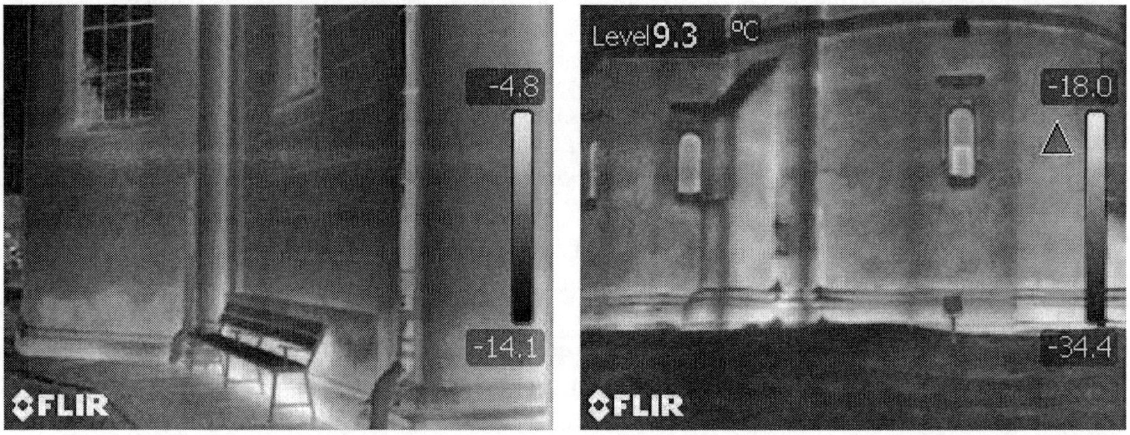

a) b)

Figure 1. a) – Thermal bridges at The Dormition of Mother of God church from Jassy
b) Thermal bridges at The Three Holy Hierarchs Monastery from Jassy

4.1. Thermal characterization of walls

IRT can be used in a qualitative way to locate the surface thermal anomalies related to different material properties, such as thermal conductivity (λ-value), heat capacity (cp), ε-value, thermal diffusivity (α), and density (ρ) [6]. Short- and long-term measurements are possible. Short-term IRT surveys are used to characterize: (i) building ages [2]; (ii) wall thicknesses [3]; (iii) material or structural components (e.g. solid or hollow bricks, concrete, thermal insulation, etc.) [4]; (iv) geometries [4] (figure 1); (v) surface properties (ε-value) [2]; and (v) energy problems related to layering techniques. They follow the standard procedures for the qualitative thermography in the building sector [2], which requires the control of whether and boundary conditions and environmental deficiencies [1]. In general, to minimize their impact, the IRT surveys should be undertaken during the cooler months, especially in evening and night-time [1]. Several multi-methodological approaches have been proposed to improve the energy assessment using the following high-depth techniques: (i) sonic trials for identifying the wall density [2]; (ii) HFM measurements for quantifying the U-value of the element [4]; (iii) ultrasonic and

IOP Conf. Series: Materials Science and Engineering **586** (2019) 012041 doi:10.1088/1757-899X/586/1/012041

electric-type micro-geophysical methods for discovering the presence of structural defects [6]; (iv) gravimetric tests for determining the moisture content 6]; (v) stratigraphic and chemical investigations for characterizing the material proprieties [5]; and (vi) computer simulation for understanding the defect through the comparison with thermal models [6]. Besides, long-term IRT surveys are used for the dynamic characterization of the α-value of walls [5], figure 2, figure 3, to consider the variation of the solar irradiance overcoming the limitations of traditional contact sensors [4]. In parallel, a novel in situ measurement procedure based on a single IR-camera and a thermal mirror has been proposed to quantify the time-shift and the temperature oscillation in thick walls, also reducing times and costs of traditional techniques [6]. Finally, IRT with periodic temperature variations has been applied to classify the percentage of moisture in traditional masonries [7].

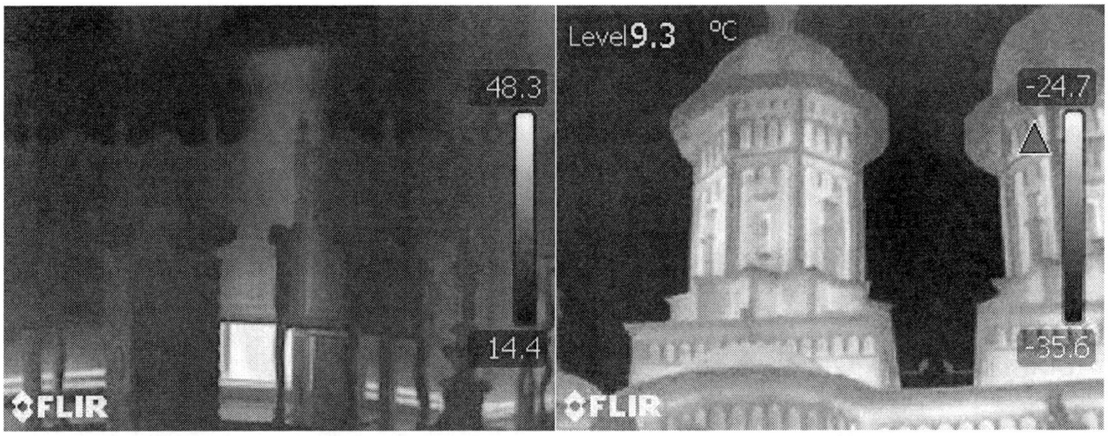

a) b)

Figure 2. a) Radiator heating element inside the Saint Peter and Pavel Monastery in Jassy, b) Heat loss in towers of the same monastery church

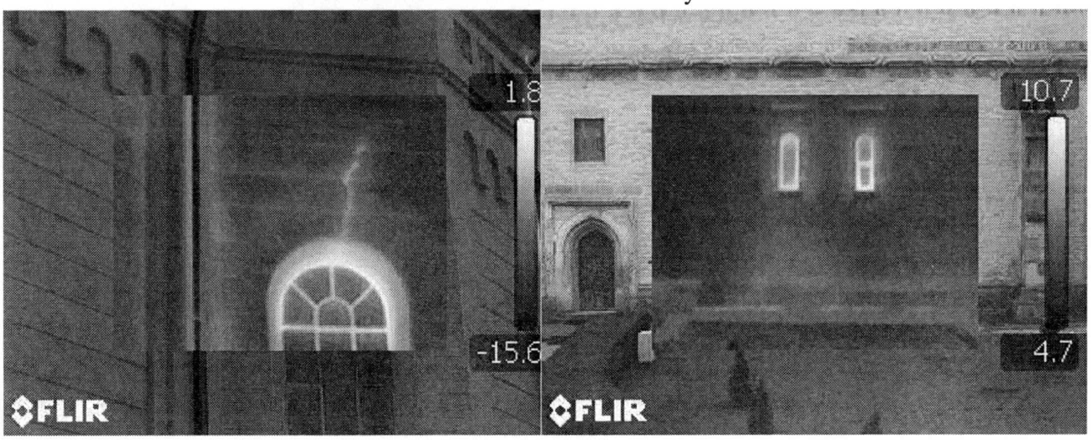

a) b)

Figure 3. a) Crack view with thermovision camera in Saint Mary Church in Jassy
b) Heat loss trough thermal bridges at junction between wall a ground

4.2. Thermal bridging and excessive heat loss areas detection

Thermal bridging is defined as additional heat losses in the building façade. It is classified into [7]: (i) "geometrical" connected to the shape of the envelope (i.e. corner of an external wall; junctions between walls, roofs or floors; adjacent walls or windows and walls); (ii) "linear" caused by discontinuities due to different materials or construction methods (i.e. corner areas between two outer walls, structural junctions between pillar and masonry, walls around windows, doors and hatches); and

(iii) "repeating" characterized by a regular pattern and distributed over a big area (i.e. mortar joints in a wall, timber studwork and beams in timber frame construction and steel wall ties in masonry cavities). IRT survey clearly identifies in a qualitative way their presence in the building façades, using standard procedures [8] (figure 4). As well, the presence of post processing techniques in the professional software permits to verify the risk of condensation and mould growth linked with the thermal bridging [9]. More recently, IRT surveys have been used in a quantitative way to investigate their impact on building envelope during in situ measurements [7] and laboratory tests [10]. Different numerical models have been proposed to overcome the perturbations caused by the climatic parameters on the external surface temperature (Ts), especially during in situ IRT surveys [10]. These models concern mainly linear thermal bridging for their diffusion in new and existing buildings. Due to the impact of boundary conditions, the studies don't consider the risk of surface condensation [11]. Furthermore, to reduce completely the impact of weather conditions, a simulation model has been defined without considering detailed meteorological observations [12]. Finally, a method for the quantitative measurement of thermal bridging performance without any supporting methods (neither numerical models nor HFM measurements) has been introduced [13]. The novelty of this approach includes the evaluation of the real heat flow rate caused by the thermal bridging and the linear thermal transmittance calculation using only IRT surveys.

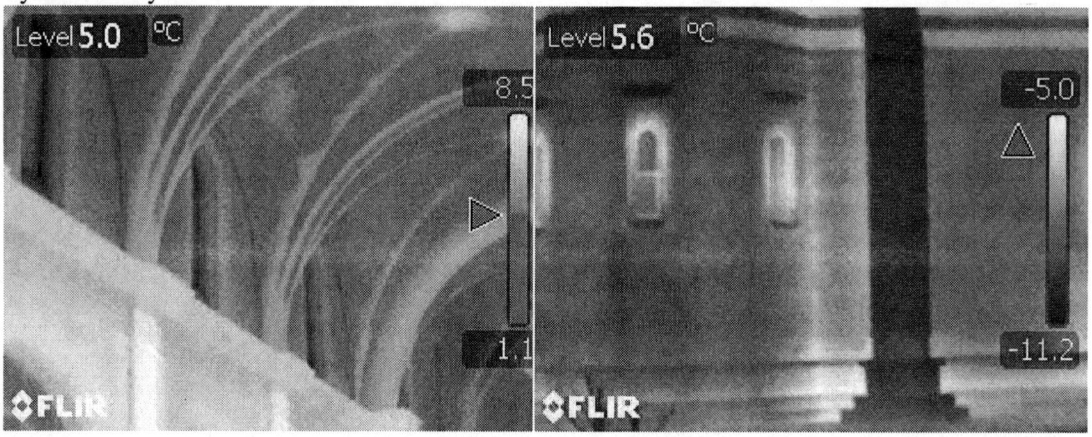

a) b)

Figure 4. a) With dark spot can be seen to heat loss in Saint Michael Church from Săbăoani b) Heat loss trough thermal bridges in The Three Holy Hierarchs Monastery from Jassy

4.3. Indoor temperature measurements

Although IR-cameras with sensors tuned at specific micrometre wave bands are used for the detection of specific gas leakages, they cannot measure the indoor air temperature [11]. Traditional IRT procedures are mainly applicable for opaque objects with substantial ε- value. Thus, the use of opaque mock targets that reach in short time the thermal equilibrium with the environment can accurately represent Ti. Mock target IRT is based on the extraction of their Ts from the radiation emitted [11], figure 4, figure 5. The interpolation of consecutive measurements provides a quick and reliable monitoring of the multidimensional spatial distribution of Ts. The thermal equilibrium of the mock targets with the environment permits to define quickly the spatial distribution of Ti. Its major advantages are the simplicity of the method and the reduced time to carry out the measurement. Several advices are suggested to easily distinguished the mock target from the environment: (i) use of IR cameras with an appropriate minimum focus distance to cut the influence of the air (e.g. 0.3 m); (ii) selection of the option picture in picture to better visualize the mock targets; and (iii) place the mock targets at least 0.5 m from walls and away from heat sources to avoid thermal reflections (e.g. radiators, windows, and so on).

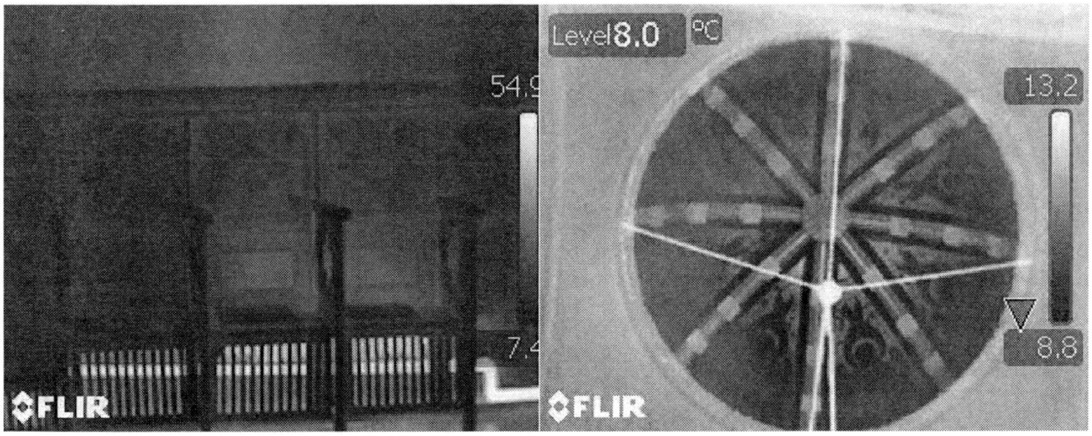

a) b)

Figure 5. a) With dark spot can be seen to heat loss in Saint Michael Church from Săbăoani b) Heat loss trough thermal bridges in The Three Holy Hierarchs Monastery from Jassy

5. Conclusion

Thermography is a non-destructive testing technology with much potential, having a great application to building materials. Research studies developed at different church revealed that emissivity is an essential parameter, since it greatly influences thermographic measurements and may restrict application of this technology to buildings. However, if the study aims for a qualitative analysis of the results, the selected emissivity value is not very important. It was also possible to confirm that thermography cannot be used to study objects in thermal or hygroscopic equilibrium, as temperature differences between the object and the environment must be significant, because of this the measure were made in winter, at -16^{0}C. Colour and thermal reflection must also be considered during thermographic testing, as they may mask the results and cause misinterpretation.

Thermography allowed studying the wetting and drying process of building materials. Temperature differences due to superficial water evaporation provided a means of recognising “wet” and “dry” areas. It was also possible to evaluate a material’s approximate drying time since small superficial temperature variations indicate that moisture is rather significant. Thermography, however, detects only superficial moisture.

Advantages of IR thermography for building study:

- the temperature measurement of the targets can be done remotely without direct contact;
- the measurement is non-destructive (it does not require sampling);
- the method is non-polluting
- the thermal image is geometrically consistent with the studied object;
- thermal, global or detailed information is obtained in real time;
- measuring accuracy is high;
- electricity consumption is insignificant;
- simple operation and relatively small gauges ensure handling, the equipment may be carried out by a single person;
- the thermal imaging system provides an image that allows a quick identification and the exact points of potential defects;
- allows association with complex recording, storage and processing equipment

Qualitative analyses are widely treated by literature, standards and practical reports. They focus on: (i) thermal characterization of walls, glazing and windows; (ii) thermal bridging and excessive heat loss areas detection; (iii) thermal insulation examination; (iv) air leakage inspection; (v) moisture and water detection; (vi) HVAC and electrical systems characterization; (vii) indoor temperature measurements; and (viii) human comfort assessment. On the contrary, quantitative studies are worthy of investigation.

Currently, they concentrate on: (i) determination of the areas with thermal anomalies; (ii) insulation level detection; (iii) U-value measurements and dynamic characterization of building elements; (iv) moisture content determination; and (v) damage evolution analysis. The literature review presents also the possible topics for the research in this field, suggesting the themes that need to be further explored. They are synthetized above:

- Comparison among different passive methodologies for the building stock analysis;
- Development of procedures and tools for:
  - U-value measurement and dynamic characterization of building components (particularly for walls and glazing systems);
  - Gas concentration determination of glass units and glazing systems;
  - Moisture content determination;
  - Damage evolution detection;
  - Long-term environmental monitoring using the mock target IRT;
  - Human comfort assessment;
  - Integration of different non-destructive testing for the building energy audit;
  - Integration of energy audit, damage, and structural assessments;
- Development of smart sensors for indoor monitoring, environmental control, and human comfort assessments.

6. References

[1] Balaras C, Argiriou A 2002 Infrared thermography for building diagnostics. Energy Build, Volume 34, pp. 171-83
[2] Barreira E, de Freitas V P, Delgado J, Ramos N M M 2012 Thermography Applications in the Study of Buildings Hygrothermal Behaviour, Infrared Thermography, Dr. Raghu V Prakash (Ed.), InTech, DOI: 10.5772/28282. Available from: https://www.i. [Online]
[3] Kylili A, Fokaides P, Christou P, Kalogirou S 2014 Infrared thermography (IRT) applications for building diagnostics: A review. Applied Energy, pp. p.531-549
[4] Pleşu R, Teodoriu G,.Ţăranu G 2012 Infrared Thermography applications for Building Investigation. Buletinul Institutului Politehnic din Iaşi
[5] 2005 American Society for Testing and Materials (ASTM) Standard guide for examining electrical and mechanical equipment with infrared thermography, E 1934. West Conshohocken: ASTM
[6] Taylor T, Counsell J, Gill S 2013 Energy efficiency is more than skin deep: improving construction quality control in new-build housing using thermography. Energy Build; 66:222–31.
[7] Öhman C 2012 Measurement in thermography. Sweden: FLIR Systems AB
[8] Cerdeira F, Vázquez ME, Collazo J, Granada E 2011 Applicability of infrared thermography to the study of the behavior of stone panels as building envelopes. Energy Build; 43:1845–51
[9] Vidasa S, Moghadam P 2013 HeatWave. A handheld 3D thermography system for energy auditing. Energy Build 2013; 66:445–60
[10] Vidasa S, Moghadam P 2013 HeatWave. A handheld 3D thermography system for energy auditing. Energy Build; 66:445–60
[11] Meola C 2007 Infrared thermography of masonry structures. Infrared Phys Technol;49:228–33
[12] Bianchi F, Pisello A, Baldinelli G, Asdrubali F 2014 Infrared thermography assessment of thermal bridges in building envelope: experimental validation in a test room setup. Sustainability; 6:7107–20
[13] Ferrarini G, Bison P, Bortolin A, Cadelano G 2016 Thermal response measurement of building insulating materials by infrared thermography. Energy Build; 133:559–64

IOP Conf. Series: Materials Science and Engineering **586** (2019) 012042 doi:10.1088/1757-899X/586/1/012042

Thermal comfort assessment for different heating system using CFD-modelling inside of an orthodox church

V Ciocan[1], E F Țurcanu[1,*], M Verdeș[1], R S Luciu[1], M C Bălan[1], S V Hudișteanu[1], A Burlacu[1]

[1]Gheorghe Asachi" Technical University of Iași, Faculty of Civil Engineering and Building Services, 700050 Iași, Romania

E-mail: turcanufemilian@gmail.com

Abstract. Thermal comfort is a general term that describes the sensation of warm or cold for the occupant, defining the state of mind of humans that expresses satisfaction. Because 90% of our time is spend inside the buildings is essential to achieve a good thermal comfort. In this paper we present two different heating system that are used in more than 90% of the church in our country. The article test two different heating system in order to evaluate the thermal comfort of the occupants and if the climate meet the requirements for the churches where the paints and the artworks must be kept in good condition for future generation.

1. Introduction

The indoor climate in historic buildings such as Metropolitan Orthodox Cathedral form Jassy (figure 1) is a crucial factor in preserving the building and its interiors. Rising of the energy cost and the threat of global climate change has put more focus on indoor climate issues in the recent years. Sustainable management of these buildings demands a balance between preserving cultural heritage and energy efficiency.[1]

Figure 1. View from Metropolitan Orthodox Cathedral form Jassy

Nowadays, we have the possibility of using methods and devices both in terms of computing capabilities and experimental techniques. We have today the opportunity to validate these models by taking into account the variation of several parameters, we also have the opportunity to correct them and to propose new models. [2]

Church has a rectangular shape with the length of approximately 42 meters, the width of 18 meters and the height of 22 meters (figure 2). The volume of air from the space exceed 16000 m^3. Now, the natural indoor environment is created in the cold season with static heaters made of cast-iron with an equivalent power of 3000 Watts. The walls are made from brick with a thick of 1.5 meters.

In cold regions, as the one in which the metropolitan cathedral is placed, with exterior temperature that can be below -18 ^{0}C, when the church is too crowded (or a humidification system is active) and certain surfaces are relatively cold (for ex. Windows, statues, frescoes, walls, ceiling) the excess of water vapour condenses on the cold surfaces. The situation is worsened by certain heating system with combustion of methane, propane, LPG etc., which produce large quantities of water vapour.[4]

Certain warm-air heating systems mitigate the decrease in relative humidity with the addition of vapour. This may do the air good, but it is not always advantageous for the surfaces of artworks, especially the cold ones (i.e. windows) or those with great thermal inertia (i.e. marble statues, walls), on which condensation may take place. [4]

When the RH level in the air and/or the moisture content in materials is high for whatever reason (e.g. crowded room and condensation, percolation, cold surfaces) a number of deterioration mechanisms may be activated. The main ones are: [4]

- oxidation and corrosion of metals (e.g., stained windows, organs, chandeliers);
- bio-decay with deterioration of organic substances (e.g., wood, leather) and/or micro-biological colonization;
- efflorescence's on bricks and plasters due to salt migration; also forced evaporation generates efflorescence.

Condensation depends entirely on two factors: the moisture content in the air (expressed in terms of mixing ratio or dew point temperature) and the temperature of the surface. Condensation can be preceded by moisture absorption if deliquescent salts are present in materials. [5]

1.1. Radiator panel heating

Radiators range from traditional cast-iron to the present pressed steel types. They have different forms: column type, single of multiple panels. Because of the affected esthetical view radiator panel heating is not popular in monumental church. They are used in small monumental churches [3].

Radiators operate as natural convectors (figure 2, figure 3). Air enters at the base of the radiator, is heated and rises for buoyancy, and finally emerges at the top or the front. In traditional radiators, hot water (90°-70°C) carries heat through pipes from the boiler to heaters. Radiators (e.g. panel, column, aluminium-convector radiators) are installed throughout the church, at some distance from, or very close to pews. Mobile radiators are electrically heated and use oil as thermo-convector fluid. Most of the heat is distributed via natural air convection and just a little via long-wave IR radiation. [5]

Benefits: The building is provided with a traditional, basic source of heat. The congregation benefits from the warm air and some people from heat radiated from radiators too. (figure 3)

Problems: Preservation: When in operation, the RH level is low, sometimes too low (figure 4). This is especially true for continuous operation in cold climates. When the system is used occasionally, or with mixed mode, damage is not limited to internal artworks and decorations, but may extends to masonry due to cycles of re-crystallization of soluble salts that follow the heating cycles. Convective motions of warm air develop above the radiators and smoke and pollutants blacken walls and ceiling during operation. Under this respect, the continuous or mixed operation is penalised. [5], [7].

1.2. Floor heating

In the late seventies of the twentieth century floor heating became popular as the main heating system in weell- insulated dwellings. Furthermore, the 'invizible' heating system was aesthetically sound and made no noise. The degree of thermal comfort was high, a high level of mean radiant temperature, a high thermal contact temperature, even with stone floor systems, and little thermal stratification. Due to these advantages is was also introduced in modern church as well as in monumental church. But due to the lack of insulation of floor heating in monumental church the building have a different behaviour. [5], [6], [7]

Benefits: Quite homogeneous heating with gradual variations. Preferably installed as a basic heating system, with continuous use, especially in mild, humid regions. In the case of continuous operation, all surfaces will be heated although in a different way: the ceiling will remain colder than walls and walls colder than the floor. Warm feet, mild body and thermal comfort. [5], [6]

Problems: Preservation: In cold regions, and in the case of continuous use, the RH may drop below the threshold of sustainability for wood and other organic materials. With intermittent or mixed operation, in the lower part of walls, where heating-cooling cycles are at highest intensity, masonry may be affected by dissolution-recrystallization cycles of soluble salts. A supply of moisture for humidification to mitigate RH may prove effective for air, but condensation and damage may occur on cold surfaces (e.g. windows, ceiling, walls) [5], [6], [7]

2. Case study

This paper described the indoor environment created when is used radiator panel heating or the floor heating system inside the same church in order to compare them. Software used is Autodesk CFD 2017 – capable of solving heat, air and water vapour simulation, the analysis where run in steady state. [7]

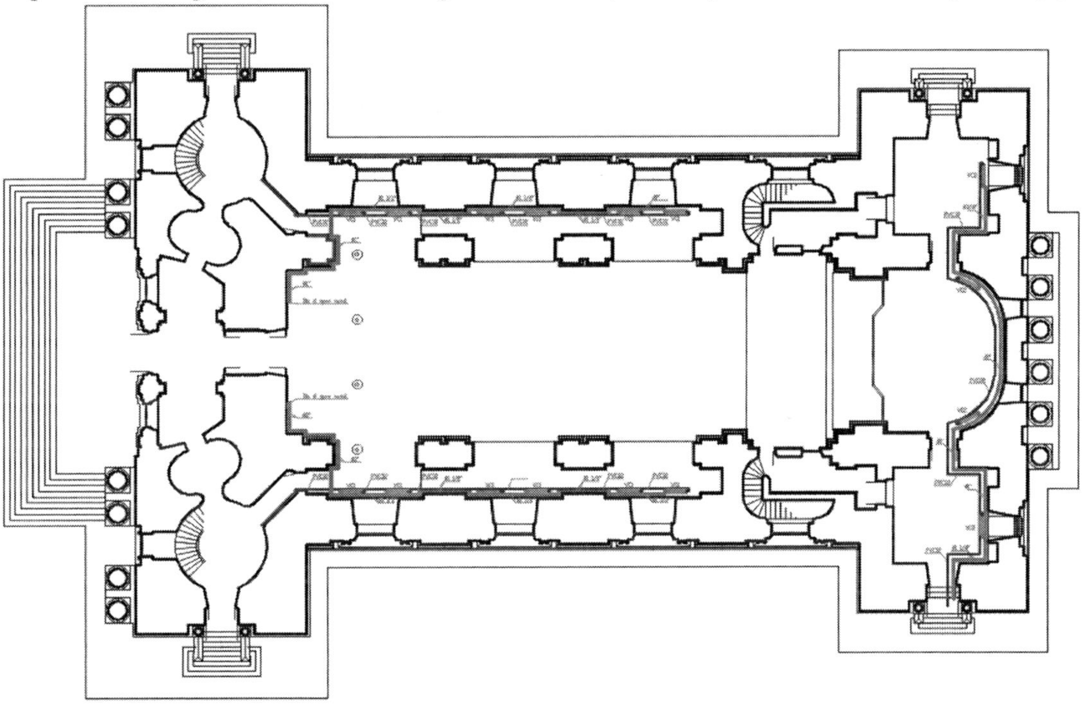

Figure 2. Metropolitan Orthodox Cathedral – Plan View with the hot water heating system

IOP Conf. Series: Materials Science and Engineering **586** (2019) 012042 doi:10.1088/1757-899X/586/1/012042

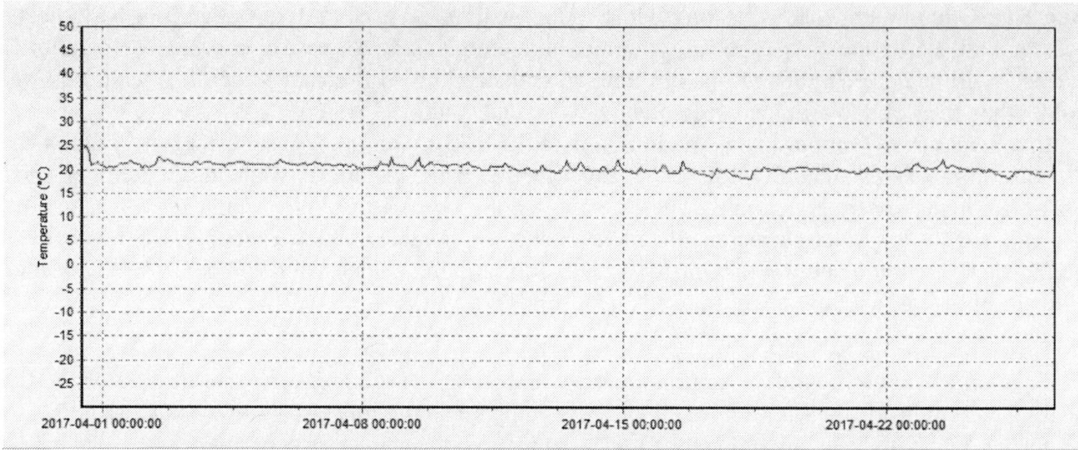

Figure 3. Air temperature variation in Metropolitan Orthodox Cathedral

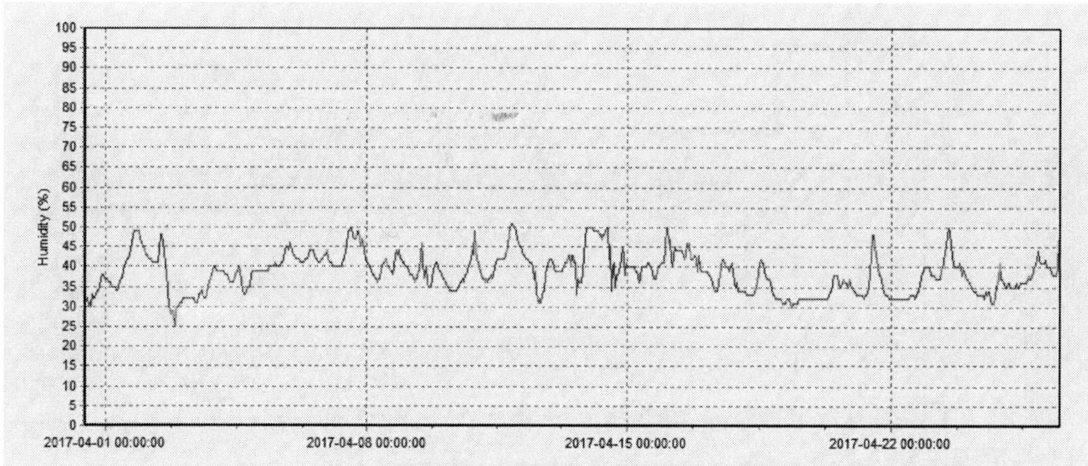

Figure 4. Air humidity variation inside the Metropolitan Orthodox Cathedral

The model has been made in Autodesk Inventor 2017 in 3d space in order to have a model closed to the investigation site. The boundary conditions imposed for walls are:

- the film coefficient for exterior walls is U=0.8 W/m^2K
- the film coefficient for interior walls is U=0.6 W/m^2K
- the film coefficient for windows is U=0.5 W/m^2K
- the film coefficient for floor is U=1 W/m^2K

The boundary condition used for static heater are taken from the site, cast-iron material with an approximated temperature of 80°C. Model discretization has a step of 0,5 meter (figure 5) in a volume of air of 16000 m^3. In the case of second modelling the static heaters were removed and replaced with floor heating system, system that had been fixed to 100 W/m^2 heat flux.

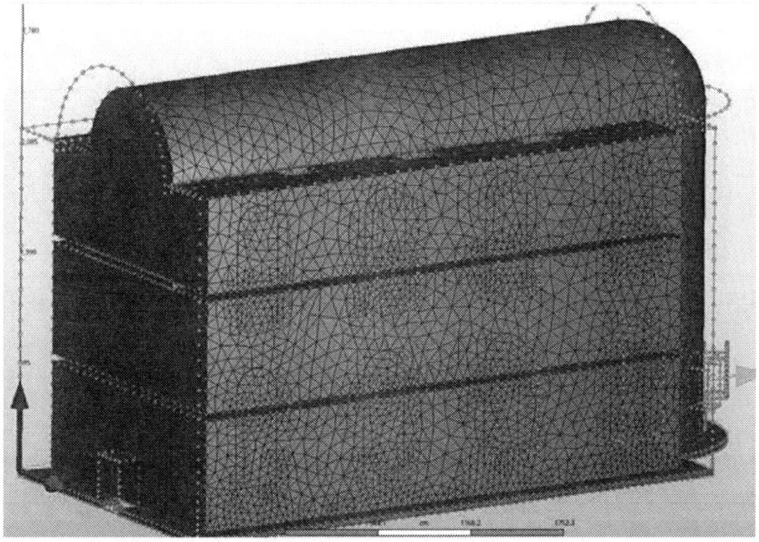

Figure 5. Metropolitan Orthodox Cathedral 3d geometry and meshing

3. Results

The CFD modelling show that the value of temperature that is create inside the church interior environment has a multi-zone variation. Locally, where static heaters are placed is warmer than the other side (figure. 6), also the air distribution in case of floor heating system can be view in figure 8, for comparison. The predicted mean vote (PMV) is in range of +0,5÷-0,5 (figure 7) and the comfort temperature of 15-20°C (figure 6).

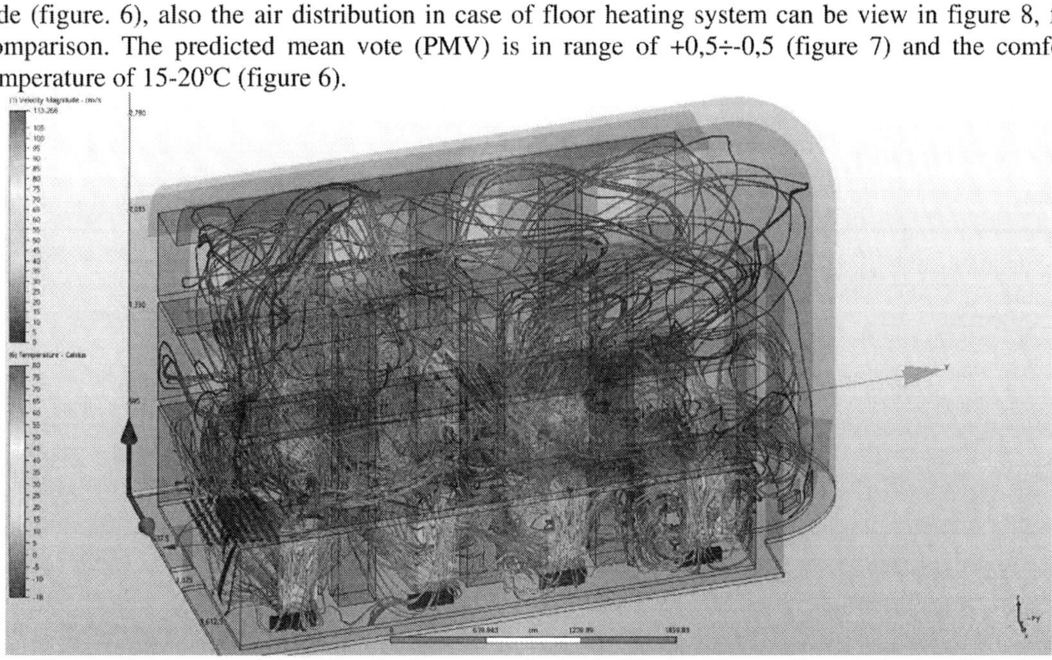

Figure 6. Air temperature distribution and velocity inside the Metropolitan Orthodox Cathedral

IOP Conf. Series: Materials Science and Engineering **586** (2019) 012042 doi:10.1088/1757-899X/586/1/012042

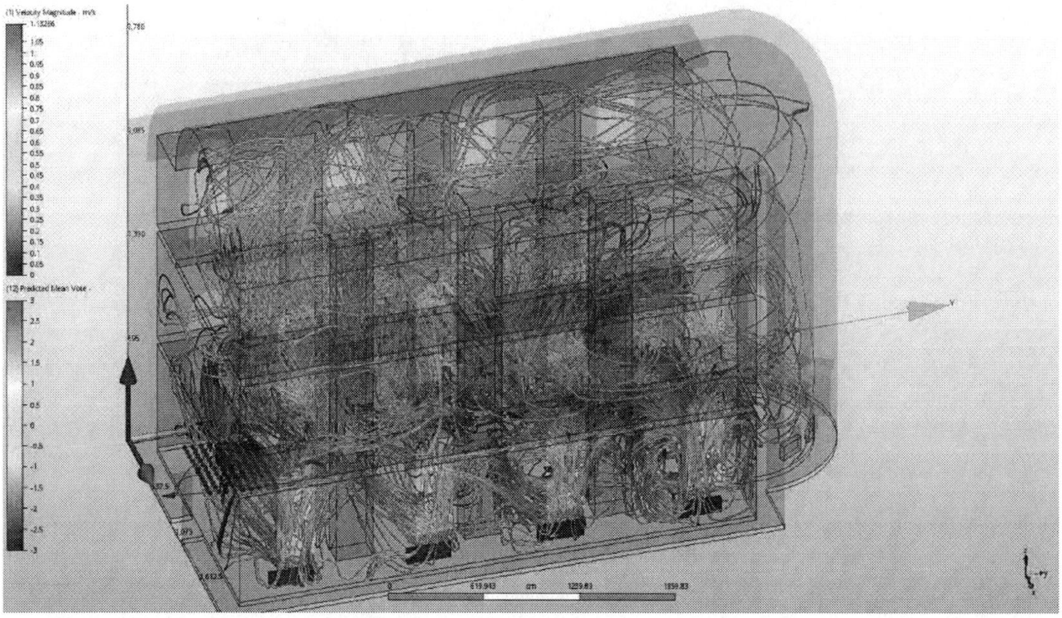

Figure 7. Air velocity and PMV viewed in cross section of Metropolitan Orthodox Cathedral

Convective rise of internal air leads to blackening of surfaces (e.g. paintings, tapestries), especially above the support brackets that trigger turbulence in the uprising airstream (figure 6). Forced air convectors blacken walls all the same, notwithstanding air filters are included (figure 7).

When in use, it generates an endless convective motion of the air above the floor, with heavy deposition of pollutants on walls (figure 8) and ceiling. The latter remains colder and is heavily blackened. Cold downdraughts formed on contact with the walls increase blackening. If possible, insulation of the roof is advisable as a preventative measure. The warm floor tends to mobilize underground vapour and in some cases it may generate a capillary rise. Ground water can be stopped by an adequate damp proof base; however, this is not always easy or possible with interventions in historic buildings.

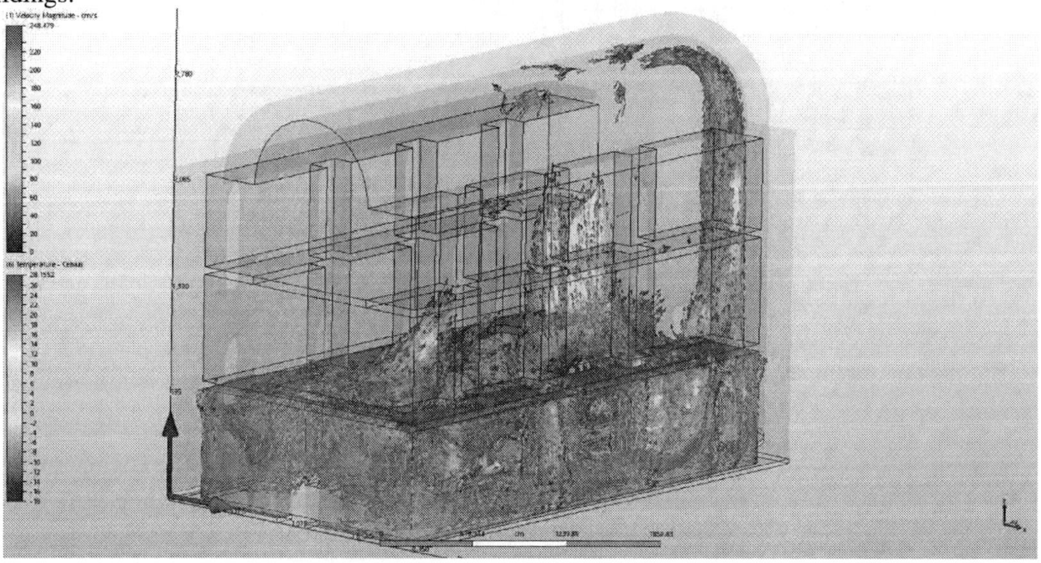

Figure 8. Floor heating system and the air volume characteristics in terms of temperature and velocity

4. Conclusion

In this case study, the local radiant heating system is a nice alternative for the "old" static heating system that heats the whole air volume of the church. First, the heating capacity of the local system is lower than that of the churches static heating system. Secondly, whereas the static heating system needs to be operated for several hours before the service starts, the local heating system needs only to be operated from 15 minutes before the service until the end of it. Therefore, the heating costs of the underfloor heating system will be importantly lower than those of the static heating system.

From the viewpoint of conservation, it is positive that the air convection is relatively small, and also in the whole church, is only heated very slowly and does not reach high temperatures. And the stratification in air temperature above a height of about 2 m, which existed when operating the hot air heating system, is not present with the local heating system. The heat is present in the zone where the people are seated.

5. References

[1] Croitoru C V, Nastase I, Bode F 2012*Thermal Comfort assessment for different ventilation strategies using a CFD approach.* Mathematical Modeling in Civil Engineering, pp. 36-44

[2] Larsen P K, Brostrom T 2011 *Climate Control in Historical Buldings. s.l.:National Museum of Denmark.*

[3] Schellen H 2002 *Heating monumental churches: indoor climate and preservation of cultural heritage.* Eindhoven: Eindhoven University of Technology

[4] Camuffo D, Pagan E, Schellen H, Limpens-Neilen D 2014 *A practical guide to the pros and cons of the various heating systems with a view to the conservation of the Cultural Heritage in Churches.* Results of the European Project Friendly Heating EVK4-CT-2001-00067

[5] Marco P, Kristian F 2018 *Indirect Investigation: Building Simulation. In: Historic Indoor Microclimate of the Heritage Buildings.* Bologna, Springer

[6] Martinez Garrido M I, Fort R, Varas Muriel M 2016 *Sensor-based monitoring of heating system effectiveness and efficiency in Spanish churches.* Indoor and Built Environment, 21(1), pp. 156-165

[7] Popovici CG, Hudişteanu SV, Chereche̦ş NC 2018 *The necessity of HVAC system for the registered architectural cultural heritage building*, E3S Web of Conferences 32, 01008

[8] Popovici CG, Hudişteanu VS 2016 *Numerical simulation of HVAC system functionality in a sociocultural building*, Procedia Technology 22, 535-542

IOP Conf. Series: Materials Science and Engineering **586** (2019) 012043 doi:10.1088/1757-899X/586/1/012043

Hygro-thermal monitoring inside Gheorghe Asachi library and their effects on heritage conservation

V Ciocan[1], E F Țurcanu[1,*], M Verdeș[1], R S Luciu[1], M C Bălan[1], S V Hudișteanu[1], A Burlacu[1]

[1]" Gheorghe Asachi" Technical University of Iași, Faculty of Civil Engineering and Building Services, 700050 Iași, Romania
E-mail: turcanufemilian@gmail.com

Abstract. The indoor environment and its dynamics is depending on the variations of the interior activity and of outdoor climate that influence the moisture in exterior walls. Heating system, along bad ventilation or air infiltration in window induce substantial fluctuations in the stability of relative humidity (RH). (RH) is the physical parameter that can induce efflorescence, drying and cracking in the timber. Gheorghe Asachi Library consist in over 1 million books, has a unique architecture with a lot of elements made from wood. The result obtained have allowed us to develop a series of recommendations that might be useful for preventive conservation.

1. Introduction

The Gheorghe Asachi library of the Technical University from Jassy is distinguish as one of the 25 most beautiful library in the world. Placed inside the Palace University and thought its classic style is decorated with wood elements. (figure 1) The foundation stone was put on 23 May 1893, and four years later King Carol I make the inauguration. Since than it known a cruel history, it was destroyed during the war and then rebuilt in 1950. [1]

Figure 1. View from Gheorghe Asachi Library of Technical University.

Library has a rectangular shape with the length of approximately 64 meters, the width of 9 meters and the height of 9 meters (figure 2). The volume of air from the space exceed 5000 m^3. The natural indoor environment is created in the cold season with static heaters made of cast-iron with an equivalent power of 2500 Watts. In the period of 08.03.2017- 13.03.2017 – there were placed sensor to monitor air temperature and humidity (figure 3, figure 4).

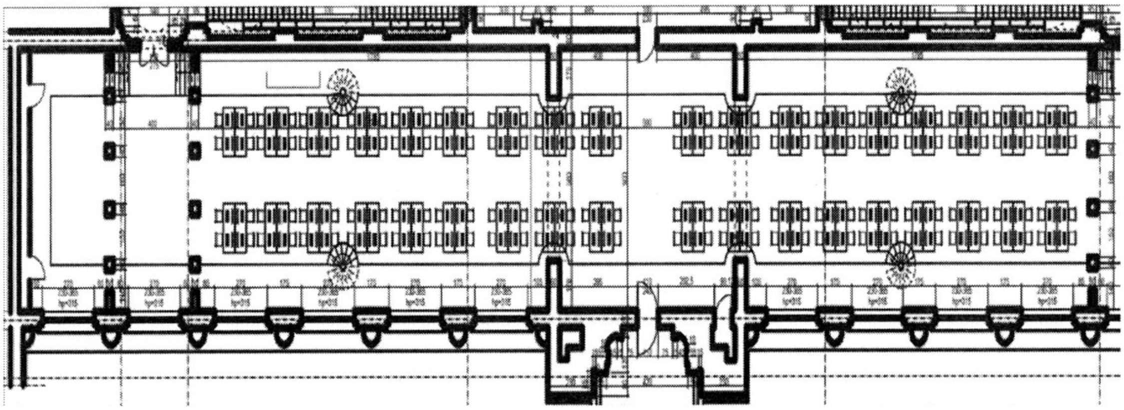

Figure 2. Plan view of the Technical University library

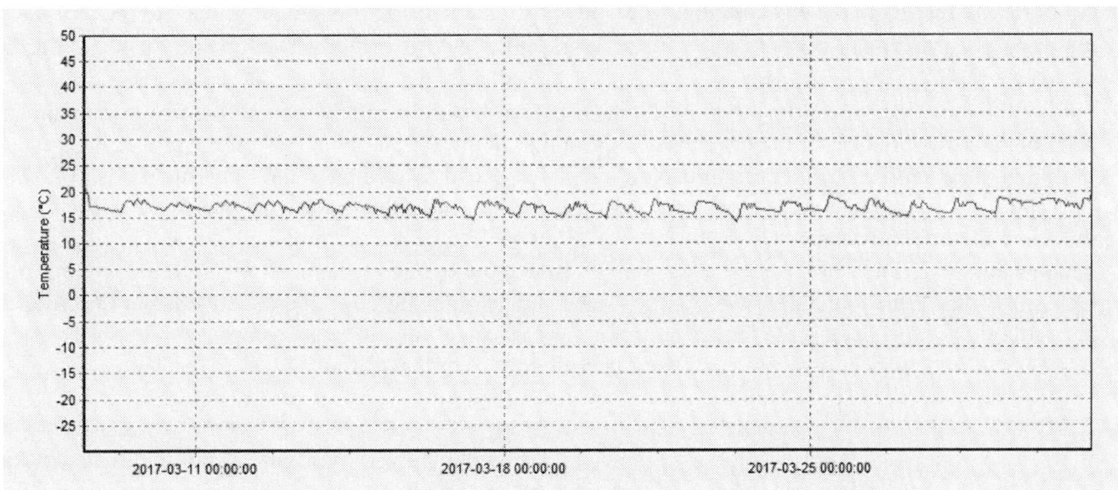

Figure 3. Air temperature variation in Technical University library

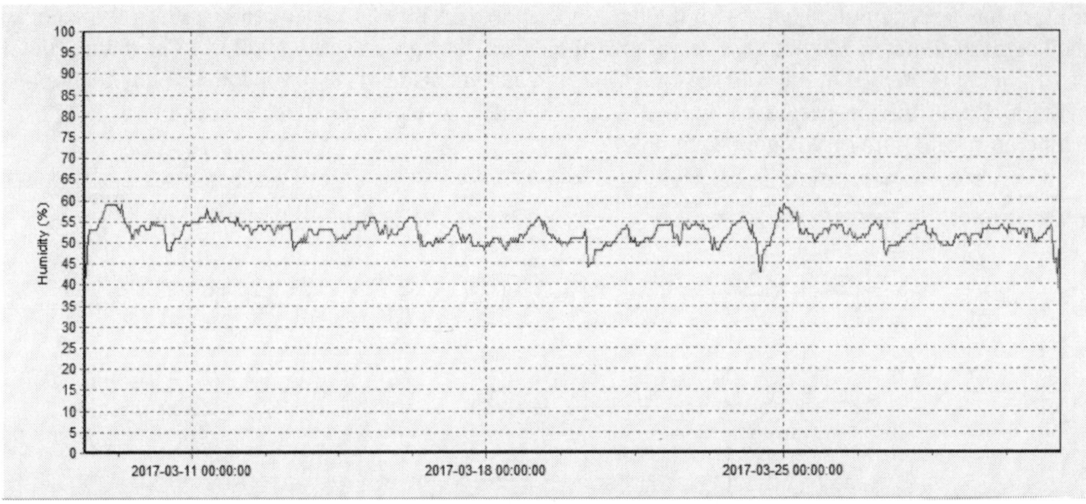

Figure 4. Air humidity variation inside the Technical University library

For a long period of time, thermal comfort, visual comfort and indoor air quality have been studied from different point of view by engineers, architect, industrial hygiene experts. The first studies on the thermal comfort with respect to the environment were made by Claude Bernard which reference to the biological regulation of the human body in 1865 in the work entitled "Introduction à l'étude de la medicine éxpérimentale".[2], [3], [4], [5]

Studies that have examined the air quality in the interior of buildings others than those of residential has been of great importance for the last years. This studies showed that the indoor air contains a wide range of pollutants like organic and inorganic components and also particulate matter and microbes. [6], [7]

The consequence of the energy crises that take place 1970 with the rising energy costs resulting from this have the effect that many building have a better sealing off to the outside air than previously had been in this case. The trend for energy saving is in use today due to the need for a global reduction in CO_2 emissions. But a more effective sealing – when unaccompanied by adequate mechanical ventilation is conducting in a considerable reduction in the air exchange rates in these rooms, with the back effect in means that the concentrations of substances emitted into the indoor air will rise. [8], [9]

2. Microclimate quality assessment

In the case of heritage building such as, Library of Technical University, figure 1, microclimate control plays a fundamental role in the deterioration process of materials and books collection preservation. [2], [4], [10]

Recent year's research make has assumed a role of great importance in the protection of work-art. So the microclimate is understand as being a synthesis of the natural environment conditions due not only weather variables such as temperature (T, °C), humidity (RH, %), solar radiation, but also to thermal exchanges with other bodies such as infrared radiation, heating and ventilation. For the preservation of artworks, an adaptive climate condition must be created to suit with micro-climate condition of conservation. [3]

3. Aim of this research

This paper described the indoor environment created when is used static heating systems, and the climate created with a HVAC heating system. The temperature and humidity measured made inside the library, are used in order to validate the modelling made in CFD environment. Software used is Autodesk CFD 2017 – capable of solving heat, air and water vapour simulation, the analysis where run in steady state. The model has been made in Autodesk Inventor 2017 in 3d space in order to have a model closed to the investigation site. Now, the heating system is a classical one with static heaters. The boundary conditions imposed for walls are:

- the film coefficient for exterior walls is U=0.8 W/m^{2}K
- the film coefficient for interior walls is U=0.6 W/m^{2}K
- the film coefficient for windows is U=0.5 W/m^{2}K
- the film coefficient for floor is U=1 W/m^{2}K

The boundary condition used for static heater are taken from the site, cast-iron material with an approximated temperature of 80oC. Model discretization has a step of 0,5 meter. (figure 5.) in a volume of air of 5000 m^3.
In the case of second modelling the static heaters were removed and placed inlet and outlet ventilation system with a volume rate flow of 160 m^3/h, system that ensure a ventilation rate of 0.5 exchanged air volume per hours.
Indoor climate simulation is performed using a computer to virtually represent a building design and perform physics-based calculations. The simulations can range from a building component to a hole indoor climate. The simulation program enables simultaneous interaction of the geometric model with outdoor conditions, occupancy, and usage of building systems to predict various loads arising in the building on an hourly basis. Basic laws of physics and energy balance equations are used for calculations.

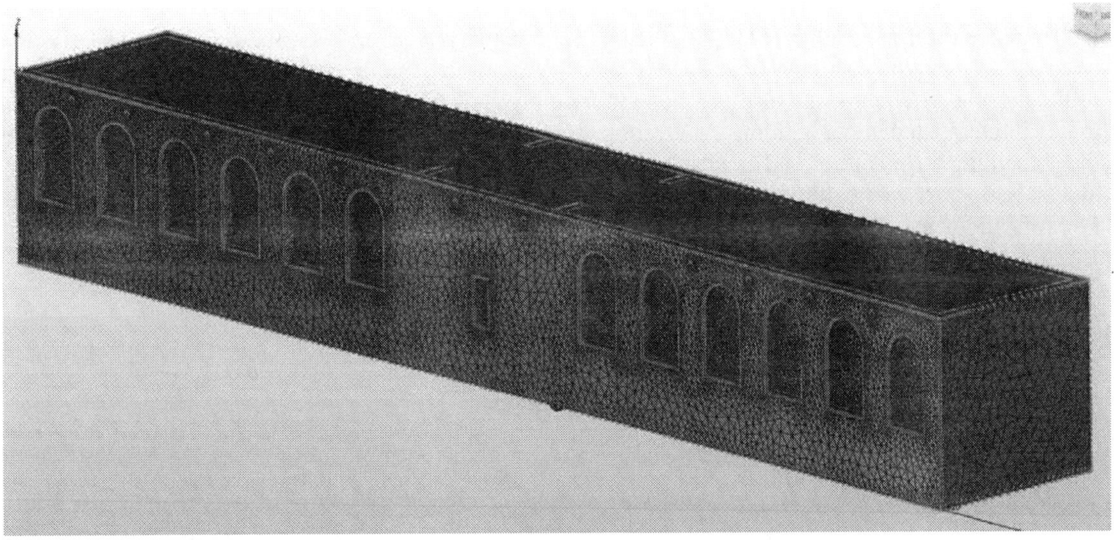

Figure 5. Library 3d geometry and meshing

4. Aim of this research

The CFD modelling show that the value of temperature that is create inside the library interior environment has a multi-zone variation. Locally, where static heaters are placed is warmer than the other side (figure 6), also the air distribution in case of HVAC system can be view in figure 8, for comparison. The predicted mean vote (PMV) is in range of +0,5÷-0,5 and the comfort temperature of 15-20°C (figure 7).

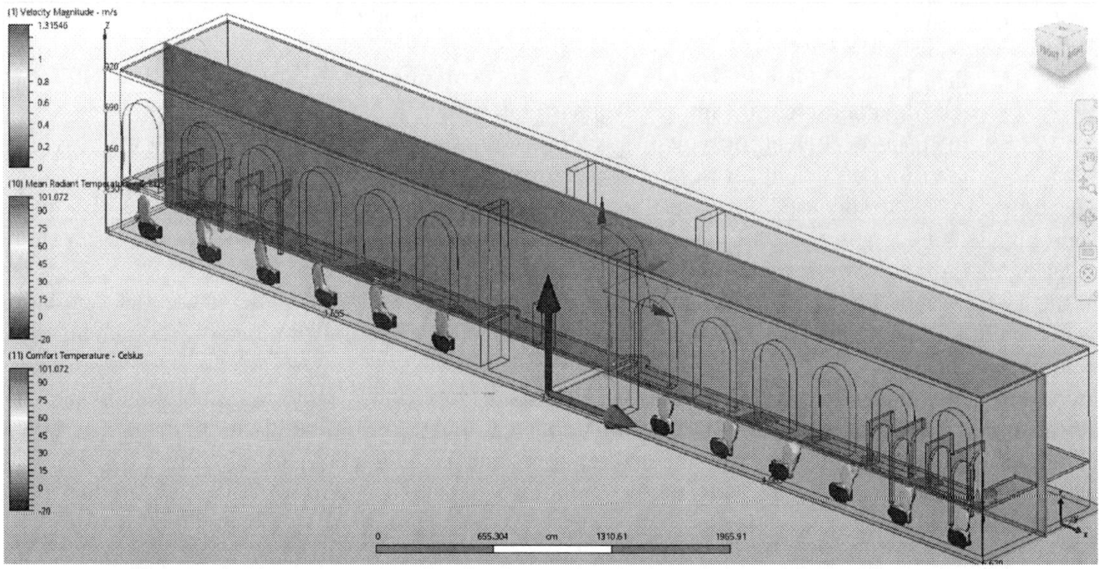

Figure 6. Air temperature distribution and velocity of air inside the library interior

There are different studies [11], [12], [13]. [14], that analysis natural ventilation as the most efficient passive cooling system and that have been used in the traditional buildings for centuries for improvement of the thermal efficiency and also for the indoor comfort. Taking into account [15], natural ventilation that is implemented into a building can become a substantial architectural design tool conduct finally to the "breathing architecture" and as a back effect this process enables the supply of fresh air into buildings and also supports the healthy breathing of the building inhabitants, provides oxygen and removes to the laws of fluid mechanics enabling analytical elaborations.

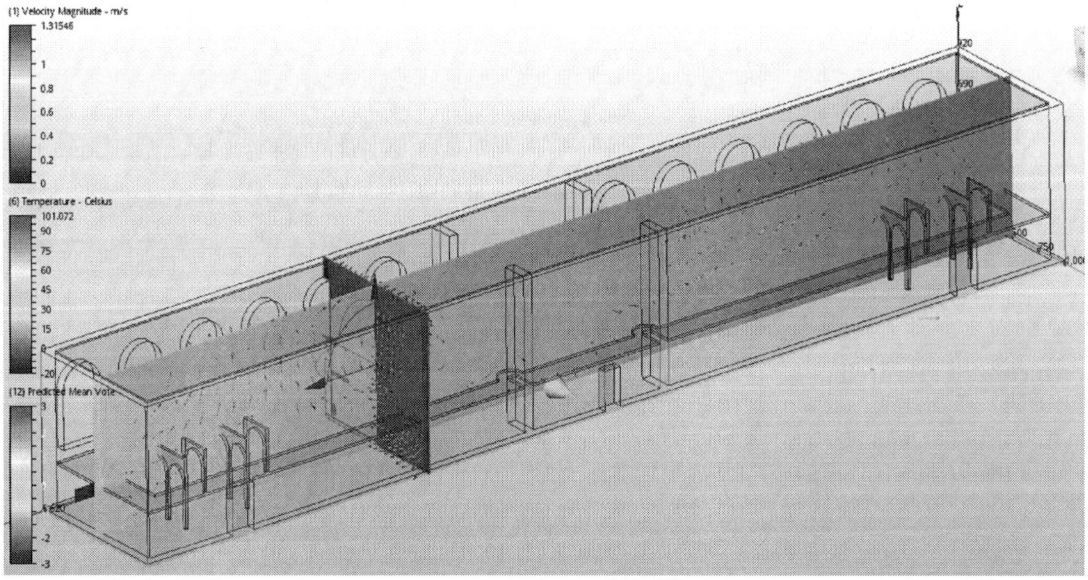

Figure 7. Range of temperature, air velocity and PMV viewed in cross section of library

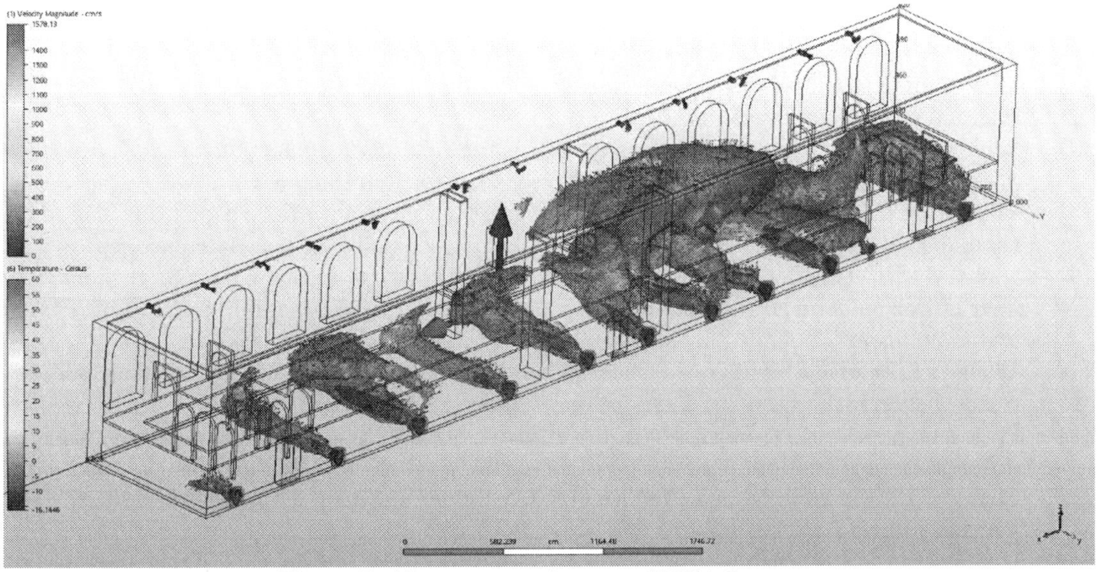

Figure 8. Air movement in case of HVAC system

5. Conclusion

The solution that uses static heating system has a slow air movement compared with the HVAC system. The temperature in case of the static heating system rise to the top of the library, beside the HVAC system where a more controlled air movement make the air, to circulate, to be cleaner and less destroying for wood elements. The moisture that can be formed in the air volume is quickly removed. For the artworks elements and for wooden, the possibility to adjust the temperature and humidity offers a better protection and good conservation. The CFD modelling and the measurements made in the library, came to the same conclusion, that if a rehabilitation process will occur in the future the chancing of the heating system must to be take into consideration.

The advantage of energy simulation over classical method is the various effects of thermophysical properties of materials and the performance of various systems under varying external and internal environmental conditions are considered in energy calculations. Most energy simulation tools do not require any special computing power as they can be run on commercially available desktop computer or laptops.

Thermal comfort is one of the main objectives of any HVAC system, including natural and hybrid ventilation system. Out of the many variables that influence, the environmental factors: the velocity of air over the person, the radiant field around the person, temperature and humidity of the surrounding air and the personal factors, clothing worn and activity of the person are the most prominent factors in the modelling thermal comfort.

6. References

[1] http://www.ziaruldeiasi.ro/stiri/sa-ne-mandrim-biblioteca-universitatii-tehnice-din-iasi-cea-mai-frumoasa-din-lume--85448.html [Online]

[2] Alterio S et al 2010 https://conservation-science.unibo.it/article/view/2319/1706 [Online]

[3] Ciocan V et al 2014 Analysis of the efficiency of ventilation systems in underground garages. *Applied Mechanics and Materials*, pp. 343-348

[4] Corgnati P S, Fabi V & Filippi M 2009 A methodology for microclimatic quality evaluation in museums: Application to a temporary exhibit, *Building and Environment*, pp. 1253-1260

[5] Nilsson H O & Holmér I 2003 Comfort climate evaluation with thermal manikin methods and computer simulation models. Indoor Air, 13(1), pp. 28-37

[6] McMullan R 2012 *Environmental science in building*, 7th ed., New York: Macmillan International Higher Education.

[7] Schellen H & Lambertus H 2002 *Heating monumental churches - indoor climate and preservation of cultural heritage*, Eindhoven(NL): Technische Universiteit Eindhoven

[8] Tian W et al 2018 Optimization on thermostat location in an office room using the coupled simulation platform in modelica buildings library: a pilot study, Cobee conference, pp. 155-165

[9] Varas-Muriel M et al 2014 Fluctuations in the indoor environment in Spanish rural churches and their effects on heritage conservation: Hygro-thermal and CO2 conditions monitoring, *Building and Environment*, Volume 82, pp. 97-109

[10] Wang H & Zhiqiang J Z 2016 Advances in building simulation and computational techniques: A review between 1987 and 2014, *Energy and Buildings*, 128(15), pp. 319-335

[11] Fu X and Wu D 2015 Comparison of the efficiency of building hybrid ventilation systems with different thermal comfort models, Energy Procedia 78, pp. 2820 – 2825

[12] Zhaia Z, El Mankibib M and Zoubirb A 2015 Review of natural ventilation models, *Energy Procedia* 78, pp 2700 – 2705

[13] Guo W, Liu X and Yuan X 2015 Study on Natural Ventilation Design Optimization Based on CFD Simulation for Green Buildings, *Procedia Engineering* 121, pp. 573 – 581

[14] Balocco C 2008 Analysis of Ancient Natural Ventilation Systems inside the Pitti Palace in Florence, *Proceedings of the COMSOL Conference* 2008

[15] Stavridou A D 2015 Breathing architecture: Conceptual architectural design based on the investigation into the natural ventilation of buildings, *Frontiers of Architectural Research* 4, pp. 127–145

IOP Conf. Series: Materials Science and Engineering **586** (2019) 012044 doi:10.1088/1757-899X/586/1/012044

Improved strategy for optimal control of Civil Engineering structures

F Păuleț-Crăiniceanu, S G Luca, C Pastia, V Florea
"Gheorghe Asachi" Technical University of Iaşi, 1 Mangeron Blvd., *Romania*

E-mail: fideliu.paulet-crainiceanu@tuiasi.ro

Abstract. Inside the Industrial Revolution 4.0, the concepts as Smart Cities and Smart Society are important parts in protection of humans against natural disasters, as strong earthquakes. Older or newer protection strategies of Civil Engineering structures (bridges, buildings) must be updated with elements of the actual technological revolution as sensors, drones, robots, nanotechnologies, AI, big data, cloud computing and so on. This paper is focused on a optimal structural active control of buildings started by many researchers during the last decades. Now, the target is to improve an optimal strategy developed by authors by proposing a more efficient use of computation. The success of this strategy leads to a faster response of computers that direct the control devices. This provides the ground for a larger range of control devices based on a larger number of sensors and, therefore, a better energy use together with a more precise anti-seismic structural protection

1. Introduction

New context of revolutionary changes, placed under Industrial Revolution 4.0 slogan, leads to concepts as Smart Cities and Smart Society. They play important roles in protection of humans against natural disasters as strong earthquakes. Therefore, older or newer protection strategies of Civil Engineering structures (bridges, buildings) must be updated with elements of the actual technological revolution as sensors, drones, robots, nanotechnologies, AI, big data, cloud computing and so on. Some early proposals were issued in [1, 2, 3] and later in [4]. Figure 1 shows two device proposals, from [4], that could lead to implementing Structural Robotics.

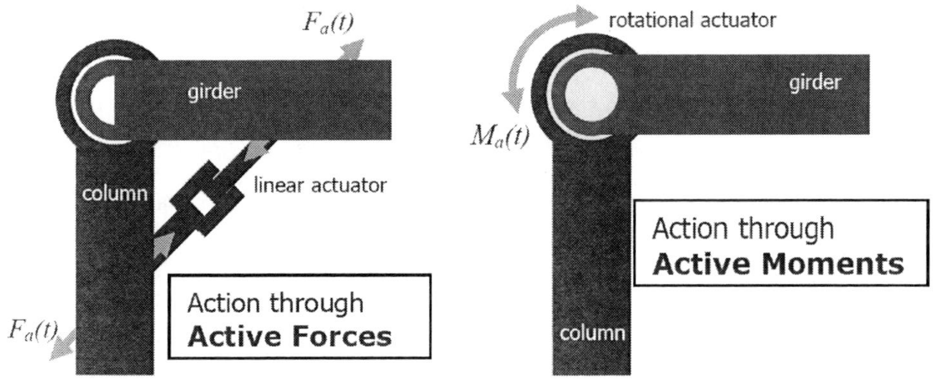

Figure 1. Structural Active Hinge, SAH – concept details [4].

During a previous work, [1], a classical optimal control approach has been used and shown as viable using 3D control systems with multiple actuators for strong earthquake effects mitigation. At that time, a pioneering strategy based on energy approach has been adopted. Due to that strategy, a methodology for limiting to only one scalar the parameters for control objective index had been issued [2, 5].

Latter, the methodology was applied to incorporate the practical problems as time delay, control noise and the introduction of estimators in the control global system. Then the methodology had been shown to be effective for decentralized disjoint and overlapping [6] control systems in Civil Engineering structures. The decentralized overlapping control model from [6] is shown in Figure 2. Other parametric studies had been done for better understanding of control devices work [7].

In this paper, an improvement of the above methodology is proposed. Starting from the previously defined matrices for the optimal objective index, a new procedure is proposed. It mainly consists in using a smaller dimension for the matrices involved in Riccati equations to the goal of a faster control computation chain.

2. Use of classical optimal control approach in Civil Engineering

In Structural Dynamics, the linear, *n* dimensional, second order differential equation with active control writes [1, 2]

$$\mathbf{M}_S\ddot{\mathbf{z}} + \mathbf{C}_S\dot{\mathbf{z}} + \mathbf{K}_S\mathbf{z} = \mathbf{f} - \mathbf{u} \tag{1}$$

where $\mathbf{M}_S$ is the *n×n* mass matrix, $\mathbf{C}_S$ is the *n×n* damping matrix $\mathbf{K}_S$ is the *n×n* stiffness matrix, $\mathbf{z}$ is the *n×1* displacement vector, $\mathbf{f}$ is the *n×1* external forces vector, and $\mathbf{u}$ is the *n×1* control forces vector. $\mathbf{z}$, $\mathbf{f}$ and $\mathbf{u}$ are all time dependent vectors.

In the case of unidirectional earthquake action, the vector of external actions might be written as

$$\mathbf{f} = \mathbf{h}_1\ddot{x}_g \tag{2}$$

where $\ddot{x}_g$ is the time dependent seismic acceleration and $\mathbf{h}_1$ is the *n×1* distribution of seismic action vector.

Also, the vector of control forces is detailed as

$$\mathbf{u} = \mathbf{L}_1\mathbf{u}_1 \tag{3}$$

with $\mathbf{L}_1$ being a *n×m* matrix showing the distribution of active forces on structural degrees of freedom, $\mathbf{u}_1$ - the *m×1* vector of active forces, and *m* is the number of active forces.

Denoting

$$\mathbf{x} = \begin{Bmatrix} \mathbf{z} \\ \dot{\mathbf{z}} \end{Bmatrix} \tag{4}$$

the classical canonical full states equation used in mitigation of earthquakes effects on Civil Engineering structures is

$$\begin{cases} \dot{\mathbf{x}} = \mathbf{A}\mathbf{x} + \mathbf{B}(\mathbf{f} - \mathbf{u}) \\ \mathbf{y} = \mathbf{C}\mathbf{x} + \mathbf{D}(\mathbf{f} - \mathbf{u}) \end{cases} \tag{5}$$

where: $\mathbf{x}$ is the time dependent *2n* dimensional vector of states, $\mathbf{y}$ is the time dependent *2n* dimensional vector of outputs. Matrices: $\mathbf{A}$ *(2n×2n)*, $\mathbf{B}$ *(2n×n)*, $\mathbf{C}$ *(p×2n)* and $\mathbf{D}$ *(p×2n)* are the canonical equation matrices. *p* is the number of available measurements.

For $\mathbf{A}$ and $\mathbf{B}$, the next is available

$$\mathbf{A} = \begin{bmatrix} \mathbf{0} & \mathbf{I} \\ -\mathbf{M}_s^{-1}\mathbf{K}_s & -\mathbf{M}_s^{-1}\mathbf{C}_s \end{bmatrix}, \quad \mathbf{B} = \begin{bmatrix} \mathbf{0} \\ \mathbf{M}_s^{-1} \end{bmatrix} \tag{6}$$

and the matrices $\mathbf{C}$ and $\mathbf{D}$ are defined depending on the practical measurement system the user is manipulating. In what follows the ideal case of full state measurement is employed in order of simplifying the content of presentation.

The linear dependency on states control low is proposed as

$$\mathbf{u}_1 = \mathbf{K}\mathbf{x} \tag{7}$$

where matrix $\mathbf{K}$ is the $m\times 2n$ control matrix. Therefore, (3) becomes

$$\mathbf{u} = \mathbf{L}_1\mathbf{K}\mathbf{x} \tag{8}$$

Using equations (8) and (2), the first equation of the system in (5) becomes

$$\dot{\mathbf{x}} = (\mathbf{A} - \mathbf{B}\mathbf{L}_1\mathbf{K})\mathbf{x} + \mathbf{B}\mathbf{h}_1\ddot{x}_g \tag{9}$$

or, in another form, denoting $\mathbf{A}_c = \mathbf{A} - \mathbf{B}\mathbf{L}_1\mathbf{K}$ and using again equation (2)

$$\dot{\mathbf{x}} = \mathbf{A}_c\mathbf{x} + \mathbf{B}\mathbf{f} \tag{10}$$

The linear quadratic control theory is imposing an optimal index to be minimized in the form [1]

$$J = \frac{1}{2}\int_0^t \left[\mathbf{x}'\mathbf{Q}\mathbf{x} + \mathbf{u}'\mathbf{R}\mathbf{u}\right]dt \tag{11}$$

where $\mathbf{Q}$ and $\mathbf{R}$ are $2n\times 2n$ and $m\times m$ parametric weighting matrices of control showing the desire of design to minimize displacements, velocities and accelerations together with a low actuators' energy consumption.

In [2], based on reducing the structural system energy, the weighting matrix $\mathbf{Q}$ was shown as

$$\mathbf{Q} = \begin{bmatrix} \mathbf{K}_S & 0 \\ 0 & \mathbf{M}_S \end{bmatrix} \tag{12}$$

and the weighting matrix $\mathbf{R}$ as

$$\mathbf{R} = diag\{r_1 \quad \cdots \quad r_m\} \tag{13}$$

or, simpler

$$\mathbf{R} = r\mathbf{I} \tag{14}$$

with r being a scalar parameter to be obtained by trial and error and $\mathbf{I}$ having $m\times m$ dimension. That is leading to a very easy way to establish the control parameters.

The first term under the integral in equation (10) can be rewritten using equation (12)

$$\mathbf{x}'\mathbf{Q}\mathbf{x} = \{\mathbf{z}' \quad \dot{\mathbf{z}}'\}\begin{bmatrix} \mathbf{K}_S & 0 \\ 0 & \mathbf{M}_S \end{bmatrix}\begin{Bmatrix} \mathbf{z} \\ \dot{\mathbf{z}} \end{Bmatrix} = \mathbf{z}'\mathbf{K}_S\mathbf{z} + \dot{\mathbf{z}}'\mathbf{M}_S\dot{\mathbf{z}} \tag{15}$$

and equation (15) is a sum of terms of energies (potential and kinetic). Therefore, the proposed $\mathbf{Q}$ matrix in equation (12) is justified and the optimized index in equation (11) is targeting a minimization of the energies in structural response.

Under the above circumstances, the solution of the optimal problem is based on solving the $2n\times 2n$ dimensional Riccati equation with constant matrices [5]

$$\mathbf{PA} - \mathbf{PBL}_1\mathbf{R}^{-1}\mathbf{L}_1'\mathbf{B}'\mathbf{P} + \mathbf{A}'\mathbf{P} + \mathbf{Q} = \mathbf{0} \tag{16}$$

where $\mathbf{P}$ is the unknown $2n\times 2n$ matrix.

Then, knowing $\mathbf{P}$ matrix, the control matrix is computed

$$\mathbf{K} = \mathbf{R}^{-1}\mathbf{L}_1'\mathbf{B}'\mathbf{P} \tag{17}$$

Further, the above methodology was applied to incorporate the practical problems as time delay, control noise and the introduction of estimators in the control global system. Then the methodology had been shown to be effective for decentralized disjoint and overlapping [6] control systems in Civil Engineering structures.

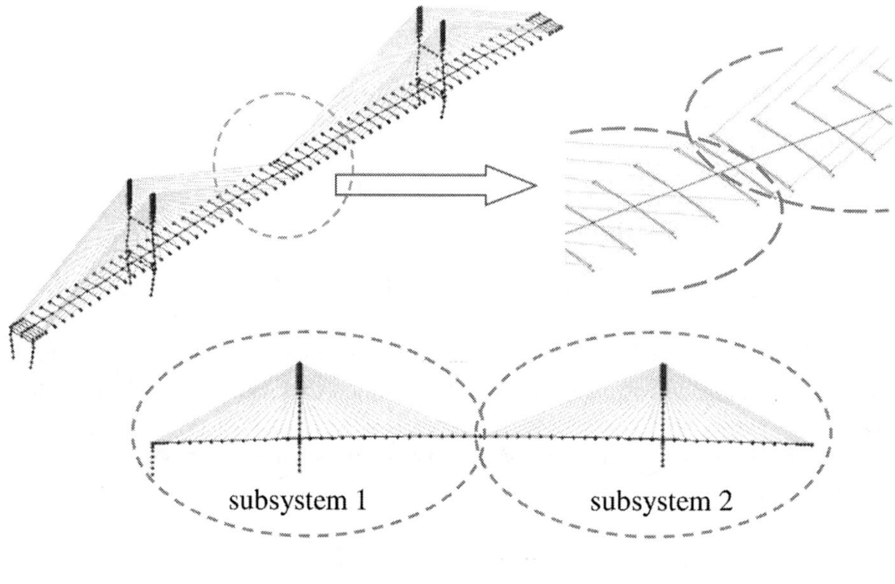

Figure 2. Bridge model and its decomposition structure for overlapping decentralized active control of a cable stayed bridge [6].

3. Proposed improvement of the strategy

The now proposed improvement is based on the observation that $\mathbf{Q}$ matrix in equation (12) is containing two zero $n\times n$ matrices and suggesting that some simplifications could be applied.

A first step is proposing a partitioning of $\mathbf{P}$ matrix into 4, $n\times n$ matrices

$$\mathbf{P} = \begin{bmatrix} \mathbf{P}_{11} & \mathbf{P}_{12} \\ \mathbf{P}_{21} & \mathbf{P}_{22} \end{bmatrix} \tag{18}$$

and, furthermore, if imposing $\mathbf{P}$ matrix as a symmetrical one

$$\mathbf{P}_{12} = \mathbf{P}_{21} \tag{19}$$

Detailing equation (17) based on equation (18), the next is obtain

$$\mathbf{K} = \mathbf{R}^{-1}\mathbf{L}_1'\mathbf{B}'\mathbf{P} = \mathbf{R}^{-1}\begin{bmatrix} \mathbf{0} & \mathbf{L}_1'\mathbf{M}_S^{-1} \end{bmatrix}\begin{bmatrix} \mathbf{P}_{11} & \mathbf{P}_{12} \\ \mathbf{P}_{21} & \mathbf{P}_{22} \end{bmatrix} = \begin{bmatrix} \mathbf{R}^{-1}\mathbf{L}_1'\mathbf{M}_S^{-1}\mathbf{P}_{21} & \mathbf{R}^{-1}\mathbf{L}_1'\mathbf{M}_S^{-1}\mathbf{P}_{22} \end{bmatrix} \tag{20}$$

which implies that only $\mathbf{P}_{21}$ and $\mathbf{P}_{22}$ $n \times n$ matrices are really needed for calculation of the $m \times 2n$ control matrix, $\mathbf{K}$.

For the present proposal, other helpful notations are

$$\mathbf{Q} = \begin{bmatrix} \mathbf{K}_S & 0 \\ 0 & \mathbf{M}_S \end{bmatrix} = \begin{bmatrix} \mathbf{Q}_1 & 0 \\ 0 & \mathbf{Q}_2 \end{bmatrix} \tag{21}$$

$$\mathbf{A} = \begin{bmatrix} \mathbf{0} & \mathbf{I} \\ -\mathbf{M}_s^{-1}\mathbf{K}_s & -\mathbf{M}_s^{-1}\mathbf{C}_s \end{bmatrix} = \begin{bmatrix} \mathbf{0} & \mathbf{I} \\ \mathbf{A}_{21} & \mathbf{A}_{22} \end{bmatrix}, \quad \mathbf{BL}_1 = \begin{bmatrix} \mathbf{0} \\ \mathbf{M}_s^{-1}\mathbf{L}_1 \end{bmatrix} = \begin{bmatrix} \mathbf{0} \\ \mathbf{B}_2 \end{bmatrix} \tag{22}$$

With all notations above, shown by equations (18), (21) and (22), the $2n \times 2n$ Riccati equation (16) becomes

$$\begin{aligned} &\begin{bmatrix} \mathbf{P}_{11} & \mathbf{P}_{12} \\ \mathbf{P}_{21} & \mathbf{P}_{22} \end{bmatrix} \begin{bmatrix} \mathbf{0} & \mathbf{I} \\ \mathbf{A}_{21} & \mathbf{A}_{22} \end{bmatrix} - \begin{bmatrix} \mathbf{P}_{11} & \mathbf{P}_{12} \\ \mathbf{P}_{21} & \mathbf{P}_{22} \end{bmatrix} \begin{bmatrix} \mathbf{0} \\ \mathbf{B}_2 \end{bmatrix} \mathbf{R}^{-1} \begin{bmatrix} \mathbf{0} & \mathbf{B}_2' \end{bmatrix} \begin{bmatrix} \mathbf{P}_{11} & \mathbf{P}_{12} \\ \mathbf{P}_{21} & \mathbf{P}_{22} \end{bmatrix} \\ &+ \begin{bmatrix} \mathbf{0} & \mathbf{A}'_{21} \\ \mathbf{I} & \mathbf{A}'_{22} \end{bmatrix} \begin{bmatrix} \mathbf{P}_{11} & \mathbf{P}_{12} \\ \mathbf{P}_{21} & \mathbf{P}_{22} \end{bmatrix} + \begin{bmatrix} \mathbf{Q}_1 & \mathbf{0} \\ \mathbf{0} & \mathbf{Q}_2 \end{bmatrix} = \begin{bmatrix} \mathbf{0} & \mathbf{0} \\ \mathbf{0} & \mathbf{0} \end{bmatrix} \end{aligned} \tag{23}$$

further developed in

$$\begin{aligned} &\begin{bmatrix} \mathbf{P}_{12}\mathbf{A}_{21} & \mathbf{P}_{11} + \mathbf{P}_{12}\mathbf{A}_{22} \\ \mathbf{P}_{22}\mathbf{A}_{21} & \mathbf{P}_{21} + \mathbf{P}_{22}\mathbf{A}_{22} \end{bmatrix} - \begin{bmatrix} \mathbf{P}_{12}\mathbf{B}_2 \\ \mathbf{P}_{22}\mathbf{B}_2 \end{bmatrix} \mathbf{R}^{-1} \begin{bmatrix} \mathbf{B}_2'\mathbf{P}_{21} & \mathbf{B}_2'\mathbf{P}_{22} \end{bmatrix} \\ &+ \begin{bmatrix} \mathbf{A}'_{21}\mathbf{P}_{21} & \mathbf{A}'_{21}\mathbf{P}_{22} \\ \mathbf{P}_{11} + \mathbf{A}'_{22}\mathbf{P}_{21} & \mathbf{P}_{12} + \mathbf{A}'_{22}\mathbf{P}_{22} \end{bmatrix} + \begin{bmatrix} \mathbf{Q}_1 & \mathbf{0} \\ \mathbf{0} & \mathbf{Q}_2 \end{bmatrix} = \begin{bmatrix} \mathbf{0} & \mathbf{0} \\ \mathbf{0} & \mathbf{0} \end{bmatrix} \end{aligned} \tag{24}$$

and

$$\begin{aligned} &\begin{bmatrix} \mathbf{P}_{12}\mathbf{A}_{21} & \mathbf{P}_{11} + \mathbf{P}_{12}\mathbf{A}_{22} \\ \mathbf{P}_{22}\mathbf{A}_{21} & \mathbf{P}_{21} + \mathbf{P}_{22}\mathbf{A}_{22} \end{bmatrix} - \begin{bmatrix} \mathbf{P}_{12}\mathbf{B}_2\mathbf{R}^{-1}\mathbf{B}_2'\mathbf{P}_{21} & \mathbf{P}_{12}\mathbf{B}_2\mathbf{R}^{-1}\mathbf{B}_2'\mathbf{P}_{22} \\ \mathbf{P}_{22}\mathbf{B}_2\mathbf{R}^{-1}\mathbf{B}_2'\mathbf{P}_{21} & \mathbf{P}_{22}\mathbf{B}_2\mathbf{R}^{-1}\mathbf{B}_2'\mathbf{P}_{22} \end{bmatrix} \\ &+ \begin{bmatrix} \mathbf{A}'_{21}\mathbf{P}_{21} & \mathbf{A}'_{21}\mathbf{P}_{22} \\ \mathbf{P}_{11} + \mathbf{A}'_{22}\mathbf{P}_{21} & \mathbf{P}_{12} + \mathbf{A}'_{22}\mathbf{P}_{22} \end{bmatrix} + \begin{bmatrix} \mathbf{Q}_1 & \mathbf{0} \\ \mathbf{0} & \mathbf{Q}_2 \end{bmatrix} = \begin{bmatrix} \mathbf{0} & \mathbf{0} \\ \mathbf{0} & \mathbf{0} \end{bmatrix} \end{aligned} \tag{25}$$

Finally, the next four, $n \times n$ dimensional Riccati equations, result as follows

- left-up equation

$$\mathbf{P}_{12}\mathbf{A}_{21} - \mathbf{P}_{12}\mathbf{B}_2\mathbf{R}^{-1}\mathbf{B}_2'\mathbf{P}_{21} + \mathbf{A}'_{21}\mathbf{P}_{21} + \mathbf{Q}_1 = \mathbf{0} \tag{26}$$

- right-up equation

$$\mathbf{P}_{11} + \mathbf{P}_{12}\mathbf{A}_{22} - \mathbf{P}_{12}\mathbf{B}_2\mathbf{R}^{-1}\mathbf{B}_2'\mathbf{P}_{22} + \mathbf{A}'_{21}\mathbf{P}_{22} = \mathbf{0} \tag{27}$$

- left-down equation

$$\mathbf{P}_{22}\mathbf{A}_{21} - \mathbf{P}_{22}\mathbf{B}_2\mathbf{R}^{-1}\mathbf{B}_2'\mathbf{P}_{21} + \mathbf{P}_{11} + \mathbf{A}'_{22}\mathbf{P}_{21} = \mathbf{0} \tag{28}$$

- right-down equation

$$\mathbf{P}_{21} + \mathbf{P}_{22}\mathbf{A}_{22} - \mathbf{P}_{22}\mathbf{B}_2\mathbf{R}^{-1}\mathbf{B}_2'\mathbf{P}_{22} + \mathbf{P}_{12} + \mathbf{A}'_{22}\mathbf{P}_{22} + \mathbf{Q}_2 = \mathbf{0} \tag{29}$$

From observing equations (26-29), the next order of solving is proposed

- solve equation (26) for $\mathbf{P}_{21}$ taking into account that $\mathbf{P}_{12} = \mathbf{P}_{21}$
- solve equation (29) for $\mathbf{P}_{22}$, replacing $\mathbf{P}_{12}$ and $\mathbf{P}_{21}$ from above.

This way, the solution of the $2n \times 2n$ dimensional Riccati equation with constant matrices in equation (16) is replaced with solving two $n \times n$ dimensional Riccati equations with constant matrices shown in equations (26) and (29).

This should lead to faster control strategies. Further numerical applications, in views for future work, are needed to prove the competitiveness of the above procedure.

4. Conclusions

In the context of smart cities, new strategies for protection of humans and structures against earthquakes are required, based on new technologies and devices.

The now proposed improvement is based on reducing the dimension of the problem involved by obtaining the control matrix. The equation used in obtaining the control matrix of an older strategy in linear quadratic linear control is replaced with two half size similar equation.

This should lead to faster control strategies. However, further numerical applications are needed to prove the competitiveness of the above procedure. This is the goal of future work.

5. References

[1] Paulet-Crainiceanu F 1997 Active Control Approach for Long Span Bridge Responses to Strong Earthquakes Ph.D. Thesis (Yokohama: Yokohama National University)

[2] Paulet-Crainiceanu F 1999 Seismic Response Control of Long Cable-stayed Bridges *Proc. 2nd World Conf. on Structural Control (Kyoto)* vol. 2 (Chichester: John Willey & Sons) p 959

[3] Paulet-Crainiceanu F 1999 Structural Robotics *Proc. 1st Romanian - American Workshop in Structural Engineering (Iasi)* ed G M Atanasiu (Iasi: “Al. I. Cuza” University Press) p 43

[4] Păuleţ-Crăiniceanu F 2006 Steps in Structural Robotics Implementation *Proc. of Computational Civil Engineering 2006 (Iasi)* ed F Păuleţ-Crăiniceanu C Ionescu *et al* (Iasi: Editura Societăţii Academice "Matei-Teiu Botez") p 531

[5] Paulet-Crainiceanu F Rodellar J and Monroy C 2001 Control Settings and Performance Analysis of an Optimal Active Control Method for Cable Bridges *Proc 2nd European Conf. on Computational Mechanics (Cracow)* (Cracow: CD-ROM) p 1

[6] Bakule L Paulet-Crainiceanu F Rodellar J and Rossell J M 2005 Overlapping Reliable Control for a Cable-Stayed Bridge Benchmark *IEEE Trans. C.S.T.* **13** 663–9

[7] Pastia C Luca S G Paulet-Crainiceanu F and Florea V 2016 Passive, Active and Semi-Active Vibration Control of a Classical Mass-Spring-Dashpot System *16th International Multidisciplinary Scientific GeoConference SGEM2016 (Sofia)* (Sofia: Stef92 Technology Ltd.) book1 vol. 3 p 575

Enhancing Civil Engineering teaching through 3D Computer Aided Design

J Gonçalves[1], B Santos[2]
[1] University of Beira Interior, Calçada Fonte do Lameiro, 6200- 358, Covilhã, *Portugal* (CITTA - Research Centre for Territory, Transports and Environment)
[2] University of Beira Interior, Calçada Fonte do Lameiro, 6200-358, Covilhã, *Portugal* (CERIS-CESUR- Civil Engineering Research and Innovation for Sustainability)

E-mail: jorge@ubi.pt

Abstract. 3D interpretation of 2D drawing is not easy for most of the Civil Engineering first-year students. Some do it naturally but most need to be trained to master this skill. In this paper, the implemented teaching methodologies aiming to facilitate the acquisition of three-dimensional project visualization skills are presented. These methodologies were developed during the last two years targeting first-year Civil Engineering students at the University of Beira Interior (Portugal). After a first contact with 2D project representation through manual drawing, students progressively elaborate computer-aided design (CAD) project starting with simple 2D drawings and culminating with a 3D project of a pre-existing building. Students are also motivated to improve their 3D graphic representation skills through a classroom contest where the winning project is printed in 3D. The training in 2D and 3D graphic representation is complemented during the second year with several in situ surveys and computer-aided drawing of topographic data. This approach proved to be very interesting for competence acquisition, qualifying students for a better 3D representation and interpretation. Students also found this methodology to be motivating.

1. Introduction

Science, technology, engineering and mathematics (STEM) are generally the least popular fields in most countries, with an average of only 24% of university graduates across OECD countries in 2016 [1]. Recent studies suggest that the number of STEM graduates at EU has remained generally stable in relation to the total number of university graduates, with a slight variation from 22.3% to 22.8% between 2006 and 2012 [2].

In this context, Shirazi A. and Behzadan A. [3] state that retaining and engaging students into the STEM fields is a major concern in the learning process. Outdated and poor teaching methods, disconnection between students and classroom technology, and lack of hands-on experiments are important reasons that keep some students away from pursuing STEM disciplines. Integrating technology into secondary and higher education, such as 3D design, can enhance student engagement and the quality of learning [4] [5] .

Portugal presents a particular situation, since Engineering is the most popular field among tertiary graduates, presenting 21% of graduates in 2016 for the field of study Engineering, Manufacturing and Construction, against a European average of 15% [6].

Over the last two decades there have been significant changes in Civil Engineering curricula in response to student, societal and industrial needs as well as professional bodies and government organisations. Some of the main factors contributing to this scenario are the easy access and the wide range of digital technology with application to project and engineering management (design and computational analysis). University curricula have responded to the growing professional appreciation of skills such as the domain of the available digital technologies and tools through a student-centred learning and teaching approach, grounded in critical thinking and problem-based learning.

According to Grand-Clement, S. [7] there is a need for digital skills to be better integrated in formal education and lifelong learning opportunities, in order to provide citizens with the tools they need to succeed in everyday life as well as in potential future careers.
Several authors and studies also refer this trend in education and in particular in the areas of Engineering, Manufacturing, Construction and Architecture [3] [8] [9] [10].
The fast development of science and technology, as well as economic and information globalization, had led to a recent education reform aiming to cultivate innovative engineers [11]. Knowledge is no longer the only measure of talent, students' ability to master new knowledge and the capacity to innovate became equally important in the demanding labour market.
This paper intends to present the implemented teaching methodologies applied in the UBI Civil Engineering education aiming to facilitate the student's acquisition of two and three-dimensional project visualization and interpretation skills using digital tools.
The paper is organized as follows: Section 2 presents a literature review of the drawing training methodologies applied in Civil Engineering education. Section 3 presents the teaching methodologies adopted in the last two academic years to motivate students in the Drawing and Topography subjects. Section 4 concludes the paper highlighting some reflections of the authors on the drawing training approaches implemented at the University of Beira Interior (UBI) and its importance in the Civil Engineering education.

2. Enhancing drawing training in Civil Engineering education

For an innovative Civil Engineering training it is important to have didactic materials to offer students recent and real problems with which they can interact, perform analyses and find solutions, promoting problem-based education (student-centred) and breaking the unilaterality of the knowledge transfer process.
Several Civil Engineering teaching research [9] [12] point-out the following difficulties in the learning process: great diversity and level of students preparation, lack of solid high school background knowledge, lack of experience in relating theoretical fundamentals to practical problems, lack of hardware and latest engineering software, lack of independent and critical thinking and lack of interest in active learning.
Among the strategies suggested to help teachers in promoting active learning are problem solving using computer-aided design. In that regard, project-based learning (PBL) has become a popular teaching method in engineering education as it provides students with opportunities to better understand the course material, gain more practical experience and learn how to apply theory to practice [5][13][14][15][16][17]. The literature indicates that PBL has several advantages over traditional methods namely as a didactic strategy: student engagement, critical thinking and team work, development of engineering identity, learning experience and theory-practice integration. Although this methodology is mainly applied to advanced year courses where students present a more mature level of knowledge, there is some evidence indicating that first-year engineering students also "generally enjoy the experience" and that PBL can also improve first-year student motivation and depth of learning [13].
Nevertheless, research [14] also suggests that a combination of traditional teaching methods with elements of PBL might be a good alternative as it allows different types of students to be satisfied: those who prefer projects and those who prefer traditional courses.
The use of digital technologies in PBL approaches help students arrive at their learning outcomes much faster. When carrying out a project, the use of graphic technologies and those relating to 3D modelling have a positive contribution towards improving the transmission of rigorous technical information and the understanding of spatial configurations. In this process scholars, rather than teaching how to use graphic technologies, use it to assist the students' learning process.
According to Shirazi A. and Behzadan A. [3], most of the students consider themselves as visual learners, acquiring better engineering and scientific topics with 2D/3D visualization, complaining about the disconnection between how they perceive technology and how technology is in fact used in

the classroom. The new generation of students, which grew up in modern technology environment surrounded by computers, videogames and smart phones is eager to use digital technologies. Technology-based instruction along with conventional teaching methods enhance students learning and enable instructors to become more familiar with new technology-based educational innovations.
Among the skills that Civil Engineering students must master project graphical representation is one of the basic. When representing any object, it can be drawn in a "flat" manner (drawing) or as a "solid" (3D model) [18]. Traditionally the results of civil engineering designs are several 2D drawings, which can be complemented by 3D models. 3D models allow untrained eyes to better understand how the project will look like when built. The 20 years of teaching experience of the authors have shown that 3D interpretation of 2D drawing is not easy for most of the Civil Engineering early years students. Some do it naturally but most need to be trained to master this skill.
According to Veide Z. and Strozheva V. [19] students' spatial ability and visualization skills may be developed with help of different training tools and methodologies: graphic courses using traditional drafting methods (pencil and drawing board), 2D/3D CAD modelling, web-based graphics applications, augmented reality based application and sketch based modelling systems. However, Duarte J. [20] points out that hand drawing is still an essential skill in traditional design education and the basis for graphical representation teaching.
Regarding 3D printing, it can be seen as belonging to a set of emerging technologies with academic advantages as a didactic tool for the learning process in graphical representation. The use of this technology will undoubtedly change engineering curricula and how to educate future engineers. Ford S. and Minshall T. [21] mention a variety of benefits that have been identified arising from the incorporation of 3D printing technology into teaching: it can facilitate learning, develop skills, increase student engagement, inspire creativity and improve attitudes towards STEM subjects and careers, while increasing teachers' interest and engagement. They also investigated the application of 3D printing in schools, universities, libraries and special education settings, and found six types of use: to teach students about 3D printing; to teach educators about 3D printing; as a support technology during teaching; to produce artefacts that aid learning; to create assistive technologies; to support outreach activities.
The disciplines of architecture and engineering were early adopters of rapid prototyping technologies to enhance teaching [5][21]. Rapid prototyping enables the production of physical models from digital ones and is used in the study of design and construction solutions. These studies can include 3D visualization of 2D designs or help to analyse more complex shapes [20].

3. Teaching experiences with 2D and 3D Computer Aided Drawing

3.1. Civil Engineering education at UBI
Currently, Portuguese Civil Engineering slowly begins to recover from the international crisis that led to an investment breakdown in construction, which led to a large decrease of local candidates for Civil Engineering courses offered by higher education institutions. To minimize this situation, in recent years UBI has invested in attracting international students, resulting in classrooms essentially composed of international students [22].
The integrated master's degree in Civil Engineering is organized according to Bologna process and comprises a set of ten academic semesters along a five years' graduation course, allowing the students to tailor their study plan on their own interests and to highlight the theoretical and practical aspects of distinct civil engineering domains. In this way, students can select one of two specialization areas (Geotechnical and Environment or Structures and Construction), have optional units available in the 1st semester of the fifth year and choose their dissertation topic in the last year [22].
The course units are grouped into 11 scientific areas: Drawing and Topography, Mathematics, Informatics, Chemistry, Mechanics and Structures, Construction, Physics, Geotechnics, Planning and Urbanism, Economics and Management, and Hydraulic and Environment [23].

IOP Conf. Series: Materials Science and Engineering **586** (2019) 012045 doi:10.1088/1757-899X/586/1/012045

The 3 course units of the scientific area Drawing and Topography, taught in the first two years of the study plan and in which computer-aided design tools are used to enhance the teaching-learning process, are presented below.

3.2. The role of 3D drawing in Civil Engineering training

It is unquestionable the relevance of technical drawing in Civil Engineering. In general, civil engineers cannot afford to test real life projects due to costs and impacts in the surrounding environment and populations. Thus, drawings are essential to visually communicate the engineer's idea to properly instruct on how to build or give a glance on how it will look. This means that not only the engineer but also the reader of the drawing must communicate with the same technical "language". This language must be learned and trained to minimize communication problems and prevent construction mistakes.
Figure 1 presents the existing relationships between real-world features and its depiction through 2D technical drawings, digital models (2D/3D) and 3D physical models, illustrating possible interconnections among these types of representation.

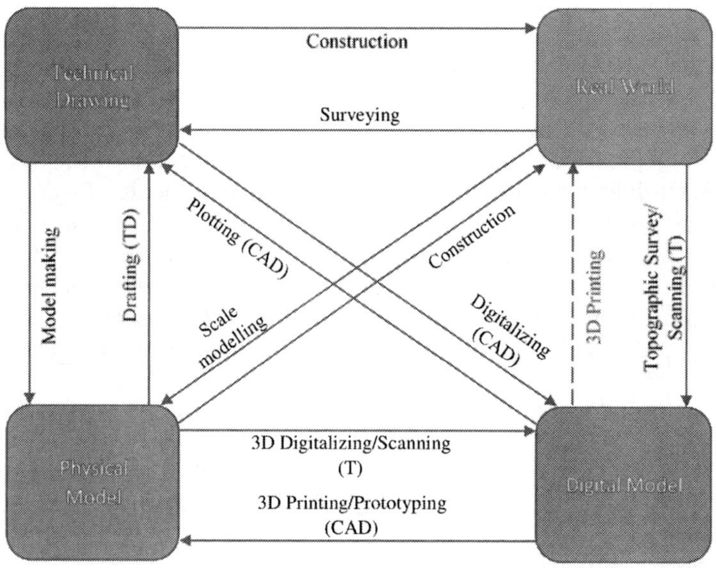

TD – Technical Drawing; CAD – Computer Aided Design; T - Topography

Figure 1. Relations between real-world features and its possible representations: Technical drawing, Digital model and Physical model.

The real-world features have traditionally been depicted by drawing and scaled models. Civil Engineering has used them since mankind started to build. The rising of the digital era brought new ways to communicate ideas and the proliferation of low-cost 3D printing devices made modelling affordable to be used as a teaching tool. There are even examples of full scale 3D printing of small buildings [24][25].
In the Civil Engineering course at UBI, the Drawing and Topography scientific area comprise 3 course units that gradually train students in basic technical drawing skills. Table 1 presents how manual drafting, computer-aided design and cartography knowledge are introduced to students in the early course years.

IOP Conf. Series: Materials Science and Engineering **586** (2019) 012045 doi:10.1088/1757-899X/586/1/012045

Table 1. Drawing skills introduced in Drawing and Topography course units.

Course Unit (Year/Semester)	Manual drafting	Computer Aided Design	Cartography
Introduction to Civil Engineering and to Technical Drawing (1Y/1S)	Technical drawing standards, 2D plans, elevation, sections	General concepts of CAD	Map reading, contour lines and profiles properties
Technical Drawing for Civil Engineering (1Y/2S)		2D drafting, 3D project	Digital Terrain Model (DTM)
Topography (2Y/1S)	Field sketches	2D drafting, terrain model	Field surveys, data georeferencing, contour lines, 3D drawing

3.3. Technical Drawing - 2D drawing training

The Technical Drawing subject is taught in the first semester of the first-year course of Introduction to Civil Engineering and Technical Drawing. It is the students' first contact with specific Civil Engineering courses and it is divided into two distinct subjects. In the first half of the semester, students are introduced to Civil Engineering history, to different types of civil engineering projects (Geotechnics, Structures, Hydraulics, Roads, Urban Planning, Construction and Environment) and perform analysis of case studies and real-life projects or construction system. In the second half, students are introduced to technical drawing basic standards, to projections, to cartography principles, orthographic projection of buildings and to civil engineering technical drawing. The students apply these knowledge with the accomplishment of an individual manual drafting work of an industrial pavilion. The lectures are organized in theoretical classes, with explanation of the program contents; and practical classes, where small practical exercises and guidance for the manual drafting of floor plans, elevations and sections of a given sketch are performed.

At the end of the semester, students must prove that they have acquired knowledge about basic civil engineering drawing skills to execute and read 2D civil engineering technical drawings. The evaluation of this knowledge includes written tests and 2 assignments: manual drawing of a terrain profile and orthographic projections of a building.

The subjects taught in the second half of the semester are essential and intend to be the grounding knowledge for students to understand and execute technical drawing in Civil Engineering domain.

3.4. Computer Aided Design - 2D and 3D drawing training

The Computer Aided Design is a subject taught in the second semester of the first-year course of Technical Drawing for Civil Engineering. It complements the introductory Technical Drawing subject of the first semester with more training in civil engineering specific design. The classes are organized in theoretical classes, with discussion of the program contents; and laboratorial classes, where students develop computer aided design projects.

During the semester, students are introduced to 2D and 3D computer-aided design using AutoCAD® software from AutoDesk. Using knowledge acquired during the course, students learn how to transform coordinates form a map, how to digitally georeference a raster image (map) and how to create a DTM (Fig. 2). In the final assignment, which aims to integrate the knowledge acquired throughout the semester, students draw a digital 2D building project, elevations and sections from printed plans. The final part of the assignment consists in preparing a printable full 3D SOLID model of the project (Fig. 3 and 4).

IOP Conf. Series: Materials Science and Engineering **586** (2019) 012045 doi:10.1088/1757-899X/586/1/012045

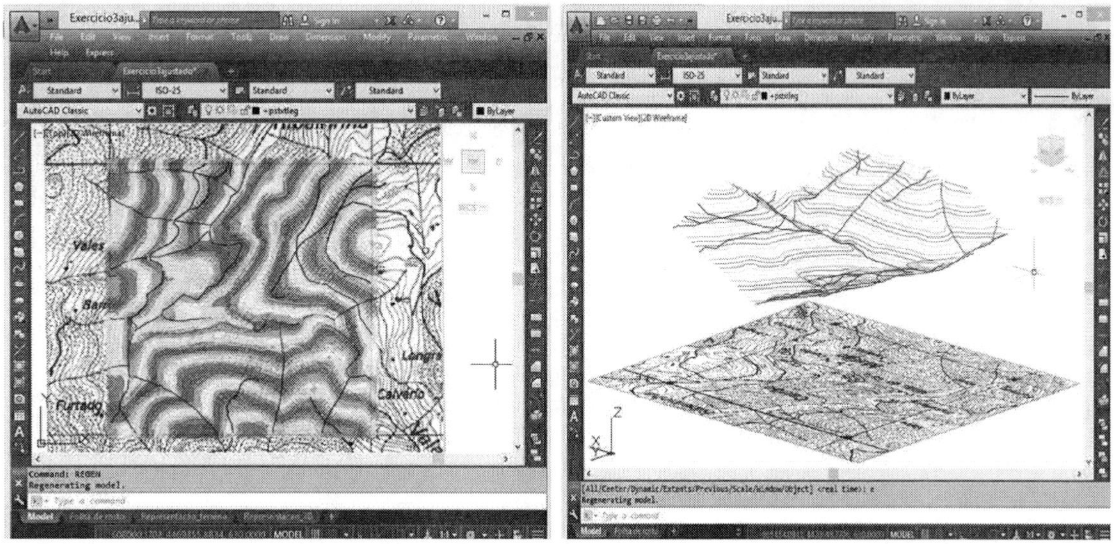

Figure 2. Digital terrain model from a georeferenced raster image and its 3D view: example of student work.

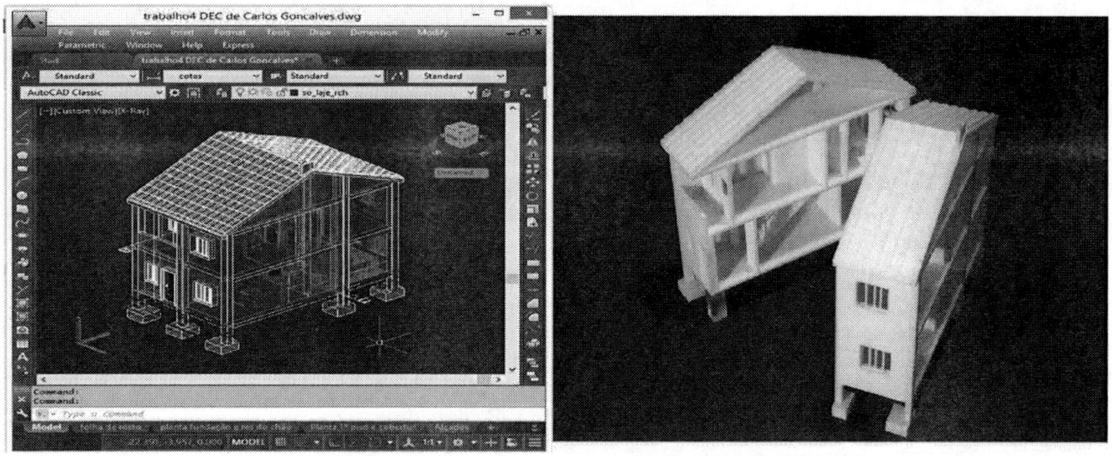

Figure 3. The 2016/2017 winning 3D model of a terraced house and its 3D printing.

The generic CAD process for 3D modelling is presented in Figure 5. Considering the acquired knowledge in manual technical drawing, students start the CAD design setting the drafting generic parameters such as layers, point type, object snapping, layouts, lettering style, line types or viewports. Due to the lack in computer-aided design background of most students, the first lessons focus on simple 2D drawings with the aim of familiarizing students with basic drawing and editing tools. The work then evolves to its 3D transformation were 3D operations are applied. After finishing the 3D design, students learn how to make the dimensioning and several ways to output the project.

At the end of the semester, students must prove that they have acquired knowledge about specific civil engineering technical drawings and skills to transform 2D technical drawings into solid 3D models. In 2016/2017 and 2017/2018, to enhance student motivation for the subject, a contest among students was proposed, awarding the best 3D project with its 3D printing (Fig. 3 and 4).

The 3D printing of the winning design was done through Fused Deposition Modelling (FDM) kit printers in laboratories at the university. This type of low-cost printers extrudes a narrow bead of hot plastic, which is selectively deposited where it fuses to the existing structure and hardens as it cools.

The 3D CAD model is transferred to the 3D printer as a standard data known as stereolithography language (STL), where the model is converted into layers that can be applied consecutively [24].

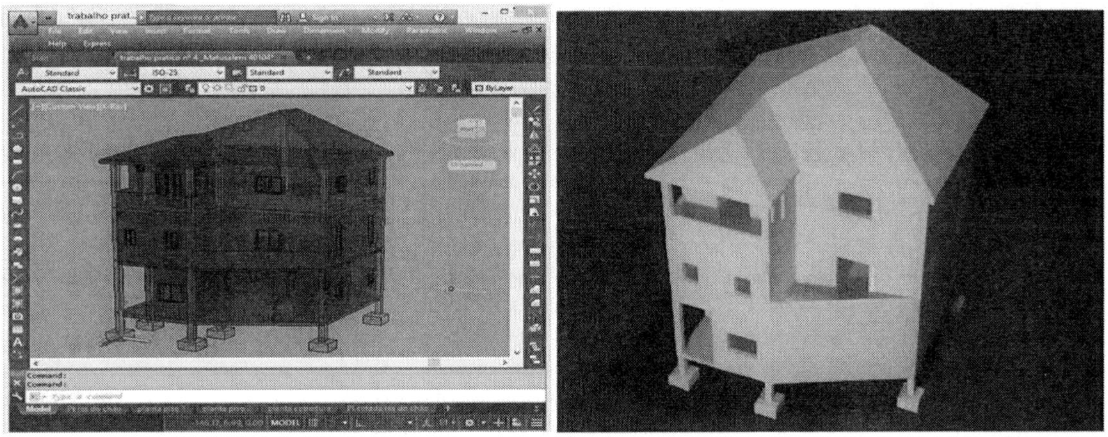

Figure 4. The 2017/2018 winning 3D model of a detached public building and its 3D printing.

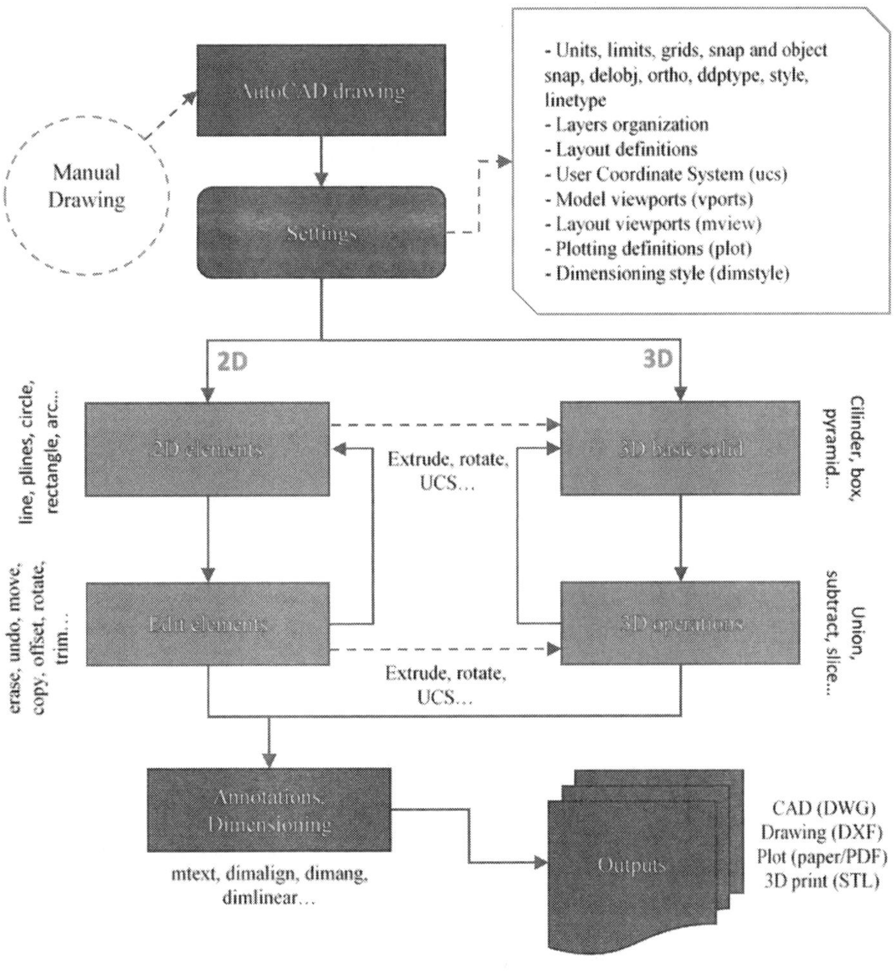

Figure 5. Generic CAD process for 2D and 3D design

Through these project-based learning methodologies students develop their 3D visualization skills, can better understand 2D interpretation mistakes and are motivated to pursue the Civil Engineering career. The evaluation of these knowledge includes written tests and the presentation of two 3D digital assignments (digital terrain model and orthographic projection of a building and its 3D project).

3.5. Topography

The Topography course is taught in the first semester of the second year and aims to introduce and develop aspects related to topography and its application to Civil Engineering. The classes are organized in theoretical lectures, with exposition of the program contents; theoretical-practical lectures, with small problems solving; and practical classes and tutorial orientations, with real practical problems solving through fieldwork and processing of collected data using computer programs.

Two-thirds of the practical and tutorials classes (20 contact hours) are dedicated to the acquisition of skills in the use of topographic instruments and a third (10 contact hours) in 2D/3D graphic representation. Students are encouraged to improve these skills with independent work using the resources available in the Drawing and Topography Laboratory.

At the end of the semester, students must have acquired enough knowledge to master topographic instruments (Digital Level and Total Station), calculations, fieldwork and digital 2D and 3D data representation. The evaluation of this knowledge includes a written test and 5 to 7 field assignments. A representation component using digital tools, where the technical drawing knowledge acquired in the first year is applied, is contemplated in the following 4 assignments:

- Digital level survey (expedite topographic survey and digital 2D representation).
- Total Station survey (rigorous topographic survey and digital 2D representation) (Fig.5).
- Precision altimeter survey using digital level and total station (constructions observation and digital 2D and 3D representation – creation of triangulated irregular network, contour lines and DTM) (Fig. 6).
- Connection of local survey to a cadastral survey network using Global Navigation Satellite System (GNSS).

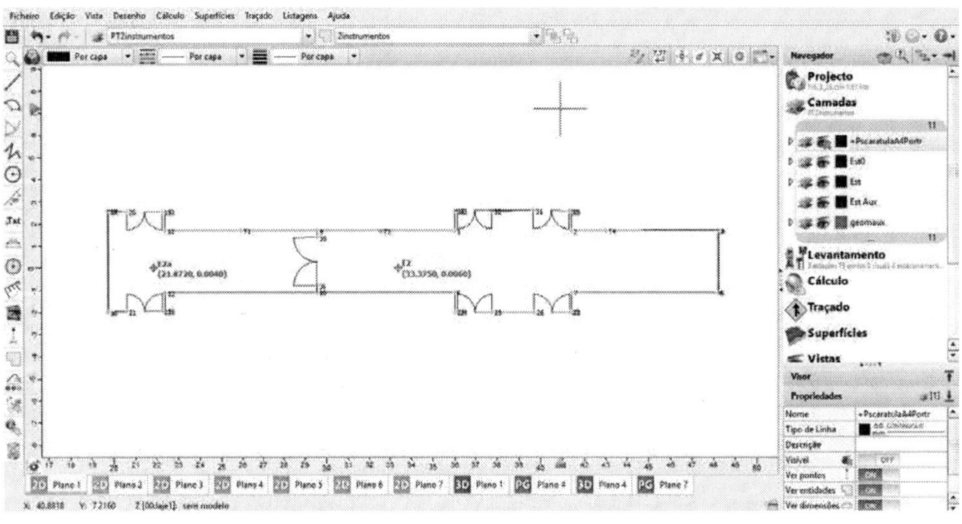

Figure 6. Rigorous topographic survey 2D representation in Cartomap®: example of student work.

The computer program used in the teaching-learning process for the fieldworks data representation, survey calculations, 2D/3D CAD design, terrain modelling, survey reports (stations, points and codes) and outputs creation is Cartomap®.

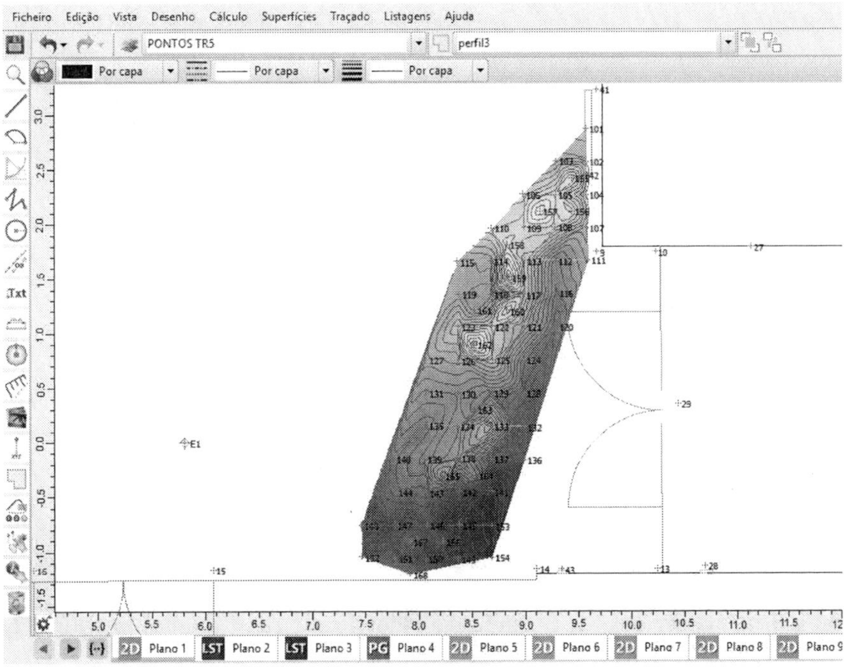

Figure 7. Precision altimetry survey representation in Cartomap®: example of student work - altimetry study of a slab deformation.

During the practical classes, students learn how to use the program in the context of topography works, but at the same time apply the technical drawing knowledge acquired in the courses of “Introduction to Civil Engineering and Technical Drawing” and “Technical Drawing for Civil Engineering”, improving their learning curve in digital representation and interpretation. These skills are extremely valuable for civil engineers, especially in today’s working world, since most jobs involve technical drawing representation and/or interpretation.

4. Final remarks

When representing any object, it can be drawn in a “flat” manner (drawing) or as a “solid” (3D model). Traditionally the results of civil engineering design are several 2D drawings representing 3D models. These drawings are essential to visually communicate the engineer’s idea to properly instruct on how to build or give a glance on how it will look. Teaching experience has shown that 3D interpretation of 2D drawing is not easy for most early year Civil Engineering students.

Among the strategies suggested in the literature to help teachers promoting active learning, project-based learning using computer-aided design combined with traditional teaching methods proved to be a good approach in Drawing and Topography subjects. This combination allows different types of students to be satisfied, those who prefer projects and those who prefer traditional courses.

In Portugal, teaching of technical drawing with 3D project is only available in 43% of Civil Engineering courses and 85% teach technical drawing and CAD.

A gradual students’ training in basic technical drawing skills has been accomplished at UBI Civil Engineering education through 3 units taught in early course years. Manual drafting, computer-aided design and cartography concepts are introduced through project-based learning approach. Students are also motivated to improve their 3D graphic representation skills through a classroom contest where the winning project is printed in 3D.

Considering the students’ multicultural background and lack of previous knowledge in drawing and 3D modelling, the contest proved to be a motivational approach in the learning process. Students have shown improvements in the skill-building process and inquires show that although most had

previously basic digital skills, the majority considered that 3D projects raised significantly their motivation and more than half considered the contest challenging. Inquiries also revealed that students need more time to develop drawing and 3D project skills.
Considering these outcomes, rearrangements have been suggested to incorporate a whole semester of technical drawing in the master's curriculum, allowing deeper introduction to 2D digital drafting. This rearrangement would allow a full second semester dedicated to 3D projects.

5. References

[1] OECD. Education at a Glance 2018: OECD Indicators. Paris: OECD; 2018. doi:10.1787/eag-2018-en.
[2] STEM Alliance. Table of of contents. STEM Educ Factsheet 2016. http://www.stemalliance.eu/publications (accessed January 28, 2019)
[3] Shirazi A, Behzadan A 2015 Content delivery using augmented reality to enhance students' performance in a building design and assembly project. Adv Eng Educ 2015;Wnter
[4] UK Department of Education 2013 3D printers in schools: uses in the curriculum
[5] Rúa E, Jimenez F, Gutierrez G, Villamizar N. 3D Printing as a Didactic Tool for Teaching some Engineering and Design Concepts. Rev Ing 2018;23:70–83. doi:10.14483/23448393.12248.
[6] Eurostat Database n.d. https://ec.europa.eu/eurostat/data/database (accessed December 20, 2018).
[7] Grand-Clement S, Devaux A, Belanger J, Manville C. 2017 Digital Learning: Education and skills in the digital age, doi:10.7249/CF369
[8] Newson T, Delatte N 2011 Case Methods in Civil Engineering Teaching. Can J Civ Eng 2011;38:1016–30, doi:10.1139/l11-023
[9] Shaaban K 2013 Practical teaching and its importance in teaching civil engineering. QScience Proc 2013, Glob Innov Conf 2013 2013:4–7, doi:10.5339/qproc.2013.gic.4
[10] Kara L. 2015 A Critical Look at the Digital Technologies in Architectural Education: When, where, and how? Procedia - Soc Behav Sci, 176:526–30, doi:10.1016/j.sbspro.2015.01.506
[11] Sun S, Su Z, Cui S 2017 Study on the continuing education innovative talents training mode of civil engineering major. IOP Conf Ser Earth Environ Sci;100. doi:10.1088/1755-1315
[12] Oliveira P, Oliveira C 2014 Integrator element as a promoter of active learning in engineering teaching. Eur J Eng Educ; 39:201–11. doi:10.1080/03043797.2013.854318
[13] Marshall J, Bhasin A, Boyles S, David B, James R, Patrick A 2018 A Project-Based Cornerstone Course in Civil Engineering : Student Perceptions and Identity Development University of Texas at Austin. Adv Eng Educ 2018:1–25
[14] Gratchev I, Jeng D 2018 Introducing a project-based assignment in a traditionally taught engineering course. Eur J Eng Educ 2018; 43:788–99. doi:10.1080/03043797.2018.1441264.
[15] Barrows H 1986 A taxonomy of problem-based learning methods. Med Educ; 20:481–6. doi:10.1111/j.1365-2923.1986.tb01386.x
[16] Kapranos P 2013 Teaching and Learning in Engineering Education – Are We Moving with the Times? Procedia - Soc Behav Sci; 102:3–10. doi:10.1016/j.sbspro.2013.10.707
[17] Googdhew P. 2012 Teaching Engineering: Can we do it better? In: Kapranos, P. Brabazon D, editor. Proc. 4th Int. Symp. Eng. Educ., p. 1–7
[18] Sampaio A, Henriques P, Cruz C 2009 Interactive models used in Civil Engineering education based on Virtual Reality technology. Proc - 2009 2nd Conf Hum Syst Interact HSI '09:170–6. doi:10.1109/HSI.2009.5090974
[19] Veide Z, Strozheva V 2015 The Visualizations Methods of Geometrical Forms in teaching of Civil Engineering Students. Proc. 10th Int. Sci. Pract. Conf., vol. 2, Rezekne, Latvia: Rezekne Higher Education Intitution, p. 312–6
[20] Duarte J 2007 Inserting new technologies in undergraduate architectural curricula: a case study. 25th ECAADe 2007:423–30

[21] Ford S, Minshall T 2019 Where and how 3D printing is used in teaching and education. Addit Manuf; 25:131–50. doi:10.1016/j.addma.2018.10.028.

[22] Santos B, Virtudes A, Gonçalves J, Amin S 2018 Multicultural Education in Civil Engineering: Planning and Transportation Approach. EDULEARN18 Conf., Palma de Mallorca, Spain: IATED 2018, p. 4991–5000.

[23] University of Beira Interior. Civil Engineering. Integr Master Degree Civ Eng n.d. http://www.ubi.pt/en/course/9 (accessed January 25, 2019).

[24] Yossef M, Chen A 2015 Applicability and Limitations of 3D Printing for Civil Structures. Conf Auton Robot Constr Infrastruct 2015:1–25.

[25] Sakin M, Kiroglu Y2017 3D Printing of Buildings: Construction of the Sustainable Houses of the Future by BIM. Energy Procedia 2017;134:702–11. doi:10.1016/j.egypro.2017.09.562

Acknowledgments

The authors acknowledge University of Beira Interior, CITTA - Research Centre for Territory, Transports and Environment (Multi/04427) and CERIS-CESUR- Civil Engineering Research and Innovation for Sustainability (ECI/04625) for supporting the performed study.

The authors also thank to Eng. Pedro Alves (MINILAB 3D) and Prof. Pedro Gabriel de Almeida (UBI) for the 3D printing of the 2016/2017 and 2017/2018 winning students' final projects.